# Clathrate Hydrates of Natural Gases

CHEMICAL INDUSTRIES

*A Series of Reference Books and Textbooks*

*Consulting Editor*
HEINZ HEINEMANN
*Heinz Heinemann, Inc.,*
*Berkeley, California*

*Volume 1:* Fluid Catalytic Cracking with Zeolite Catalysts, *Paul B. Venuto and E. Thomas Habib, Jr.*

*Volume 2:* Ethylene: Keystone to the Petrochemical Industry, *Ludwig Kniel, Olaf Winter, and Karl Stork*

*Volume 3:* The Chemistry and Technology of Petroleum, *James G. Speight*

*Volume 4:* The Desulfurization of Heavy Oils and Residua, *James G. Speight*

*Volume 5:* Catalysis of Organic Reactions, *edited by William R. Moser*

*Volume 6:* Acetylene-Based Chemicals from Coal and Other Natural Resources, *Robert J. Tedeschi*

*Volume 7:* Chemically Resistant Masonry, *Walter Lee Sheppard, Jr.*

*Volume 8:* Compressors and Expanders: Selection and Application for the Process Industry, *Heinz P. Bloch, Joseph A. Cameron, Frank M. Danowski, Jr., Ralph James, Jr., Judson S. Swearingen, and Marilyn E. Weightman*

*Volume 9:* Metering Pumps: Selection and Application, *James P. Poynton*

*Volume 10:* Hydrocarbons from Methanol, *Clarence D. Chang*

*Volume 11:* Foam Flotation: Theory and Applications, *Ann N. Clarke and David J. Wilson*

*Volume 12:* The Chemistry and Technology of Coal, *James G. Speight*

*Volume 13:* Pneumatic and Hydraulic Conveying of Solids, *O. A. Williams*

*Volume 14:* Catalyst Manufacture: Laboratory and Commercial Preparations, *Alvin B. Stiles*

*Volume 15:* Characterization of Heterogeneous Catalysts, *edited by Francis Delannay*

*Volume 16:* BASIC Programs for Chemical Engineering Design, *James H. Weber*

*Volume 17:* Catalyst Poisoning, *L. Louis Hegedus and Robert W. McCabe*

*Volume 18:* Catalysis of Organic Reactions, *edited by John R. Kosak*

*Volume 19:* Adsorption Technology: A Step-by-Step Approach to Process Evaluation and Application, *edited by Frank L. Slejko*

*Volume 20:* Deactivation and Poisoning of Catalysts, *edited by Jacques Oudar and Henry Wise*

*Volume 21:* Catalysis and Surface Science: Developments in Chemicals from Methanol, Hydrotreating of Hydrocarbons, Catalyst Preparation, Monomers and Polymers, Photocatalysis and Photovoltaics, *edited by Heinz Heinemann and Gabor A. Somorjai*

*Volume 22:* Catalysis of Organic Reactions, *edited by Robert L. Augustine*

*Volume 23:* Modern Control Techniques for the Processing Industries, *T. H. Tsai, J. W. Lane, and C. S. Lin*

*Volume 24:* Temperature-Programmed Reduction for Solid Materials Characterization, *Alan Jones and Brian McNicol*

*Volume 25:* Catalytic Cracking: Catalysts, Chemistry, and Kinetics, *Bohdan W. Wojciechowski and Avelino Corma*

*Volume 26:* Chemical Reaction and Reactor Engineering, *edited by J. J. Carberry and A. Varma*

*Volume 27:* Filtration: Principles and Practices, second edition, *edited by Michael J. Matteson and Clyde Orr*

*Volume 28:* Corrosion Mechanisms, *edited by Florian Mansfeld*

*Volume 29:* Catalysis and Surface Properties of Liquid Metals and Alloys, *Yoshisada Ogino*

*Volume 30:* Catalyst Deactivation, *edited by Eugene E. Petersen and Alexis T. Bell*

*Volume 31:* Hydrogen Effects in Catalysis: Fundamentals and Practical Applications, *edited by Zoltán Paál and P. G. Menon*

*Volume 32:* Flow Management for Engineers and Scientists, *Nicholas P. Cheremisinoff and Paul N. Cheremisinoff*

*Volume 33:* Catalysis of Organic Reactions, *edited by Paul N. Rylander, Harold Greenfield, and Robert L. Augustine*

*Volume 34:* Powder and Bulk Solids Handling Processes: Instrumentation and Control, *Koichi Iinoya, Hiroaki Masuda, and Kinnosuke Watanabe*

*Volume 35:* Reverse Osmosis Technology: Applications for High-Purity-Water Production, *edited by Bipin S. Parekh*

*Volume 36:* Shape Selective Catalysis in Industrial Applications, *N. Y. Chen, William E. Garwood, and Frank G. Dwyer*

*Volume 37:* Alpha Olefins Applications Handbook, *edited by George R. Lappin and Joseph L. Sauer*

*Volume 38:* Process Modeling and Control in Chemical Industries, *edited by Kaddour Najim*

*Volume 39:* Clathrate Hydrates of Natural Gases, *E. Dendy Sloan, Jr.*

*Volume 40:* Catalysis of Organic Reactions, *edited by Dale W. Blackburn*

Additional Volumes in Preparation

Fuel Science and Technology Handbook, *edited by James G. Speight*

Oxygen in Catalysis, *A. Bielanski and Jerzy Haber*

Industrial Drying Equipment: Selection and Application, *Kees van't Land*

# Clathrate Hydrates of Natural Gases

**E. Dendy Sloan, Jr.**
Colorado School of Mines
Golden, Colorado

Marcel Dekker, Inc. 1990 New York and Basel

Marcel Dekker, Inc. and the author make no warranty with regard to the *Clathrate Hydrates of Natural Gases* software, its accuracy, or its suitability for any purpose other than specified in the volume. This software is licensed solely on an "as is" basis. The only warranty made with respect to the *Clathrate Hydrates of Natural Gases* diskette is that the diskette medium on which the software is recorded is free of defects. Marcel Dekker, Inc. will replace a diskette found to be defective if such defect is not attributable to misuse by the purchaser or his agent. The defective diskette must be returned within ten (10) days to:

Customer Service
Marcel Dekker, Inc.
Post Office Box 5005
Cimarron Road
Monticello, New York 12701
(914) 796-1919

Comments, suggestions, or bug reports concerning *Clathrate Hydrates of Natural Gases* are welcome and should be sent to:

E. Dendy Sloan, Jr.
Colorado School of Mines
1500 Illinois Street
Golden, Colorado 80401

**Library of Congress Cataloging-in Publication Data**

Sloan, E. Dendy, Jr.
Clathrate hydrates of natural gases.

(Chemical industries; v. )
Includes bibliographical references.
1. Gas, Natural--Hydrates. 2. Clathrate compounds.
I. Title. II. Series.
TN880.S55 1989 665.7 89-23371
ISBN 0-8247-8296-8 (alk. paper)

This book is printed on acid-free paper.



MARCEL DEKKER, INC.
270 Madison Avenue, New York, New York 10016

Current printing (last digit)
10 9 8 7 6 5 4 3 2 1

PRINTED IN THE UNITED STATES OF AMERICA

# Preface

Since each reader brings a different background and perspective to a monograph, it may be worthwhile initially to provide both an apologia and a guideline for reading this book. The book was written with the intention of being of some use to engineers and scientists in practice and in research, and readers of different backgrounds may wish to follow different paths through the chapters.

The primary reason the book was written was to tie together the different concepts which have arisen in the literature (primarily over the last half-century) regarding hydrates of natural gas. These unique compounds are more properly called clathrate hydrates to distinguish them from the stoichiometric hydrates commonly found in inorganic chemistry. Our understanding of these compounds has changed dramatically in recent decades and the new discoveries have provided fresh perspectives on past theories and data. It was hoped that such an overview would yield new understanding and insights for both the author and the reader, and that future directions might be suggested for research and practical applications in these unusual natural gas solids.

The second reason for the book was to provide a balance between experimental and theoretical perspectives. The monograph was intended as a single document which recorded much if not all of the thermodynamic and transport property data obtained since 1934, the time of discovery of hydrates in pipelines. As noted in Chapter 1, 1934 marks the transition from hydrates as an academic curiosity to hydrates as an industrially important compound. Too often the comparative availability and low cost of computers cause the mistaken elevation of theory over experiment. In the field of hydrates, however, the most significant advances in knowledge have been made by those researchers who have performed painstaking hydrate experiments, guided by intuition and theory. Experiments have provided the physical understanding upon which to base and correct theories. In almost every case, the most marked theoretical advances, such as those of van der Waals and Platteeuw (1959), have been based upon significant experimental advances, such as the determination of the hydrate crystal structures by von Stackelberg and coworkers, Claussen, and Pauling and Marsh in the preceding two decades.

The final reason for the book was to provide a vehicle for understanding and use of the IBM-PC compatible computer programs on the disk in the frontpapers. The programs and the book should serve as complements to each other. The main program on the floppy disk is an abbreviated version of the one written in this laboratory and marketed in various revisions during the decade ending January 1, 1988. Normally such programs, which are based on fairly complex statistical thermodynamics, cannot be easily written and "debugged" from the published literature without substantial time and effort. It is not necessary to understand the theory (Chapter 5) in order to use the computer program to perform several hydrate formation calculations; the reader should simply follow the directions and examples in Appendix B. However, without the computer program the theory may otherwise remain somewhat sterile. At the same time, the book may serve as a forum for a more thorough exposition of the program's principles than can be normally provided in documents accompanying a program.

Both the industrial practitioner and the researcher may wish to read Chapter 1 which provides a historical perspective of clathrate hydrates. One cannot deal with hydrates without some knowledge of the all-important crystal structures, provided in Chapter 2. Chapter 3, on the kinetics of hydrate formation, is an effort to provide the "how" of hydrate formation to the "what" of the hydrate crystals in Chapter 2. As such, it will be of primary interest to the researcher, and possibly of interest to the practitioner. Chapter 4, which provides an overview of hydrate phase equilibria, is the last chapter which should be of common interest to both practitioners and researchers. A summary of these recommendations and those pertaining to the remainder of the book is given in the following table:

**A Suggestion on How to Read This Book**

| Reader's Background | Practitioner | Researcher |
|---|---|---|
| Applicable Sections | | |
| Chptr 1 - History | All Sections | All Sections |
| Chptr 2 - Structures | 2.1.1,2.1.2,2.1.3 | All Sections |
| Chptr 3 - Kinetics | 3.1,3.2.1 | All Sections |
| Chptr 4 - Phase Equilibria | 4.1,4.3,4.5,4.6 | Applicable Sections |
| Chptr 5 - Statistical Thermodynamics | None | All Sections |
| Chptr 6 - Experimental Methods and Data | 6.1<br>6.2 as needed | 6.1<br>6.2 |
| Chptr 7 - Hydrates in the Earth | 7.1 | All Sections |
| Chptr 8 - Processing and | As needed | None |

| | | |
|---|---|---|
| Production | | |
| Appdx A - Thermodynamic Relations | A.2 | A.1 |
| Appdx B - Program Manual and Examples | All Sections | All Sections |

There are many limitations to the book. First, it was intended to apply primarily to clathrate hydrates of significant components in natural gas. Although other hydrate formers (such as olefins) and gases heavier than normal butane are largely excluded, the principles of crystal structure and kinetics in Chapters 2 and 3 will still apply. Secondly, primarily due to language inability and literature access, the book has a Western Hemisphere perspective. Two translations (Schroeder (1927) and Makogon (1985)) were made in preparation for this manuscript. Discussions at some length were held with Professor Makogon and with Professor Berecz and Ms. Balla-Achs, whose earlier hydrate monographs were published in Russian in 1974 and in Hungarian in 1977, respectively, before their publication in English in 1981 and 1983. Yet this book falls short of a truly worldwide perspective, as witnessed by the bibliography of approximately 1000 Soviet references recently received from Professor Makogon, which were excluded due to time and space constraints. Finally, as in all single author manuscripts, the book is the limited product of a single perspective which has been shaped by past workers and present colleagues.

Particularly noteworthy were the contributions by my colleague and former mentor, Professor Kobayashi of Rice University, who shaped my initial thinking on the subject and continues to shape ideas in hydrate research through the exacting experimental effort in his laboratory. Dr. W.R. Parrish of Phillips Petroleum Company has encouraged and contributed to the work from our laboratory for almost a decade. Dr. K.A. Kvenvolden of the U.S. Geological Survey has been generous with his publications and discussions regarding in situ hydrates and their geochemistry. My academic colleagues Professor Holder (University of Pittsburgh) and Professor Bishnoi (University of Calgary) have graciously shared their recent theoretical and experimental results which are of central importance to our current hydrate understanding.

My colleagues at the Colorado School of Mines, Professors F.H. Poettmann, M.S. Selim, and T.R. Wildeman, have been a constant source of inspiring and challenging ideas, many of which are incorporated here. The graduate students and postdoctoral fellows have done most of the work and much of the thinking which has evolved from this laboratory. These young minds have served to maintain a fresh perspective (and at times a nudge) for the author: S. Adisasmito, Dr. B. Al-Ubaidi, M. Amer, Dr. G.B. Asher, M.S. Bourrie (deceased), S.B. Cha, Dr. P.B. Dharmawardhana, Dr. D.D. Erickson, A. Giussani, T. Kotkoskie, J.J. Johnson, S. Mann, L. McClure, P. D. Menten, B. Müller-Bongartz, T. Nguyen, H. Ouar, Dr. R.M. Rueff, Dr. K.A. Sparks, J.W. Ullerich, B.E. Weiler, S. Yamanlar, Dr. M.H. Yousif, and Dr. C. O. Zerpa.

Finally, the intrinsic joy of learning about clathrate hydrates has in itself been a pleasure which I hope will be communicated through these pages to younger workers in the field. The survival of a research area, like that of a civilization, depends on whether the young see learning as a worthwhile goal. In noting that these pages doubtless contain several mistakes, the author invokes the acute observation of Francis Bacon,[1] "Truth emerges more readily from error than from confusion."

Because our lives are enabled and shaped by those around us, I dedicate this book to my parents, for their kind forebearance, to Marjorie, for her magnanimous patience and good humor, and to Trey and Mark, who compose the future.

E. Dendy Sloan, Jr.

[1] "Novum Organum" Vol. VIII of "The Works of Francis Bacon," ed. J. Spedding, R.L. Ellis, and D.D. Heath, p210, New York, 1969

# Contents

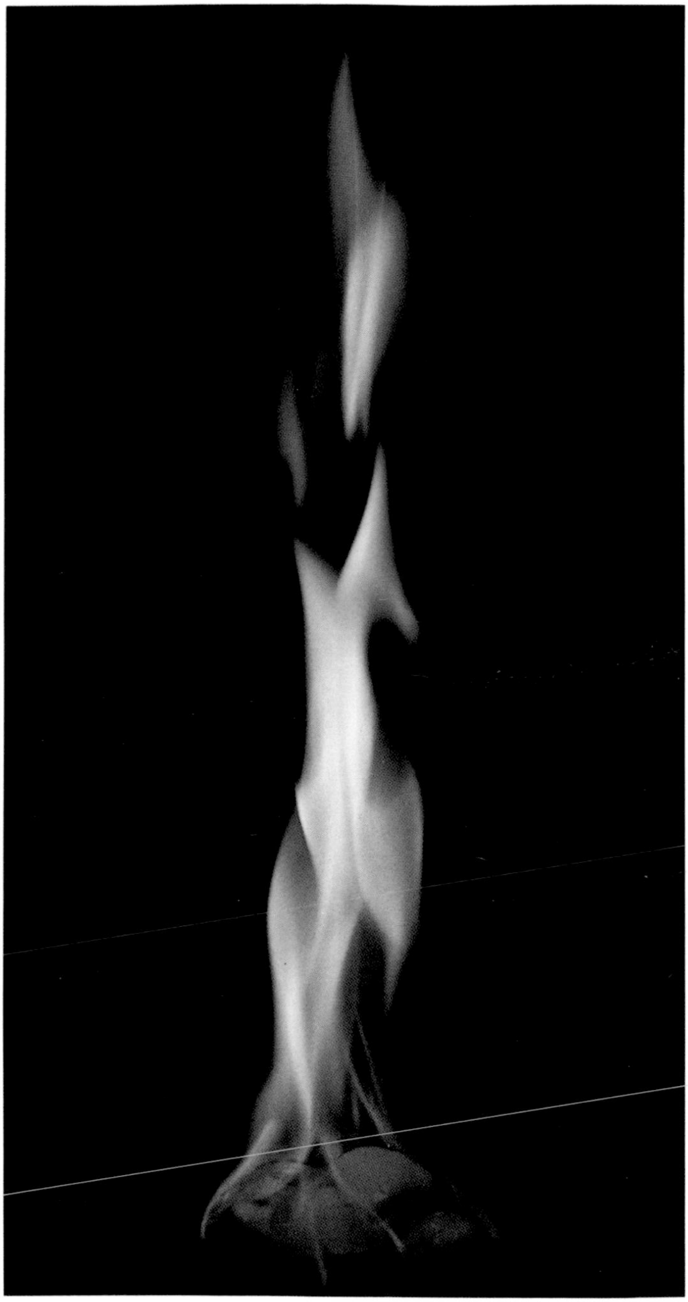

The Burning Snowball: Picture of Methane Hydrate Mass Supporting Its Own Combustion (Courtesy United States Department of Energy)

# 1

# Historical Perspective

The history of natural gas hydrates has evolved over three major periods, as follows:

- The first period, continuing from their discovery in 1810 until the present, concerned gas hydrates as a scientific curiosity which caused a combination of both gas and water into a solid phase.
- The second period, continuing from about 1934 until the present, predominantly concerned natural gas hydrates as a man-made substance which was a hindrance to natural gas producers and processors.
- The third major period, from the mid 1960's until the present, came with the discovery that nature had predated man's fabrication of hydrates by millions of years, **in situ** in both the deep oceans and permafrost regions as well as in extraterrestrial environments.

As a result, the present is a culmination of the three periods, representing the most fascinating time in the history of natural gas hydrates. During the first century after their discovery the number of hydrate publications totalled approximately 40; recently, the number of publications both in the popular and the technical press has increased dramatically with over 80 publications in 1982 alone. The purpose of this chapter is to review the three above periods, to provide an overall historical perspective. The major concepts will be briefly discussed; the detailed investigations will be presented in the appropriate chapters which follow.

## 1.1 Hydrates as a Laboratory Curiosity

Natural gas hydrates were discovered by Sir Humphrey Davy (1811, p.30), with these brief comments on chlorine (then called oxymuriatic) gas:

> It is generally stated in chemical books, that oxymuriatic gas is capable of being condensed and crystallized at low temperature; I have found by several experiments that this is not the case. The solution of oxymuriatic gas in water freezes more readily than pure water, but the pure gas dried by muriate of lime undergoes no change whatever at a temperature of 40 below $0^0$ of Fahrenheit."

Over the following one and one-quarter centuries, researchers in the field endeavored with two major goals in mind. As a first priority, there was a concern to identify all of the compounds which formed hydrates. Secondly, there was a wish to quantitatively describe the compounds, their compositions, and their physical properties. Table 1-1 provides an summary of the research over this period. Thorough details of the research during the first 125 years of hydrates are presented in the reviews of Schroeder (1927), Davidson (1973), or Berecz and Balla-Achs (1983).

In Table 1-1, the following pattern is often repeated: (a) the discovery of a new hydrate is published by an investigator, (b) a second researcher confirms the existence of the hydrate, but disputes the composition proposed by the original investigator, (c) a third (frequently more) investigator refines the measurements made by the initial two investigators, and proposes slight extensions. As a typical example, in the case of chlorine hydrate, after Davy's discovery in 1810, Faraday confirmed the hydrate (1823) but proposed that there were 10 water molecules per molecule of chlorine. Then Ditte (1882), Mauméné (1883), and Roozeboom (1884) refined the ratio of water to chlorine. Other notable work of Table 1-1 was done by Cailletet (1877) and Cailletet and Bordet (1882) who first measured hydrates with mixtures of two components.

**Table 1-1 Hydrates from 1810 - 1934**

| Year | Event |
|---|---|
| 1810 | Chlorine hydrate discovery by Sir Humphrey Davy |
| 1823 | Corroboration by Faraday - formula $Cl_2 \cdot 10H_2O$ |
| 1882,83 | Ditte and Mauméné disputed the composition of chlorine hydrates |
| 1884 | Roozeboom confirmed the formula $Cl_2 \cdot 10H_2O$ |
| 1828 | Bromine hydrates discovered by Löwig |
| 1876 | $Br_2$ hydrates corroborated by Alexeyeff as $(Br_2 \cdot 10H_2O)$ |
| 1829 | $SO_2$ hydrates found by de la Rive - $SO_2 \cdot 7H_2O$ |
| 1848 | Pierre determined the formula $SO_2 \cdot 11H_2O$ |
| 1855 | Schoenfield measured the formula as $SO_2 \cdot 14H_2O$ |
| 1884,85 | Roozeboom first postulated upper and lower hydrate quadruple points, using $SO_2$ as evidence |
| 1856-8 | $CS_2$ hydrate composition disputed by Berthelot ('56), Millon('60), Duclaux('67), Tanret('78) |
| 1885 | Chancel and Parmentier determined chloroform hydrates |

| | |
|---|---|
| 1882 | Wroblewski measured carbon dioxide hydrates |
| 1877,82 | Cailletet, and Cailletet and Bordet first measured mixed gas hydrates from $CO_2$ + $PH_3$, and from $H_2S$ + $PH_3$ |
| 1882 | De Forcrand suggested formula $H_2S\cdot(12\text{-}16)H_2O$ measured 30 binary hydrates of $H_2S$ and with a second component such as $CHCl_3$, $CH_3Cl$, $C_2H_5Cl$, $C_2H_5Br$, $C_2H_3Cl$. He indicated all compositions as $G\cdot 2H_2S\cdot 23H_2O$ |
| 1888 | Villard obtained the temperature dependence of $H_2S$ hydrates |
| 1888 | De Forcrand and Villard measured temperature dependence of $CH_3Cl$ and $H_2S$ hydrates |
| 1888 | Villard measured hydrates of $CH_4$, $C_2H_6$, $C_2H_4$, $C_2H_2$, $N_2O$ |
| 1890 | Villard measured hydrates of $C_3H_8$, and suggested that the temperature of the lower quadruple point is decreased by increasing the molecular mass of guest; Villard suggested hydrates were regular crystals |
| 1896 | Villard measured hydrates of Ar, hypothesized nitrogen and oxygen hydrates; first used heat of formation data to get water/gas ratio |
| 1897 | De Forcrand and Thomas sought double (w/ $H_2S$ or $H_2Se$) hydrates; found mixed (other than $H_2S_x$) hydrates of numerous halohydrocarbons mixed with $C_2H_2$, $CO_2$, $C_2H_6$ |
| 1902 | De Forcrand first used Clausius-Clapeyron relation for $\Delta H$ and compositions; tabulated 15 hydrate conditions |
| 1919 | Scheffer and Meyer refined Clausius-Clapeyron technique |
| 1923,25 | De Forcrand measured hydrates of krypton, xenon |

### 1.1.1 Hydrates of Hydrocarbons Distinguished from Ice

Two French workers, Villard and de Forcrand were the most prolific researchers of this period. Villard (1888) first determined the existence of methane, ethane and propane hydrates. De Forcrand (1902) tabulated equilibrium temperatures at one atmosphere for 15 components, including those of of natural gas, with the exception of iso-butane (first measured in 1954 by von Stackelberg and Müller). The early period of hydrate research is marked by a tendency to set an integral number of water molecules per guest molecule, owing to the existing knowledge of stoichiometric hydrates which are of a substantially different nature. It gradually became clear that the newer clathrate hydrates distinguished themselves by being both non-stoichiometric and crystalline; at the same time they differed from normal hexagonal ice because they had no effect on polarized light.

### 1.1.2 Determining the Hydrate Composition

The work represented in Table 1-1 witnesses one of the early research difficulties of gas hydrates, which is still present to a limited extent - that of the experimental determination of the composition of the hydrate phase. Whereas many solids such as carbon dioxide precipitate in a relatively pure form, or a form of fixed composition, the nature of gas hydrates is such that their composition is variable with temperature, pressure, and composition of the associated fluid phases. While the composition measurement of either the gas or the water phase is tractable, measurement of the hydrate composition is difficult.

On a macroscopic basis, it is difficult to remove all of the excess water and hydrate former from the hydrate mass; this causes a substantial decrease of the accuracy of hydrate composition measurements. The hydrate formations often occlude water or a guest within the solid in a metastable configuration, and thus invalidate the composition obtained via dissociation. Of the macroscopic methods to determine hydrate composition, one of the most accurate was the indirect method proposed by Villard (1896) via the heat of formation, both above and below the ice point. Miller and Strong (1946) provided another method, as discussed in Section 4.7.2. Recently spectroscopic investigations have provided a resolution to the direct measurement of composition; Davidson et al. (1983) were able to determine some of the details of the guest molecule occupancy of each type of cage.

### 1.1.3 Phase Diagrams Provide Classification Scheme

Roozeboom (1884, 1885) defined both an lower hydrate quadruple point $Q_1$ ($I-L_W-H-V$)[1] and an upper quadruple point $Q_2$ ($L_W-H-V-L_{HC}$) shown in the pressure-temperature plot of Figure 1-1 for several components of natural gases. In the figure, H is used to denote hydrates, V for vapor, and $L_W$ and $L_{HC}$ for aqueous and hydrocarbon liquid phases, respectively. These quadruple points are unique and invariant for each hydrate former, so that their values provide a quantitative classification, given in Table 1-2 for hydrate components of natural gas. Each quadruple point occurs at the intersection of four three phase lines. The lower quadruple point is marked by the transition of $L_W$ to I, so that with decreasing temperature, $Q_1$ denotes where hydrates cease to be formed from vapor and liquid water and begin forming from vapor and ice. The early researchers took $Q_2$, the approximate point of intersection of line $L_W-H-V$ with the vapor pressure of the hydrate former, to represent an upper temperature limit for hydrate formation from that component. Since the vapor pressure critical temperature can be too low to allow such an intersection, some natural gas components such as methane and nitrogen have no upper quadruple $Q_2$, and consequently they have no

[1]Listed in order of decreasing water concentration

upper temperature limit for hydrate formation. Phase diagrams are discussed in detail in Chapter 4.

## 1.2 Hydrates as a Problem to the Natural Gas Industry

In the mid-1930's Hammerschmidt studied the 1927 hydrate review of Schroeder (Katz, 1988), to determine that natural gas hydrates were blocking gas transmission lines, frequently at temperatures above the ice point. This discovery was pivotal in causing a more pragmatic interest in gas hydrates, and shortly thereafter led to the regulation of the water content in natural gas transmission. Many workers including Hammerschmidt (1939), Deaton and Frost (1946), Bond and Russell (1949), Kobayashi et al. (1951), and Woolfolk (1952) investigated the effects of inhibitors on hydrates. In particular, many chloride salts, such as those of calcium, sodium, and potassium were considered, along with others. Methanol gradually became one of the most popular inhibitors, due to its property of becoming concentrated in the free water phase after being vaporized into the gas at some upstream point. Inhibitors are considered in detail in Chapters 4 and 8.

**Table 1-2: Natural Gas Component Quadruple Points**

| Component | T,P at $Q_1$ K, MPa | T,P at $Q_2$ K, MPa |
|---|---|---|
| Methane | 272.9, 2.563 | No $Q_2$ |
| Ethane | 273.1, 0.530 | 287.8, 3.39 |
| Propane | 273.1, 0.172 | 278.8, 0.556 |
| iso-Butane | 273.1, 0.113 | 275.0, 0.167 |
| n-Butane | does not form hydrates as single hydrocarbon | |
| Carbon Dioxide | 273.1, 1.256 | 283.0, 4.499 |
| Nitrogen | 271.9,14.338 | No $Q_2$ |
| Hydrogen Sulfide | 272.8, 0.093 | 302.7, 2.239 |

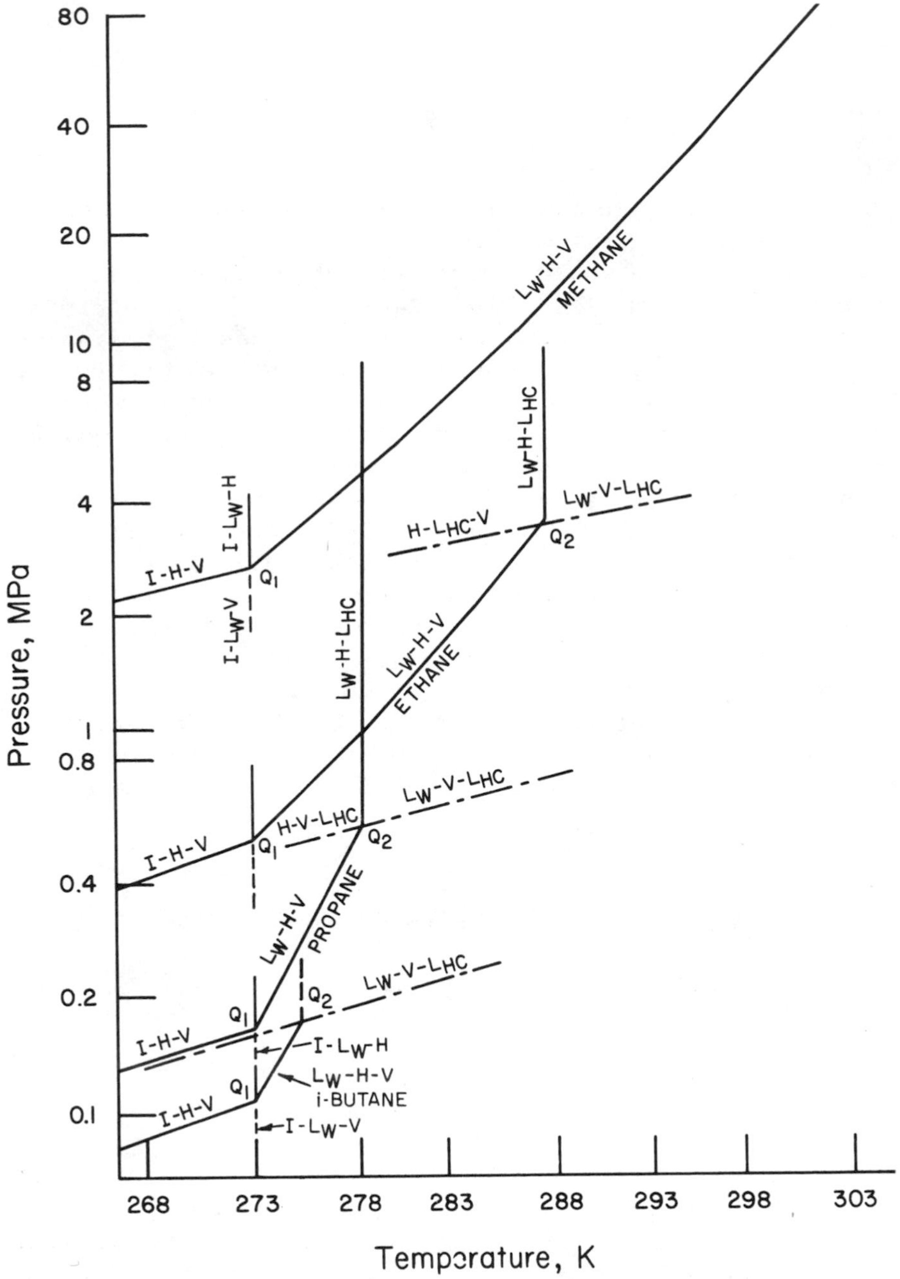


1-1 Phase Diagrams for Some Simple Natural Gas Hydrocarbons which Form Hydrates
Modified, with permission, from the <u>Handbook of Natural Gas Engineering</u> (Katz et al., 1959), by the McGraw-Hill Bk. Co.

### 1.2.1 Initial Experimental Studies on Natural Gases

After Hammerschmidt's initial discovery, the American Gas Association commissioned a thorough study of hydrates at the U.S. Bureau of Mines. In an effort spanning World War II, Deaton and Frost (1946) experimentally investigated the formation of hydrates from pure components of methane, ethane, and propane, as well as their mixtures with heavier components in both simulated and real natural gases. Many of the current predictive techniques are still compared to the data of the fundamental study by Deaton and Frost. It should be remembered however, that while this study was both painstaking and at the state-of-the-art, the data were of somewhat limited accuracy, particularly in the measurements of gas composition. As will be seen in Chapter 4, small inaccuracies in gas composition can dramatically affect hydrate formation temperatures and pressures. Yet for example, Deaton and Frost were unable to distinguish between normal butane and isobutane and so considered the sum of their mol fractions. Composition measurement methods such as chromatography did not come into common use until late in the 1950's.

### 1.2.2 Initial Correlation Work

When Hammerschmidt (1934) initially determined hydrates in pipelines, he published a correlation summary of over one hundred hydrate formation data points. Shortly afterward, Professor D.L. Katz and his students at the University of Michigan commenced both an experimental and correlative study of the field. Because it was impractical to perform experiments to measure hydrate formation conditions for every gas composition, two correlative methods were determined. The second (and simplest available) method, generated by Professor Katz (1945) and students in a graduate class, is presented in Figure 1-2. The plot enables the user to estimate a hydrate formation pressure, given a temperature and gas gravity (molecular weight of gas divided by that of air.) The original work also enabled the determination of the hydrate formation limits to Joule-Thomson expansions of natural gases, as in throttling gas through a valve. This method and its limitations are discussed in detail in Section 4.3.1, but it may be considered to be a useful first heuristic for estimating hydrate formation conditions.

The initial method by Wilcox, Carson, and Katz (1941) resulted in the generation of distribution coefficients (sometimes called $K_i$ values) for hydrates on a water-free basis. With a substantial degree of intuition, Katz determined that hydrates were a solid solution which might be treated similarly to an ideal liquid solution. Establishment of the $K_i$ value (defined as the component mol fraction in the gas relative to that in the hydrate phase) for each of a number of components enabled the user to determine the pressure and temperature of hydrate formation from mixtures. These $K_i$ value charts were generated well in advance of the determination of the hydrate crystal structure, and were labelled "tentative" in the publication title - yet they continue to be of high utility to the natural gas industry half a century later. The method is discussed in detail in Section 4.3.2.

The above two predictive techniques, and the accompanying experimental work from Katz's laboratory, provided a watershed in hydrate research. For the first time, industry had reasonably accurate methods to predict hydrate formation conditions for gas mixtures, without having to do costly experimental measurements. Consequently hydrate research centered on the determination of the hydrate crystal structure(s). Further refinements of the $K_i$ values were determined by Katz and co-workers in the Handbook of Natural Gas Engineering (1959), by Robinson and co-workers (Johari and Robinson, 1965, Robinson and Ng, 1976) and by Poettmann (1984).

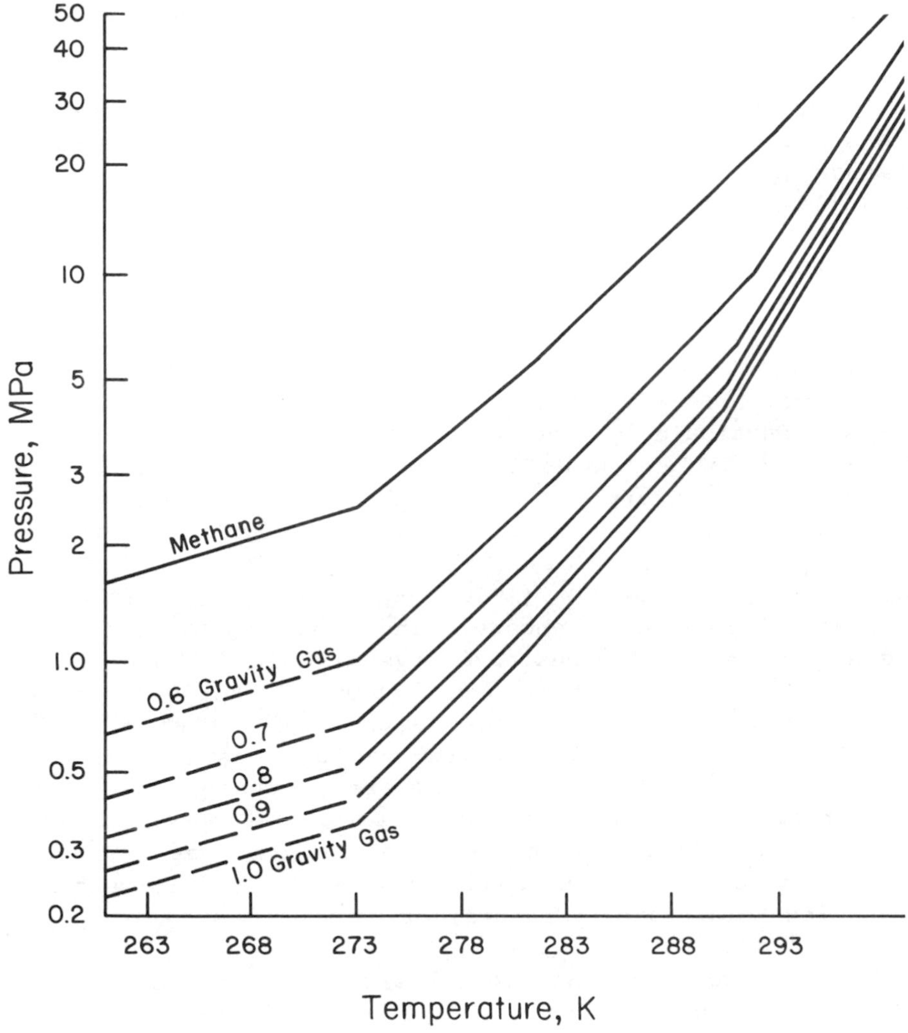


1-2 Gas Gravity Chart for Prediction of Three-Phase ($L_w$-H-V) Pressure and Temperature
(Reproduced, with permission, from Transactions AIME (Katz, 1945), by the American Institute of Mining, Metallurgical, and Petroleum Engineers.)

### 1.2.3 Hydrate Crystal Structure and Type Definitions

In the 1940's and 1950's von Stackelberg and co-workers performed the fundamental X-ray diffraction experiments on hydrates. The interpretation of these early diffraction experiments by von Stackelberg and Müller (1949, 1951a,b, 1952), Claussen(1951a,b), and Pauling and Marsh (1952) led to the determination of the two common hydrate crystal structures, shown in Figure 1-3, which are detailed in Chapter 2. During the period from 1959 to 1967 an extensive series of crystallographic studies were performed on clathrate hydrates by Jeffrey and co-workers. Particularly useful are the studies of each structure (McMullan and Jeffrey, 1965, Mak and McMullan, 1965) and the summary reviews (Jeffrey and McMullan, 1967; Jeffrey, 1984). These studies showed hydrates to be members of the class of compounds labelled "clathrates" by Powell (1948) - after the Latin "clathratus," meaning to encage. The two hydrate structures are composed of repetitive crystal units, shown in Figure 1-3, which have almost spherical "cages" of hydrogen-bonded water molecules; each cage contains at most one guest molecule, held with van der Waals bonds. The hydrate crystalline structures are discussed, along with their mechanical properties in Chapter 2. Throughout this book the common name "natural gas hydrate(s)" may be used interchangeably with the correct designation "clathrate hydrate(s) of natural gas."

Von Stackelberg and co-workers also classified hydrates in a scheme which is still used in the hydrate literature. "Mixed" hydrates is the term reserved for hydrates of more than one component, in which cages of the same kind are occupied by two types of molecules, with the restriction of at most one molecule per cage. "Double" hydrates initially was reserved for structure II hydrates in which one component is hydrogen sulfide or hydrogen selenide; it has come to mean hydrates in which each size cage is primarily occupied by a different type molecule. Von Stackelberg proposed that double hydrates

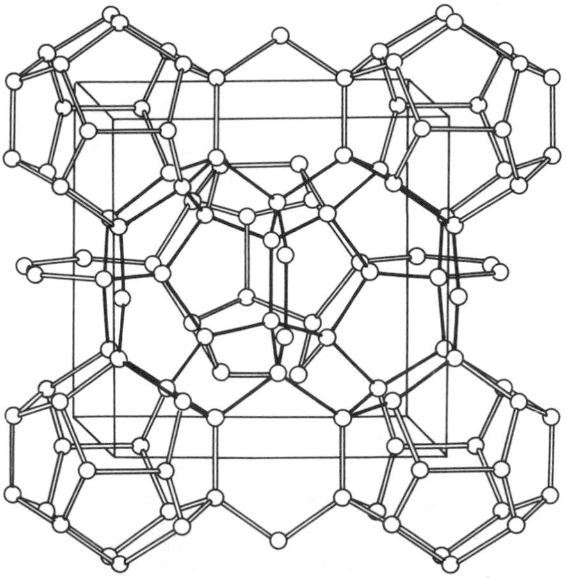

(a)

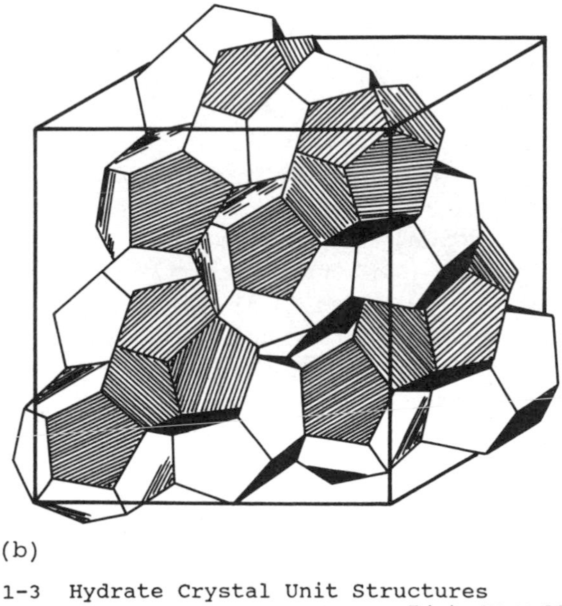

(b)

1-3 Hydrate Crystal Unit Structures
I(a) (McMullan and Jeffrey, 1965)
and II(b) (Mak and McMullan, 1965)
(Both figures reproduced with permission from the Journal of Chemical Physics by the American Institute of Physics.)

were stoichiometric due to their almost invariant composition; Platteeuw and van der Waals (1959) suggested this invariance was caused instead by the azeotropic nature of such hydrates. A hilfsgase or "help gas" hydrate is composed of components such as oxygen or nitrogen which would normally require high pressures to form hydrates but increase the hydrate forming ability of a second, more strongly hydrated component. To complete these common definitions, Davidson (1973) proposed that the term "simple" hydrate denote the presence of only one guest species; this definition is most common currently.

### 1.2.4 Basis for Current Thermodynamic Models

With the determination of hydrate structure, more rigorous predictive methods were formulated for hydrate thermodynamic property predictions. Barrer and Stuart (1957) initially suggested a statistical thermodynamic approach to determining gas hydrate properties. In a similar yet more successful approach, van der Waals and Platteeuw (1959) proposed the foundation of the method currently used. This method is perhaps the best modern exemplar of the use of statistical thermodynamics to enable the prediction of macroscopic properties such as temperature and pressure, using microscopic properties such as intermolecular potentials. It represents one of the few, routine uses of statistical thermodynamics in industrial practice. The advantage of the method, in addition to accuracy, is that in principle, it enables the user to predict properties of mixtures from parameters of simple hydrate formers. Since there are only eight natural gas components (yet virtually an infinite number of natural gas mixtures) which form hydrates, the method represents a tremendous saving in experimental effort for the natural gas industry. The basic van der Waals and Platteeuw method, along with the below modifications, are detailed in Chapter 5.

Child (1964) and McCoy and Sinanoglu (1963) refined the van der Waals and Platteeuw method using different intermolecular potentials such as the Kihara potential. Workers such as Marshall et al. (1964) and Nagata and Kobayashi (1966) at Rice University first fit simple hydrate parameters to experimental data for methane, nitrogen, and argon, thereby significantly reducing the prediction error for mixtures of those gases. Parrish and Prausnitz (1972) showed in detail how this method could be extended to all natural gases. Further extensions to the van der Waals and Platteeuw method have been made by many workers. Efforts to improve the original assumptions by van der Waals and Platteeuw are detailed in a recent review by Holder et al. (1988). Erbar and co-workers (Wagner, et al. 1985) and Anderson and Prausnitz (1986) presented improvements to inhibition prediction. Robinson and co-workers introduced guest interaction parameters into their prediction scheme, as summarized by Nolte et al. (1985). Researchers in this laboratory (Sloan and Parrish, 1983, Sloan et al., 1987) presented yet another modification of the theory for hydrate phase predictions. These last three groups of researchers have made computer programs commercially available for the prediction of hydrate properties. An abbreviated IBM-PC compatible version of such a program is found in the frontpapers of this book. The above methods are presented in detail in Chapter 5.

### 1.2.5 Studies of Hydrate Kinetics

In the mid-1960's, driven by the promise of natural gas hydrates as a substantial energy resource, a large experimental effort was begun in the U.S.S.R. in a research group led by Makogon (1974, 1985) at the Gubkin Petrochemical and Gas Industry Institute. The area of hydrate kinetics and thermodynamics have had priority in the Soviet research program, because the same physics can be applied to problems of hydrate formation in transmission/processing equipment as well as those of **in situ** hydrates, the third major area of hydrate study. The Soviet studies from the 1960's are unique in being among the first to place emphasis on the kinetics of hydrate formation, both in the bulk phases and in porous environments.

Lubas (1978) and Bernard et al. (1979) investigated the formation of hydrates in gas wells. Bishnoi and co-workers (Vysniauskas, 1983a,b; Kim,1985; Englezos et al., 1987a,b,c; Kalogerakis et al. 1988a,b) have investigated the macroscopic kinetics of hydrate growth such as might be encountered in oceans or in transmission lines which might be blocked with hydrates. Other recent studies in kinetics are by Holder and co-workers (Angert et al., 1982a,b; Godbole, 1982; Kamath et al.,1984) and workers from this laboratory (Selim and Sloan, 1985,1987,1989; Ullerich et al. 1987; Yousif et al., 1988, 1989). The details of hydrate kinetics are reviewed in Chapter 3.

The onset of kinetic measurements marks the maturing of hydrates as a field of research. Typically research efforts begin with the consideration of time-independent thermodynamic equilibrium properties due to relative ease of measurement. As an area matures and phase equilibrium thermodynamics become better defined, research generally turns to time-dependent measurements such as kinetics and transport properties. The paucity of transport data (Section 2.2.3) and kinetic data (Chapter 3) indicates that these areas are at the state-of-the-art for current research.

### 1.2.6 New Gas Production and Processing Problems

Since 1970 hydrate research has followed production and processing problems in unusual environments, such as on the North Slope of Alaska, in Siberia, in the North Sea, and in deep ocean drilling. Kobayashi and co-workers (Sloan et al., 1976, Song and Kobayashi, 1983, 1984) and workers in this laboratory (Sloan et al., 1986, 1987) have measured concentrations of water in hydrate-forming fluid phases in equilibrium with hydrates (when there is no free water phase present) for application in cryogenics or cold regions such as the North Slope. Ng, Robinson, and co-workers (1983, 1984, 1985a,b, 1986,

1987) have performed the most comprehensive measurements of aqueous phase concentrations of methanol and glycols needed to inhibit hydrates formed from both the gas and condensed hydrocarbon phases, such as may be present in North Sea transmission lines. Recently there have been reports by Barker and Gomez (1987) about hydrate formation in deep ocean drilling operations; current experimental work in this laboratory addresses these phenomena.

As mentioned above, hydrate formation in deepwater pipelines as in the North Sea is another substantial problem. Pipelines which transport water and condensed hydrocarbons such as gas condensate or crude oil have limited possibilities for removing hydrates once plugs have formed. Earlier work by Scauzillo (1956) which suggests that formation may be inhibited by the input of hydrocarbon liquids, cannot be confirmed by thermodynamic calculations, and one must resort to kinetics or to fluid mechanics experiments. Thus the construction of large-scale fluid flow pipelines are currently underway to model hydrate formation in subsea pipelines.

### 1.2.7 Hydrates in Storage and Separation

Several researchers have considered the hydrate properties causing the separation and concentration of gas and water. In particular, Barduhn and co-workers (Rouher and Barduhn, 1962, Barduhn et al., 1962) worked in desalination of seawater. The first Soviet hydrate work available in the western literature was that of Nikitin (1936,1937,1939,1940) who developed a method for separating rare gases with $SO_2$ hydrates; Nikitin was also the first to suggest a guest/host lattice structure for hydrates. Tsarev et al. (1979), Trofimuk et al.(1983a,b) suggested other hydrate separations of light gas components. Miller and Strong (1946), Parent (1948), and Dubinin and Zhidenko (1979) investigated the storage and transportation of natural gas in hydrate form.

## 1.3 Hydrates as an Energy Resource

A world atlas, giving sites with evidence of hydrate deposits, both on-shore and off-shore, is presented in Figure 1-4. Estimates of hydrate reserves, given in chapter seven are very high but uncertain. Since each volume of hydrate can contain as much as 184 volumes of gas (STP), hydrates are currently considered a potential unconventional energy resource. Due to the potential magnitude of this resource, the United States Department of Energy funded the compilation of well-established hydrate data in the Handbook of Gas Hydrate Properties and Occurrence (Lewin and Associates and Consultants, 1983).

### 1.3.1 In Situ Hydrates in the Soviet Union

An initial overview of the Soviet hydrate literature, with particular emphasis on natural occurrences, was recently published (Krason and Ciesnik, 1985). Makogon (1974) reported the first three Russian publications from more obscure primary references: Strizhev (1946), Mokhnatkin (1947), and Palvelev (1949) first hypothesized that hydrates existed in rocks in the northern regions of that country, suggested **in situ** formation mechanisms, and discussed the possibility of hydrate formation associated with coals, respectively. The major portion of the Soviet work has been done on hydrate thermodynamics and kinetics since the 1960's. The research emphasis on hydrate kinetics in the Soviet literature seems to be motivated, in large part, by the idea that the kinetics are key to accurate determinations of hydrate reserves (Barkan and Voronov, 1982).

The most commonly cited global hydrate reserve estimations are those of Trofimuk et al. (1981) with $57 \times 10^{12}$ $m^3$ of gas in continental hydrates, and $5\text{-}25 \times 10^{15}$ $m^3$ of gas in hydrates in the oceans. While there is substantially more verification for the estimates on land

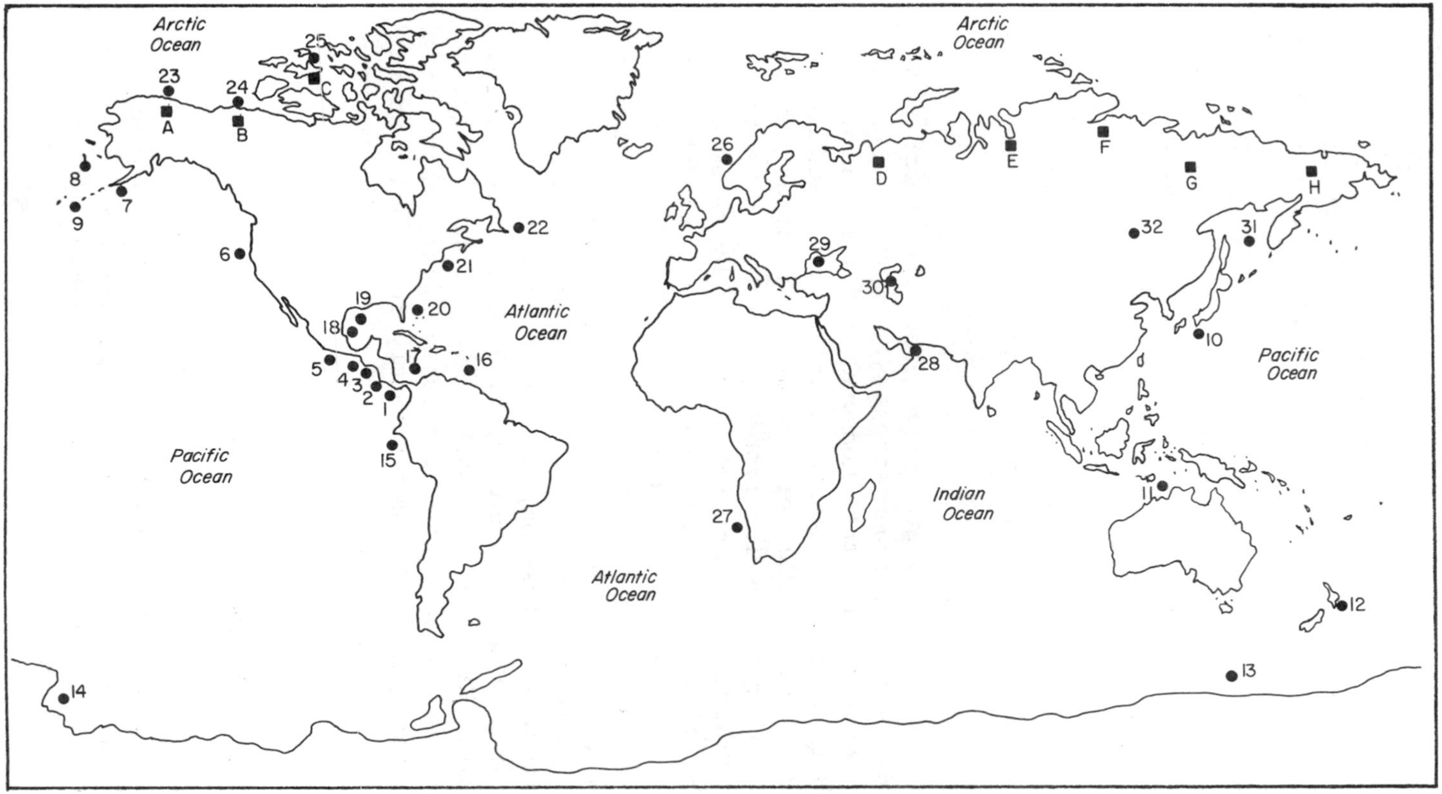


1-4 Map of In-Situ Hydrate Locations (See Section 7-1 for nomenclature)
(Reproduced, with permission, from <u>Chemical Geology</u> (Kvenvolden, 1988a), by Elsevier Science Publishers, B.V.)

than in the oceans, these estimates are controversial. However, recent estimates of world reserves by different methods in the eastern hemisphere (Makogon, 1988b) and in the western hemisphere by Kvenvolden (1988a) are of the same order of magnitude, which show the hydrate resource to be twice as great as the combined fossil fuel energy reserve. Calculations have been made, but there is no systematic experimental investigation of marine hydrates in the Soviet literature. For hydrates beneath the permafrost however, there are two major Soviet hydrate deposits associated with gas fields, at Messoyakha and Vilyui, upon which testing and/or production have been done.

In 1967, the Soviets discovered the first major hydrate deposit in the permafrost (Makogon, 1987). The hydrate deposit, in the Messoyakha field has been estimated to involve at least one-third of the entire gas reservoir, with depths of hydrates as great as 900 meters. During the period from 1969 to 1983 more than $5 \times 10^9$ $m^3$ of gas has been released from hydrates in the Messoyakha field (Makogon, 1984). The information in the Soviet literature on the production of gas from the Messoyakha field is discussed in Chapter 7. The majority of the Soviet publications are by eight authors, listed here in decreasing order with respect to number of publications: Y.F. Makogon, N.V. Cherskii, V.P. Tsarev, A.A. Trofimuk, V.A. Khoroshilov, S. Byk, and V.A. Fomina, and Nikitin. Krason and Ciesnik, (1985) indicate that no other Soviet author has published more than five articles; Makogon (1988a) has recently furnished a Soviet Bibliography which cites a number of hitherto unknown hydrate references.

### 1.3.2 Hydrates in the Western Hemisphere

In regard to **in situ** hydrate deposits in permafrost regions of the Western Hemisphere, in 1972 a core of hydrates was recovered from the Arco-Exxon Northwest Eileen Well Number Two in West Prudhoe Bay, Alaska

(Collett, 1983). Also in 1972, hydrates were found in drilling an Imperial well in Canada's MacKenzie Delta (Bily and Dick, 1974). Instances such as the above have caused further geologic investigations of permafrost hydrates in Russia, Canada, and Alaska. Collett (1983) investigated possible hydrate occurrances in 125 wells in the North Slope of Alaska. Weaver and Stewart (1982) and Franklin (1982) indicated well log responses in several wells in the Arctic Archipelago region. Judge (1982) and Collett, Kvenvolden, et al. (1988) summarized instances of hydrates reported in Canada and in the National Petroleum Reserve in Alaska, respectively.

The Deep Sea Drilling Project (DSDP), currently the Ocean Drilling Program (ODP) of the National Science Foundation, has undertaken the most systematic evaluation of ocean hydrate deposits. The DSDP has recovered hydrate cores in the deep oceans from both coasts of the U.S.A., from the Mid-America Trench off Guatemala and off the coast of Peru. Oceanic hydrate cores have also been recovered from the Gulf of Mexico and three Soviet water bodies. Even with these core recoveries, most of the evidence of hydrates in the oceans has been inferential. Bottom Simulating Reflectors (BSR's) have been used to infer the existence of hydrates in most of the locations shown in Figure 1-4. Much of the geochemical and geophysical research for these determinations has been done by Kvenvolden and co-workers (1980, 1981, 1982, 1983, 1984, 1985a,b) and Brooks and co-workers (1984, 1985 a,b,c, 1987). Oceanic and permafrost hydrates are discussed in Chapter 7.

### 1.3.3 Properties for Discovery and Recovery

The determination of **in situ** hydrates spawned an entire wave of research to determine the hydrate properties needed for geological research and gas recovery. Several workers measured sonic velocity and thermal conductivity of hydrates in sediments. Other workers measured the calorimetric properties needed to determine dissociation. Davidson (1983) summarized hydrate properties as being similar to ice, with a few notable exceptions.

Along with the new measurements of hydrate properties, came several studies to determine the recoverability of gas from hydrates beneath the permafrost. Soviet researchers indicated that thermal stimulation from above ground was not economically viable, (Trofimuk et al., 1982) and suggested alternatives of pressure reduction, inhibitor injection, geothermal stimulation, or **in situ** combustion techniques. Recovery techniques modelled in the Western Hemisphere concerned either pressure reduction or thermal stimulation. The first of these was by McGuire (1982), followed by Holder et al.(1983), Kamath and co-workers (1984, 1987), Burshears et al. (1986), and workers in this laboratory (Selim and Sloan, 1985, 1987, 1989; Yousif et al., 1988,1989). Such studies suggest that the recovery of gas from hydrates is viable economically, and work to recover this resource has been accomplished since 1971 in the U.S.S.R. Details of these studies are given in Chapter 7.

### 1.3.4 Hydrates in Unusual Environments

Other "natural" hydrates have been suggested during the last two decades. Miller (1974) concluded that hydrates of carbon dioxide were on Mars, and Pang et al. (1983) indicated that the E rings of Saturn were hydrates. Delsemme and Miller (1970), Mendis (1974), and Makogon (1987) suggested that hydrates exist in comets; in particular that carbon dioxide and water in Halley's comet were combined in the form of hydrates.

Shoji and Langway (1982) described air hydrates found with ice cores off Greenland, while Tailleur and Bowsher (1981) indicated the presence of hydrates associated with coals in permafrost regions. Rose and

Pfannkuch (1982a) have considered the applicability of the "Deep Gas Hypothesis" to the origin of methane in hydrates. Pauling (1961) suggested that hydrates were formed by gases during the process of anesthesia, while Miller (1961) concluded that hydrates were not present in anesthesia, but gas molecules caused substantial structure in the water without forming actual solids, as indicated in Section 3.1.2. Ripmeester et al. (1987), suggested that a new hydrate structure H, which can contain very large molecules, e.g. methyl cyclohexane, may be found in nature in the future.

## 1.4 Relationship of This Chapter to Those Which Follow

With the overview provided by the hydrate history of the current chapter, the remainder of this monograph provides a detailed development of research done in each of the above three areas, together with examples of industrial interest. The following chapter presents the chemical structure of hydrates and, by inference, begins to consider molecular and macroscopic properties with emphasis on similarities and differences from ice. Chapter 3 discusses kinetics with an objective of understanding how hydrate formation occurs. Chapter 4 provides phase diagrams and simple prediction schemes for each of the hydrate phase diagram regions. Chapter 5 details the statistical thermodynamic prediction method by van der Waals and Platteeuw (1959), and its refinements and extensions. Chapter 6 provides a listing of the hydrate thermodynamic and transport property data since 1934 for natural gas pure components, mixtures, and inhibitors together with the common experimental techniques. Chapter 7 discusses **in situ** hydrates in the oceans and permafrost, as well as recovery schemes. Finally, Chapter 8 considers some common industrial problems (and solutions) concerning hydrates in processing and production. The appendices and the computer programs in the frontpapers deal with the predictions initially discussed in Chapters 4 and 5.

# 2

# Molecular Structure and Similarities to Ice

All of the common natural gas hydrates belong to either of two crystal structures, named structure I (12 Å cell) and structure II (17.3 Å cell), as shown in Figure 1-3. The determination of the structures is due primarily to the x-ray diffraction work over a score of years by von Stackelberg and his students (1932, 1949, 1954 a,b,c,d,e, 1956), notably Müller (1951a,b) and Meuthen (1958). The proposal for crystal structure II was made by Claussen (1951a), who also proposed structure I (1951b) at the same time as Pauling and Marsh (1952) and Müller and von Stackelberg (1951b) based on x-ray diffraction data. Jeffrey (1984) and his colleagues have done the most notable confirmation. The first portion of this chapter details the molecular hydrate structures, as well as that of ice - a related common water crystal.

Because the hydrate structures consist of 85% water (minimum) on a molecular basis, many of the hydrate mechanical properties resemble those of ice Ih. Among the exceptions to this heuristic are thermal expansivity and thermal conductivity, the only transport property of interest. The final portion of this chapter examines mechanical, electrical, and transport properties with emphasis on those properties which differ from ice.

## 2.1 Crystalline Structures of Ice Ih and Natural Gas Hydrates

In Section 2.1.1 two common condensed phases of water and the hydrogen bond are considered. With these bases relating to knowledge of a more common substance, the departures of hydrates from ice are considered in the hydrate cavity structures of Section 2.1.2. The cavities are then assembled to comprise each hydrate structure in Section 2.1.3, before a summary is presented in Section 2.1.4.

### 2.1.1 Ice Ih, Water, and the Hydrogen Bond

In this section, the structures of ice, water, and the hydrogen bond are based upon the classical works of Bernal and Fowler (1933), Pauling (1935), and Bjerrum (1952), as well as the reviews of Frank (1970), and Stillinger (1980). These subjects are treated in comprehensive detail in the seven volume series edited by Franks (1972-1982), to which any student of water compounds will wish to refer.

#### 2.1.1.a. Ice

The most common solid form of water is known as ice Ih, with the molecular structure as shown in Figure 2-1, (Durrant and Durrant, 1962). In ice each water molecule (shown as a circle) is hydrogen bonded (solid lines) to four others in essentially tetrahedral (Lonsdale, 1958) angles. These angles have the highest stability because there is almost no geometrical distortion; that is the O-O-O angle is 109.5°. Stillinger (1980) suggests that the tetrahedral coordination represents the most feasible way of packing molecules about a central water molecule to permit fully developed hydrogen bonds.

The water molecules in ice are bound in this tetragonal hydrogen-bonded structure so that ice forms in the non-planar "puckered" hexagonal rings as shown, rather than sheets. In ice Ih the typical distance between oxygen nuclei is 2.76 Å ; covalent-bonded protons are about 1 Å from an oxygen nucleus, and the hydrogen bond length comprises the remaining 1.76Å. Only one proton lies on each line connecting adjacent oxygen atoms in a hydrogen bond. The ice structure then appears as normal water molecules hydrogen-bonded in a solid lattice. Since water molecules are similarly bonded in hydrates, both water molecules and hydrogen bonds are considered briefly in the following two sections.

Other properties of the ice structure are common to hydrates, as detailed in Section 2.3. One of the most frequently cited molecular properties is that of Bjerrum crystal defects and their propagation. In his classic paper, Bjerrum (1952) considered the possible reasons for low temperature disorder in the ice lattice. The normal electrostatic interactions caused by the orientations of the oxygens and the protons are necessary but insufficient to account for the total disorder. In addition, as shown in Figure 2-2, thermally excited water molecules cause proton position faults in the ice lattice. There are two types of defects shown; the D defect has two protons between oxygen nuclei, and the L defect has none. These defects cause the surrounding water molecules to pivot about one hydrogen bond to resolve the unfavorable defect energy strain.

The faults act as catalysts to promote dipole turns, with one fault for every $10^6$ molecules, corresponding to a turn rate of $10^{-12}$ $sec^{-1}$ at an orientation fault site. A second type fault, that of proton jumps from one molecule to a neighbor, accounts for the

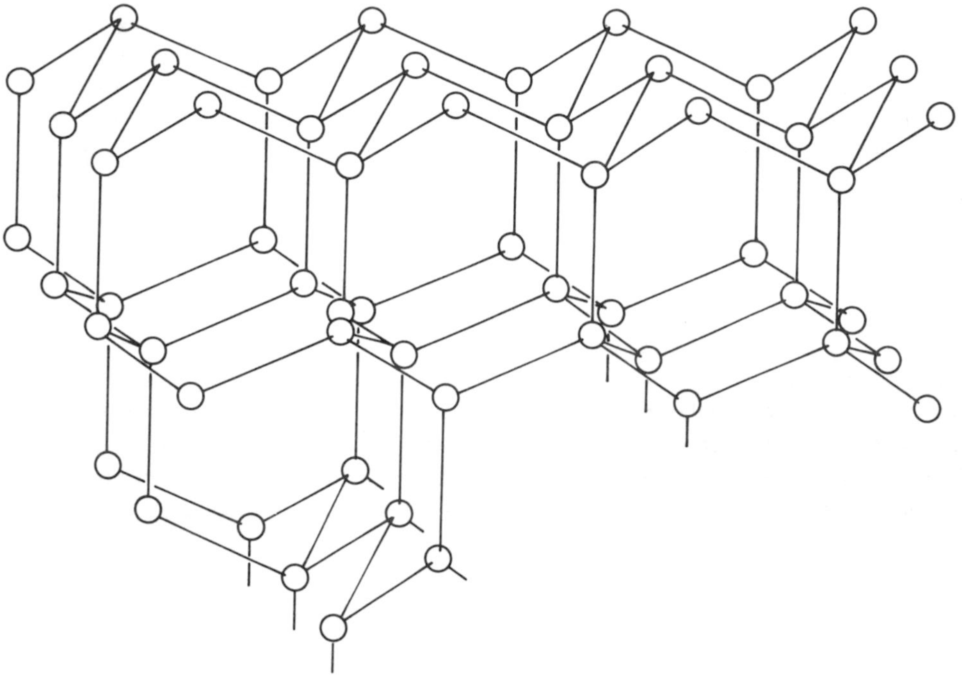

2-1 Basic Crystal Structure for Ice Ih (Reproduced, with permission, from Introduction to Advanced Inorganic Chemistry (Durrant and Durrant, 1962), by John Wiley and Sons, Inc.)

electrical conductivity of ice, but were shown to be insufficient to account for the high dielectric constant of ice. Molecular turns at orientation fault sites are responsible for the dielectric properties of ice. Further discussion of such ice faults is provided by Franks (1972), and by Onsager and Runnels(1969) who note that interstitial migration may be a more likely self diffusion mechanism. Similar defects are later seen to occur in hydrates, particularly when a polar guest species is present.

### 2.1.1.b. The Water Molecule

Figure 2-3a,b shows the model of Bernal and Fowler (1933) for the water molecule. The molecular geometry has been well determined (Benedict et al., 1956) from rotational and vibrational spectra. The oxygen atom has eight electrons which occupy the $1s^2 2s^2 2p^2$ configuration. Each hydrogen atom has an electron occupying the $1s^1$ configuration; these electrons are shared with two bonding electrons of oxygen to obtain water.

Four valence electrons form "lone-pair" orbitals which have been determined (Pople, 1951) to point above and below the plane formed by the three nuclei of the molecule. The shared electrons with the protons give the molecule two positive charges, and the lone pair electrons give the molecule two negative charges. The result is a molecule with four charges and a permanent electric dipole (McClellan, 1963) of 1.84 Debye. While the rotational energy levels of individual water molecules are

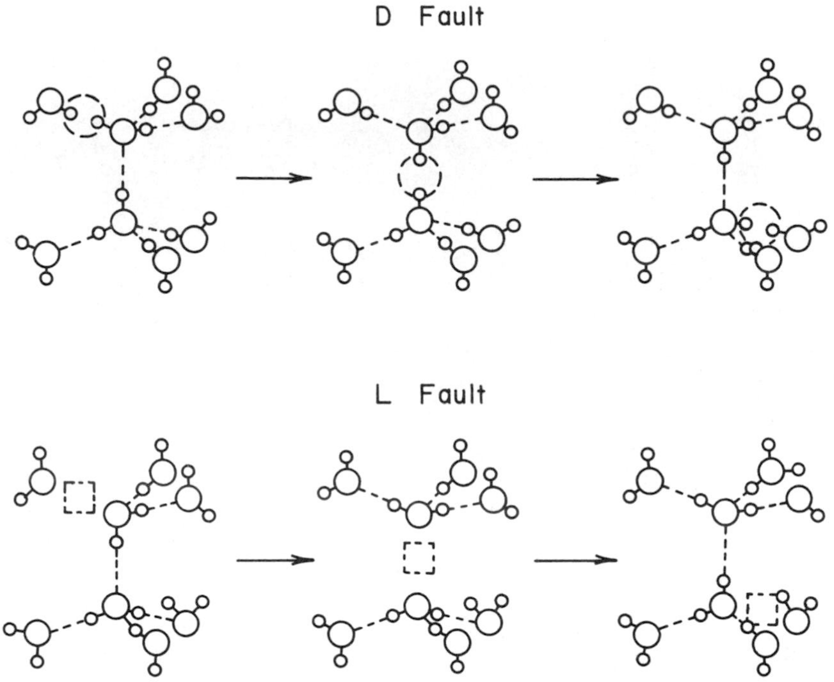


2-2 Two Types (D and L) of Bjerrum Faults (Reproduced, with permission, from the Journal of Chemical Physics (Onsager and Runnels, 1969), by the American Institute of Physics.)

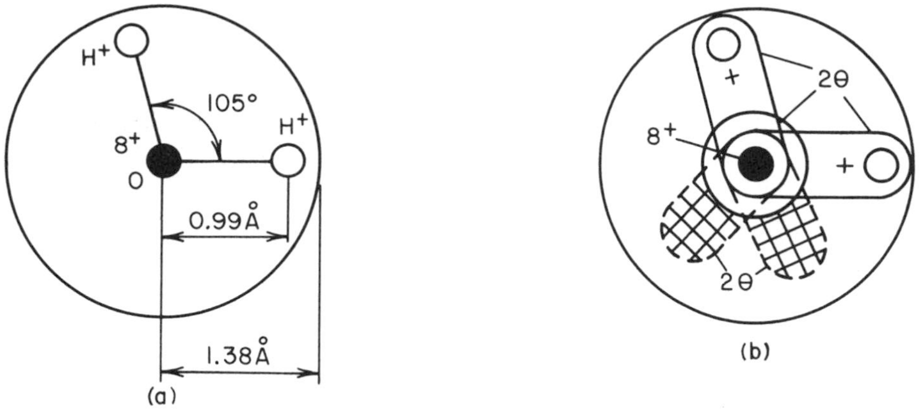


2-3 Water Model of Bernal and Fowler
(Reproduced, with permission, from Hydrates of Natural Gas, (Makogon, 1974) by the PennWell Publishing Co.)

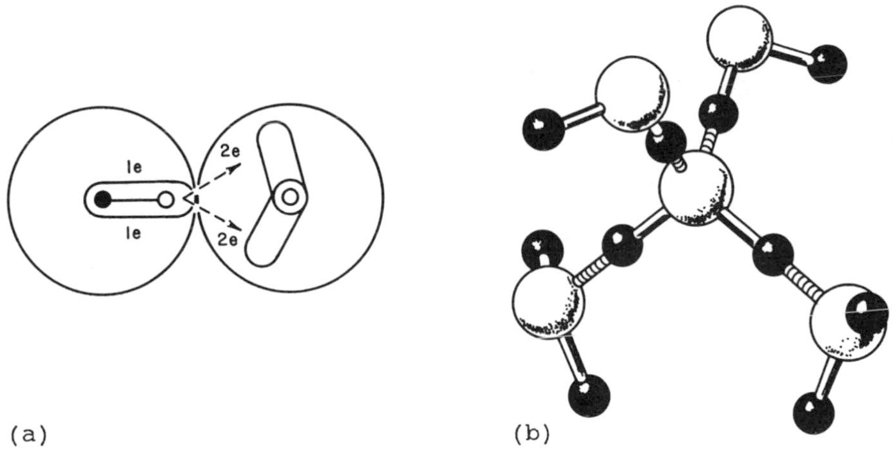


2-4 Hydrogen Bonding (Hydrogen Bonds are Crosshatched)
(a) Between Two Molecules (Reproduced, with permission, from Hydrates of Natural Gas (Makogon, 1974), by the PennWell Publishing Co.)
(b) Between Four Molecules (Reproduced, with permission, from Water: A Comprehensive Treatise (Franks, 1973: vol. 1, pg. 172), by Plenum Publishing Corp.)

spaced so that they need be considered, the vibrational energy spacings are too far apart to be of concern in ice or hydrate structures. Only the ground electron states, which give the molecule its four charges, need be considered.

### 2.1.1.c. The Hydrogen Bond

In 1920 two young men working in the G.N. Lewis' laboratory at Berkeley proposed the hydrogen bond (Latimer and Rodebush, 1920), using a simplified electrostatic point charge model of the water molecule. More recent work by Kollman (1977) indicated that the simplified model was satisfactory due to a cancellation of two other energy components. In the preceeding section the water molecule was described as having two positive and two negative poles. A water hydrogen bond, shown in Figure 2-4a, is caused by the attraction of the positive pole on one molecule to the two negative poles on a neighboring molecule.

Through this mechanism, each water molecule is attached to four others, donating two and accepting two hydrogen bonds, as illustrated in Figure 2-4b. Each proton of the molecule is attracted to the two negative poles of a neighboring molecule, thereby providing two bonds. Also the two negative poles on the initial molecule attract two positive poles from two other water molecules. The four surrounding molecules are arranged tetrahedrally around the central molecule. Along with the attraction of the hydrogen bond, the fact that these strong bonds separate the water molecules rigidly causes the solid density to be less than that of the liquid. In ice only 34 per cent of the volume is occupied by water molecules, in contrast to 37 per cent volume occupation in liquid water; this explains the unusual property of a decrease in density on freezing, or the fact that ice floats.

The energy required to break one hydrogen bond (ca. 5 kcal/mol) is substantially greater than a typical van der Waals bond (ca. 0.3 kcal/mol) such as would attract two non-polar molecules in a fluid. On the other hand, the hydrogen bond energy is not nearly as large as that of a covalent chemical bond (102 kcal/mol) such as exists between hydrogen and oxygen within an individual water molecule (Cottrell, 1958). Due to these bond strengths, when hydrates form or dissociate, only hydrogen bonds are considered between neighboring molecules. Van der Waals forces are present, but are insignificant relative to hydrogen bonds. Chemical bonds between hydrogen and oxygen are never perturbed within a water molecule when hydrates are formed.

Hydrogen bonds are also present to a large extent in liquid water. Luck (1973) notes that the hydrogen bonds inhibit the vaporization of liquid water, with the result that its normal boiling point is 260 K higher than that of methane, a compound with a similar molecular weight. Stillinger (1980) indicates that many of the unique properties of water are caused by the hydrogen bond, such as a maximum in liquid density at 277 K, an isothermal compressibility minimum at 319 K, an anomalously high heat capacity, etc.

In the solid-liquid transition, Pauling (1945, p. 304) has compared the heat of sublimation to the heat of fusion of ice to estimate that only 15% of the hydrogen bonds split on the melting of ice. This calculation supported the Frank and Evans (1945) "iceberg" model of liquid water molecules in a hydrogen-bonded network. Franks (1975) explained that the term "iceberg" is often misused, because such clusters were originally described as very short-lived or flickering and not measurable on a macroscopic time scale. Stillinger (1980) notes that disconnected icebergs should not be present, but that a more likely model is that of a random, three dimensional network of hydrogen bonds, rather than long-lived clusters of molecules. Similarly, random networks of hydrogen bonded water molecules are present to a large degree when hydrates form or dissociate. These mechanisms are discussed in relation to their effect on the nucleation of hydrates in the following chapter.

### 2.1.2 Hydrate Crystalline Cavities and Structures

The information on water structures in this section is derived, almost entirely, from the excellent reviews of Jeffrey and McMullan (1967) and of Jeffrey (1984), with the addition of substantial work on molecular motions reviewed by Davidson and Ripmeester (1984). Of more than 100 compounds which form clathrate hydrates with water molecules, all form either structure I or structure II, with four exceptions: 1) bromine (Allen and Jeffrey, 1963), 2) dimethyl ether (Gough et al. 1974, 1975), 3) ethanol (Brownstein et al., 1967, and Calvert and Srivastava, 1969), and 4) molecules approximately the size of methylcyclohexane or larger (Ripmeester et al., 1987). Ripmeester (1988) notes that ethanol probably is not a true clathrate, and that structure H clathrate requires at least 15 hydrocarbon molecules. However, since none of these exceptions involve natural gas compounds, they are not included here; the reader is referred to the reviews by Davidson (1973), Davidson and Ripmeester (1978), Jeffrey (1984), and the paper of Ripmeester et al. (1987) for other structures.

#### 2.1.2.a The Three Basic Cavities in Hydrates

The hydrate structures of Figure 1-3 are composed of the three polyhedra formed by hydrogen-bonded water molecules shown in Figure 2-5, with properties tabulated in Table 2-1. Jeffrey (1984) suggested the nomenclature discription ($n_i^{m_i}$), for these polyhedra, where $n_i$ is the number of edges in the i type of face and $m_i$ is the number of faces of i type. Thus the basic 12-hedra (dodecahedral cavity) of Figure 2-5a is denoted $5^{12}$ because it has twelve pentagonal faces. The 14-hedra (tetrakaidecahedral cavity) of Figure 2-5b is denoted $5^{12}6^2$ because it has 12 pentagonal and two hexagonal faces. Finally, the 16-hedra (hexakaidecahedral cavity) in Figure 2-5c is denoted $5^{12}6^4$ because, in addition to 12 pentagonal faces, it contains 4 hexagonal faces. Tabushi et al. (1981) have suggested that the 15-hedra is absent (in all clathrates except bromine) due to an unfavorable strain relative to the other three cavities.

Jeffrey noted that the 12-, 14-, and 16-hedra are not stable in a pure water structure. Since the three normal cavities are larger than ice cavities, hydrate cavities are prevented from collapse under their own attractive forces only by the presence of a guest molecule, either in the cavity itself, or in a large percentage of the neighboring cavities. Clathrate and ice densities are 45% (minimum) and 57% respectively, of the densities of closely packed spheres of water.

The Pentagonal Dodecahedra. The basic building block of both hydrate structures is the $5^{12}$ pentagonal dodecahedra cavity of Figure 2-5a, which is present as the small cavity in both hydrate structures. Because the pentagonal face is fundamental to the $5^{12}$ cavity, it is worth consideration.

Small clusters such as pentamers are difficult to examine experimentally, but a substantial amount of information has been determined via computer simulation. Molecular dynamics studies have been able to simulate unusual water attributes such as the density maximum, the isothermal compressibility minimum, and diffraction patterns. Such studies by Stillinger and Rahman (1974) suggest that the highest fraction of water molecules at 283

**Table 2-1 Geometry of Cages**

| Hydrate Crystal Structure | I | | II | |
|---|---|---|---|---|
| Cavity | Small | Large | Small | Large |
| Description | $5^{12}$ | $5^{12}6^{2}$ | $5^{12}$ | $5^{12}6^{4}$ |
| Number of Cavities/Structure | 2 | 6 | 16 | 8 |
| Average Cavity Radius, Å | 3.91 | 4.33 | 3.902 | 4.683 |
| Variation in Radius[1], % | 3.4 | 14.4 | 5.5 | 1.73 |
| Coordination Number[2] | 20 | 24 | 20 | 28 |

1. Variation in distance of oxygen atoms from center of cage.
2. Number of oxygens at the periphery of each cavity

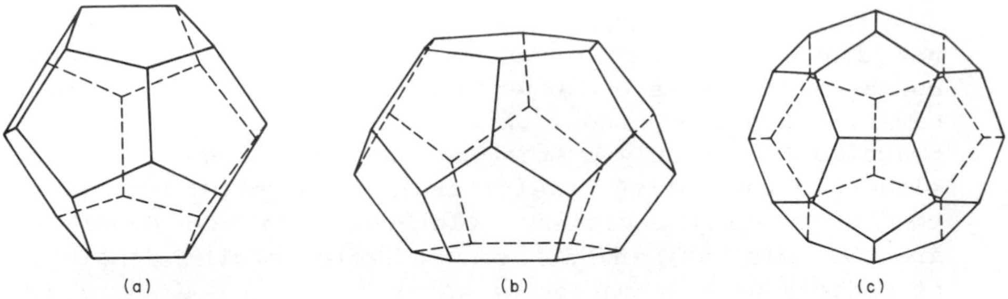


2-5 Three Cavities in Gas Clathrate Hydrates
(a) Pentagonal Dodecahedron ($5^{12}$)
(b) Tetrakaidecahedron ($5^{12}6^{2}$)
(c) Hexakaidecahedron ($5^{12}6^{4}$)
(Reproduced, with permission, from <u>Progress in Inorganic Chemistry</u> (Jeffrey and McMullan, 1967), by John Wiley and Sons)

K is hydrogen-bonded to two others. They also calculated that the pentamer is the most likely structure to spontaneously arise in the liquid at many temperatures (Rahman and Stillinger, 1973), followed in frequency by hexamers.

In a review of water, Frank (1970) noted that closed rings of bonds are always more stable than the most stable open chains of the same cluster number, due to the extra energy of the hydrogen bond. Through molecular dynamic studies of many five-molecule clusters, Plummer and Chen (1987) have shown that the cyclic pentamer which comprises all of the $5^{12}$ faces is effectively the only stable five member cluster above 230 K.

When twelve of the pentamers are combined the $5^{12}$ structure results. All 60 of the molecules in the 12 pentamers are not required for the structure; because pentamers share sides, only 20 molecules are required to make a $5^{12}$ cavity. Euclid (fl.c. 300 B.C.) proved that the structure is the largest of the five strictly regular convex polyhedra; it is the only hydrate cavity which has (almost) planar faces that have both equal edges and O-O-O angles. There is only $1.2^{0}$ departure of the O-O-O angles from the tetrahedral angles of ice Ih, and the O-O bond lengths exceed those in ice by only 1%. Angell

(1981) indicated that unstrained $5^{12}$ polyhedra arise naturally within the random hydrogen-bonded network in supercooled water.

Chen (1980, pg. 109) suggested that the $5^{12}$ cavity seems geometrically favored by nature because it maximizes the number of bonds (30) to molecules (20) along the surface, when compared to similar cavities. Holland and Castleman (1980) studied a number of clusters and determined that the $5^{12}$ cluster had a hydrogen bond advantage over ice, and that it was less strained than other clathrate clusters. The $5^{12}$ cavity (as well as the other hydrate cavities below) obeys Euler's theorem (Lyusternik, 1963) for convex polyhedra (F+V=E+2); the 12 faces (F) plus the 20 vertices (V) equals the 30 edges (E) plus 2. The $5^{12}$ cavity appears so geometrically appealing that it was once erroneously used to suggest both a clathrate model for water (Pauling, 1957) and a clathrate theory of anaesthesia (Pauling, 1961).

As shown in Table 2-1, the $5^{12}$ type cavity is almost spherical with a radius of 3.91 Å and 3.902 Å in structure I and II, respectively. This small dimensional difference determines the size of the occupation molecule. Until recently, it was thought (Davidson, 1973) that the smallest hydrate guest molecules occupied[1] the $5^{12}$ cavity of structure I. Davidson et al.(1984) crystallographically confirmed the suggestion of Holder and Manganiello (1982) that guest molecules of pure argon (3.83 Å diameter) or krypton (4.04 Å diameter) occupied the $5^{12}$ cavities of structure II. More recently Davidson et al. (1986) and Tse et al.(1986) determined that nitrogen and oxygen, respectively, occupied the $5^{12}$ cavity of structure II. Methane and hydrogen sulfide, with diameters of 4.36Å and 4.58Å respectively, occupy the $5^{12}$ cavity of structure I. Helium, hydrogen, and neon, the smallest molecules with diameters less than 3.0 Å, cannot stabilize any cage, but are able to pass through the "bars" formed by hydrogen bonds; therefore they do not form hydrates.

[1] To "occupy" means to fit into a cavity; molecules which occupy small cavities will also occupy the large cavities of the structure.

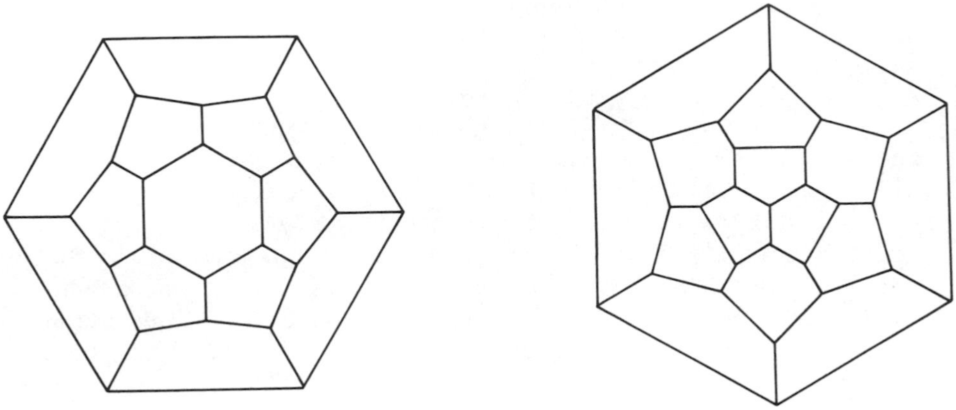

(a) (b)

2-6 Schlegel Diagrams for (a) Tetrakaidecahedron ($5^{12}6^2$)
(b) Hexakaidecahedron ($5^{12}6^4$)
(Reproduced, with permission, from the Journal of Chemical Physics (Allen, 1964), by the American Institute of Physics)

The Tetrakaidecahedron. The semi-regular solid geometries of the 14-hedra ($5^{12}6^2$) and the 16-hedra ($5^{12}6^4$) are detailed by Allen (1964), who presented Schlegel diagrams of these cavities, to complement the diagrams of Figure 2-5b,c. In these two-dimensional diagrams, shown in Figure 2-6 a,b for the $5^{12}6^2$ and $5^{12}6^4$ cavities respectively, one has the perspective of looking into the cavity interior from one of the hexagonal faces. The periphery of each figure is hexagonal, and the placement of the other hexagonal faces serve to locate the pentagonal faces.

The cavity follows Euler's theorem, with 14 faces plus 24 vertices to give 36 edges plus 2. Figure 2-6a shows the 14-hedra to be formed by two facing hexagons, with 12 connecting pentagons. Each hexagon has six pentagons attached to its edges, resulting in two "cups" formed by hexagons at the base with pentagons at the sides. The two cups join at the periphery of the pentagon edges to form the cavity. The orthogonal view of the $5^{12}6^2$ cavity, shown in Figure 2-5b, represents the most non-spherical of the cavities. The four crystallographically different types of oxygen sites are 4.25 Å, 4.47 Å, 4.06Å, and 4.645 Å from the 14-hedra cavity center, giving a 14% variation from sphericity, resembling an oblate ellipsoid. This cavity also has the largest O-O-O angle variation ($5.1^0$) from the tetragonal angle preferred by water.

With an average radius of 4.33 Å, the $5^{12}6^2$ cavity is large enough to contain molecules smaller than 6.0Å in diameter. Ripmeester (1988) indicates that the 14-hedron is the preferred cage for almost all structure I hydrates, in which it plays the main stabilizing role. The oblate nature of this cage causes the shape of the guest molecule to play a particularly large role in its occupation. Of the natural gas components, only ethane and carbon dioxide will stabilize this cavity as simple hydrates. If the $5^{12}$ cavities of structure I are stabilized by smaller molecules, those molecules may also occupy the $5^{12}6^2$ cavities, as discussed in Section 2.1.2.

<u>The Hexakaidecahedron</u>. Figures 2-5c and 2-6b provide the orthogonal view and the Schlegel diagram for the 16-hedra. The latter diagram allows Euler's theorem to be verified; 16 faces plus 28 vertices equals 42 edges plus 2. The $5^{12}6^4$ notation for this cavity indicates four hexagonal faces; the hexagonal faces are symmetrically arranged so that their normals form the vertices of a tetrahedron. Each hexagonal face is surrounded entirely by pentagonal faces. No two hexagons share a common edge. The radii from the cavity center to the crystallographically different oxygen sites does not vary more than 1.7%; therefore the 16-hedra is the most spherical cavity of the three types.

The $5^{12}6^4$ cavity has the capacity to contain molecules as large as 6.6 Å. Consequently, when the larger components of natural gas such as propane and isobutane form simple (single guest species) hydrates, they stabilize this cavity alone in structure II but the smaller cavities remain vacant. Wu et al. (1976) experimentally proved that n-butane, a much larger molecule, can be enclathrated over a limited range of temperature and pressure, when a help gas such as methane is present to occupy smaller cavities in structure II. Davidson et al. (1977) similarly formed a hydrate of n-butane and hydrogen sulfide. While no one has proven microscopically that n-butane forms a structure II hydrate (Ripmeester, 1988), it is universally assumed to occupy the $5^{12}6^{4}$ cavity due to its size. Trimethylene oxide' (Hawkins and Davidson, 1966), cyclopropane (Hafemann and Miller, 1969, Majid et al., 1969) and ethylene sulfide (Ripmeester, 1988) are three molecules which can form in either the 14-hedra of structure I or the 16-hedra of structure II as simple hydrates.

### 2.1.2.b Hydrate Crystal Cells - Structures I and II

Jeffrey (1984) listed seven of a series of hydrate inclusion compounds in terms of the possible associations of polyhedra such as those in the last section. He noted however, that a "study of all the possible three-dimensional four-connected structures based on polyhedra with twelve or more vertices is a formidable topological problem." Indeed the most recently determined structure (Ripmeester et al., 1987), for hydrates of large molecules such as methylcyclohexane and a help gas, does not seem to be one of the seven listed by Jeffrey.

The basic crystal properties of both structures are given in Table 2-2. This table allows the contrast of structures obtained by linking the basic $5^{12}$ cavity in two different ways. All modes of associating pentagonal dodecahedra lead to either fivefold or sixfold coordination of water molecules, except the following two, which yield fourfold hydrogen bonds: (1) by linking the vertices of dodecahedra, or (2) by sharing common faces of adjacent dodecahedra. Structure I is an example of vertex linking of the 12-hedra in three dimensions, while structure II illustrates face sharing of the 12-hedra in three dimensions.

**Table 2-2 Hydrate Crystal Cell Structures**

| | | |
|---|---|---|
| Structure | I | II |
| Crystal System | Cubic | Cubic |
| Space Group | Pm3n(#223) | Fd3m(#227) |
| Lattice Description | Body Centered | Diamond |
| Lattice Parameter, Å | 12 | 17.3 |
| Ideal Unit Cell Formula[1] | $6X \cdot 2Y \cdot 46H_2O$ | $8X \cdot 16Y \cdot 136H_2O$ |
| Atomic Positions, number and site symmetry | X(d) 6, $\bar{4}2m$ | X(b) 8, $\bar{4}3m$ |
| | Y(a) 2, $m^3$ | Y(c) 16, $\bar{3}m$ |
| | O(c) 6, $\bar{4}2m$ | O(a) 8, $\bar{4}3m$ |
| | O(i) 16, 3m | O(e) 32, 3m |
| | O(k) 24, m | O(g) 96, m |
| | ($^1/_2$ H)(i) 16, 3 | 2($^1/_2$ H)(e) 32, 3m |
| | 3($^1/_2$ H)(k) 24, m | 3($^1/_2$ H)(g) 96, m |
| | 2($^1/_2$ H)($\ell$) 48, $\ell$ | ($^1/_2$ H)(i) 192, m |

1. X and Y refer to guest molecules in large voids and in 12-hedra, respectively

(Table modified from Jeffrey (1984, p. 150))

Davidson and Ripmeester (1984) discuss the mobility of water molecules in the host lattices, based upon NMR and dielectric experiments. Water mobility comes from molecular reorientation and diffusion, with the former being substantially faster, in contrast to water mobility in ice. Dielectric relaxation data suggests that Bjerrum defects in the hydrate lattice, caused by guest dipoles, may enhance water diffusion rates. Spectroscopic evidence and further discussion of these phenomena are given in Section 2.2.1.

<u>Structure I</u>. Definitive X-ray diffraction data on structure I was obtained by McMullan and Jeffrey (1965) for ethylene oxide hydrate, as presented in Table 2-2. The common pictorial view of structure I is presented in Figure 1-3a. In that figure, the front face of a 12 Å

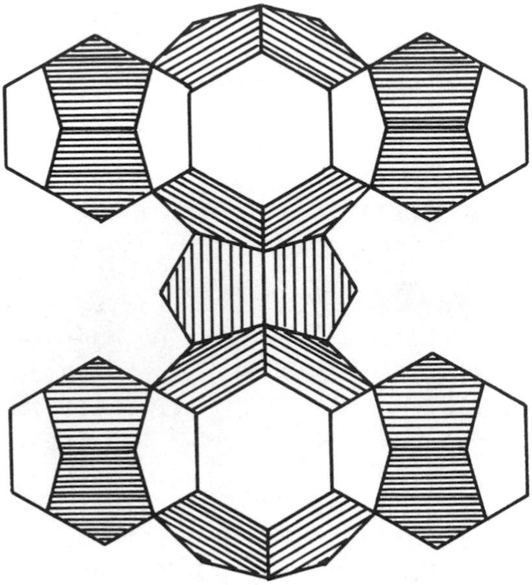

(a)

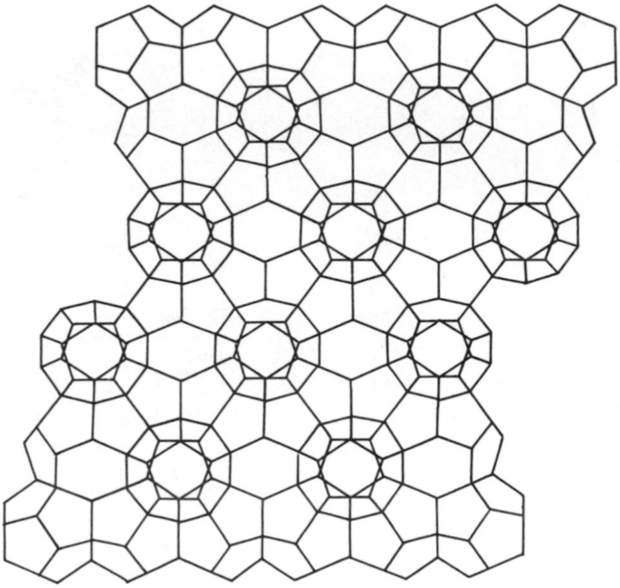

(b)

2-7 (a) Linking of Five $5^{12}$ Polyhedra by Two $5^{12}6^{2}$ Polyhedra to Form Structure I

(b) Two Dimensional View of Face Sharing of $5^{12}$ Polyhedra to Form $5^{12}6^{4}$ Polyhedra Voids for Structure II

(Reproduced, with permission, from Progress in Inorganic Chemistry (Jeffrey and McMullan, 1967), by John Wiley and Sons)

cube is shown, with two complete 14-hedra (emphasized hydrogen bonds) connecting four 12-hedra.

An alternative, less frequent view is found in Figure 2-7a, which shows five 12-hedra connected by two of the 14-hedra; this is a primary illustration of the linking of pentagonal dodecahedra through additional water molecules, which form the 14-hedra. Jeffrey (1984, p. 153) suggests that the structure is defined by the 14-hedra, in the following description:

> "There is no direct face sharing between the 12-hedra. The structure can be constructed from the vertices of face-sharing 14-hedra arranged in columns with the 14-hedra sharing their opposing hexagonal faces. These columns are then placed in contact so as to share a pentagonal face between each pair of 14-hedra, with the column axes along (x,1/2,0) of a 12 Å cubic cell. The remaining space is then the 12-hedra in a body-centered cubic arrangement of their centers."

While Figure 1-3a appears to contain many water molecules, there are only 46 water molecules inside the structure I cubic cell and there are only eight polyhedra totally within the cube. Each of the six faces contains two halves of a 14-hedra, for a total of six 14-hedra within the cell. Each of the eight vertices of the cube contains one-eighth of a 12-hedra which, added to the 12-hedra in the center of the cube, gives a total of two 12-hedra per cell.

<u>Structure</u> <u>II</u>. After crystal structure II was deduced, a definitive X-ray diffraction study of tetrahydrofuran/hydrogen sulfide hydrate was undertaken by Mak and McMullan (1965), two of Jeffrey's colleagues. They confirmed the crystal to be a diamond lattice, which fits within a cube 17 Å on a side, with parameters as given in Table 2-2 and shown in Figure 1-3b. In direct contrast to the properties of structure I this figure il-

lustrates how a crystal structure may be completely defined by the vertices of the smaller 12-hedra. Because the 12-hedra outnumber the 16-hedra in the ratio 16:8, only the 12-hedra are clearly visible in Figure 1-3b.

A second view of a layer of structure II is presented in Figure 2-7b, which is a two dimensional view of the way many face-sharing 12-hedra are arranged so that the residual voids are 16-hedra (with hexagonal faces). This is only a partial view of the crystal cell; such layers are stacked in a staggered pattern ABCABC so that the centers of the 16-hedra form a diamond lattice within a cube, with shared hexagonal faces. Jeffrey (1984, p. 156) suggests that the voids formed by connected 16-hedra could accommodate much larger guests (perhaps sharing voids) than those normally found in gas hydrates, but such clathrates have yet to be found. When the layer in Figure 2-7b forms the pattern ABABA a hexagonal, rather than cubic, cell is formed; this structure is not found in natural gas hydrates, but is present with such molecules as isopropylamine, $(CH_3)_2CHNH_2 \cdot 8H_2O$.

### 2.1.3 Characteristics of Guest Molecules

A second classification of hydrates is obtained through consideration of the guest molecules. Such a classification is a function of two factors: 1) the chemical nature of the guest molecule and 2) the size, and to a minor extent the shape, of the guest. The size of the guest molecule is directly related to the hydrate number and, in most cases, to its non-stoichiometric value.

#### 2.1.3.a. Chemical Nature of Guest Molecules

Two classifications of the chemical nature of the guest molecule have been proposed. The first scheme by von Stackelberg and co-workers (1956) was a combination of both size and chemical nature as discussed in Section 1.2.3. The second scheme by Jeffrey and McMullan (1967) characterized the guest molecules in one of four groups: 1) hydrophobic compounds, 2)water soluble acid gases, 3)water soluble polar compounds, and 4)water soluble ternary or quaternary alkylammonium salts.

Jeffrey (1984) clearly summarized chemical nature-based classification schemes by stating that the guest molecule must not contain either a single strong hydrogen-bond group, or a number of moderately strong hydrogen bonding groups. The molecules of natural gas components are not involved in hydrogen bonding, and so their chemical nature is not a delimiter. Most of the natural gas molecules which form hydrates are hydrophobic, with the notable exceptions of hydrogen sulfide and carbon dioxide, so that they fall within the first two catagories of Jeffrey and McMullan (1967). In the past, experimentalists have capitalized on the complete aqueous miscibility of the cyclic ethers, such as ethylene oxide in forming structure I and tetrahydrofuran in forming structure II, to provide easy access to hydrate crystal structures. While these cyclic ethers are easy to enclathrate, they promote a disproportionate number of Bjerrum defects in the hydrate (Davidson and Ripmeester, 1984).

Restrictions on Guest Motions. The physio-chemical nature of the guest molecule, once enclathrated, has been studied in some detail. X-ray or neutron diffraction results are insufficient for the separation of effects of orientation disorder and thermal oscillatory motion (Jeffrey, 1984), so that dielectric and nuclear magnetic resonance (NMR) techniques have been successfully applied by Davidson (1971) and Davidson and Ripmeester (1984). The translational degree of freedom of the guest is limited, but potential energy calculations indicate that most guest molecules prefer to be located off-center near the cavity wall, particularly at temperatures below the normal boiling point of nitrogen.

Davidson (1971) determined that the most important rotation inhibition interactions between guest molecules (in adjacent cages) were the dipole-dipole interactions, but even they were of minor importance. Within a single cage, both non-polar and polar guest molecules such as ethylene oxide, tetrahydrofuran, and acetone have only small barriers to rotational freedom, which approximates that in the vapor phase. The rotational freedom is probably due to the fact that the sum of the cage water dipoles effectively cancel near the center of each cage, and the quadrupolar fields are relatively small.

### 2.1.3.b. Geometry of the Guest Molecules

In a review of the motions of guest molecules in hydrates, Davidson (1971) indicated that all molecules between the sizes of argon (3.8Å) and cyclobutanone (6.5Å) can form hydrates, if the above restrictions of chemical nature are obeyed. Ripmeester (1988) notes that the largest simple structure II formers are m- and p-dioxane and carbon tetrachloride, each with a molecular diameter of 6.8Å. The structure I and II water lattice parameters of 12Å and 17.3Å show no significant distortion by any guest, as determined by von Stackelberg and Jahns (1954).

Size of Guest Molecules. To determine the size upper limit of each cavity available for a guest, Davidson suggested subtracting the van der Waals radius of the water molecule (1.45Å) from the "average cavity radius" values given in Table 2-1. To determine the upper and lower limits to guest size, it is instructive to consider the diameter ratios of the guest molecule to each cavity for simple (no other guest species present) hydrate formers.

Table 2-3 presents the diameter ratios of some guests (including natural gas components) relative to those of each cavity in both structures. Also presented are two unusual molecules, cyclopropane and trimethylene oxide, which can form simple hydrates of either structure; discussion of the hydrates of these molecules is deferred until the next subsection. The ratios denoted with an "ℱ" are those occupied by the simple hydrate former. The values of the table demonstrate a ratio lower bound of about 0.77, below which the molecular attractive forces cannot contribute to cavity stability. Above the upper bound ratio of about 1.0, the guest molecule cannot fit into the cavity without distortion. Species capable of entering the $5^{12}$ cavity of either structure will also enter the large cavity of that structure.

For simple hydrate-forming components of natural gas, ethane is shown to occupy the $5^{12}6^{2}$ cavities of structure I with a ratio of 0.955, while propane and isobutane each occupy the $5^{12}6^{4}$ cavities of structure II with ratios of 0.971 and 1.005, respectively. N-butane does not form a simple hydrate, because even with the given diameter of the **gauche** molecular configuration, the diameter ratio is far beyond the upper limit of 1.00. Nitrogen has recently been determined (Davidson et al., 1986) to stabilize the $5^{12}$ cavity of structure II. Molecules of methane, and hydrogen sulfide can occupy the $5^{12}$ cavities of structure I with size ratios between 0.89 and 0.93.

In Table 2-3 it is interesting to note that the simple hydrate of methane always occupies the small cavity of structure I (diameter ratio = 0.886) while the diameter ratio of methane to the small cavity of structure II is 0.889, apparently a very small difference. Ripmeester (1988) indicates that for small, simple hydrate formers the transition from structure II to structure I is brought about by the additional stability gained by the guest molecule in the $5^{12}6^{2}$ cavity. When the restriction of a simple hydrate is removed, the addition of a small amount of a second, larger hydrocarbon has a dramatic effect. In Chapter 5 it is predicted that the addition of a small amount (≥ 0.5 mol %) of propane to methane will cause the hydrate structure to change to structure II, as evidenced by a substantial decrease in hydrate formation pressure at a given temperature. Since propane can only fit into the large cavity of structure II, and since the diameter ratio for methane in the $5^{12}$ cavity of each

**Table 2-3 Ratios of Molecular Diameters[1] to Cavity Diameters[2] for Some Molecules Including Natural Gas Hydrate Formers**

| | | (Molecular Diameter)/(Cavity Diameter) | | | |
|---|---|---|---|---|---|
| | | Structure I | | Structure II | |
| | Cavity Type ⇒ | $5^{12}$ | $5^{12}6^{2}$ | $5^{12}$ | $5^{12}6^{4}$ |
| Molecule | Guest Dmtr(Å) | | | | |
| He | 2.28 | 0.463 | 0.396 | 0.465 | 0.353 |
| $H_2$ | 2.72 | 0.553 | 0.470 | 0.555 | 0.421 |
| Ne | 2.97 | 0.604 | 0.516 | 0.606 | 0.459 |
| Ar | 3.8 | 0.772 | 0.660 | 0.775[†] | 0.599[†] |
| Kr | 4.0 | 0.813 | 0.694 | 0.816[†] | 0.619[†] |
| $N_2$ | 4.1 | 0.833 | 0.712 | 0.836[†] | 0.634[†] |
| $O_2$ | 4.2 | 0.853 | 0.729 | 0.856[†] | 0.649[†] |
| $CH_4$ | 4.36 | 0.886[†] | 0.757[†] | 0.889 | 0.675 |
| Xe | 4.58 | 0.931[†] | 0.795[†] | 0.934 | 0.708 |
| $H_2S$ | 4.58 | 0.931[†] | 0.795[†] | 0.934 | 0.708 |
| $CO_2$ | 5.12 | 1.041 | 0.889[†] | 1.044 | 0.792 |
| $C_2H_6$ | 5.5 | 1.118 | 0.955[†] | 1.122 | 0.851 |
| c-$C_3H_6$ | 5.8 | 1.178 | 1.007[†] | 1.182 | 0.897[†] |
| $(CH_2)_3O$ | 6.1 | 1.240 | 1.059[†] | 1.244 | 0.943[†] |
| $C_3H_8$ | 6.28 | 1.276 | 1.090 | 1.280 | 0.971[†] |
| i-$C_4H_{10}$ | 6.5 | 1.321 | 1.128 | 1.325 | 1.005[†] |

| | | (Molecular Diameter)/(Cavity Diameter) | | | |
|---|---|---|---|---|---|
| | | Structure I | | Structure II | |
| | Cavity Type ⇒ | $5^{12}$ | $5^{12}6^{2}$ | $5^{12}$ | $5^{12}6^{4}$ |
| Molecule | Guest Dmtr(Å) | | | | |
| $n\text{-}C_4H_{10}$ | 7.1 | 1.443 | 1.232 | 1.447 | 1.098 |

ǂ indicates the cavity occupied by the simple hydrate former

1. molecular diameters obtained from von Stackelberg and Müller (1954), Davidson (1973), Davidson et al.(1984,(1986), or Hafemann and Miller (1969).
2. cavity radii from Table 2-1 minus 1.45Å water radii

structure differs less than 0.5%, this change of structure is not surprising. Molecules such as natural gas components of hydrogen sulfide and nitrogen with cavity diameter ratios similar to methane, might be expected to undergo similar structural changes in the presence of larger molecules.

Of the natural gas components which form simple hydrates, nitrogen, propane and isobutane are known to form structure II. Methane, ethane, carbon dioxide, and hydrogen sulfide all form structure I as simple hydrates. Yet because the larger molecules of propane and isobutane will only fit into the large cavity of structure II, natural gas mixtures containing propane and isobutane will usually form as hydrate structure II. In the following section, the discussion of guest size continues, particularly with regard to the affect of size on hydrate number.

Shape of Guest Molecules. The shape of guest molecules plays a minor part in the hydrate structure and properties. Davidson and co-workers (1977,1984) have performed NMR spectroscopy and dielectric relaxation measurements, where applicable, in order to estimate the barriers to molecular reorientation for all natural gas hydrate components, except carbon dioxide. If substantial barriers to rotation occur, they would affect such properties as hydrate heat capacity.

For structure I formers, essentially no barriers were found for methane and hydrogen sulfide, while the average barrier for ethane was 1.2 kcal/mol, indicating a rotational restriction due to its shape in the oblate 14-hedra. The average barriers for structure II formers were 0.6, 1.2, and 1.4 kcal/mol for propane, isobutane and n-butane (double hydrate with hydrogen sulfide) respectively. These barriers, while measurable, are still small, on the order of a van der Waals bond.

Due to shape restrictions, n-butane is enclathrated (as a mixed hydrate) in the **gauche** isomer (Davidson et al., 1977) rather than the **trans** isomer which is preferred in the gas phase. At most temperatures of interest to the natural gas processor, above 100K the guest molecules have very small restrictions to reorientation. For guest molecules of intermediate sizes, such as cyclopropane and trimethylene oxide, small changes in size caused by thermal stimulation of rotational and vibrational energies may be sufficient to determine the occupied cavity as discussed in the following section.

### 2.1.3.c. The Filling of Hydrate Cages

In both hydrate structures each cavity can contain at most one guest molecule; there is no published instance of occupancy of a cavity by more than one guest. Figure 2-8, a revision of a figure originally by von Stalkelberg (1949), presents the sizes of simple gas hydrates relative to each cavity. This figure, explained further in the following sections, summarizes the individual cavity discussions in Section 2.1.2.

Ideal Hydrate Numbers. In Figure 2-8, note the fact that new X-ray and neutron diffraction results (Davidson et al. 1984, 1986) for argon, krypton, oxygen and nitro-

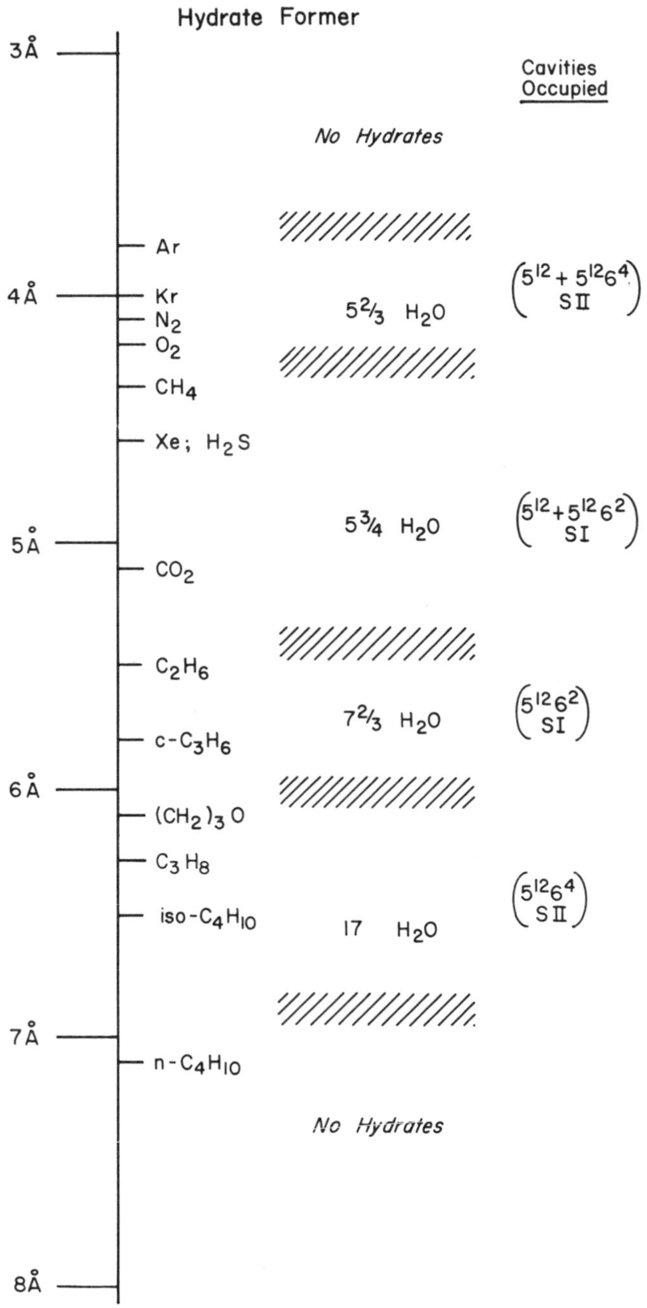


2-8 Comparison of Guest Molecule Sizes and Cavities Occupied as Simple Hydrates
(Modified, with permission, from Naturwissenchaften (von Stackelberg, 1949), by Springer-Verlag.)

gen indicate that these simple hydrates not only stabilize the 16 smaller cavities of structure II but also occupy the 8 larger cavities of that structure, for an ideal guest/water ratio of $24G:136H_2O$ or $G:5^2/_3H_2O$. If guest molecule G is too large for the small cavity of structure II, the ideal ratio will be $8G:136H_2O$ or $G:17H_2O$.

The ideal guest/water ratio is $G:5^3/_4H_2O$ for molecules which can occupy both cavities of structure I, and $G:7^2/_3H_2O$ for occupants of only the large cavity of structure I. As indicated in Figure 2-8 molecules of transitional size (shaded region) such as cyclopropane (Majid et al., 1969) and trimethylene oxide (Hawkins and Davidson, 1966), which have diameters of 5.8Å and 6.1Å respectively, may form either structure.

The ideal hydrate numbers validate the notion that a substantial amount of hydrocarbon is present in the hydrate. For example, if all the cavities of structure I are filled, each volume of hydrate may contain 189 volumes of gas at standard temperature and pressure. This ratio shows hydrated gas density to be equivalent to a highly compressed gas, but somewhat less than the density of a liquid hydrocarbon. The similarity of hydrates to a highly compressed gas suggests their use for storage, or as an unconventional gas resource, where they occur **in situ** in the deep oceans or permafrost.

Because it is impossible for all of the cavities to be occupied (an analog would be a perfect crystal) simple hydrates always have more water molecules than the ideal composition. Usually the ratios range from $G:5^3/_4H_2O$ to $G:19H_2O$, with typical fractional occupancies of the smaller cavities of 0.7 to 0.9, based upon size restrictions. This variation causes clathrate hydrates to be called "non-stoichiometric hydrates", to distinguish them from stoichiometric salt hydrates.

Hydrate Non-Stoichiometry. The cause of the non-stoichiometric properties of hydrates has been considered. Evidence for the view that only a fraction of the cavities need be occupied, is obtained from both the experimental observations of variation in composition, and the theoretical success of the statistical thermodynamic approach of van der Waals and Platteeuw (1959) (see Chapter 5).

Davidson (1971) indicated that occupancy of the smaller cages is incomplete for molecules less than 5.0 Å in diameter, and that the larger cages of each structure seem to be almost completely occupied. Via the use of NMR (currently the only experimental technique for determining molecular occupation of cavities) Ripmeester and Davidson (1981) have determined that for $^{129}Xe$, the occupancy of the small cages of structure I is 0.74 times that of the large cages.

Glew (1959) suggested that the most non-stoichiometric guest molecules are those for which the size of the guest approaches the upper limit of the free volume of a cavity. For two molecules which approach the size limit of cavities, Glew and Rath (1966) presented experimental evidence that the non-stoichiometry of hydrates of both chlorine and ethylene oxide was due to the composition of the phase in equilibrium with the hydrates. A systematic determination of both hydration number (Cady, 1983a) and relative cage occupancies (Davidson and Ripmeester, 1984) shows that molecules such as $CH_3Cl$, $SO_2$, and $Cl_2$ are the most non-stoichiometric. While theoretical calculations using the van der Waals and Platteeuw model provides some rationale for the non-stoichiometry, experimental quantification of non-stoichiometry as a function of guest/cavity size ratio has yet to be determined.

Of particular interest to the question of structure are the simple hydrates of cyclopropane, and trimethylene oxide because they can form in either the $5^{12}6^2$ cavity of structure I, or the $5^{12}6^4$ cavity of structure II, as a function of formation conditions. Note that these simple hydrates should be considered as examples of structure change, because they are essentially stoichiometric in structure II, but somewhat less stoichiometric in structure I. Here, for purposes of illustration, we consider cyclopropane, first determined to form each structure by Hafemann and Miller (1969). Figure 2-9 is a corrected phase diagram by Majid et al. (1969), similar to that of Figure 1-1. Note in this figure that most of the hydrate area is occupied by structure I, except for a smaller area between 257.1 K and 274.6 K where structure II hydrate forms, This provides four quadruple points: the normal lower point $Q_1$ ($I-L_W-H_{II}-V$) at about the ice point, the normal upper point $Q_2$ ($L_W-H_I-V-L_{CP}$) at 289.4 K, a new low point $Q_3$ ($I-H_I-H_{II}-V$) at 257.1 K, and a new intermediate point $Q_4$ ($L_W-H_I-H_{II}-V$) at 274.6 K. Between the new quadruple points is line $\overline{Q_3}\overline{Q_4}$ where both hydrate phases coexist with vapor; the line may be a unique example of the way molecular size affects the cavity occupied.

Table 2-4 is a subsection of Table 2-3 which lists ratios of molecular diameters to cavity diameters for cyclopropane and trimethylene oxide. Both molecules are within the previously determined size stability ratio for the $5^{12}6^2$ cavity of structure I and the $5^{12}6^4$ cavity of structure II. However both guest molecules appear to be just at the upper stability limit of the $5^{12}6^2$ cavity. Majid et al. (1969) note that there are appreciably more Bjerrum defects in the water molecules of the cage, presumably caused by the distortion of the $5^{12}6^2$ cavity by the guest cyclopropane molecule. Ripmeester (1988) indicates that recent $^2H$ NMR measurements show that the motion of cyclopropane in the 14-hedron is quite anisotropic.

From the initial work by Hafemann and Miller (1969) and Majid et al.(1969), together with the ratios of Table 2-4, one may infer a hypothesis for hydrate structure change which may be related to non-stoichiometry, particularly in in structure I. It seems reasonable to suggest that an increase in rotational and vibrational energy causes the guest molecule to enlarge with temperature at low pressure, so that it becomes too large for the $5^{12}6^2$ cavity of structure I, but it will fit the larger $5^{12}6^4$ cavity of structure II. At temperatures lower than 257.1 K the guest expansion may not be great enough to

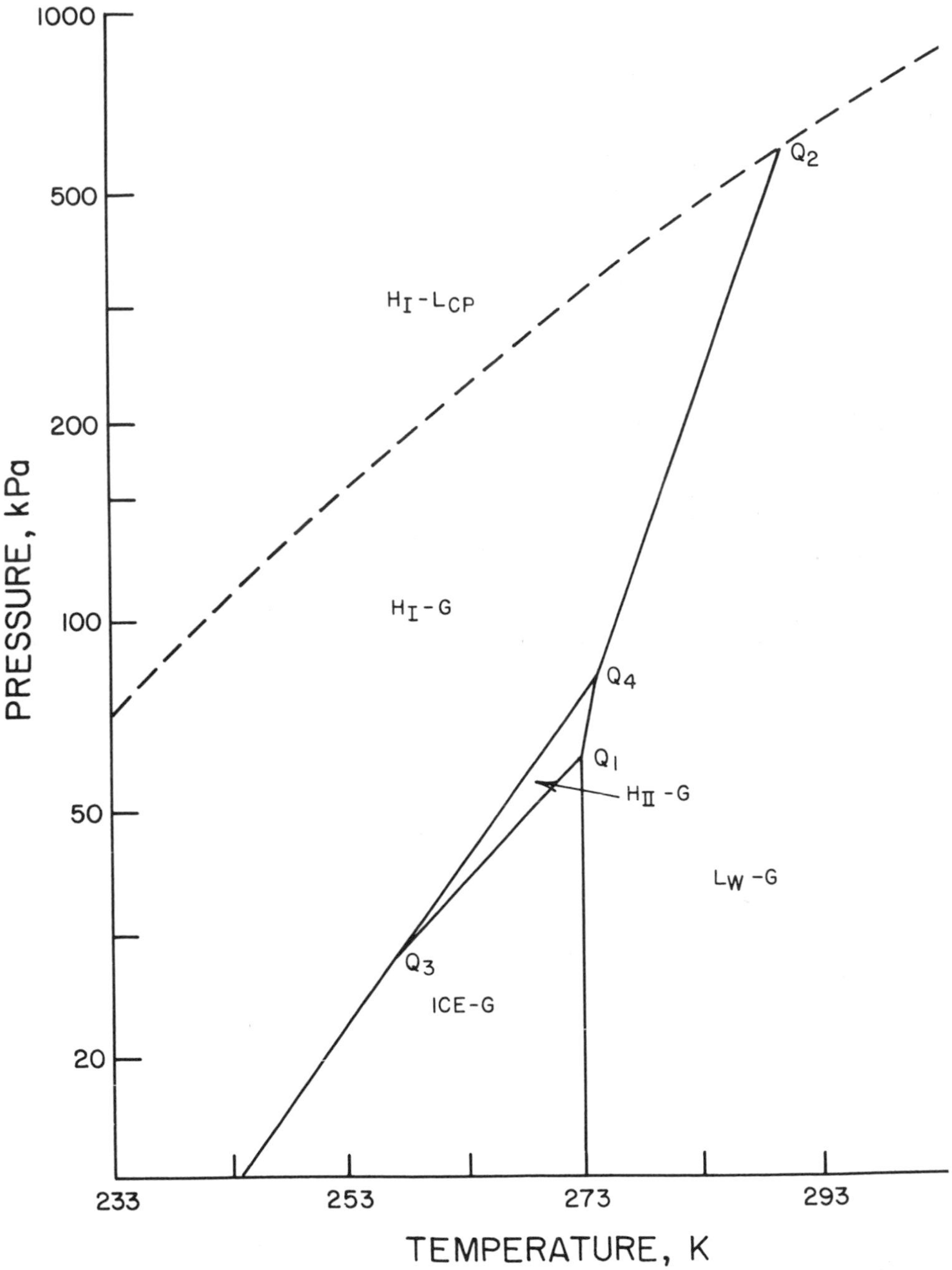


2-9 Pressure-Temperature Phase Diagram for Cyclopropane (Reproduced, with permission, from the Canadian Journal of Chemistry (Majid et al., 1969), by the National Research Council of Canada.)

**Table 2-4. Ratios of Molecular Diameters[1] to Free Cavity Diameters[2] for Cyclopropane and Trimethylene Oxide**

| | | (Molecular Diameter)/(Cavity Diameter) | | | |
|---|---|---|---|---|---|
| | | Structure I | | Structure II | |
| Cavity Type ⇒ | | $5^{12}$ | $5^{12}6^{2}$ | $5^{12}$ | $5^{12}6^{4}$ |
| Molecule | Guest Dmtr(Å) | | | | |
| $c\text{-}C_3H_6$ | 5.8 | 1.178 | 1.007ƒ | 1.182 | 0.897ƒ |
| $(CH_2)_3O$ | 6.1 | 1.240 | 1.059ƒ | 1.244 | 0.943ƒ |

ƒ indicates the cavity occupied by the simple hydrate former

1. molecular diameters from Hafemann and Miller (1969) and from Davidson (1986),
2. cavity radii from Table 2-1 minus 1.45Å water radii

cause the structure change, while at temperatures greater than 274.6 K the pressure may be sufficient to mitigate the change in hydrate structure. The thermal expansions of the cages, discussed in Section 2.2 do not appear to be large enough to increase their relative dimensions appreciably.

Following Glew's suggestion about the relationship between cavity diameter and non-stoichiometry, it would seem reasonable to infer from Table 2-4, that both cyclopropane and trimethylene oxide should exhibit much less stoichiometry in structure I than in structure II. Davidson (1973, page 151) presents experimental evidence of the non-stoichiometry of cyclopropane with four hydrate numbers ranging from 7.4 to 8.1. Unfortunately, the fact that trimethylene oxide is miscible causes it to have a very dissimilar phase diagram to either Figure 1-1 or 2-9, so that a corresponding analysis is inhibited.

### 2.1.4 Summary Statements for Hydrate Structure

Based upon the analyses presented in Section 2.1 the following statements represent a heuristic summary of hydrate structure:

1. With few exceptions clathrate hydrates form in either body centered cubic structure I, or in diamond cubic structure II.

2. The hydrogen bond is the basis for the interactions of the water molecules, bonding in tetrahedral structures similar to that of ice. The pentagonal water cluster formed by hydrogen bonds is fundamental to the entire water structure; hexagonal clusters exist as cavity faces at regular, but relatively infrequent intervals in the structure.

3. The basis of both hydrate structures is the pentagonal dodecahedra $5^{12}$ cavity, which is linked to other $5^{12}$ cavities through its vertices to form structure I, or through face sharing in three dimensions to form structure II.

4. The spaces between the spherical $5^{12}$ cavities are the larger, oblate $5^{12}6^{2}$ cavities in structure I, or the spherical $5^{12}6^{4}$ cavities in structure II.

5. The percentage occupation of each type cavity, with at most one molecule per cavity, must be high in order for the hydrate structure to be stable. This heuristic only applies to the cavities which are of sufficient size to encase a guest molecule, i.e. due to its size, one would expect propane to have a high occupancy in the $5^{12}6^{4}$ cavity but not the $5^{12}$ cavity of structure II.

6. Occupation of hydrate cavities is determined to a large degree by the size, and to a lesser degree by the chemical nature and shape, of the guest molecule.

7. Water molecules retain some degree of reorientation in the crystal lattice, but substantially less diffusion

occurs of solid water molecules in hydrate than in ice. Guest molecules within each cavity retain a surprisingly high degree of rotational freedom.

8. Hydrate non-stoichiometry appears to be related to the ratio of the guest molecule diameter to the free cavity diameter; non-stoichiometry increases as that ratio approaches unity. See Example 5-1 for the relationship between the hydration number, cage occupancies, and the chemical potential of water.

## 2.2. Comparison of Properties of Hydrates and Ice

If all of the cavities of structure I or structure II were occupied as a simple hydrate, for example with xenon or argon, the minimum number of water molecules (5.75 and 5.67 respectively) would be obtained per guest molecule. Both of these values yield a structure which is 85 mol per cent water. As discussed in Section 2.1, due to the non-stoichiometric nature of hydrates, the mol fraction of water is invariably higher than 0.85.

With such high water contents, it is useful as a first approximation to consider some properties of hydrates as variations from those of ice. Davidson (1973) provides a second, microscopic rationale for this comparison, by noting that hydrate hydrogen bonds average only 1% longer than those in ice, and the O-O-O angles differ from the ice tetrahedral angles by 3.7° and 3.0° in structures I and II respectively.

Table 2-5 is a slight modification of a summary by Davidson (1983) of such properties for both ice and hydrates; microscopic properties (from spectroscopy) and macroscopic properties are listed. While the values of the table were generally measured or estimated for methane or propane hydrates, the contribution of the guest molecule (other than causing the structure to exist) may be considered small for these properties, to a first approximation. In many of the properties which are derived from structure, the differences between the hydrate crystal structures are not appreciable. One might intuitively expect properties based on the water crystal structure to exhibit less variation between hydrate structures than between those properties of ice, in view of the fact that the $5^{12}$ cavity is common to each hydrate structure.

### 2.2.1 Spectroscopic Implications

Proton Nuclear Magnetic Resonance (NMR) spectroscopy and dieletric constant measurements provide evidence about the motion of the water molecules in crystal structures, as reviewed by Davidson and Ripmeester (1984). At very low temperatures (< 50K) molecular motion is "frozen in" so that the hydrate lattices become rigid and the hydrate proton NMR second moments (mean square line width) are essentially those of ice. At temperatures greater than 200-250 K, however, water motion becomes appreciable in the solid and NMR analysis suggests that the first order contribution to motion is due to reorientation of water molecules in the structure; the second order contribution is due to translational diffusion.

This is one distinguishing feature between hydrates and ice; water molecules diffuse two orders of magnitude slower in hydrates than in ice. As shown in Table 2-5, water molecules in ice diffuse almost an order of magnitude faster than they reorient about a fixed position in the crystal structure. In direct contrast, hydrate water molecules reorient 20 times faster than they diffuse. As for all solids, however, water diffusion rates in either solid structure is still several orders of magnitude slower than that in a vapor or liquid.

The dielectric constant values in Table 2-5 also suggest that, while hydrate water molecules reorient rapidly relative to molecules in other solids, those reorientation rates are only one-half those in ice. The hydrate value is lower than that of ice due to the lower density of hydrogen bonded water molecules. Far infrared

**Table 2-5 Comparison of Properties of Ice and Hydrates**
Modified from Davidson (1983)

| Property | Ice | Gas Hydrates Struct. I | Struct. II |
|---|---|---|---|
| **Spectroscopic** | | | |
| Crystallographic Unit Cell | | | |
| Space Group | $P6_3/mmc$ | Pm3n | Fd3m |
| No. $H_2O$ molecules | 4 | 46 | 136 |
| Lattice Parameters at 273 K | a=4.52<br>c=7.36 | 12.0 | 17.1 |
| Dielectric Constant at 273 K | 94 | ~58 | 58 |
| Far infrared spectrum | Broad Peak at 229 $cm^{-1}$ | Peak at 229$cm^{-1}$ with others | |
| NMR rigid lattice second moment of $H_2O$ protons ($G^2$) | 32 | 33±2 | 33±2 |
| Water molecule reorientation time at 273 K (μsec) | 21 | ~10 | ~10 |
| Diffusional jump time of water molecules at 273K (μsec) | 2.7 | >200 | >200 |
| **Mechanical** | | | |
| Isothermal Young's modulus at 268 K ($10^9$ Pa) | 9.5 | $8.4^{est}$ | $8.4^{est}$ |
| Poisson's Ratio | 0.33 | ~0.33 | ~0.33 |
| Bulk Modulus (272 K) | 8.8 | 5.6 | NA |
| Shear Modulus (272 K) | 3.9 | 2.4 | NA |
| Velocity Ratio (Compress/Shear) (272 K) | 1.88 | 1.95 | NA |

| Property | Ice | Gas Hydrates Struct. I | Gas Hydrates Struct. II |
|---|---|---|---|
| Linear thermal expansion at 273K ($K^{-1}$) | $56\times10^{-6}$ | $104\times10^{-6}$ | $64\times10^{-6}$ |
| Adiabatic bulk compressibility at 273 K ($10^{-11}$ Pa) | 12 | $14^{est}$ | $14^{est}$ |
| Speed of longitudinal sound at 273 K (km/sec) | 3.8 | 3.3 | 3.6 |
| Transport | | | |
| Thermal Conductivity at 263K (W/(m-K) | 2.23 | 0.49±.02 | 0.51±.02 |

spectral data by Bertie and co-workers (1975, 1978) suggest however, that the strength of the hydrogen bonds in hydrates is very similar to that in ice.

### 2.2.2 Elastic Properties

There is a paucity of reliable, consistent data for hydrate elastic properties. However, since these properties depend on crystal structure, many can be estimated reliably.

Whalley (1980) presented a theoretical argument to suggest that both the thermal expansivity and Poisson's ratio should be similar to that of ice. With the above two estimates, Whalley calculated the compressional velocity of sound in hydrates with a value of 3.8 km/s, a value later confirmed by Whiffen et al.(1982) via Brillouin spectroscopy. Kiefte et al.(1985) performed similar measurements on simple hydrates to obtain values for methane, propane, and hydrogen sulfide of 3.3, 3.7, and 3.35 km/s respectively, in substantial agreement with calculations by Pearson et al. (1983).

Pandit and King (1982) and Bathe et al.(1984) presented measurements using transducer techniques, which are somewhat different from the accepted values of Kiefte et al. The reason for the discrepancy of the sonic velocity values from those in Table 2-5 and above is not clear. Pandit and King (1982) also presented values for the bulk modulus, the shear modulus, and the ratio of compressional to shear velocity, with comparisons to ice values as given in Table 2-5; no other measured values are available for comparison of these properties.

### 2.2.3 Thermal Conductivity of Hydrates

Stoll and Bryan (1979) at the Lamont-Doherty Observatory first measured the thermal conductivity of propane hydrates to be a factor of about 5 less than the value of ice (2.23 $W \cdot m^{-1} \cdot K^{-1}$), which was measured by Ross, Andersson, and Backstrom (1978). Several measurement techniques by Cook and Laubitz (1981), Ross and Andersson (1982), Cook and Leaist (1983), and by Asher (1987) confirmed the low thermal conductivity of hydrates, as well as the similarities of the values for each structure, shown in Table 2-5. The thermal conductivity of the solid hydrate (0.5 $W \cdot m^{-1} \cdot K^{-1}$) more closely resembled that of liquid water (0.605 $W \cdot m^{-1} \cdot K^{-1}$) measured by Venart, Prasad, and Stocher (1979). A pictorial summary of the relative thermal conductivites of water structures, including those in Ottawa sand, was presented by Asher (1987) in Figure 2-10.

Ross, Andersson, and Backstrom (1981) also determined that tetrahydrofuran hydrate thermal conductivity had a proportional temperature dependence, but no pressure dependence. Ross and Andersson (1982) suggested that this behavior, which was never before reported for crystalline organic materials, was associated with the properties of glassy solids. The early work on this anomalous property led to the development of a thermal conductivity needle probe (Asher, 1987) as a possible means of **in situ** discrimination of hydrates from ice in the permafrost; further discussion of thermal conduction

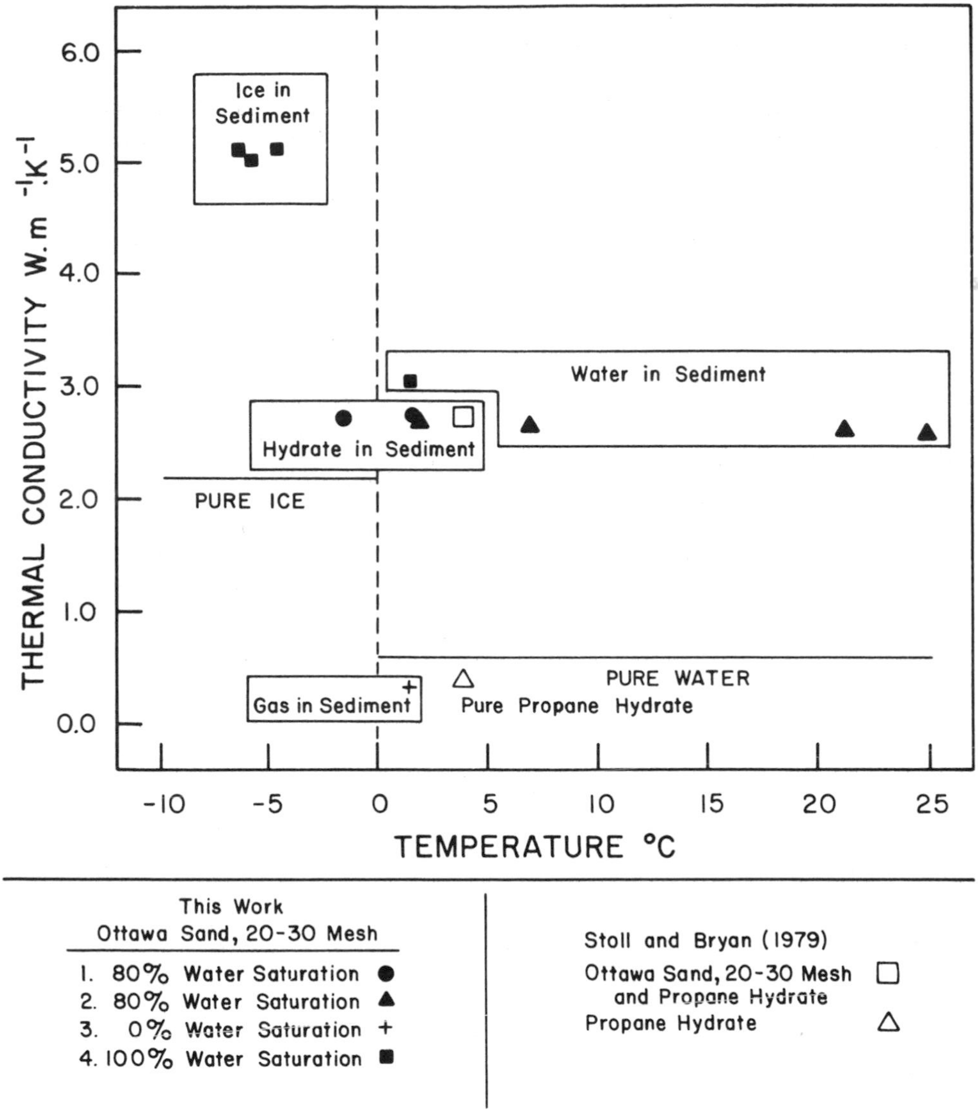


2-10 Thermal Conductivity of Gas, Water, Ice, and Hydrates both With and Without Unconsolidated Sediment (Asher, 1987)

of hydrates in sediments is deferred until Chapter 7 on **in situ** properties and resource recovery.

Dharma-wardhana(1983) calculated the unusual hydrate thermal conductivity properties by equating the phonon mean free path to the lattice constant for each hydrate crystal. His analysis was scaled to thermal conductivity values for ice, using disorder in the lattice. The effects of guest molecules were discounted.

A second reason for the anomalous hydrate thermal conductivity has come from the molecular dynamic studies of Tse, Klein, and McDonald (1983a,b, 1984, 1987a), whose calculation method gave results which comply with the far-infrared spectroscopic measurements of Bertie and Jacobs (1982). In their study of several guest molecules in structure I hydrate, Tse et al. (1984) noted that the low frequency translational and rotational motions of the guest molecule scatter quasi-particles associated with lattice vibration quanta, called phonons, which allow heat to be conducted through the hydrate.

In the hydrate lattice structure, the water molecules are largely restricted from translation or rotation, but they do vibrate about a fixed position in order to transmit energy. The vibration potential of a particle in a crystal lattice is a function of distance from other lattice particles, as shown in Figure 2-11. For small vibrational departures from the minimum, the potential may be approximated as a parabola (shown by the dashed line); for somewhat larger vibrations, departure from the parabolic potential develops, giving anharmonicity. This anharmonicity provides a mechanism for the scattering of phonons, providing a lower thermal conductivity.

In hydrates, the frequencies of the guest molecule translational and rotational energies are similar to those of the low frequency lattice (acoustic) modes, so that frequency coupling increases the crystal anharmonicity. The results of Tse et al.(1983b) show, in

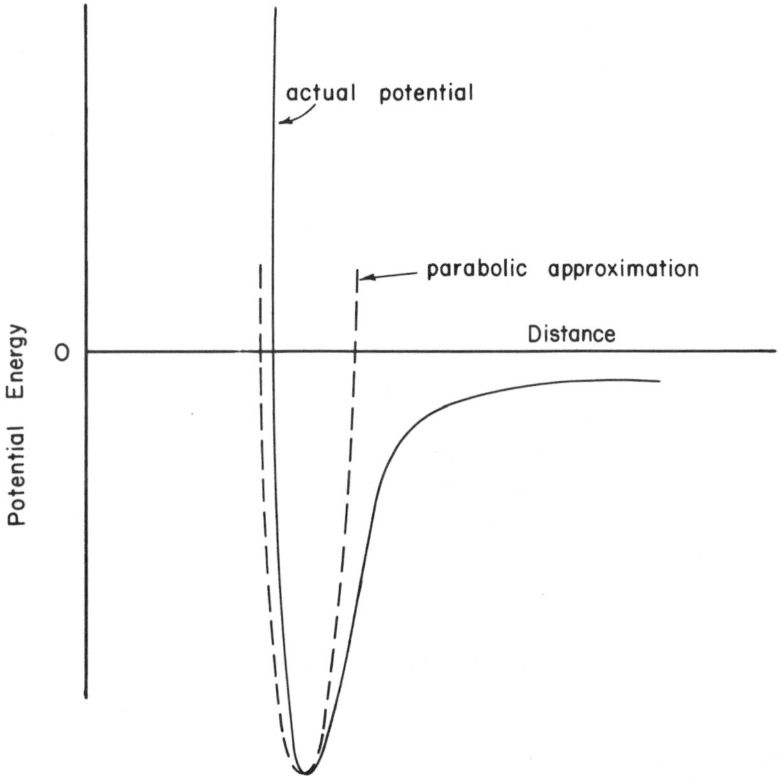


2-11 Anharmonic (solid) and Harmonic (dashed) Potentials for Vibration of a Particle

Figure 2-12 for both empty and filled structure I, that some of the crystal phonon frequencies are changed, and the phonon density distribution is changed (see peaks 1, 2, and 3) by the inclusion of the guest molecule. This weak coupling between the guest and host lattice does not noticeably affect most structural thermodynamic and mechanical properties, but it has a marked effect on the transport of heat. Tse et al. (1984) also used this hypothesis to explain the unusual temperature dependence of thermal conductivity. Molecular dynamic studies of structure II hydrate by Tse and Klein (1987) have confirmed the phonon densities to be similar to those of structure I, thereby explaining the similarity in thermal conductivity.

Yet a third reason for the anomalous value of hydrate thermal conductivity has recently been proposed by Ahmad and Phillips (1987) who suggested that proton disorder gives rise to tunnelling states which causes phonon scattering. It is unclear as to which of the three above explanations (or combination thereof) provides the actual explanation of the phenomena of thermal conductivity. Tse (1988) recently indicated that the scaling hypothesis was not able to explain the slope change of the thermal conductivity with temperature, while the tunnelling effect should give a higher rather than lower thermal conductivity.

### 2.2.4 Thermal Expansion of Hydrates and Ice

Linear thermal expansion coefficients ($d\ell/\ell dT$) of both hydrate structures and ice have recently been determined through dilatometry by Roberts et al. (1984) and through X-ray powder diffraction by Tse (1987b), and Tse et al. (1987). These coefficients, shown in Table 2-5, are taken from the work of Tse and should be considered accurate to about ±10%. The values for hydrate are approximately $10\times10^{-6}$ $K^{-1}$ (THF) to $50\times10^{-6}$ $K^{-1}$ (EO) larger than those of ice (Touloukian, et al.(1977)). The property differences become more marked at low temperatures.

Two factors may be possible for the differences in thermal expansion: (a) structural differences between ice and hydrates, and (b) interactions between the guest molecules and the water lattice, such as those indicated in Section 2.2.3. Through constant pressure molecular dynamic calculations for thermal expansion of ice and of empty structure I, Tse et al.(1987) determined that structural difference only contributed a small part to the larger hydrate thermal expansion.

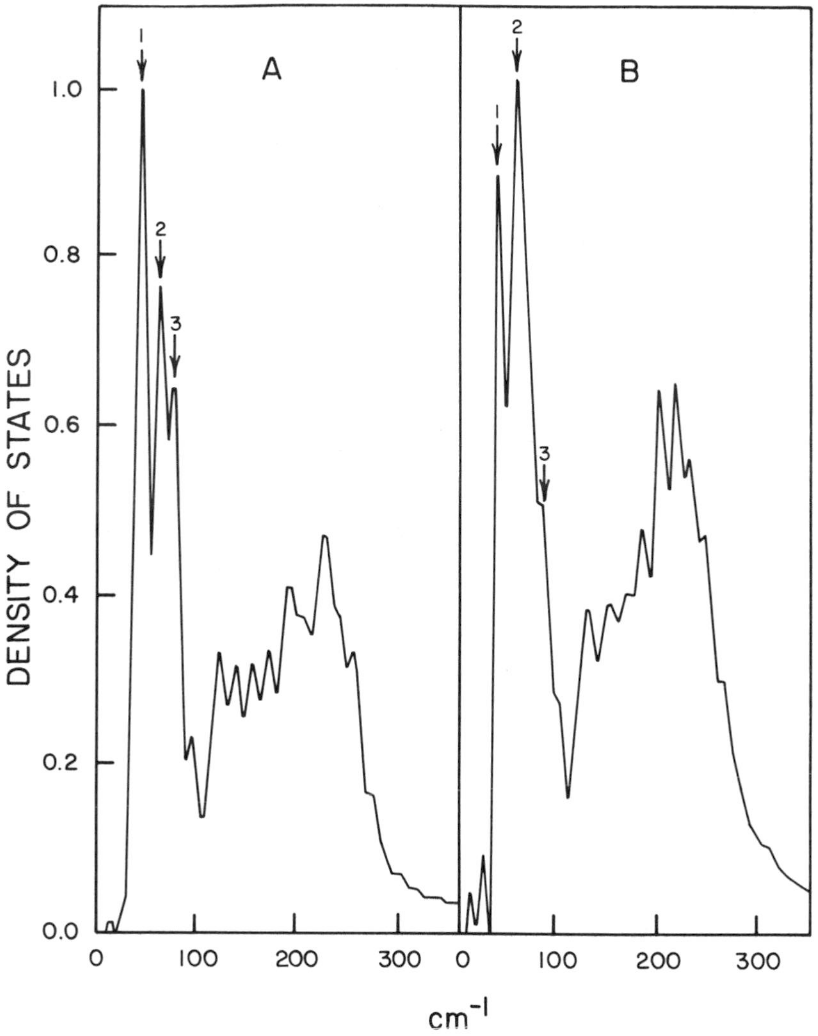


2-12 Translational Frequency Spectra for (A) Empty Hydrate (205K) and (B) Methane Hydrate (214K); Numbered Peaks as in Text (Reproduced, with permission, from the Journal of Physical Chemistry (Tse et al., 1983b), by the American Chemical Society.)

## 2.3 The What and the How of Hydrate Structures

With the conclusion of the present chapter, the reader should have a firm notion of what the molecular structures are and how these structures compare to that of ice. The physical properties of ice and hydrates have been compared and contrasted.

If water will normally form ice in the absence of a solute molecule, the question arises about the mechanism for forming a clathrate with an exact structure, when the solubility of hydrocarbon molecules in liquid water is known to be small (or negligible in ice), relative to the amount of hydrocarbon needed for hydrates. Thus, along with the definition of what the hydrate structures are, comes the logical question of how these structures form. Such a question, while entirely natural, has a detailed answer which is only beginning to be realized. The microscopic mechanisms, together with the macroscopic kinetics of hydrate formation, are the major considerations of Chapter 3.

# 3

# Kinetics of Hydrate Formation

Perhaps one of the most intriguing questions regarding the hydrate molecular structure concerns kinetics. An understanding of the kinetics of hydrate formation is integrated with the hydrate structures of the previous chapter. Experimental study of formation represents a challenge due to the high degree of metastability observed during the nucleation of the crystal to a stable size for continued growth. The result is that there is a paucity of reliable quantitative kinetic data in the literature. As a consequence, the study of formation phenomena represents one of the most promising areas for current hydrate research. On the other hand, hydrate dissociation phenomena are fundamentally different, because they do not exhibit such a high degree of metastability; therefore dissociation kinetics are relegated to Chapter 7 on Gas Recovery.

The knowledge of formation kinetics is bifurcated. The first area concerns the primary nucleation process, during which the hydrate nuclei grow and disperse in an attempt to achieve critical size. This nucleation step is a microscopic phenomenon involving tens to hundreds of molecules, and is difficult to observe experimentally. Support for current hypotheses of primary hydrate nucleation is provided by the phenomena of water freezing, the dissolution of hydrocarbons in water and computer simulations of both these liquid phenomena. There are however, only a small number of data on the primary nucleation process itself. The experimental data, calculations, and hypotheses of primary hydrate nucleation are examined in Section 3.1.

After the step in which critical nuclei sizes are obtained, progress in the quantification of crystal growth (or dissociation) has been made by Miller and Smythe (1970), Falabella (1975), and Bishnoi and co-workers (Vysniauskas (1980), Kim (1985), and Englezos et al. (1987a,b, 1988)). A new secondary nucleation hypothesis has recently evolved from the vacuum cryogenic studies of both hydrates and ice by Devlin and co-workers (1983, 1984 a,b, 1985 a,b, 1987, 1988). Vysniauskas and Bishnoi (1983a) indicated that much of the Soviet kinetic work, including that of Babe et al.(1972), Malenko (1974), and Selznu and Stupin (1977), provided significant qualitative information on hydrate growth kinetics. Secondary nucleation concepts are quantified in Section 3.2.

## 3.1 Concepts of Hydrate Nucleation

Nucleation thermodynamics were originated by Gibbs (1928) at the end of the last century. Volmer and Weber (1926) indicated that kinetics played a major role in nucleation, by the growth and decay of clusters of molecules. Nucleation phenomena are much more difficult to measure than are the macroscopic, time independent properties of equilibrium thermodynamics. The experimental difficulties of hydrate nucleation measurements include metastability, microscopic scales, and time dependence of kinetics. As a result, most of the knowledge about hydrate nucleation is conceptual, based on calculation, or at best on fragmentary experimental evidence.

In order to achieve some understanding of the nucleation of hydrate crystals, it is useful to first study the mechanism of water supercooling and freezing in Section 3.1.1. Two hypotheses of water nucleation exist. The first is founded on the liquid water flickering "iceberg" theory of Frank and co-workers (1945, 1957, 1968), as summarized by Makogon (1974) in relation to hydrates. The more modern hypothesis, is based on the liquid water "hydrogen-bonded network" model of Stillinger and Rahman (1973,1974), with support both in the supercooled water data and simulations of Angell (1982) and Speedy (1987). The most recent explanation of the modern model has been summarized by Dore (1988), which portrays the hydrogen-bonded network as having gel-like properties in the subcooled region.

A second, more comprehensive understanding of hydrate nucleation is gained by a study of hydrocarbon solubility in water, in Section 3.1.2. This second mechanism for hydrate formation is also supported through both experimental observations and computer simulations. Apparently this mechanism has yet to be expressed explicitly by researchers in hydrate kinetics, yet it evolves naturally from the current work on aqueous solutions of non-polar molecules.

While there is a notable lack of data on the nucleation of hydrate themselves, the determination of the critical nuclei size is a key ingredient to any comprehensive theory. An analysis of the nucleation induction period in the data of Falabella (1975) for hydrate formation from ice, provides a hypothesis for the effect of solute size upon the nucleation process. These concepts are discussed in Section 3.1.3.

### 3.1.1 Mechanisms for Nucleation of Water Molecules

The hydrogen bond physical chemistry, discussed in Section 2.1.1, enables the hypothesis of two mechanisms for water freezing which are unique among liquids: (1)the flickering cluster mechanism, and (2) the hydrogen-bonded "network-cluster" mechanism.

Frank and co-workers (1945, 1957, 1970) and Nemethy and Scheraga (1962a) provided a model of water as an equilibrium mixture of short-lived ($10^{-10}$ second) hydrogen-bonded clusters, together with a non-hydrogen-bonded dense phase. Makogon (1974) incorporated these concepts into a hydrate nucleation mechanism, indicating that the clustering of water molecules provides the beginning of hydrate nuclei. Makogon used X-ray diffraction data to indicate short-term clustering with a decrease in temperature, to provide some experimental verification of pre-freezing phenomena via "flickering clusters" of liquid water.

Figure 3-1 depicts a progression of portions of short-lived water clusters, from two molecules at a temperature just below the critical point of liquid water, to the formation of a partial cavity attachment at lower temperatures. If a guest molecule were in the vicinity of a partial hydrate cage such as Figure 3-1f, and if the resultant mass were to attract a second partial cavity such as Figure 3-1g, then a $5^{12}$ cavity might form as the beginning of a hydrate crystal.

If the thermodynamic conditions were conducive to the addition of other water clusters and guest molecules, then a basic crystal structure I or II might occur, which could grow until a critical nucleus was obtained to enable the next stage of hydrate crystallization. Makogon (1974) suggested that the radius $r_c$ of the critical hydrate nucleus is given by:

$$r_c = \frac{2\sigma T_e}{\Delta H(\Delta T)} \qquad (3.1)$$

where ΔH is the latent heat of crystallization, σ is the "nuclei's specific surface energy" and ΔT is the degree of subcooling at a given three phase ($L_W$-H-V) equilibrium pressure.

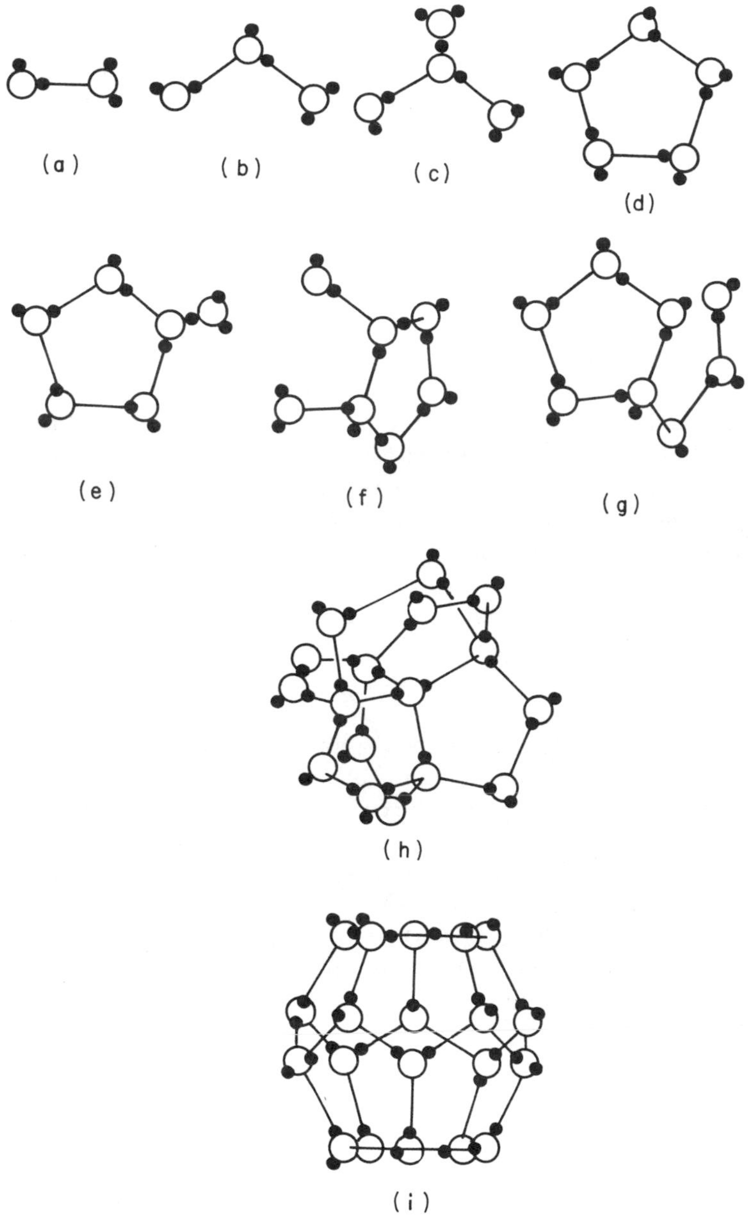


3-1 Short Lived Water Molecule Aggregates with Progression from High Temperatures (a) to low temperatures (h); (i) Pentagonal Dodecahedra ($5^{12}$) (Reproduced, with permission, from Hydrates of Natural Gas (Makogon, 1974), by PennWell Publishing Co.)

While the cluster hypothesis is attractive in its simplicity, it has lately been somewhat discredited by experimental evidence, in some instances by the very founders of the "flickering cluster" hypothesis (Frank, 1970). As Narten and Levy (1969) pointed out, for interpretation of X-ray data on water, the only recourse presently available is to calculate such data from proposed models for comparison with experiment. Such interpretation is necessary but not sufficient for the verification of a model, since more than one model may lead to a similar comparison with the data.

Narten and Levy interpreted small angle X-ray scattering data to indicate that there are no regions larger than a few molecular diameters for which the density differs more than a negligible extent from that of the bulk. Raman scattering results on stretching frequencies of HDO by Wall and Hornig (1965) also suggest that water cannot be a mixture of large clusters and non-bonded molecules. An alternative to the "flickering cluster" hypothesis is the "network-cluster" model.

Stillinger (1980) described water as a "macroscopically connected (three dimensional) random network of hydrogen bonds, with frequent strained and broken bonds, that is continually undergoing topological reformation." A schematic of such a random network, which includes polygons, is provided in Figure 3-2. Stillinger based this model on the calculated results of molecular dynamics studies recently made possible through the use of digital computers.

Molecular dynamics is a technique to simulate water structures, whereby an accurate water atomic potential function, is used to enable Newton's equations of motion to be solved for a small (typically $\leq$ 250) ensemble of molecules. Alternatively, a solute molecule in water may be simulated with the addition of an interatomic potential for the guest-host interactions. The total energy (kinetic and potential) and a velocity average (temperature) of the system are obtained, from which all other

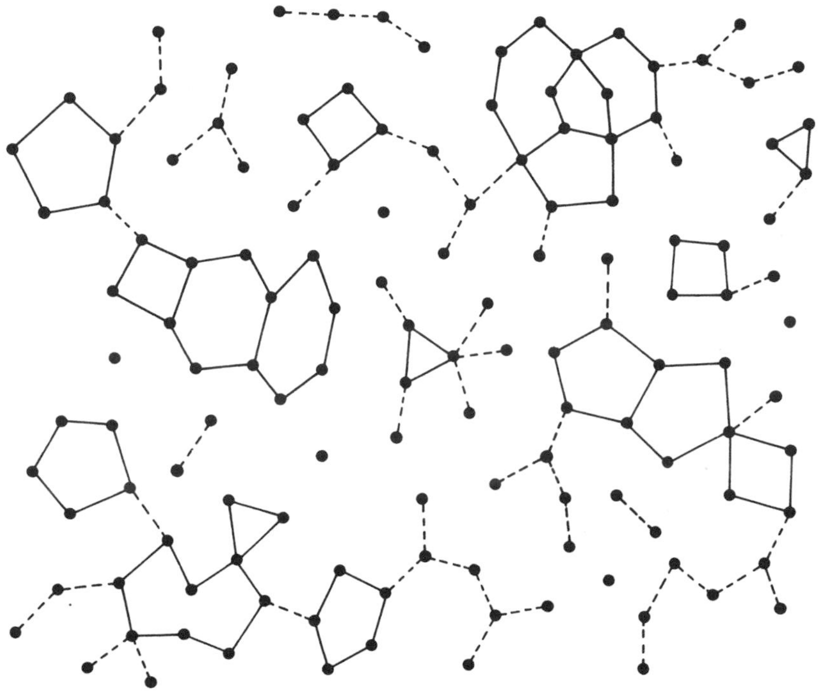

3-2 Networks of Water Molecules from Molecular Dynamics (Reproduced, with permission, from the Journal of the American Chemical Society (Rahman and Stillinger, 1973))

thermodynamic properties may be calculated for a constant volume system. This technique also enables the study of the time-dependent properties, such as nucleation.

The molecular dynamic simulation technique has been validated through comparison of calculated properties with experimental thermodynamic water data, such as the density maximum, the high heat capacity, and diffraction patterns (Stillinger and Rahman, 1974) as well as the hydrate infrared (vibrational) spectra data by Bertie and Jacobs (1977, 1982). With acceptable comparisons of many computed and experimental properties of water structures, there is little doubt that a substance very similar to water has been simulated.

Rahman and Stillinger (1973) showed that the most common portions of a water molecular network were five and six member polygons. The calculations of Speedy et al. (1987) reinforced the idea that pentagons and hexagons prevail when water is cooled below 277K. Chen (1980) studied the molecular dynamics of eight possible networks of five water molecules (Plummer and Chen, 1987), and of the $5^{12}$ cavity (Plummer and Chen, 1983) for pure water. Chen found that the most stable pentamer at temperatures above 230 K is the ring structure, which may exist up to 315 K. When a monomer molecule collides with either a ring pentamer or a $5^{12}$ cavity, Chen found that a hydrogen-bonded surface adsorption of the monomer always occurs (without cavity or ring disruption), due to the presence of many degrees of atomic freedom in the cluster, which acts as a dissipating heat reservoir.

Measurements of the properties of supercooled water by Angell (1982) and Speedy and co-workers (1987(2)) indicated that the vital parameter was the concentration and spatial distribution of several unstrained hydrogen-bonded polyhedra, such as the $5^{12}$ cavity, embedded within and linked to the random network of molecules. The hydrogen bond angles in these polyhedra are such that it is thermodynamically favorable for them to share edges and faces without the introduction of strain, rather than attach themselves to a strained polyhedra. As a result unstrained polyhedra are shown to exhibit an attraction for each other. The sharing of polyhedra faces and edges in supercooled water is suggestive of pre-nucleation phenomena for hydrates.

### 3.1.2 Dissolution in Water and Formation Mechanism

Normally the solubility of non-polar gases in water is very small. Miller and Hildebrand (1968) indicate that the solubility of gases in water is at least an order of magnitude lower than their corresponding solubility in cyclohexane. The Gibbs Free Energy ($\Delta G$) of solution, the thermodynamic potential for the solution at constant temperature and pressure, is small and positive in accordance with a small solubility. ΔG is determined primarily by a negative entropy of solution (Franks and Reid, 1972).

Since ΔG ≡ ΔH - TΔS, a large negative (unfavorable) entropy of solution overcomes the negative (favorable) enthalpy of solution to determine the value of ΔG. Large negative entropy changes and large positive heat capacities changes are unique to aqueous solutions of gases. Values for solubilities, as well as infinite dilution enthalpies, entropies, and heat capacities of solution at 298 K are given in Table 3-1. It should be noted that the entropies of solution in Table 3-1 would be much more negative were they determined at infinite dilution, rather than at a standard state of one atmosphere. Himmelblau (1959) noted a dramatic decrease in entropy of solution for hydrate formers relative to such small nonhydrate formers as He, $H_2$, and Ne; this is an indication of some special solubility phenomenon associated with size.

Entropy is taken as an indication of disorder on a molecular scale. The large negative entropy of solution in Table 3-1 is generally considered as evidence of the creation of structure within the body of water. The large heat capacity changes also indicate the structuring effect of the solute on the water molecules. The size of the solute molecule has a substantial effect on the solubility. A summary by McAuliffe (1966) of the solubility data for 17 paraffins shows a good correlation of the solubility with the molar volume of the hydrocarbon, with the notable exceptions of those molecules which form hydrates, particularly methane, ethane, and propane.

In a review of the thermodynamics of water, Franks and Reid (1973) present Figure 3-3, to show that the optimum molecular size range for maximum solubility is very similar to that for hydrate stability (compare with Figure 2-8). Franks and Reid note, "this is not intended to imply that long-lived clathrate structures exist in solu-

**Table 3-1 Solution Properties of Natural Gas Hydrate Components at 298K**

| Component | Solubility[1] $10^5 x_2$ | $-\Delta H_{soln}$[2] KJ/mol | $-\Delta S_{soln}$[3] J/(K-mol) | $\Delta Cp_2$[4] (KJ/(K-mol) |
|---|---|---|---|---|
| Methane | 2.48 | 13.26 | 44.5 | 55 |
| Ethane | 3.10 | 16.99 | 57.0 | 66 |
| Propane | 2.73 | 21.17 | 71.0 | 70 |
| Iso-Butane[5] | 1.69 | 25.87 | 86.8 | NA |
| n-Butane | 2.17 | 24.06 | 80.7 | 72 |
| Nitrogen | 1.19 | 10.46 | 35.1 | 112 |
| Hydrogen Sulfide[6] | NA | 26.35 | 88.4 | 36 |
| Carbon Dioxide[7] | 60.8 | 19.43 | 65.2 | 34 |

1. Solubility at 101.3 KPa from Miller and Hildebrand (1968) except $i\text{-}C_4H_{10}$, $H_2S$, and $CO_2$ as indicated below

2. $\bar{H}^{\circ}_{L_2} - \bar{H}^{*}_{G}$, transfer from gas to infinite dilution liquid, from Franks and Reid (1973)

3. $\bar{S}_{L_2} - \bar{S}^{*}_{G}$, transfer from gas to liquid, standard state, fugacity = 101.3 KPa

4. $C_p$ from Alexander et al.(1971) except $CH_4$, $C_2H_6$, and $H_2S$ from D'Orazio and Wood (1963)

5. $i\text{-}C_4H_{10}$ properties calculated from Wetlaufer et al.(1964)

6. $H_2S$ properties calculated from Selleck et al.(1952), and hypothetical $\Delta H_{soln}$, $\Delta S_{soln}$ extrapolated below $Q_2$ (302K)

7. $CO_2$ properties calculated from Alexander et al. (1971)

tion, only that the stabilization of (the) water structure by apolar solutes resembles the stabilization of water in a clathrate lattice." Glew (1962) noted that, within experimental error, the heat of solution for ten hydrate formers (including methane, ethane, propane, and hydrogen sulfide) was the same as the heat of hydrate formation from gas and ice, thereby suggesting the coordination of the aqueous solute with surrounding water molecules.

The most popular explanation of such experimental evidence stems from the hypothesis of Frank and Evans (1945), that the presence of the solute molecule causes an increase in the order of the bulk water molecules, with the caution that such ordering should not be considered as long-lived or complete in any sense. Over the last four decades this basic hypothesis has been extended and given various degrees of mathematical justification by eminent workers such as Frank and Quist (1961), Nemethy and Scheraga (1962b), and Frank and Franks (1968). The approach of Ben-Naim (1980) differs from the others in detail, yet the basic conclusions appear the same, regarding the effect of the solute molecule in structuring the solvent. In experimental studies of aqueous solutions of alcohols, the apolar portion of the alcohol molecule has been suggested to affect the water solvent in much the same manner as strictly apolar solutes, by Alexander and Hill (1969), Arnett at al.(1969) and Krishnan and Friedman (1969).

Geiger, Stillinger, and Rahman (1979) performed computer simulations on dissolved non-polar solutes in water, which indicated that the water network includes rearrangements in the form of a clathrate-type cage. In a similar molecular dynamic (MD) comparison of the $5^{12}$ cavity and of 8 nitrogens dissolved in 192 water molecules, Dang (1985) found that the hydration number was 17, which approaches the coordination number of 20 for the $5^{12}$ cavity. In Monte Carlo computer studies Owicki and Scheraga (1977) determined that the average coordination number of water molecules around dissolved methane was 23, while Swaminathan et al. (1977) determined the value to be closer to 20.

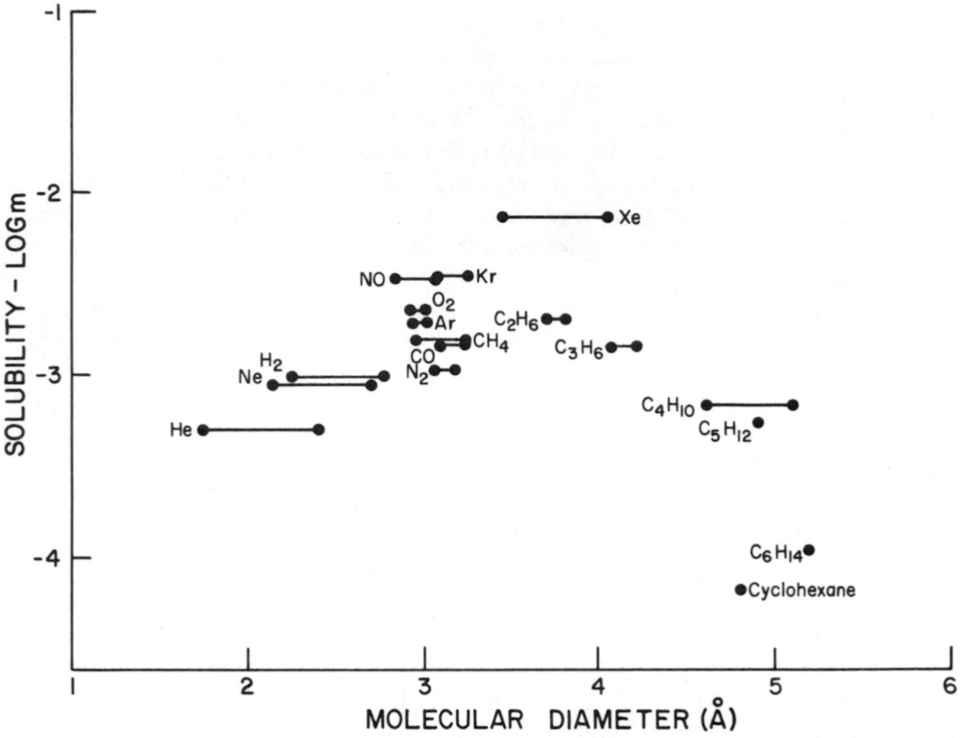


3-3 Maximum in Water Solubility for Hydrate-Forming Molecules (Reproduced, with permission, from Water: A Comprehensive Treatise (Franks and Reid, 1973) by Plenum Publishing Corp.)

When water rearranges its structure around a dissolved hydrocarbon, Stillinger (1980) indicates that any non-bonded hydrogens always point outward from the convex structure, thereby encouraging further bonding. The stereographic pictures in Figure 3-4 from the Monte Carlo studies of dissolved methane by Swaminathan et al. (1977) seem to confirm the clathrate hypothesis of dissolution. During the reorientation of water molecules to accommodate the solute molecule, the bonds are calculated as stronger in the solvation cages, which possess many of the geometrical properties of the $5^{12}$ cavity.

Franks (1975) and Ben-Naim (1980) review a substantial body of experimental evidence for the so-called

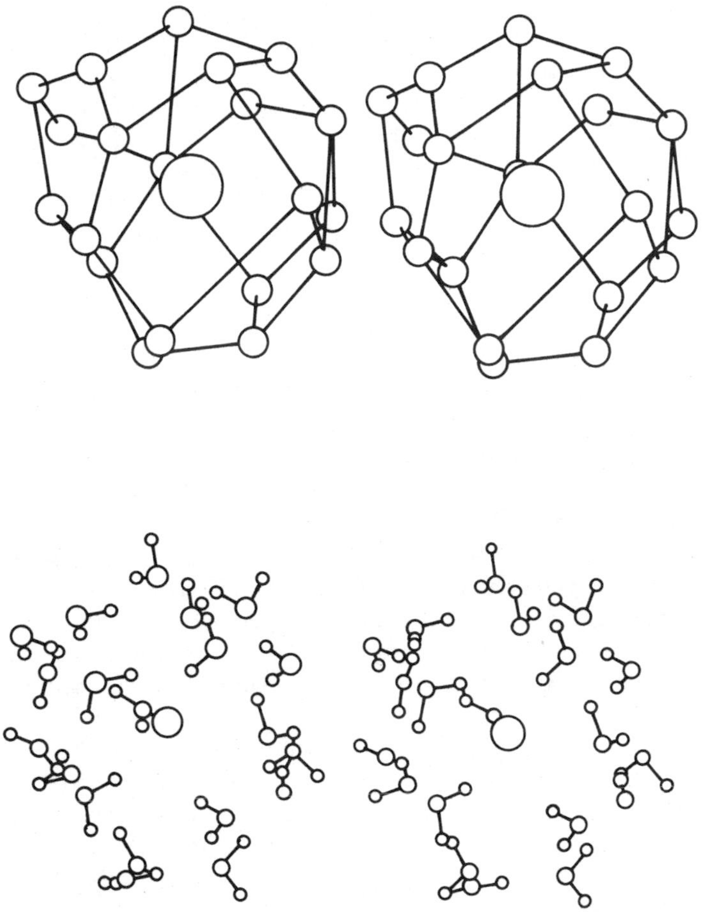

3-4 Alignment of Water Molecules into Clusters by a Dissolved Apolar Molecule (Large Circles) from Monte Carlo Studies. Top Figure Shown with Lines Connecting Water (Reproduced, with permission, from the Journal of the American Chemical Society (Swaminathan et al., 1978))

"hydrophobic-bond", in which non-polar dissolved molecules and their solvent clusters are attracted to each other. While there are no actual bonds involved, the sharing of imperfect cluster faces is viewed as a thermodynamic tendency to minimize the negative entropies of solution. The tendency for face or edge sharing of individual solvation clusters, Stillinger (1980) points out, is the same as the tendency for clustering of pure supercooled water.

This mechanism of nucleation of a hydrate structure is clearly hypothesized in the above descriptions of water network-clusters, supercooled water, and dissolved hydrocarbons in liquid water. In view of the degree of sophistication of current water theory, it would appear that this mechanism of hydrate nucleation would be an integral portion of the theory of water. The incremental step from (a) a short-lived liquid cluster of hydrogen-bonded water molecules around a dissolved hydrocarbon, with several irregular and broken hydrogen bonds, to (b) a fixed position of well-formed hydrogen bonds in a solid clathrate cage around a guest molecule, appears evident.

If one accepts the hypothesis of the increased solubility due to water clustering, three central hydrate concepts (solubility, metastability and the establishment of critical nuclei) appear to be part of a spectrum of a single physical phenomenon. As discussed above, the solubility of apolar molecules in the water seems to be due to: (a) normal van der Waals forces to disperse the molecules randomly in the water, and (b) the clustering of the water molecules to provide the dissolved structure. The water molecular clustering itself appears to provide enhanced solubility.

Holder (1988) and coworkers recently considered the thermodynamic solubility of methane around the hydrate formation point; for equilibrium conditions, they determined that the solubility was not enhanced beyond the measured values of Culberson and McKetta (1950, 1951). In Bishnoi's laboratory Englezos et al.(1988) measured the consumption of methane as a function of time prior to secondary nucleation, which they called the hydrate nucleation or turbidity point. They determined that the methane concentration was supersaturated in the solution as a function of time, before the onset of secondary nucleation. One may interpret the kinetic data of Englezos et al. (1989) as an indication that the end of the period for increased solubility marks the time at which the critical nucleus size is determined, at which time secondary nucleation begins. In a broader interpretation, the time required for enhanced solubility or clustering and the establishment of the critical nucleus size is synonymous with the time of hydrate metastability. In this way one can join the concept of time required for enhanced solubility with the time needed for establishment of a critical nucleus, to the concept of the end of the metastable period before secondary nucleation.

Lingelem and Majeed (1989) attempted to quantify the time for natural gas hydrate crystals to appear in a sapphire cell at high pressure. The cell was agitated at 120 rpm with water which had not been used to form hydrates previously. The visual observation of hydrate crystals in the cell was taken as the endpoint of the primary nucleation period. A correlation was constructed, as in Figure 3-5, between the time required for the appearance of nuclei and the degree of cooling below the equilibrium temperature at a given pressure. A single datum of pressure decrease in the figure indicates that the time to visually observe hydrate appearance may be appreciably greater than the secondary nucleation time marked by a decrease in pressure (increased gas encapsulation). If such a correlation were indeed valid, it would be of great utility to both researchers and practitioners. should be noted however that several laboratories have unsuccessfully attempted to reproduce this correlation.

Englezos et al. (1989) suggested that the nuclei grow to critical size due to the encapsulation of gas molecules at the crystal-water interface within the volume of aqueous solution; however the mixing rates used in the experiments were very high (600 rpm). In other work Englezos and Bishnoi (1988b) measured the nucleation

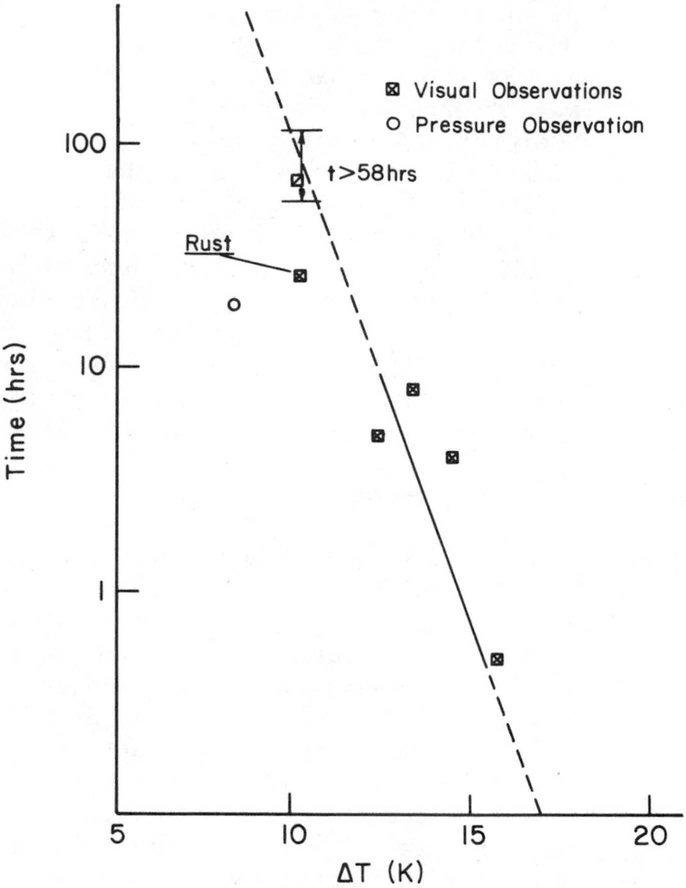


3-5 Hydrate Metastability Time as a Function of Subcooling (Reproduced, with permission, from the Proc. 68th Annual GPA Convention (Lingelem and Majeed, 1989), by the Gas Processors Association.)

point to be a factor of twenty less than the calculated thermodynamic limit to methane solubility; the calculated limit to methane solubility itself was a factor of five less than the methane concentration required in the hydrate phase.

Due to the large amount of gas encapsulated within hydrate relative to that dissolved in the liquid, it would seem that the most likely place for the crystals to grow is at the point of contact of the crystal-water interface with the high concentrations of gas, at the boundary of the macroscopic aqueous liquid phase. The high mixing rates may in fact cause the interfacial gas-liquid-crystal structures to be mixed within the volume of the liquid for semi-homogeneous dispersion of nuclei.

Regardless of the mechanism of nucleation, however, the fixing of the solid structure is a function of the critical radii of the crystal nuclei. It is the nuclei of critical size (or larger) which will continue to grow at the expense of shrinking nuclei which are not of critical size. Next consider the quantification of critical nuclei sizes.

### 3.1.3 Size of the Critical Nuclei

Recently, Bishnoi co-workers (Englezos et al. (1987a,b)) have determined an expression for the radius of the critical nucleus, using the Gibbs Free Energy per unit volume of product formed $\Delta g$, in a modification of Equation (3.1) as:

$$r_{cr} = - \frac{2\sigma}{\Delta g} \tag{3.2}$$

and

$$(-\Delta g) = \frac{RT}{v_m} \left[ \sum_1^2 \theta_j \ln\left(\frac{f_j}{f_{\infty,j}}\right) + \frac{n_w v_w (P-P_\infty)}{RT} \right] \tag{3.3}$$

where the surface tension $\sigma$ is the value of ice in water, $v_m$ and $v_w$ are the molar volumes of hydrate and water, respectively, $\theta_1$ and $\theta_2$ are the mol fractions of guests in the hydrate cages on a water free basis, $f_j$ and $f_{\infty,j}$ are the experimental and equilibrium fugacities, respectively, of component j at temperature T, $(P-P_\infty)$ represents the overpressure beyond the equilibrium value, and $n_w$ is the number of water molecules per gas molecule.

Chen (1980, pg. 7) suggested that the use of bulk phase properties, such as those in Equations (3.1) through (3.3) may be satisfactory only in qualitative analyses. Microscopic critical clusters contain several tens to hundreds of molecules, and as such have a spectrum of sizes and properties, which may be difficult to quantify with a single number on a macroscopic scale. Because the hydrate nucleation phenomena are very difficult to quantify experimentally, recent research efforts have centered on computer simulation, using realistic interatomic potentials for water and the guest molecule, as indicated in the previous section.

#### 3.1.3.a. Clustering for Hydrate Reformation

Makogon presented data to support his concept that, on dissociation hydrates do not totally decompose but leave a partial structure which will form more readily with a further decrease in temperature. Chen's molecular dynamic studies seem to confirm Makogon's data which indicate that both the pentamer ring and the residual structure are stable up to 315 K. Vysniauskas and Bishnoi (1983b) measured the data in Table 3-2, which relates the induction time for hydrate formation to the thermal history of the water. The result of this effect is shown in the successively decreasing hysteresis of the

**Table 3-2**
**Effect of Thermal History of Water on the Induction Delay for Hydrate Formation**

| Water Description | Mean Induction Delay, minutes |
|---|---|
| Hot Tap Water (about 333 K) | 18.13 |
| Double Distilled Water | 11.75 |
| Cold Tap Water (283-288 K) | 4.95 |
| Dissociated Hydrate (left overnight) | 2.50 |
| Thawed Ice (used immediately) | 0.75 |
| Dissociated Hydrate (used immediately) | 0.0 |

cooling curves of Figure 3-6 of Schroeter et al., (1983) on hydrates formation. To obtain the lines in this figure, a container with gas and liquid water was isochorically cooled into the metastable region until hydrates formed in the portion of the curve labelled $S_1$. The container was then heated and hydrates dissociated along the liquid water-hydrate-vapor ($L_W$-H-V) line until point H was reached, where the last hydrate crystal was depleted. Successive cooling curves $S_2$ and $S_3$ show decreased metastability from line $VL_WH$, because residual hydrate structure is present in the water.

A summary of nucleation heuristics, together with a hypothesis generated in the following section, is in Section 3.1.5.

### 3.1.4 Size, Nucleation, and Induction

As the concluding portions of a five part series on hydrates, Barrer and his students studied the kinetics of hydrate formation in the liquid nitrogen region of temperatures. Barrer and Ruzicka (1962) experimentally observed the kinetics of argon, krypton, and xenon hydrate formation from ice using a glass ball-mill apparatus, designed to mitigate any inhibition effect caused by a layer of hydrates on the ice surface. They were concerned with the effect of inerts as help gases when chloroform or tetrahydrofuran was present to stabilize the larger cavities. A qualitative Gibbs Free Energy analysis was used to explain the sensitivity of nucleation to the degree of subcooling.

Barrer and Edge (1967) substituted stainless steel balls for the glass balls in a study of the formation of simple hydrates of the same three inert gases from ice. They demonstrated that, while argon and xenon formed hydrates rapidly, an induction period of about one hour was observed for krypton. During the induction period the conversion to hydrate was negligible, but the period was followed by a rapid growth of hydrate and a gradual decline, caused by exhaustion of the ice. The induction

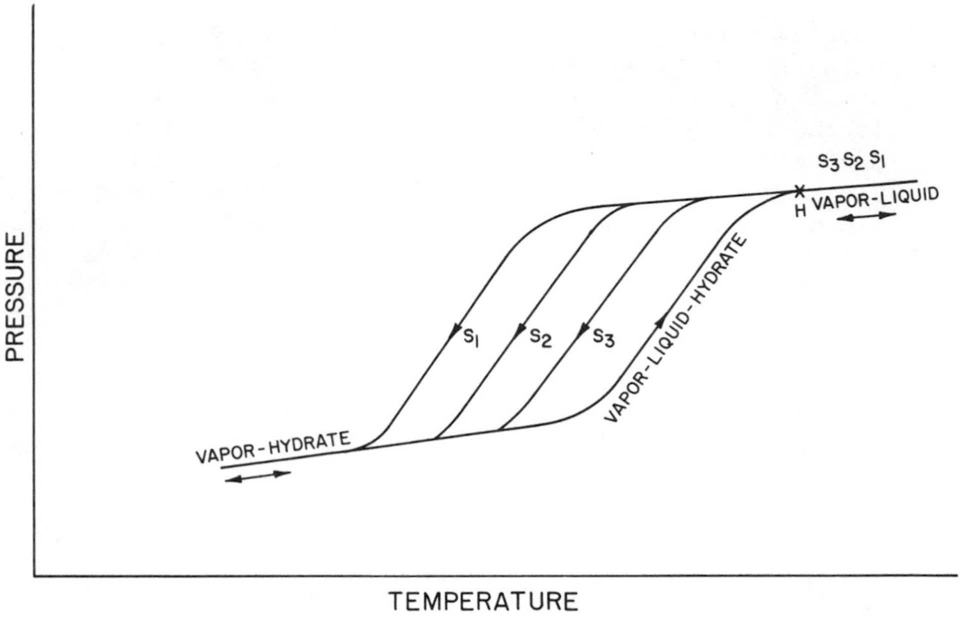


3-6 Successive Cooling Curves for Hydrate Formation with Successive Runs Listed as $S_1 < S_2 < S_3$
(Reproduced, with permission, from Industrial and Engineering Chemistry Fundamentals (Schroeter, et al., 1983), by the American Chemical Society.)

phenomenon appeared to be similar to that of subcooled solutions in the absence of a seed crystal. Barrer and Edge offered no explanation as to why krypton alone exhibited the induction period, and they did not attempt to quantify the reaction rate.

Almost a decade later, Falabella (1975) determined both the equilibrium and kinetic properties of hydrates of methane, ethane, ethylene, acetylene, carbon dioxide, argon, and krypton at temperatures from 148 to 240 K and pressures below atmospheric. Falabella used an apparatus similar to Barrer and Edge and duplicated their results for krypton. While replications of his own experiments do indicate a small amount of scatter, Falabella was able to obtain two types of kinetic results: (a) one type with a clearly defined induction period for methane and krypton, with typical data shown in Figure 3-7, and (b) a second type for other gases, which exhibited no induction

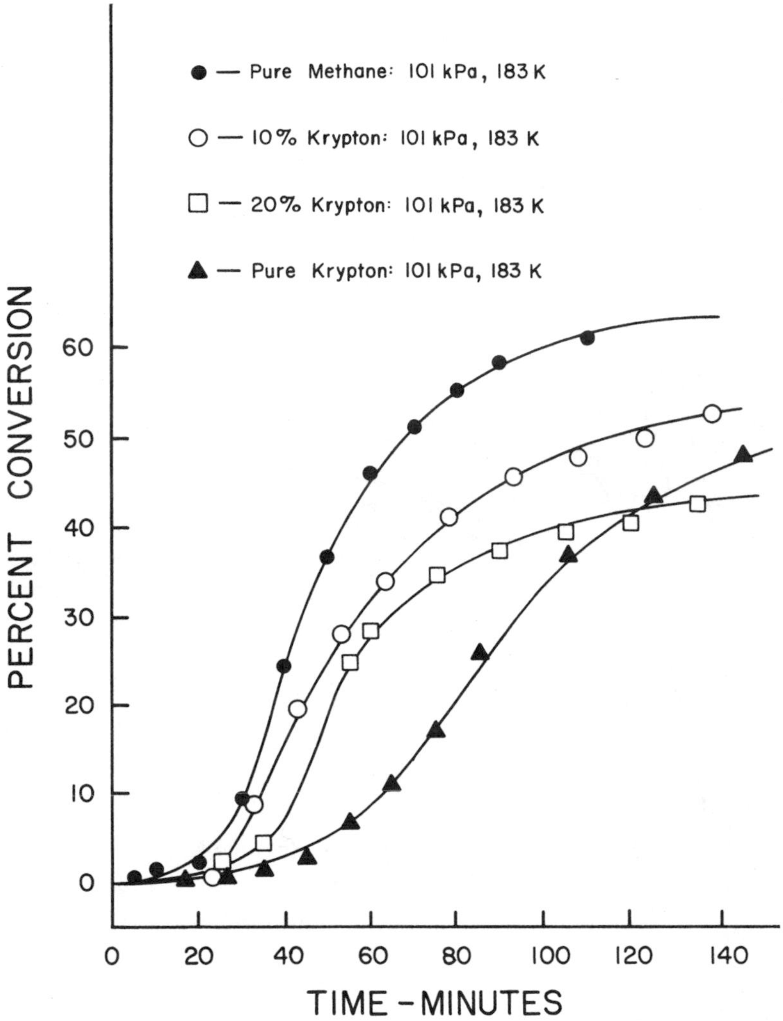


3-7 Kinetics of Methane and Krypton Hydrate Formation from Ice With Induction Period
(Reproduced with permission from B.J. Falabella (Falabella, 1975))

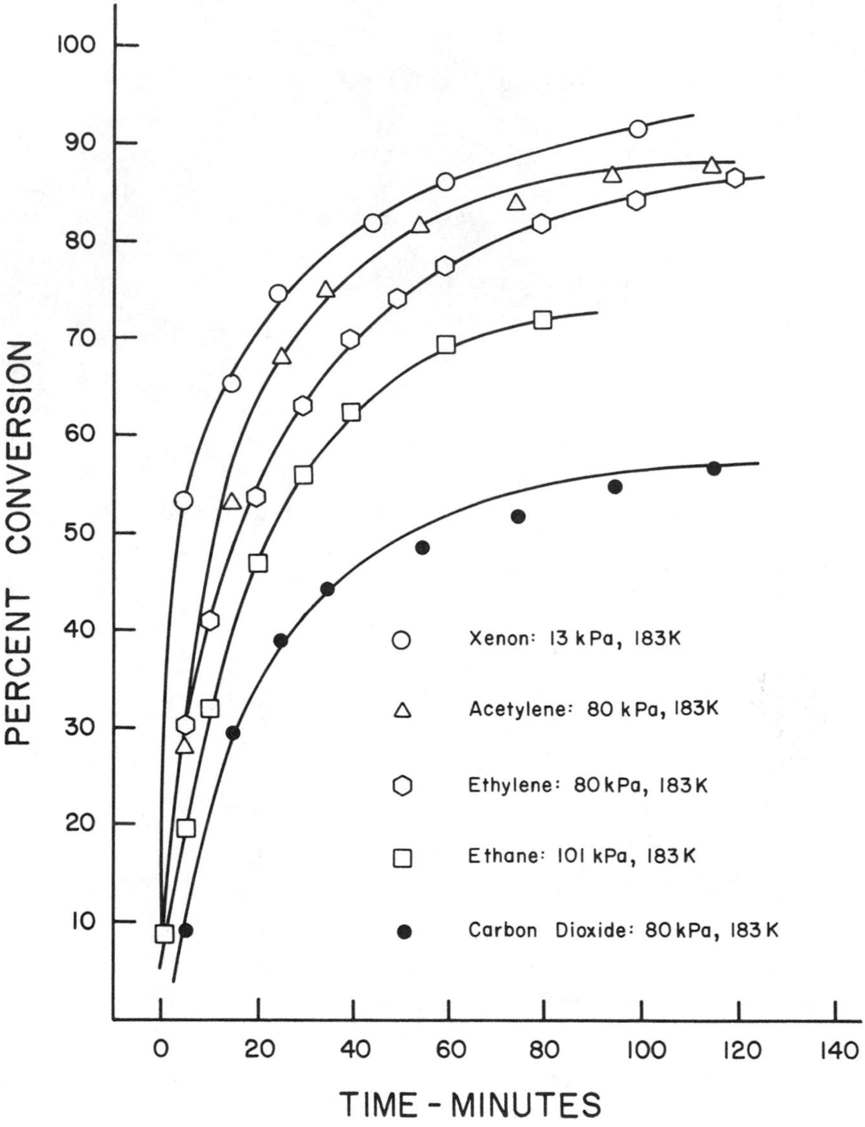


3-8 Kinetics of Some Hydrate Formations from Ice Without Induction Period (Reproduced with permission from B.J. Falabella (Falabella, 1975))

period, shown in Figure 3-8. No explanation was provided for the induction period of either methane or krypton hydrate.

An examination of the induction period in Figure 3-7 for krypton and methane can provide some insight into the formation of critical nuclei for crystal growth. In the discussion accompanying Table 2-3, it was noted that methane (4.36 Å diameter) normally stabilizes the $5^{12}$ cavity of structure I, while the guest:cavity size ratio of the methane molecule was within 0.5 per cent of that of the $5^{12}$ cavity of structure II. Also, although Falabella believed that krypton (4.04Å diameter) stabilized the $5^{12}$ cavity of structure I, almost a decade later Davidson et al. (1984) experimentally determined that krypton stabilized the $5^{12}$ cavity of structure II. When the smaller cavities are occupied in a simple hydrate, guest molecules also enter the larger cavities.

The size difference between methane and krypton is just sufficient to discriminate between the slight differences in each $5^{12}$ cavity diameter, to cause a change of hydrate structure. These differences in guest and host cavity sizes may provide one explanation of the induction period observed in the kinetics of methane and krypton hydrate. As discussed in Section 2.1.3 on hydrate number, Glew (1959) first suggested that when the size of a guest molecule approached the limiting size of a cavity, a high degree of thermodynamic non-stoichiometry was obtained; Glew's notion seems to be validated by the non-stoichiometry of molecules such as cyclopropane, which can stabilize the large cavities of either structure, as a function of temperature or pressure. An analogous behavior on a kinetic level is suggested by the induction period of methane and krypton.

As a microscopic hypothesis for the cause of induction, consider the initial formation of simple hydrate nuclei of methane or krypton in each type of $5^{12}$ cavity, which are very similar in size, so that little discrimination is obtained. In Section 2.1.2 it was determined that the $5^{12}$ cavity is the "basic building block" of each hydrate structure, connected by its vertices to form structure I, or connected by its faces to form structure II. It is suggested that the induction period for the simple hydrates of methane and krypton, may represent a period of oscillation between the $5^{12}$ cavities of structure I and II, before structure I achieves the critical nuclei radius stability for methane hydrate, or before krypton hydrate attains the critical size for stability for structure II. Ripmeester's (1988) indication of substantial stability within the $5^{12}6^{2}$ cavity for guests such as methane suggests that the large cavity may play a deciding role for structure I formers.

Such an hypothesis is supported from two sources. First, when simple hydrates of other guest molecules (Ar, $C_2H_6$, $C_2H_4$, $C_2H_2$, Xe, or $CO_2$) were formed from ice in a ball mill, no kinetic induction period was observed, as shown in Figure 3-8. The lack of induction period may be due to the structural stability brought about by a higher discrimination between the cavity diameters; this is confirmed by calculation of the dimensional guest:cavity ratio shown in Table 2-3 for the molecules studied by Falabella. Molecules smaller than krypton, such as argon (3.8 Å in diameter), would more readily stabilize smaller $5^{12}$ cavity in structure II. Molecules larger than methane, such as xenon (4.58 Å diameter), are sufficient in size to provide substantial stability to the $5^{12}$ cavity of structure I. For such small molecules, the stability of the $5^{12}$ cavities provide the major stabilizing effects on hydrate nucleation; the inclusion of the small molecules in larger cavities provide only a minor portion of the stability to each crystal structure.

Secondly, as indicated in Section 2.1.3, the addition of a small amount of propane to methane will cause the entire hydrate structure to change on formation, because propane can fit only the $5^{12}6^{4}$ cavity of structure II. In an analgous manner, the addition of small amounts of ethane to krypton may cause the hydrate structure to change from structure II for pure krypton, to structure I for the mixture, because ethane provides substantial stability to the $5^{12}6^{2}$ cavity. While no crystallographic data are available to support this suggestion, Falabella's kinetic data in Figure 3-9 clearly show that the addition of 5% ethane to krypton eliminated the induction period observed in pure krypton. Similarly, the addition of 10% ethane to methane eliminated the induction period as shown in Figure 3-10, again probably due to the high stability of ethane in the $5^{12}6^{2}$ cavity, but without hydrate structure change.

The above hypothesis, based on the induction period for the formation of simple hydrates of methane and krypton from ice, then suggests consideration of a new nucleation rate parameter, namely the size ratio of guest molecules to cavities. If the above hypothesis is valid, molecules such as nitrogen and oxygen with sizes intermediate between krypton and methane (see Table 2-3) should show a nucleation induction period greater than that of krypton. Such kinetic induction is displayed in the data of a single kinetic run in Figure 3-10 for mixtures of methane and air.

In recent work Wright (1986) was unable to measure hydrate nucleation rates from ice at temperatures around the ice point either from water or from ice when the surface of the water/ice was not agitated. As reported by Hwang et al. (1989), the melting ice appeared to provide a template for the formation of hydrates. They reported that hydrate formation diminishes and often ceases once a protective film of hydrate forms at the surface. Falabella (1975) suggested that hydrate formation from ice at low temperature has the advantage of high molecular surface mobility of the water molecules. Hwang et al. (1989) propose that the mechanism of hydrate formation at temperatures around the ice point may differ from that at lower temperatures.

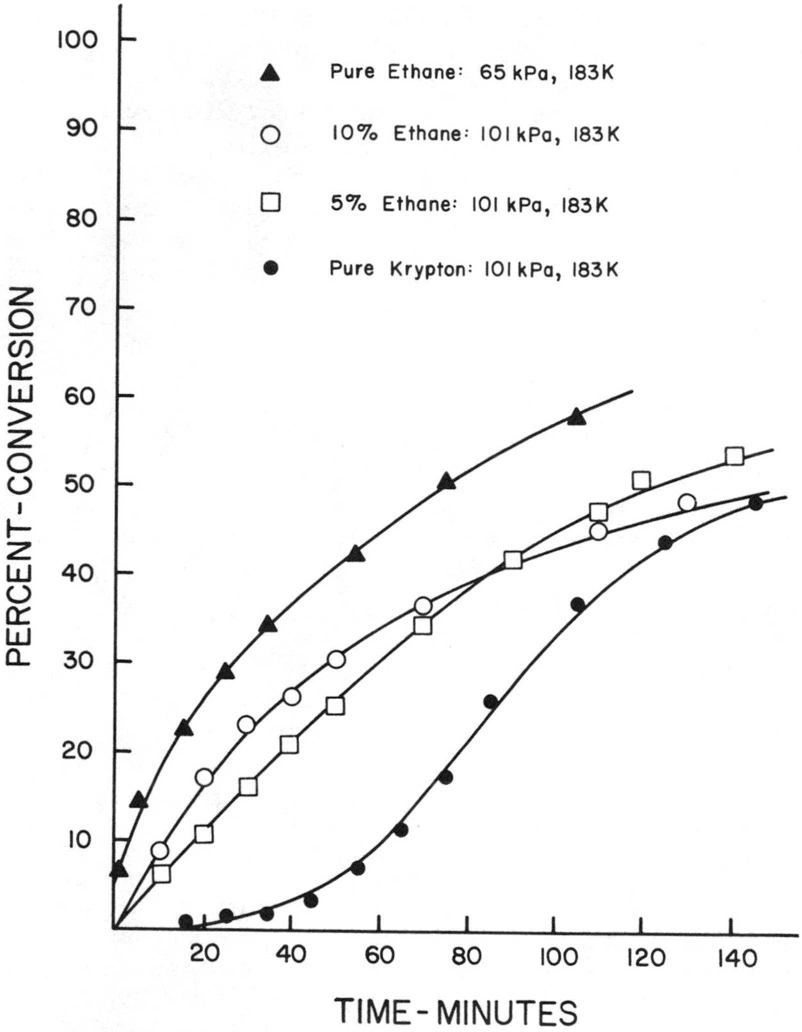


3-9 Effect of Ethane on Induction Kinetics of Krypton Hydrate (Reproduced with permission from B.J. Falabella (Falabella, 1975))

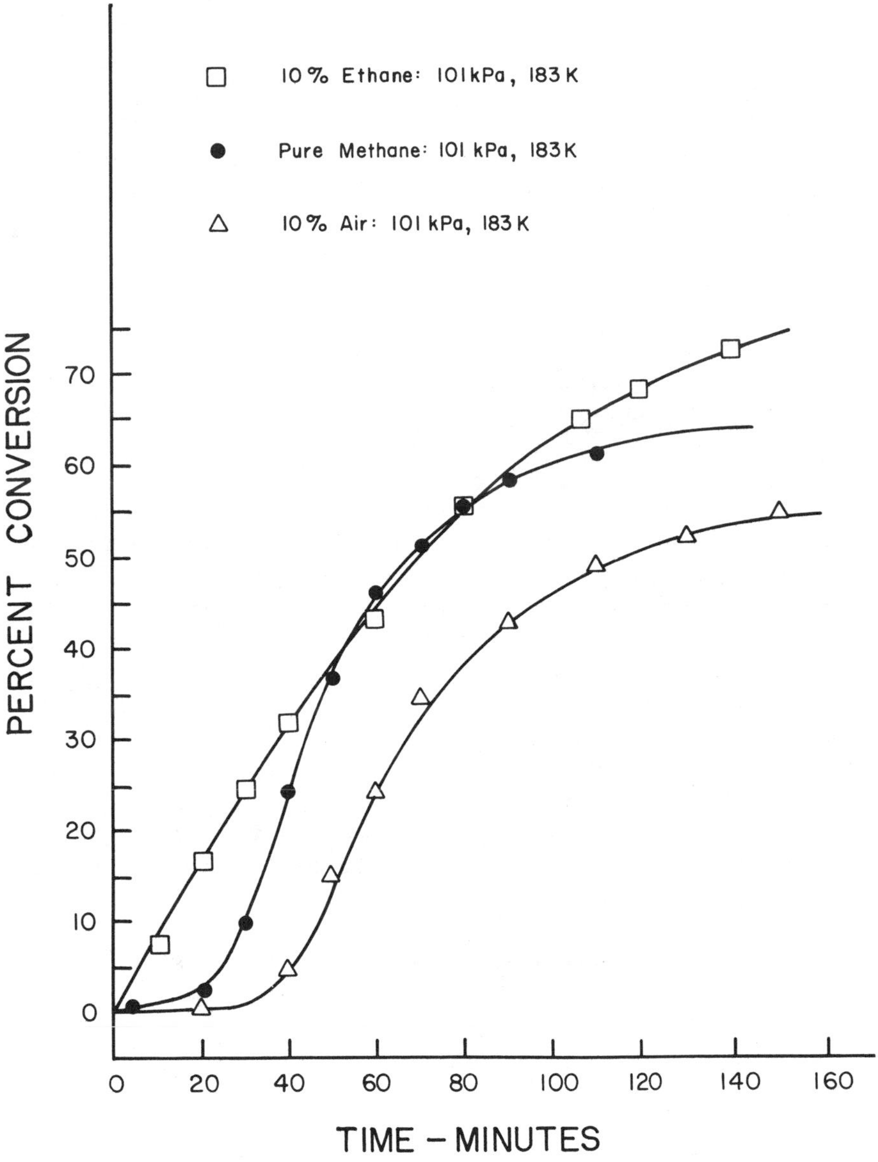


3-10 Effect of Addition of Ethane or Air Upon Methane Hydrate Induction Kinetics (Reproduced with permission from B.J. Falabella (Falabella, 1975))

### 3.1.5 Summary of Gas Hydrate Nucleation Phenomena

The nucleation phenomena listed below are modified from those of Makogon's (1974) qualitative summary of Soviet work, supplemented by the material presented in this section. The phenomena are qualitatively described as follows:

1. Most frequently, the formation of hydrate nuclei occurs at an interface of hydrocarbon fluid and liquid water. This interface may be broadly interpreted to be either a free gas-liquid interface, a liquid hydrocarbon-liquid water interface, or a fluid hydrocarbon-adsorbed water interface.

2. Hydrate formation within the body of a hydrocarbon phase is inhibited by the low solubility of water in a hydrocarbon (usually less than a few molecules per thousand). Conversely, hydrate formation within a body of water is unlikely due to the low solubility of hydrocarbon in water. As shown in Section 2.1.3, a very substantial amount of hydrocarbon is encapsulated on hydrate formation. Both of these notions indicate the interface as the most probable place for hydrate formation to occur, due to high concentrations of both water and guest. Under unusual conditions of saturation, crystallization centers can form directly within the hydrocarbon fluid or within the bulk of liquid water.

3. The hydrate nucleation rate is promoted by subcooling (or general displacement from equilibrium conditions), composition, surface area, and by the availability of the guest and host species to the point of formation. Liquid water molecules are arranged around a dissolved solute molecule in a "prehydrate" structure, with essentially the correct coordination number. Thermodynamics favors the edge and face sharing of both clathrate-like clusters of supercooled water and of neighboring clusters of solvated non-polar atoms.

4. On a non-equilibrium, time-dependent basis, it is possible to supersaturate the liquid with guest molecules

prior to rapid hydrate growth. The time required for supersaturation is hypothesized to coincide with both the establishment of critically sized hydrate nuclei and the time of metastability of hydrate formation, prior to secondary nucleation.

5. While the host molecules have electrostatic attraction for each other the hydrate structure has electrostatic neutrality. The guest molecule interacts with the water cage through van der Waals attractive forces due to a cancellation of water forces. The translation and rotation of the guest molecule may be inhibited by inclusion, but the volumetric properties of the host lattice are substantially unchanged by the presence of the guest. The basic $5^{12}$ cavity is sturdy and will absorb the kinetic energy of impinging water molecules to grow rather than erode.

6. In special cases, notably for methane and krypton, the induction period of hydrate formation from ice may be caused by a vacillation between structures before a critical nucleus size is achieved. This initial nucleation phenomenon may be due to an approximation to the optimum fit of the methane and krypton molecule in either type $5^{12}$ cavity. The induction period may be prevented by the inclusion of a second species of hydrate guest.

7. Once hydrates have been formed and decomposed, there is some residual crystal-like network in the liquid water below 300K. Above 315K the residual network is almost completely eroded. The residual structure discourages metastability upon reformation of hydrates. The residual agglomerate population varies inversely with temperature and time.

## 3.2 Hydrate Crystal Growth Kinetics

After the nucleation of hydrate crystals of at least critical radii, the hydrate growth rate becomes more straightforward and therefore more quantifiable. However, there persists a paucity of kinetic data on crystal growth rate after primary nucleation, during the period called secondary nucleation. Most of the primary nucleation parameters (displacement from equilibrium through subcooling or overpressurization, geometry and surface area, and gas composition) are also important in secondary nucleation. The results of macroscopic laboratory kinetic studies will be presented before considering the microscopic models for hydrate crystal growth.

### 3.2.1 Macroscopic Models of Crystal Growth

Knox, et al. (1961) studied the crystallization rate of propane hydrates in a stirred slurry pilot plant reactor for a commercial desalination process. Their data indicated that an optimum degree of subcooling was achieved at 2-3 °C, a value which is reinforced by that of Makogon (1981 p. 57). While the exact particle size was not determined, Knox and co-workers indicated that very small crystals are obtained with 2-3 °C of subcooling indicating rapid growth, and larger particles occurred with slower growth rates at 1°C.

In unpublished work, Pinder studied the kinetics of hydrate formation from carbon dioxide and double hydrates of tetrahydrofuran(THF)-hydrogen sulfide, with anomalous results. The unsatisfactory results were caused by a formation of hydrate crystals on the water interface, which shielded the water from further reaction with the gas; this concept has been expressed by many researchers in both kinetic and equilibrium studies. To avoid such a surface hindrance, Pinder (1964,1965) studied the simple hydrate kinetics and slurry rheology of THF, a water miscible component. Noting that THF and water may form several hydrogen-bonded complexes, he concluded that rate kinetics within the liquid volume were diffusion dependent, so that no analogy could be made between his results and gas-water reaction rates.

Miller and Smythe (1970) measured kinetic data for the formation of carbon dioxide hydrates from ice. In their analysis the isothermal hydrate growth rate was determined by a typical exponential decrease in pressure of carbon dioxide, as it was encaged in hydrate, given by:

$$\ell n \left[\frac{P_o - P_\infty}{P - P_\infty}\right] = kt \tag{3.4}$$

where $P_o$, $P_\infty$, and P are the initial pressure, the equilibrium pressure, and the pressure as a function of time t, in hours. The first order rate constant k had values of 0.22, 0.71, 1.4, and 2.2 $hour^{-1}$ at 152 K, 162 K, 167 K, and 172 K, respectively. An Arrhenius plot of these constants gave an activation energy of 5.9 kcal/gmol carbon dioxide. No details of the apparatus were given, but it is presumed to be similar to that of Hafemann and Miller (1969).

Via macroscopic measurements of hydrate growth from ice, Falabella (1975) determined the kinetics of eight simple hydrates, and mixtures of methane+ethane, methane+air, methane+krypton, and krypton+ethane from ice between 90K and 203 K. He used essentially the same experimental equipment and isothermal-isobaric procedure as described by Barrer and Edge (1967), who had studied the simple hydrate kinetics of argon, krypton, and xenon. The ball mill apparatus provided both surface renewal and agitation, but it inhibited the determination of an accurate surface area.

In the studies by Barrer and Edge and by Falabella, an identical induction period was noted for krypton; a molecular hypothesis for such nucleation seeding is presented in Section 3.1.4 of this work. By neglecting the induction period for both krypton and methane, Falabella was able to obtain a satisfactory empirical fit to the data of all his simple hydrates as well as the data of Barrer and Edge using second order kinetics, based upon isothermal-isobaric fractional conversion of ice, as:

$$\frac{d(1-X)}{dt} = -k(1-X)^2 \tag{3.5}$$

to obtain an integral expression of

$$\frac{X}{(1-X)} = kt \qquad (3.6)$$

where: $X \equiv \dfrac{\text{(moles of ice converted to hydrate)}}{\text{(moles of ice available to form hydrate)}}$

$k$ = reaction rate constant, minutes$^{-1}$

$t$ = elapsed time, minutes.

Falabella was not able to control such variables as ice crystal size and distribution in his ball-mill apparatus, so they were discounted in his analysis. He notes that, apparently due to the fact that some ice was occluded, only 90% (approximately) of the ice was available for conversion. He was able to analyze his methane hydrate data for pressure kinetic effects. At 183 K, kinetics at four pressures were fit to the following modification of Equation (3.5):

$$\frac{d(1-X)}{dt} = -k(1-X)^2(P-P_\infty)^{1/2} \qquad (3.7)$$

By analyzing his 183 K methane data at the maximum rate point (15% conversion), Falabella was able to discount the induction period. He determined the rate constant k to be 0.67±0.2 min$^{-1}$ at 56, 66, and 76 cm Hg. The highest pressure of 94 cm Hg gave an anomalous k of 0.83 min$^{-1}$, possibly because the pressure was too far beyond the equilibrium value of 42 cm Hg. An Arrhenius analysis indicated the methane data could not be analyzed for temperature effects.

Kamath (1984) determined both a surface and a volume model for hydrate formation from melting ice at 274K. Both models may be summarized by the below equation:

$$\frac{dx}{dt} = K\,(1-x)^y \left[\frac{P_o}{Z_oRT} - \frac{P_e}{Z_eRT} - \frac{M_ox}{V_RM_wn}\right] \qquad (3.8)$$

where x is the fractional ice conversion, t is time, $M_W$ is the molecular weight of water, $M_O$ is the initial amount of ice, $V_R$ is the volume of the pressure vessel, n is the hydrate number, and subscripts o and e represent original and equilibrium values. The exponent y is 2/3 for the surface model and unity for the volume model. The constant $K = k_1 S_O V_R M_W n/M_O$ for the surface model and $k_2 n$ for the volume model, where $k_i$ is a fitted rate constant (1=surface; 2=volume) and $S_O$ is the initial surface area. The value of the lumped parameter $k_1 S_O$ between 0.0125 $hr^{-1}$ and 0.038 $hr^{-1}$ gave good agreement with experimental data in the surface model, while a value of $k_2$ = 2.439 cc/gmol/hr for the volume fit the data to within 8.8%.

Vysniauskas and Bishnoi (1983a,b, 1985) presented kinetic experimental studies above the ice point (274-284 K) on the growth of methane and ethane hydrates after critical nuclei are established. Their isothermal, isobaric experiments, performed in a semi-batch stirred tank reactor, determined the rate of hydrate formation by the consumption of gas into the stirred water-hydrate mixture. Because the slurry eventually became too viscous for good mixing, they only converted 5% of the initial 300 $cm^3$ water charge to hydrates. The area of the reaction was assumed to be the contact surface area (106 $cm^2$) of the water with the gas. The overall gas consumption rate was determined to be a function of the interfacial area, the temperature and pressure, and the concentration of the critically-sized clusters of nuclei.

For each isotherm, a logarithmic plot of the reaction rate as a function of pressure was linear except for an initial convex portion, which was attributed to the formation of critically sized clusters, as indicated in Section 3.1. Vysniauskas and Bishnoi fit their data to the reaction rate expression:

$$r = A\, e^{-\Delta E_a/RT}\, e^{-a/\Delta T^b}\, P^{\gamma} \qquad (3.9)$$

where A is the pre-exponential constant, $\Delta E_a$ is the energy of activation, and $\gamma$ is the overall order of reaction with pressure. As shown in Figures 3-11 and 3-12 for methane and ethane respectively, the above model with parameters shown in Table 3-3 fits the data very well. Also in Table 3-3 are natural gas kinetic parameters by Topham (1984) from his analysis of earlier data by Bishnoi and co-workers (1980a,b); to obtain the values in the table, parameters a and b were held constant, $\gamma$ was constrained to be approximately 3, and A and $\Delta E$ were fitted to the data.

In Table 3-3 a negative energy of activation is indicative of a physical process, in which both the reorganization of the water molecules into a hydrate lattice and the energy on guest-host interaction is considered. Such negative temperature correlations have been previously determined for systems in which reactants with low energy or temperature react faster than those with high energy (Mozurkewich and Benson, 1984). Geometry, particularly that of the nuclei, is recognized as a major variable in all crystallization processes. In the work of Vysniauskas and Bishnoi, a single apparatus was used to generate all of the data at an optimized stirring rate, with no rippling of the vortex surface. As a consequence, no study is available concerning the translation of the data to hydrate formation conditions with other geometries.

Hwang et al. (1989) fit the data of Wright in the secondary nucleation region for growth from ice from spinning ice disks and determined that the growth rate was inversely proportional to the thickness of the hydrate layer, when no agitation was provided. They also determined that the rate constants were a function of the geometry of the system and were not generalizable parameters.

The final comment in the previous paragraph is applicable to all kinetic work to date. That is, one must obtain dissociation or formation rate constants which are independent of geometry, experimental method, and mass transfer effects in order to have a satisfactory kinetic

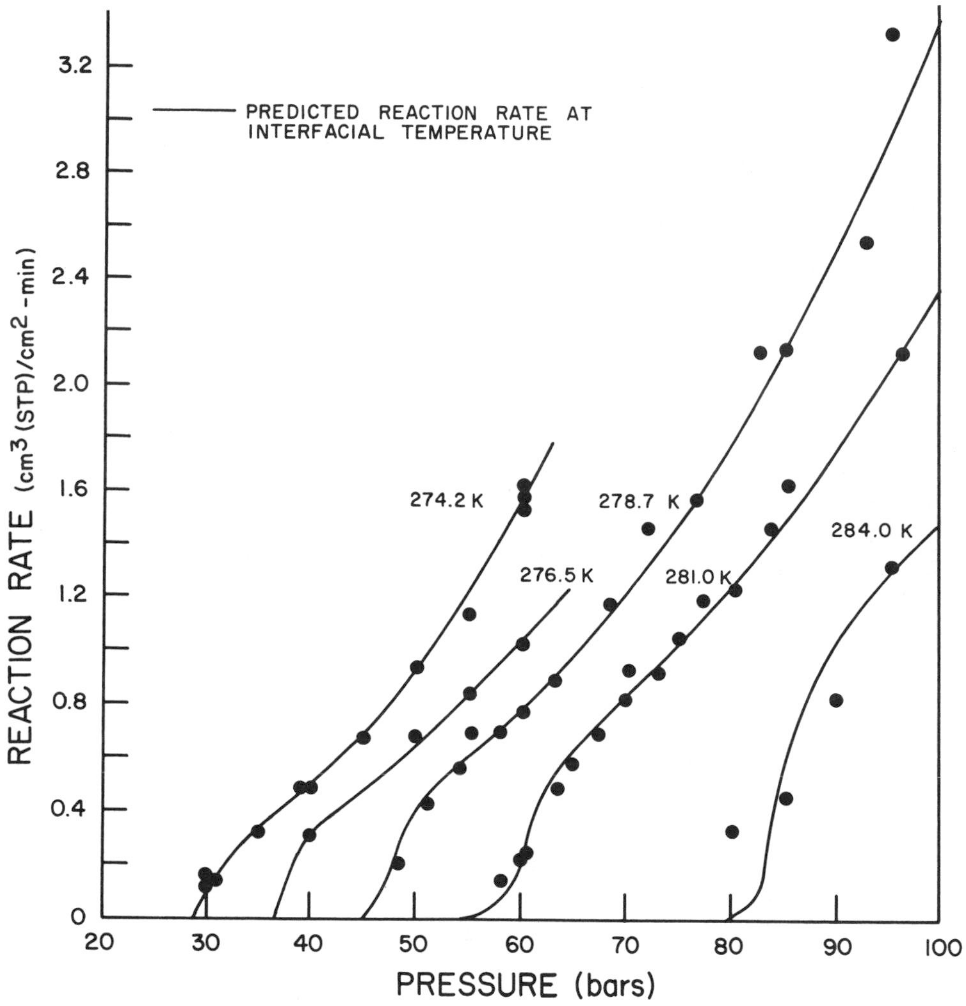


3-11 Isothermal-Isobaric Data for Methane Hydrate Formation (Reproduced, with permission, from Chemical Engineering Science (Vysniauskas and Bishnoi, 1983b), by Pergamon Press)

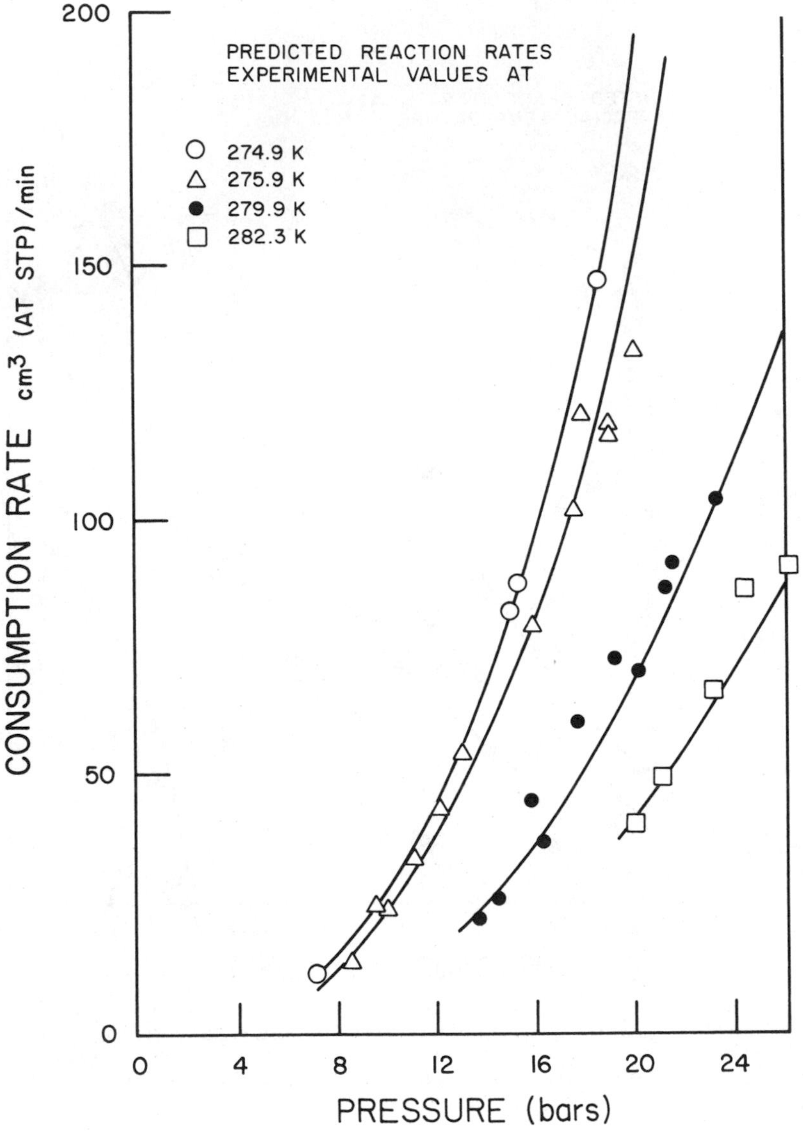


3-12 Isothermal-Isobaric Data for Ethane Hydrate Formation (Reproduced, with permission, from Chemical Engineering Science (Vysniauskas and Bishnoi, 1985), by Pergamon Press.)

**Table 3-3**
**Parameters in Equation (3.9)**

| Parameter | Methane | Ethane | Mixture I[1] | Mixture II[2] |
|---|---|---|---|---|
| A, m/sec(MPa)$^{\gamma}$ | $7.262\times10^{-27}$ | $3.827\times10^{-30}$ | $4.294\times10^{-42}$ | $1.547\times10^{-44}$ |
| ΔE, kJ/gmol | -106.20 | -133.01 | -191.35 | -204.74 |
| $\gamma$ | 2.986 | 2.804 | 3.235 | 2.90 |
| a | 0.0778 | 0.0778 | 0.0778 | 0.0778 |
| b | 2.411 | 2.411 | 2.4 | 2.4 |

1. Mixture I is 87.34 (mol)% $CH_4$, 9.18% $C_2H_6$, 3.00% $C_3H_8$, and 0.48% i-$C_4H_{10}$

2. Mixture II is 83.1% $CH_4$, 10.19% $C_2H_6$, 4.01% $C_3H_8$, 0.85% i-$C_4H_{10}$, 0.64 n-$C_4H_{10}$, 0.22 i-$C_5H_{12}$, 0.20 n-$C_5H_{12}$, 0.39 n-$C_6H_{14}$, 0.29 $N_2$, and 0.11 $CO_2$

model. Questions such as these are very much at the current state-of-the-art. Much more remains to be done in this area. To overcome some of these restrictions a mechanism of hydrate formation should be considered without regard to geometries.

### 3.2.2 Microscopic Models for Hydrate Crystal Growth

Barrer and Ruzicka (1962) studied the rate of crystal growth of double hydrates of argon and tetrahydrofuran, particularly with respect to argon inclusion. They argued that the rate of crystal growth was limited by minima in Gibbs Free Energy as a function of guest occupation. The minima were regulated by the rate of transport of guest molecules from the bulk solution to the hydrate face, given by the Fick's second law:

$$\frac{dn}{dt} = A\ \mathcal{D}\ \frac{\Delta C}{\delta} \tag{3.10}$$

where dn/dt is the hydrate growth with time, A is the surface area of the crystal, $\mathcal{D}$ is the diffusion coefficient of the guest through the bulk solution, $\Delta C$ is the difference in concentration across the boundary layer, and $\delta$ is the thickness of the fluid boundary layer surrounding the hydrate crystal. Once guest molecules cannot be supplied to the face of the crystal at a rate to satisfy the optimum cavity occupation, a Gibbs Free Energy minimum was obtained causing crystal growth to cease. Their work gives the qualitative suggestion that nucleation rate decreases rapidly with increasing temperature.

Barrer and Edge (1967) postulated that the mechanism for crystal growth of hydrates from ice was that of condensation of water molecules into a hydrate structure; at 90K the formation rate of argon hydrate was $1.88\times10^{18}$ water molecules•gram$^{-1}$•sec$^{-1}$. Falabella (1975) noted that the Knudsen sublimation rate of ice at 90 K is too low ($1.02 \times 10^{7}$ water molecules•gram$^{-1}$•sec$^{-1}$) to supply sufficient water for such hydrate formation rates. The surface mobility of ice, as indicated by the low diffusional jump time (2.7 μsec) listed in Table 2-5, might be an important part of the hydrate kinetic mechanism from ice.

In a series of low pressure cryogenic studies of co-condensation of thin films of hydrate crystals from the gas phase, Devlin and co-workers (1983, 1984 a,b, 1985 a,b,1987 and 1988) showed that hydrates of ethylene oxide may be formed from the gas phase, and nonpolar molecules can be co-condensed with water if only a few percent of ethylene oxide is present. Noting that previous researchers show that the polar ether promotes a series of defects in the hydrate lattice, Devlin and co-workers (1987) suggested that hydrate crystallization was controlled by the concentration of mobile orientation

(Bjerrum L type) defects. In addition they indicated that the propagation of the crystal structure is limited by crystal growth through an amorphous phase. If the residual defects propagate through the crystal body to the surface in an annealing process, then the crystal will stabilize and grow. On the other hand, if the residual defects increase, the crystal integrity will be lost. In effect, the mobile defects are required for minimization of disorder within the body of the crystal. This work represents a new hypothesis for hydrate crystal growth from the vapor phase, with a firm theoretical basis that is yet to be developed.

In work originally published here, we (Selim and Sloan, 1987) suggest a molecular mechanism for the secondary nucleation of simple hydrates of methane in a liquid film surrounding a saturated gas-liquid interface, as depicted in Figure 3-13. In that figure, the three dimensional hydrate cages are shown in two dimensions for simplicity. Four steps were proposed for the formation of hydrates at the gas-liquid-solid interface, as follows, with letters in reference to Figure 3-13:

(1) Water molecules diffuse to the surface of the hydrate nuclei A. This step offers negligible resistance due to the high concentration of water.

(2) Water molecules adsorb onto the initial cage and use that cage as an orientation template for a partial cage formation, B. This step is taken as a first order physical reaction with the stoichiometry:

$$A + H_2O \xrightarrow{k_1} B \tag{3.11a}$$

(3) Methane is adsorbed into the partial cage structure B, from the gas-liquid interface to obtain component C. Instantaneously, the water bonds of the partial cage reorient slightly to accommodate the van der Waals forces imposed by the methane guest in the partial cage. The resulting methane molecule adsorbed in the adjusted partial cage is called component D in Figure 3-13, with the stoichiometry:

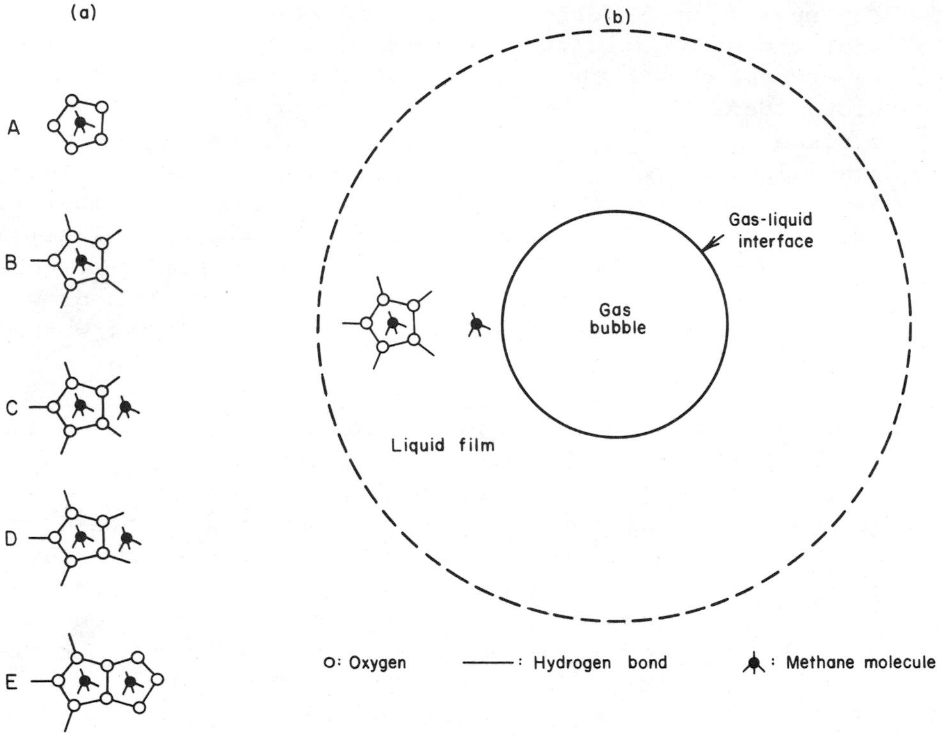


3-13 A Mechanism for the Molecular Growth of Hydrate Crystals
(a) with Successive Growth of Subnuclei as A→B→C→D→E
(b) in Liquid Film Surrounding Gas-Liquid Interface
(Selim and Sloan, 1987)

$$B + CH_4 \xrightarrow{k_2} D \tag{3.11b}$$

(4) Water molecules diffuse to the partial cage with the adsorbed methane. Again, this step has negligible resistance due to high water concentration. The water molecules adsorb in a formation capable of hydrogen bonding plus accommodating the methane molecule already in the cage. The new encapsulated cage is called component E, which is very similar to the initial reactant A, and is thus assumed proportional to A in concentration. The final stoichiometry is:

$$D + H_2O \xrightarrow{k_3} E \qquad (3.11c)$$

Since the reactions take place in an abundance of water, the water concentration was assumed constant. Further, since the reaction takes place at the gas-liquid interface, the methane molecules are likely to be present in abundance so that their concentration was also assumed constant. With these assumptions, the first order rate equations become:

$$\frac{d[A]}{dt} = -k_1[A] \qquad (3.12a)$$

$$\frac{d[B]}{dt} = k_1[A] - k_2[B] \qquad (3.12b)$$

$$\frac{d[D]}{dt} = k_2[B] - k_3[D] \qquad (3.12c)$$

$$\frac{d[E]}{dt} = k_3[D] - k_1[E] \qquad (3.12d)$$

with the following solution for [E]:

$$[E] = a_1 e^{-k_1 t} + a_2 e^{-k_2 t} + a_3 e^{-k_3 t} + a_4 e^{-k_1 t} \qquad (3.13)$$

where the a's are computed from the following set of linear equations:

$$\sum a_i = 0 \qquad (3.14a)$$

$$\sum k_i a_i = 0 \qquad (3.14b)$$

$$\sum k_i^2 a_i = 0 \qquad (3.14c)$$

$$\sum k_i^3 a_i = \left( \prod_{j=2}^{4} k_j \right) [A]_0 \qquad (3.14d)$$

where $[A]_0$ is the initial concentration of A. In the unlikely case for which all the k's are equal, the solution becomes:

$$\left[\frac{E}{A}\right]_0 = \frac{(kt)^3}{3!} e^{-kt} \qquad (3.15)$$

which may be recognized as a Poisson distribution. The simplified case in the above equation confirms our intuition that the concentration of species E (with two cages) will increase with time, and then decrease as the three cage structures of hydrate crystals begin to form. Experimental examination of this mechanism is currently being carried out.

In more detailed secondary nucleation experiments than performed by Vsyniauskas(1980), Bishnoi and coworkers determined a mechanistic model for hydrate formation, based upon a two step process for crystal growth. Following experiments on formation of simple hydrates of methane and ethane, Englezos et al.(1987b) were able to show that mixtures formed a linear combination of the rates for each gas. They eliminated the first step of diffusion of dissolved gas to the interface by vigorous agitation, and accounted for the second step of adsorption of the guest onto the crystal, for a simple hydrate crystal growth expression of:

$$\left(\frac{dn}{dt}\right) = K^* A_p \; (f - f_\infty) \tag{3.16}$$

where n is the number of moles of guest incorporated into the crystal structure over time t, $A_p$ is the surface area of a spherical particle, and the rate constant $K^*$ represents adsorption. A binary mixture of methane and ethane was shown to have kinetics of a linear combination of the components as:

$$\left(\frac{dn}{dt}\right)_p = \sum_1^2 \left(\frac{dn_j}{dt}\right) = \sum_1^2 K_j^* \; A_p \; (f - f_\infty) \tag{3.17}$$

The radius of the critical nuclei was determined by Equations (3.2,3). Visual observation of the turbidity point was taken as the indication that the nuclei were of critical size. After the turbidity point, the global rate $R_y(t)$ of secondary nucleation was assumed to be proportional to the second moment $\mu_2$ of the particle size distribution $\phi(r,t)$, as given by:

$$R_y(t) = 4\pi\mu_2\sum_1^2 K_j^* \ (f - f_\infty)_j \qquad (3.18)$$

where:

$$\mu_2 = \int_0^\infty r^2\phi(r,t)\,dr \qquad (3.19)$$

The second moment of the particle size distribution is physically interpreted as an indication of the width of the particle size spectrum. For a Dirac function of $\phi$, the global rate R is proportional to the volume of uniform spheres; for $\phi$ of zero, there is no global rate. With the parameters $K_j^*$ from pure component data, Englezos et al. used the above model to predict the mixture data to within 11.3% for most runs over the temperature range of 273 to 284K and the pressure range of 0.68 to 5.60 MPa. Only values of $K_j^*$ and the solubilities of gas at the turbidity point are needed; no other adjustable parameters are in the model.

Englezos et al. (1988) determined a rate expression for hydrate decomposition by multiplying the expression in Equation (3.16) by a negative on each side. The rate constant $K^*$ was taken as $K_d$ and fit to the equation, with good comparison to experimental data.

## 3.3 Thermodynamics of Hydrates as the End Point of Kinetics

This chapter should serve to convince the reader that, while our knowledge has advanced a small step in the last 180 years, we have only begun to understand hydrate kinetic phenomena. Questions regarding the rate of hydrate formation in processing equipment or pipelines, in the deep oceans, or in porous media remain largely unknown and are the object of current investigations. In addition, kinetic studies from unusual liquid compositions, such as with high amounts of inhibitors, with clays, or from three phase ($L_W$-V-$L_{HC}$) regions are at the state-of-the-art for researchers.

In the absence of quantitative kinetic information, researchers and practitioners alike have turned to thermodynamics as the end point of kinetics. That is, given a large amount of time, the system will equilibrate to the thermodynamic conditions, and the kinetic rate of hydrate formation will become negligible. Because we cannot accurately determine the time for hydrate formation, we instead estimate the appropriate temperatures, pressures, and compositions (the thermodynamic properties) of hydrate formation. If one knows the thermodynamic temperature and pressure of hydrate formation for a given gas, the prevention of hydrates is ensured by maintaining a higher temperature or a lower pressure. At present the knowledge of the thermodynamics of hydrate formation is much more complete than the kinetics, so that the thermodynamics may be used to provide a bound to hydrate calculations.

As an example, consider a gas pipeline in which free water is known to exist. Optimally, the pipeline operator might wish to know how long the pipeline can be in service without becoming blocked with hydrates; this however is a kinetic question which is presently beyond the state-of-the-art. Instead, using thermodynamic methods, the pipeline operator could determine the temperatures and pressures at which hydrates were possible, to determine whether his pipeline operation was in jeopardy. If the conditions of pipeline operation were outside the thermodynamic stability region of hydrates, no hydrate formation will occur regardless of the time involved. Alternatively, the operator might wish to know how much hydrate inhibitor to inject into the gas, in order to keep the pipeline in the fluid phase; this is another thermodynamic question. Such questions of thermodynamics are dealt with in Chapters 4, 5, and applied in Chapter 8.

# 4

# Fundamental Phase Equilibrium Calculations for Natural Gas Hydrates

The phase equilibria of natural gas hydrates represents the most important set of properties which differ from those of ice. As indicated in Chapter 3, hydrate phase equilibria is relatively well-defined and determines a boundary to the kinetic problem, which is incompletely understood. The goal of this chapter is to provide a qualitative understanding of clathrate phase equilibria through a description of the different regions of the phase diagram, together with the simplest methods of calculation of those equilibria. Techniques for initial estimates are provided to enable the use of more sophisticated computer techniques, such as the ones in the following chapter, to judge their reliability for an application. In addition, an understanding of the concepts of the current chapter is necessary for applications - to hydrates in the earth (Chapter 7) and to hydrate problems in production and processing (Chapter 8).

The phase diagrams for water-hydrocarbon systems provide a convenient overview of the types of phase equilibrium calculations. These diagrams differ substantially from the normal hydrocarbon mixture phase diagrams primarily due to the hydrogen bond phenomena indicated in Chapter 2. The phase diagrams discussed in Section 4.1 serve as a reference point for the calculation methods in the remainder of this chapter and the next.

Section 4.2 addresses the solubility of hydrocarbons in an aqueous phase, a phenomenon which finds many applications in hydrate formation. For example, in the previous chapter a kinetic heuristic was given indicating the nucleation and growth of hydrates at the interface of free water and gas. The small aqueous solubility of the hydrocarbon provides an explanation for the low probability of hydrate formation in the bulk phase. The question of the equilibrium solubility of hydrocarbons in an aqueous phase, particularly around the hydrate point, thus appears to be integral to the study of hydrate kinetics.

Section 4.3 concerns the most useful conditions of hydrate equilibria. Determination of these equilibria will enable the answer to the question, "Given a gas composition, what are the temperature and pressure conditions at which hydrate forms from gas and free water?" In this section two historical methods, namely the gas gravity method (Section 4.3.1) and the $K_i$-value method (Section 4.3.2), for calculating the pressure-temperature equilibrium of three-phases (liquid water-hydrate-vapor, $L_W$-H-V)[1] are discussed. For the method of Section 4.3.2 a user-friendly IBM-PC compatible program is provided on the floppy disk in the frontpapers, and its use is defined. These methods yield initial estimates for the calculation and provide a qualitative understanding of the equilibria. A statistical thermodynamic method provides yet a more accurate three-phase calculation; however because it is both more comprehensive and detailed, it is relegated to a separate chapter immediately following this one.

---

[1] Phase (and abbreviation) nomenclature given in order of decreasing water concentration in each phase

The two calculation methods in Section 4.3 are for the three-phase ($L_W$-H-V) region extending between the two quadruple points $Q_1$ and $Q_2$ in Figure 1-1. Section 4.4 provides a method to use the techniques of Section 4.3 to locate both quadruple points on a pressure-temperature plot. Section 4.4 also qualitatively discusses the equilibrium of three condensed phases (aqueous liquid-hydrate-hydrocarbon liquid ($L_W$-H-$L_{HC}$)). Determination of equilibria from condensed phases provides an answer to the question, "Given a liquid hydrocarbon and a free water phase, at what pressure and temperature conditions will hydrates form?"

The inhibition of three-phase hydrate formation is discussed in Section 4.5. The determination of this phase equilibrium enables the answer of such questions as, "How much methanol (or other inhibitor) should be placed in the free water phase so that hydrates will not form at the pressures and temperatures of interest?" Classical empirical techniques such as that of Hammerschmidt (1934) are suitable for hand calculation and provide a qualitative understanding of the effect of inhibitors.

The calculation of two-phase (hydrate-fluid hydrocarbon) phase equilibrium is discussed in Section 4.6. The question, "To what degree should the hydrocarbon gas or liquid be dried in order to prevent hydrate formation?" is addressed through these equilibria. In this section the older water content charts are shown to be in substantial error, and a hand calculation method is presented as a better estimate.

The final technical portion of the chapter, Section 4.7, concerns the relationship of phase equilibria to other hydrate properties. The hydrate application of the Clapeyron equation is discussed with regard to calculating heats of formation and the hydrate number. Other techniques for determining the hydrate number are also discussed here.

Throughout this chapter and the next, it should be borne in mind that our physical understanding is incomplete, and thus the calculation methods should be taken only as estimates. Statements of accuracy limits are provided for each method, where possible. Since these methods are only as good as available experimental data, when questions of veracity arise, the examined data should be taken as reliable and the calculation method questioned. If it were economical to obtain accurate experimental data for each case under consideration by the user, that would be the preferred course.

As an example, the determination of three-phase ($L_W$-H-V) formation data for simple hydrates (one guest) has been experimentally established with a high degree of certainty. Since an infinite number of composition possibilities exist for binary hydrocarbon gas mixtures, a plethora of possible mixtures exist for the eight known hydrate formers in natural gas. In the three-phase region, only the effort and expense of obtaining hydrocarbon mixture hydrate data justify the construction of a means of interpolation and careful extrapolation of data sets.

## 4.1 Diagrams of Phase Behavior for Hydrocarbon+Water Systems

The phase behavior of hydrocarbon+water mixtures differs significantly from that of normal hydrocarbon mixtures. Differences from more common phase diagrams arise from two effects, both of which have their basis in hydrogen bonding. First, the hydrate phase is a significant part of all hydrocarbon+water phase diagrams for hydrocarbons with a molecular weight lower than n-butane. Secondly, water and hydrocarbon molecules are so different that, in the condensed state, two distinct liquid phases form, each with a very low solubility in the other.

Almost all of the concepts of hydrocarbon+water phase equilibria may be found in the diagrams of a binary hydrocarbon+water mixture. For simplicity, consider first the system propane+water, one of the few binaries with comparatively plentiful experimental data for the construction of such diagrams. Similar phase diagrams exist for binary water mixtures with the natural gas components of ethane and isobutane. Modifications are observed for binary systems of water with methane or nitrogen, discussed in Section 4.1.2. Modifications for water-soluble hydrate formers such as carbon dioxide, and hydrogen sulfide are also discussed in Section 4.1.2. Phase diagrams for multicomponent hydrocarbons+water also differ slightly, as discussed in Section 4.1.3.

### 4.1.1 Phase Diagrams of the Propane+Water Binary

The diagrams of the propane+water binary are taken from an analysis (Harmens and Sloan, 1989) which represents an extension and in some cases a correction, to those published earlier by Kobayashi (1951), Katz et al. (1959) and Bourrie and Sloan (1986).

#### 4.1.1.a. The Propane+Water P-T Diagram

An ideal presentation of the phase diagrams for propane+water would use the three dimensions of pressure, temperature, and composition. Due to the restrictions of the printed page, we must first consider a two-dimensional pressure-temperature (P-T) diagram which compresses the composition axis (perpendicular to the page) followed by several temperature-composition (T-x) isobars of the next subsection. The P-T diagram of Figure 4-1, while condensed, provides an overview and guide to the T-x diagrams.

Figure 4-1 contains all of the major features of the propane+water system, and contains the smaller region of Figure 1-1. Figure 4-1 is shown with a somewhat distorted scale; due to the large ranges of pressure and temperature, many of the intricate features would be highly compressed and indistinguishable with a rigorous quantitative scale. The relative positions of the lines and points on the diagram are maintained in order to provide a qualitative overview. The structure of this diagram

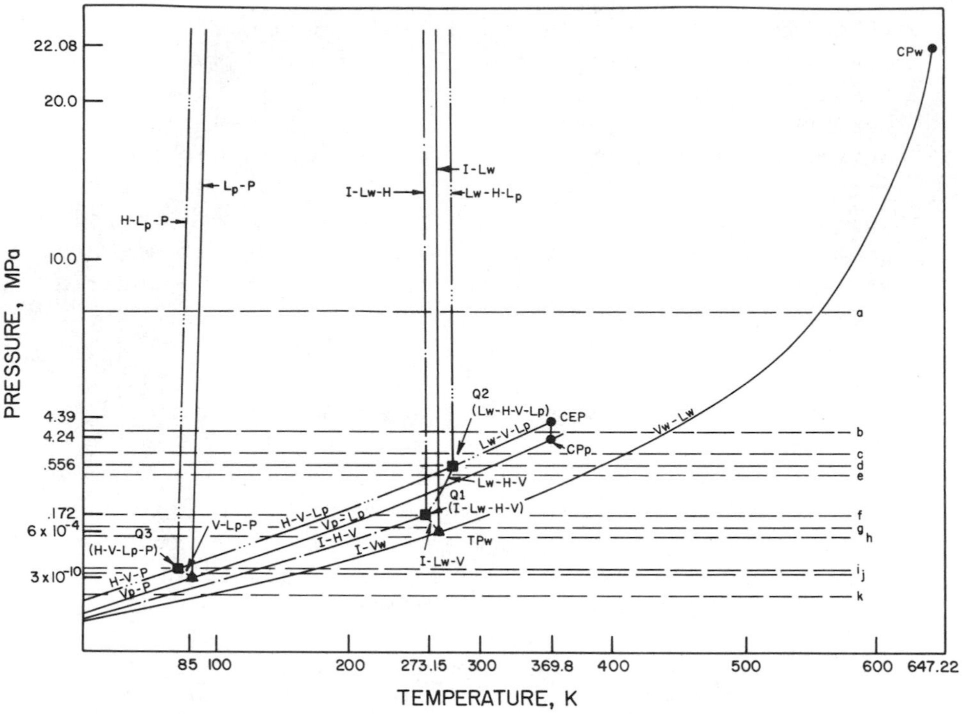


4-1 Pressure-Temperature Diagram for Propane and Water. Isobars (Dashed lines a through k) Correspond to T-X Diagrams in Figure 4-2

rests on experimental data for characteristic points and phase boundaries, and on the Gibbs Phase Rule (Gibbs, 1928, p.96). In both the P-T and the T-x diagrams the following nomenclature is used:

H,V represent the hydrate and vapor phases, respectively
P,I represent pure solid propane and pure ice,
$V_P, V_W$ represent pure propane and pure water vapor, and
$L_P, L_W$ represent liquid propane and aqueous liquid, both as pure liquids and with slight miscibilities.

The initial reference points of Figure 4-1 are the triple points (▲) for the pure components propane and water ($TP_P$ and $TP_W$) at $3.0 \times 10^{-7}$ KPa, 85.5 K and 0.618 KPa, 273.15 K respectively, and the pure component critical (●) points ($CP_P$ and $CP_W$) at 4.24 MPa, 369.8 K and 22.08 MPa,647.2 K, respectively. The P-T lines for each pure component are shown as three types of solid lines: (1) connecting the triple points and the critical points for the vapor pressures (lines $V_P$-$L_P$ and $V_W$-$L_W$), (2) extending below the triple points for the sublimation pressures (lines $V_P$-P and I-$V_W$), and (3) extending almost vertically from the triple point, off the scale of the diagram, for the fusion lines ($L_P$-P and I-$L_W$).

Together with the triple points of the pure components, the three quadruple points (■) $Q_1$, $Q_2$, and $Q_3$ in Figure 4-1 represent the only points of line intersections. Makogon (1981,p.8) notes that there should be a fourth quadruple point $Q_4$ at pressures approximating 1000 MPa; however such a quadruple point has never been measured and will not be discussed here. By the Phase Rule, the four-phase points for the binary are invariant and provide reference points on the P-T projection. The same quadruple points shown in Figure 1-1 are given in the two upper quadruple points $Q_1$ (I-$L_W$-H-V), and $Q_2$ ($L_W$-H-V-$L_P$) located at 0.172 MPa, 273.15 K, and at 0.556 MPa, 278.85 K, respectively. The lower quadruple point, $Q_3$ (H-V-$L_P$-P), at a pressure and temperature only fractionally greater and less respectively than the triple point of propane, has never been experimentally measured, but is determined through phase analysis (Katz et al., 1959, Figure 5-5).

In the P-T projection, non-horizontal broken lines represent all three-phase equilibria, which are univariant by the Phase Rule. The only three-phase line which has an upper critical end point is the $L_W$-V-$L_P$ line which terminates. at conditions only slightly variant (4.39 MPa, 369.7 K) from the propane critical point. Other three-phase lines terminate at either quadruple points or triple points, except the almost vertical lines, whose terminals have yet to be measured or predicted.

The Use of the P-T Plot. To demonstrate the utility of the P-T diagram, consider the region around $Q_2$. The position of the $L_W$-H-V line between $Q_1$ and $Q_2$ has been the object of the vast majority of propane (and other natural gas) hydrate phase equilibrium studies. Once the vapor hydrocarbon composition is determined, such studies only concern the pressure and temperature of hydrate formation from liquid water and vapor, without regard for either the composition of the condensed phases or the water content of the hydrocarbon vapor.

The two-phase $L_W$-V region forms to the right of the $L_W$-H-V line, as two phases common to the bounding lines $L_W$-H-V and $L_W$-V-$L_P$. As the temperature is lowered to the left of the $L_W$-H-V line, one fluid phase disappears, leaving either a two-phase H-V region between $L_W$-H-V and H-V-$L_P$, or a $L_W$-H region between $L_W$-H-V and I-$L_W$-H; the excess fluid phase is the phase which persists into the two-phase region. Similar two-phase regions may be determined in the remainder of the diagram as those areas which are common to the three-phase lines which bound them.

Emanating from $Q_2$ is the line $L_W$-H-$L_P$, which is almost vertical (Verma, 1974) due to three incompressible phases. Due to the vertical nature of $L_W$-H-$L_P$, the location of $Q_2$ is taken as the upper temperature of hydrate formation of propane+water; Section 4.4 discusses this point in detail. The P-T locus of the $L_W$-H-$L_P$ line is of interest to those concerned about the hydrate formation of liquid hydrocarbons from a free aqueous phase.

### 4.1.1.b. T-x Isobaric Diagrams for Propane+Water

The series of T-x isobaric diagrams of Figure 4-2 point out other distinguishing features of the binary system. The abscissa of each of these diagrams represents the mol fraction of propane, increasing in the positive direction, with decreasing mol fraction of water. Since mutual solubilities, as well as some temperature differences, are very small, the T-x diagrams cannot be drawn to scale. Some essential but small scale features are over-emphasized, so that the diagrams are distorted, but qualitatively correct. The details of these diagrams are shown in translation between them and the P-T diagram, as in the following two subsections.

Translation Between P-T and T-x Diagrams. Because these diagrams may at first appear difficult to interpret, eleven horizontal isobars (dashed lines) have been drawn on Figure 4-1, with letters matching the captions of the T-X isobars of Figure 4-2. Any T-x diagram may be constructed from the appropriate isobar on the P-T diagram. It may be useful to consider the heuristics shown below for diagram translations by regarding the T-x diagrams of Figure 4-2 simultaneously with the P-T diagram of Figure 4-1:

1. Composition is compressed onto the P-T plane of Figure 4-1; therefore the expansion of the composition dimension in the T-x isobars should incorporate the following concepts:

a. All solid phases are shown as vertical lines on the T-x diagrams. The solid ice (I) and the solid hydrocarbon (P) phases are pure; the solid mixture hydrate (H) composition was taken as constant since it varies less than 1% from 0.0556 mol fraction propane with temperature and pressure.

b. The single phase regions on the T-x diagrams for aqueous liquid ($L_W$) and the hydrocarbon liquid ($L_P$) are almost pure (typically $0.999^+$ mol fraction (Kobayashi, 1951)) due to the high degree of immiscibility of each component, caused by hydrogen bonding.

c. The components are miscible in all proportions in the vapor (V) or supercritical fluid phase (either V or $L_P$). The vapor or supercritical regions represent the upper single-phase regions shown on the T-x diagrams.

d. For simplicity, only single-phase regions (areas V, $L_W$, and $L_P$) and vertical lines (I,P,H) are marked on the T-x diagrams. Other areas on the diagrams represent

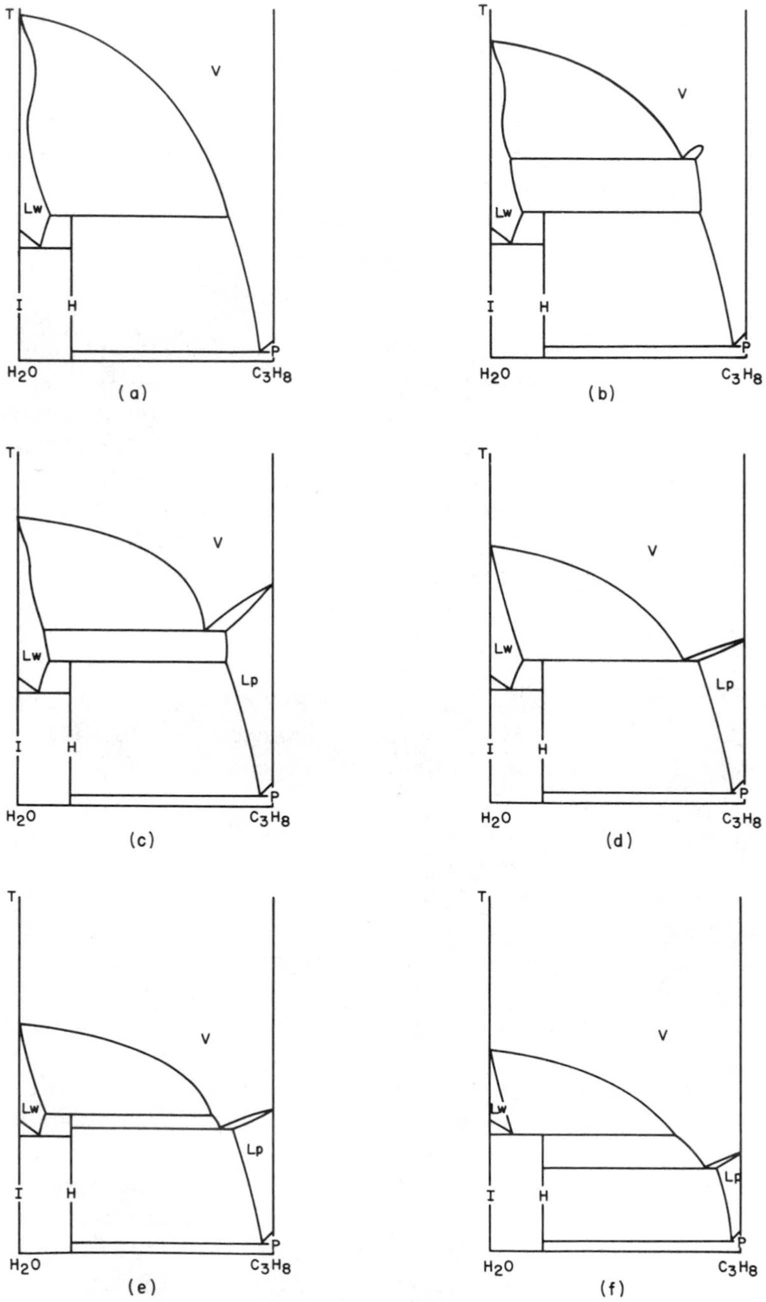


4-2 Temperature-Composition Diagrams for Propane and Water. Isobars (Subfigures a through k) Correspond to Horizontal Dashed Lines Indicated in Figure 4-1

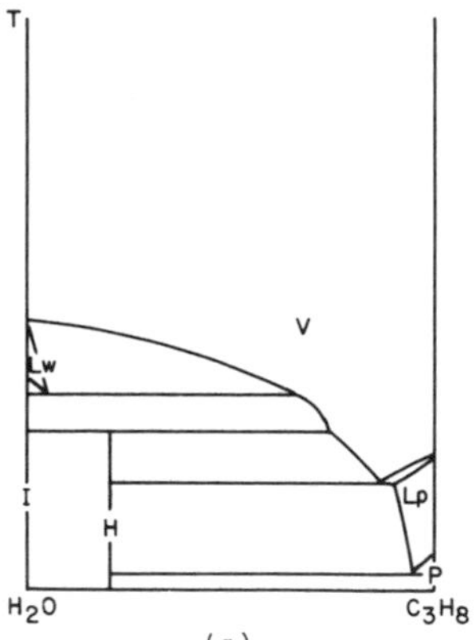
T
V
Lw
I
H
Lp
P
$H_2O$
$C_3H_8$

(g)

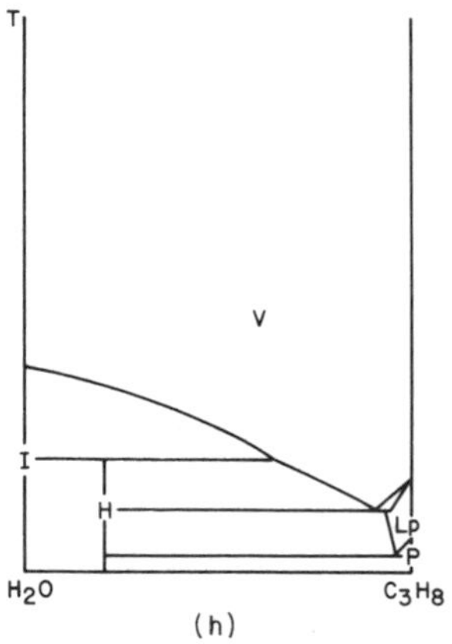
T
V
I
H
Lp
P
$H_2O$
$C_3H_8$

(h)

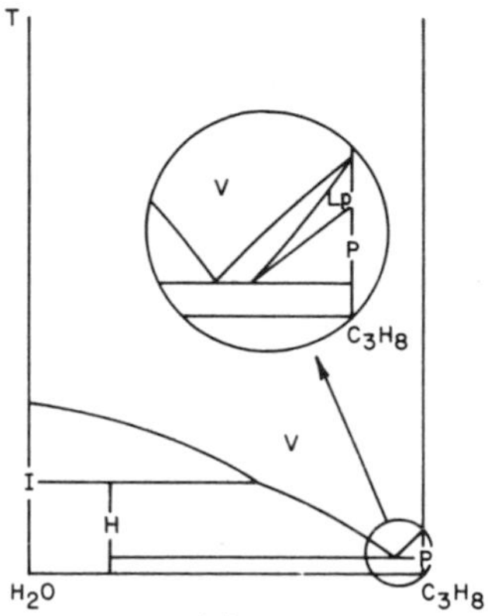
T
V
Lp
P
$C_3H_8$
V
I
H
P
$H_2O$
$C_3H_8$

(i)

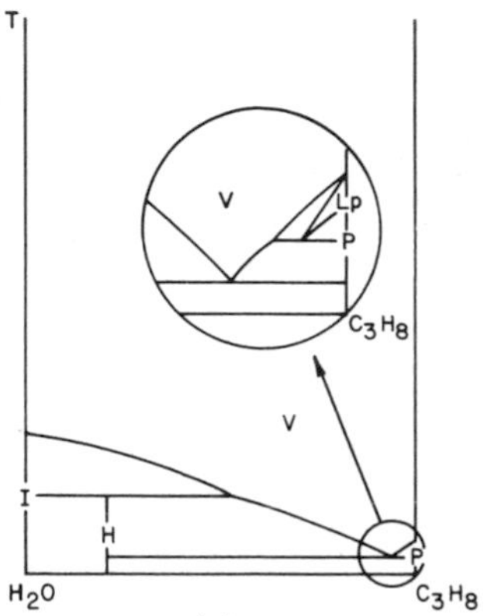
T
V
Lp
P
$C_3H_8$
V
I
H
P
$H_2O$
$C_3H_8$

(j)

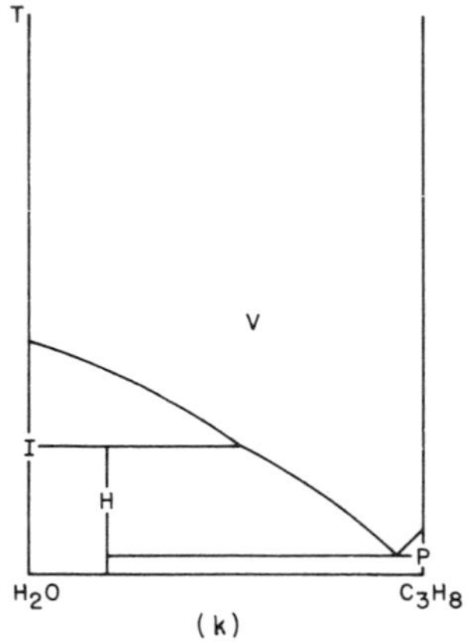
T
V
I
H
P
$H_2O$
$C_3H_8$

(k)

two-phase equilibria, bounded by either the above single phase areas or the three-phase lines, indicated in Item 5 below.

2. Since pure component triple points and critical points on the P-T figure are invariant, they will appear on a T-x diagram (as points along the abscissa extremes) only if the isobar of that diagram coincides with the pressure of the triple point or critical point (e.g. at a pressure just above that of Figure 4-2h, or at a pressure just below that of Figure 4-2j).

3. A pure component two-phase line on the P-T diagram marks the upper temperature of a mixture two-phase area. On the T-x diagram a pure component two-phase point will appear at either end of the abscissa of a T-x diagram (e.g. Figures 4-2a through 4-2k for water, and Figures 4-2c through 4-2k for propane).

4. The intersection of a three-phase line with an isobar on the P-T diagram maps into a horizontal isotherm on the T-x diagram (e.g. the $L_W$-V-$L_P$ lines on Figures 4-2b,c or the H-V-$L_P$ lines on Figures 4-2e,f,g,h).

5. While the three-phase lines do not intersect except at quadruple points, the regions between three-phase lines represent areas of two-phase equilibria for phases which are common to both three-phase lines. For example, at any isobar (such as lines b and c) between 0.556 and 4.4 MPa on Figure 4-1, the $L_W$-$L_P$ area is observed on the P-T figure between the two rightmost three-phase lines ($L_W$-V-$L_P$ and $L_W$-H-$L_P$). The $L_W$-$L_P$ area is defined on T-x Figures 4-2b,c.

6. Quadruple points on the P-T diagram will appear on the T-x diagram (as horizontal isotherms) only if the isobar coincides with the invariant quadruple pressure (e.g. Figures 4-2d,f, and i.)

An Example of Phase Equilibria on the T-x Diagram. Any of the eleven T-x diagrams in Figure 4-2 may be constructed from the appropriate isobar on the P-T diagram in an analogous manner, with regard to the above heuristics. This may be easily done via a thought experiment, in which a high temperature vapor mixture is placed in an inverted cylinder with a weight on the top enclosing piston. The closed assembly is then installed in a temperature bath and the temperature is reduced at constant pressure, from a high temperature such as 600 K to a temperature just below 85 K. It will be necessary to access both Figures 4-1 and 4-2 (almost simultaneously) to comprehend the relations of these two diagrams, discussed in the below paragraphs.

**Example 4.1: Generating Isobar e in Figure 4-2 from Figure 4-1.**

Consider isobar e on Figure 4-1, corresponding to Figure 4-2e, with a weight atop the above inverted cylinder, just sufficient to exert a constant pressure at a value between quadruple points $Q_1$ and $Q_2$. As the temperature is lowered from around 600K in the single vapor phase, the water vapor pressure temperature is reached on isobar 4-1e, signaling the beginning of the aqueous liquid-vapor two-phase region, marked on the ordinate of Figure 4-2e. The $L_W$-V region on Figure 4-2e expands with increasing propane concentration as the temperature is lowered, and terminates at the three-phase $L_W$-H-V line on each diagram. The overall composition of the original vapor mixture determines the temperature at which the liquid phase initially appears.

In the $L_W$-V region, as in any two-phase region of the T-x diagram, the overall initial gas composition may also be used with the Lever Rule to determine the relative amount of each phase present. Unfortunately such a quantitative analysis is prevented in these diagrams by their qualitative, illustrative nature; however, more quantitative T-x diagrams may be drawn to determine phase amounts.

For extremely propane-rich initial mixtures, the $L_W$-V region on isobar 4-1e and 4-2e will not be obtained; instead a vapor-liquid propane (V-$L_P$) region will represent the first two-phase region encountered with a decrease in temperature. Just above the temperature of the $L_W$-H-V line, the propane vapor pressure temperature is obtained at $V_P$-$L_P$ on isobar e of both diagrams, indicating the beginning of the vapor-liquid hydrocarbon region. This two-phase area also increases in Figure 4-2e as the temperature is lowered and terminates at the three-phase H-V-$L_P$ temperature on both diagrams.

As the temperature is lowered below the three-phase $L_W$-H-V line, two two-phase regions appear. To the right of the hydrate line in Figure 4-2e a two-phase hydrate-vapor (H-V) region appears for systems which are propane-rich in total composition. To the left of the hydrate line, an aqueous liquid-hydrate ($L_W$-H) area appears for total compositions which are richer in water. The H-V area is terminated at the three-phase H-V-$L_P$ boundary, and the $L_W$-H region concludes at the I-$L_W$-H three-phase line on both Figures 4-1 and 4-2e.

When the temperature is decreased to the (almost vertical) fusion line for water on the P-T diagram, an ice-aqueous liquid two-phase area appears at 273.15K on the left ordinate of the T-x diagram , for mixtures with a high, water-rich total composition. The I-$L_W$ region grows in propane content in Figure 4-2e as the temperature is lowered only about 0.05 K, until its terminus at the I-$L_W$-H boundary. Similarly, for mixtures very propane-rich as the temperature is lowered below the propane fusion condition ($L_P$-P) at about 85 K, a two-phase ($L_P$-P) region appears on the right ordinate of the T-x diagram. The hydrocarbon liquid phase of the $L_P$-P region increases in water composition until the temperature is lowered to the three-phase (H-$L_P$-P) boundary, which is essentially vertical on the P-T diagram. On the T-x diagram, the H-$L_P$-P line acts as a lower bound to the two-phase H-$L_P$ area, which initially occurred at the higher three-phase (H-V-$L_P$) temperature.

At the very lowest temperatures, mixtures which are initially water-rich will equilibrate in the I-H two-phase region of Figure 4-2e, common to the Figure 4-1 three-phase lines (I-$L_W$-H) and (I-H-V). Also at low temperatures, propane rich mixtures will equilibrate in the Figure 4-2e two-phase H-P region, common to the Figure 4-1 three-phase lines (H-V-P) and (H-$L_P$-P).

---

The reader may wish to construct other T-x diagrams from the P-T diagram following a technique similar to that used in the above example. For purposes of hydrate experiments and calculations, the region of primary interest on the T-x diagrams is the three-phase $L_W$-H-V line, which can be seen to emanate from the lower quadruple point $Q_1$ of Figure 4-2f, and terminate at the upper quadruple point $Q_2$ of Figure 4-2d. The composition points along the $L_W$-H-V isotherms have only been of minor interest, perhaps because trace amounts limited their experimental accessibility. The P-T locus indicates thermodynamic conditions of hydrate formation or dissociation from aqueous liquid and from a given hydrocarbon vapor composition, on a water-free basis. Section 4.3 will concern two calculation methods for the $L_W$-H-V line, and Section 4.5 will deal with methods of calculating the temperature depression of this line through use of inhibitors, such as salts or methanol.

### 4.1.2 P-T-x Phase Equilibria for Other Binary Systems

The natural gas components methane+water and nitrogen+water have phase diagrams which resemble each other; yet both systems differ from the above diagrams for propane+water. A phase analysis of the methane+water system, analogous to that above for propane+water, has been detailed by Kobayashi and Katz (1949). For simplicity, only the qualitative differences between the methane+water system and the propane+water system are shown here. The methane+water system is the simpler of the two, primarily because the low critical temperature of methane (190.6 K) eliminates a high temperature portion of the $L_W$-V-$L_{HC}$ line, so that an intersection with the $L_W$-H-V line is impossible. Without an intersection of $L_W$-V-$L_{HC}$ and $L_W$-H-V the upper quadruple point $Q_2$ does not occur. Therefore no upper hydrate formation temperature

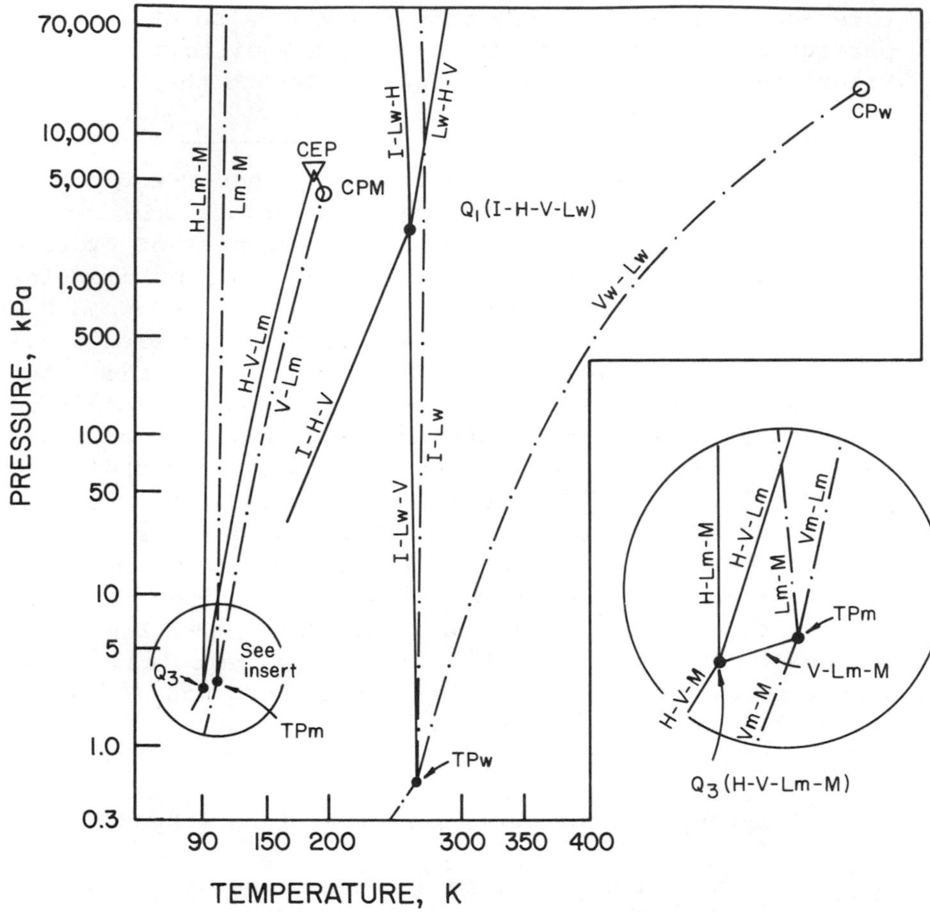


4-3 Pressure-Temperature Diagram for Methane and Water (Reproduced, with permission, from Transactions AIME (Kobayashi and Katz, 1949), by the American Institute of Mining, Metallurgical, and Petroleum Engineers.)

has been measured for methane (or for nitrogen for similar reasons), and the $L_W$-H-V line extends to very high pressures.

Methane+water analogies for the propane+water P-T Figure 4-1 and for the T-x Figures 4-2a,b are presented in Figures 4-3 and 4-4a,b respectively, by Kobayashi and Katz (1949). The methane+water binary has a middle quadruple (I-$L_W$-H-V) point $Q_1$, and a lower quadruple (H-

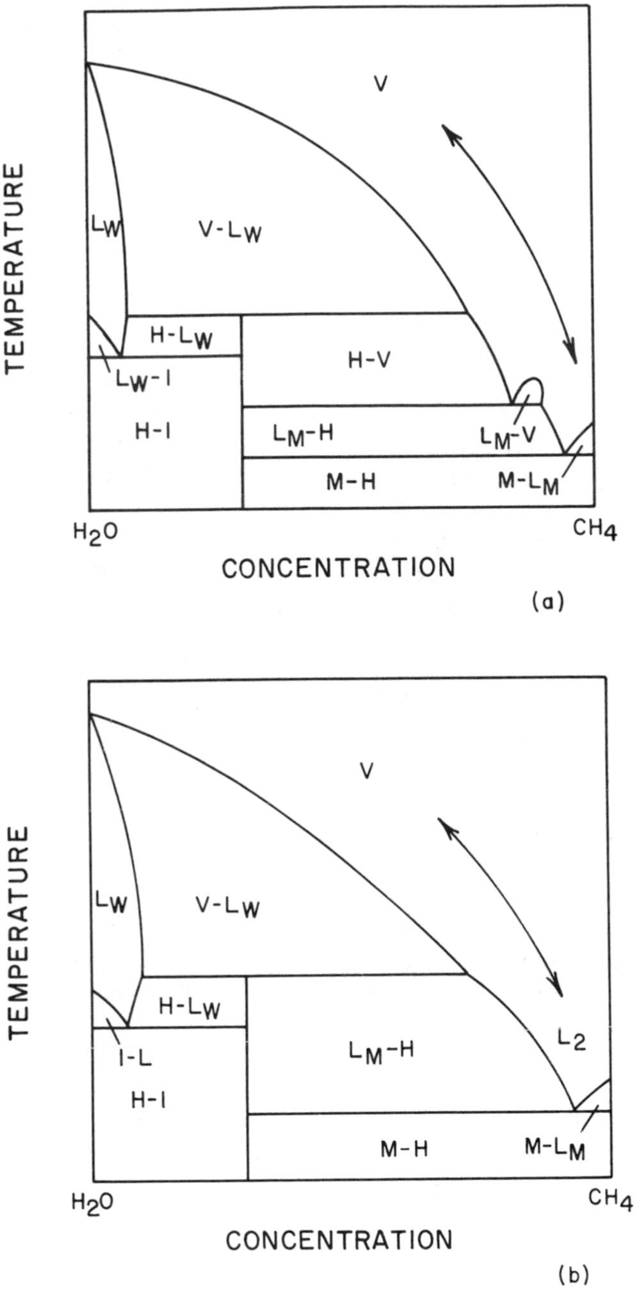


4-4 Temperature-Composition Diagrams for Methane and Water (Reproduced, with permission, from Transactions AIME (Kobayashi and Katz, 1949), by the American Institute of Mining, Metallurgical, and Petroleum Engineers.)

V-$L_M$-M) point $Q_3$ analogous to those for propane+water in the T-x Figure 4-2f and Figure 4-2i, respectively. Figures 4-2f,g for propane+water are qualitatively correct for similar isobars of methane+water at $Q_1$, and at pressures between $Q_1$ and $Q_3$, respectively. These conditions encompass the range of temperatures and pressures of major interest in the study of hydrates.

For water-soluble components such as carbon dioxide or hydrogen sulfide, Thies (1988) noted that the phase diagrams of Figures 4-1 and 4-2 must also be changed slightly. The pure $CO_2$ vapor pressure lies above the three-phase lines on either side of $Q_2$ on the P-T diagram; for the T-x diagrams that means that the small $CO_2$-rich vapor-liquid region on the right point downward instead of upward. The pure $H_2S$ vapor pressure crosses the H-V-$L_{H_2S}$ line a little to the left of $Q_2$ on the P-T projection; at high temperature $H_2S$ behaves like $CO_2$ and at lower temperatures it behaves like propane.

### 4.1.3. Phase Equilibria for Multicomponent Hydrocarbons+Water

Robinson et al. (1987) provided Figure 4-5 for hydrate formation in the presence of a multicomponent hydrocarbon mixture+water. The straight line labelled $L_W$HV represents the hydrate formation region equivalent to the region between the middle quadruple point $Q_1$ (I-$L_W$-H-V) and the upper quadruple point $Q_2$ ($L_W$-H-V-$L_P$) for the propane+water P-T diagram, Figure 4-1. The difference comes about because the analog of the $L_W$-V-$L_P$ line for propane+water has become, in the multicomponent+water system, a vapor-liquid area defined by the curve ECFKL shown superimposed on the hydrate formation line. Consequently the upper quadruple point evolves into a line ($\overline{KC}$) for the multicomponent system.

The line $\overline{KC}$ may not be straight in the four-phase region (see Example 8-5) but is drawn that way for illustration. The location of the lower point K is determined by the intersection point of the phase envelope ECFKL

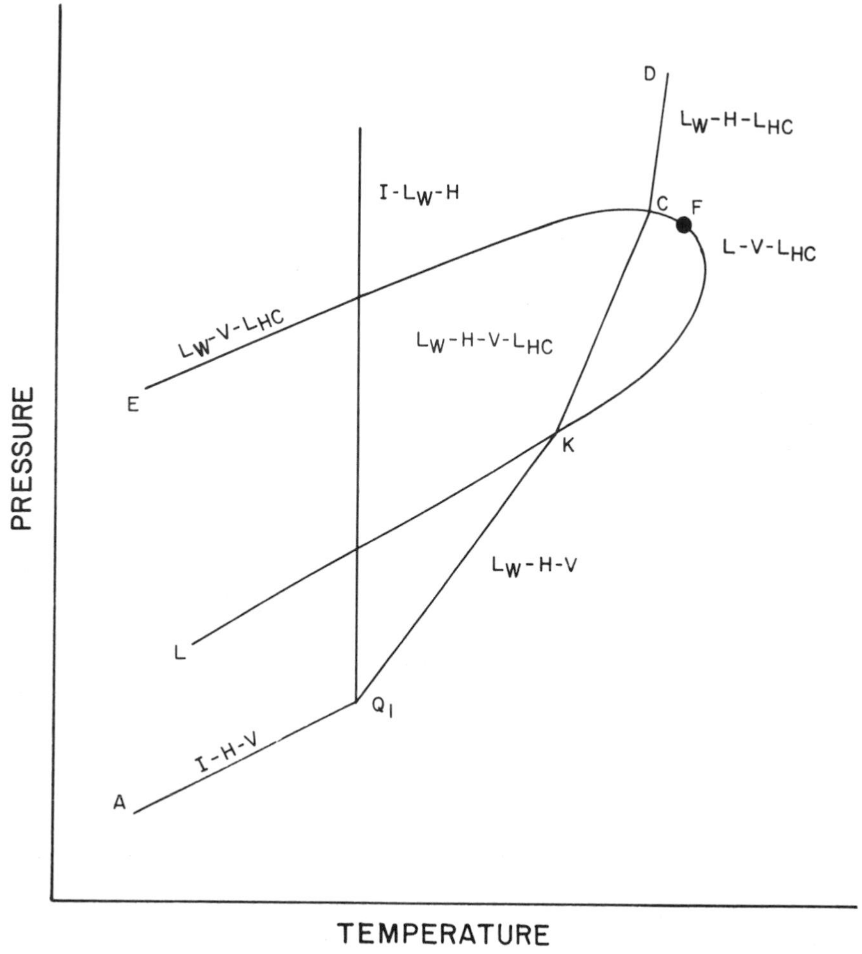


4-5 Pressure-Temperature Diagram for a Multicomponent Hydrocarbon Mixture and Water
(Reproduced, with permission, from the Proc. 66th Annual GPA Convention (Robinson, et al., 1987), by the Gas Processors Association.)

with the $L_W$-H-V line, determined by the methods of Section 4.3 or those of the Chapter 5. To determine the upper point C, first a vapor-liquid equilibrium calculation is performed, assuming the liquid phase is the composition of the vapor used to calculate K; the resulting equilibrium vapor is used with the normal $L_W$-H-V calculation methods to determine the upper intersection with the phase envelope ECFKL. A more thorough treatment of the calculation of multicomponent equilibria involving a condensed hydrocarbon phase is given in Section 8.1.5.

## 4.2 Solubility of Hydrocarbons in the Aqueous Phase

The solubility of hydrocarbons in the aqueous phase determines the width of the region $L_W$ marked on the T-x diagrams of Figures 4-2 and 4-4. Of all gases, the alkanes have received the most experimental attention, as indicated in a chronological summary of studies in Table 4-1. A comprehensive review of the solvent properties of water is found in the treatise by Franks (1972, Vol. 2, Chapter 1 and Chapter 5 (with Reid)).

The discussion of the molecular properties of dissolved hydrocarbons in water has been treated in detail in Section 3.1.2 so there is no need to repeat it here. The reader is referred to that section for numerical values of solubilities, enthalpies, and entropies of solution. Recall however, the essential detail that water molecules form networks with short-lived partial hydrogen-bonded clusters around solute molecules, giving a maximum in solubility for those molecules with diameters which form hydrates. In Section 3.1.2 it was suggested that the kinetic or time-dependent supersaturation was identical with the concepts of hydrate primary nucleation and metastability of hydrate formation.

Questions have arisen however regarding the time-independent liquid phase solubility close to the hydrate point. The equilibrium solubility has been experimentally verified in the gas-aqueous liquid region in several studies. In the hydrate literature however, there is a concern about the aqueous solubility at the three-phase ($L_W$-H-V) condition at which hydrates first form.

### 4.2.1 Aqueous Solubility in the ($L_W$-H-V) Region

The few studies which have been made of gas solubility in the aqueous phase at hydrate conditions below 300K

**Table 4-1 Studies of Solutions of Natural Gas Components in Water**

| Date | Investigators | System (other than water) |
|---|---|---|
| 1937 | Butler | methane, ethane, butane |
| 1944 | Dodson, Standing | natural gas |
| 1944 | Reamer et al. | n-butane |
| 1952 | Reamer et al. | n-butane |
| 1950 | Culberson, Horn, McKetta | methane, ethane |
| 1950 | Culberson, McKetta | ethane |
| 1951 | Culberson, McKetta | methane |
| 1951 | Brooks, Gibbs, McKetta | n-butane |
| 1952 | Claussen, Polgase | methane, ethane, propane, and n-butane |
| 1952 | Morrison | methane, ethane, propane, n-butane |
| 1952 | Morrison, Billett | methane, ethane, propane, n-butane, nitrogen,carbon dioxide |
| 1952 | Selleck et al. | hydrogen sulfide |
| 1953 | Kobayashi, Katz | methane, ethane, propane, nitrogen |
| 1956 | Dodds, et al. | carbon dioxide |
| 1958 | Gjaldbaek, Niemann | nitrogen, ethane |
| 1959 | Himmelblau | nitrogen, methane |
| 1961 | Wehe, McKetta | propane |
| 1963 | McAuliffe | methane, ethane, propane |
| 1966 | | n-butane, iso-butane |
| 1964 | Wetlaufer, et al. | methane, ethane, propane, n-butane, iso-butane |
| 1968 | Nosov, Barlyaev | iso-butane |
| 1969 | Razaryan, Ryabtsev | iso-butane, n-butane |
| 1974 | Ben-Naim, Yaacobi | methane,ethane |
| 1976 | Rice, Gale, Barduhn | n-butane |
| 1982 | Gillespie, Wilson | methane, carbon dioxide, hydrogen sulfide |

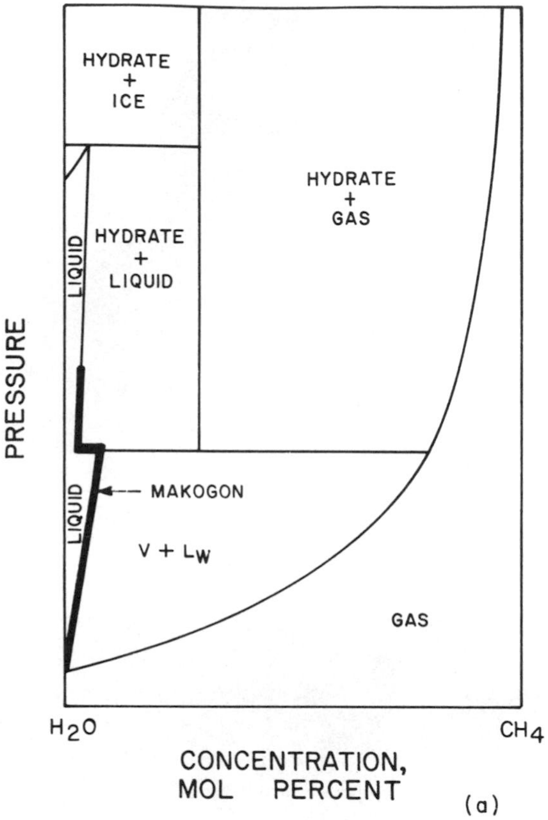


4-6 Methane and Water Pressure-Composition Diagram with Aqueous Solubility Results
(a) (Makogon and Koblova, 1972)
(b) (Reproduced, with permission, from Petroleum Transactions AIME (Culberson and McKetta, 1951), by the Society of Petroleum Engineers, Inc.)

provide two different concepts of solubility. The more recent data set, determined by Makogon and Koblova (1971), enabled them to postulate a new theory for the formation of hydrates from the body of solution. However, the initial measurements by McKetta and co-workers appear to both contrast with the Soviet results and fit the phase diagrams presented in Section 4.1 more closely.

Makogon and Koblova (1971) experimentally determined the aqueous solubility of methane at temperatures of 283

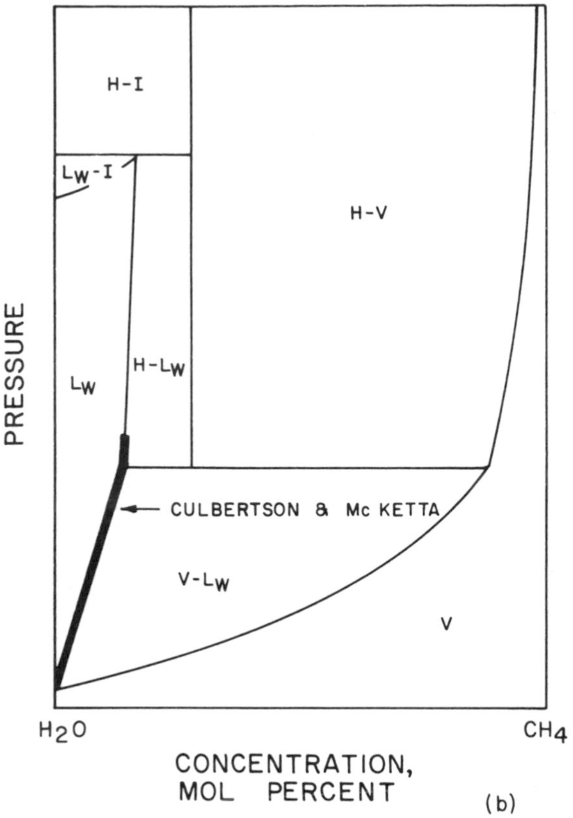


(b)

K, 287 K, and 293 K in the pressure range below 11.7 MPa. As the pressure increased beyond the hydrate formation condition at 283 K, a dramatic decrease in gas solubility was observed, as plotted in Figure 4-6a. This decrease of gas solubility, according to Makogon (1974), allows excess gas to be released from solution, for hydrate formation from the bulk water. While this theory has been adopted by some workers in the Western Hemisphere (e.g. Claypool and Kaplan (1974); Finley and Krason (1986a)), other Soviet researchers have expressed doubts. Barkan and Voronov (1982) studied gas saturated water in geologic environments; they determined that only an insignificant amount of dissolved gas was transferred to hydrates as they were formed in three-phase equilibrium.

The initial set of solubility studies by Culberson, et al. (1950), and Culberson and McKetta (1951) provided measurements of the solubility of methane at 298K over the pressure range from 2 to 69 MPa. The majority of solubility measurements were in the gas-liquid region, but each study measured two data points in the hydrate region at high pressure. Holder and co-workers (1988) measured the solubility of gas in liquid water around the hydrate point, both in water which had formed hydrates and in water with no residual structure; their results show no dramatic decrease in pure component solubility at the three-phase ($L_W$-H-V) condition. Those results do indicate a slight increase beyond that achieved by a mol fraction addition of Henry's constants for the combined solubility of binary and multicomponent gases in water at the hydrate point.

The phase diagrams of the previous section provide a convenient means for qualitative evaluation of the above two concepts. For this purpose it is convenient to choose the isothermal P-x phase diagram, which has the effect of inverting the isobaric T-x diagrams of Figures 4-2 and 4-4. The darkened line in Figure 4-6a is a qualitative illustration of the data of Makogon and Koblova, while a similar line in Figure 4-6b qualitatively illustrates the two data sets of McKetta and co-workers and that of Holder. Figure 4-6b shows the solubility line to have a discontinuous slope, in accord with normal phase analysis. Figure 4-6a however, indicates an offset decrease in the solubility at the hydrate point; this disruption of concentration is unanticipated in the previous phase diagram analyses. Even at the quadruple point $Q_1$ (see Figure 4-2f) some 20K lower, no such dramatic fissure is found in the plot of the mol fraction of the hydrocarbon.

The models of clusters of solvent molecules around the solute (discussed in Section 3.1.2) would have to change dramatically at the hydrate point, were they to accommodate the Soviet solubility data. Professor Makogon (1988a) provided an explanation of the above point, by indicating that his experimental solubility data were obtained with a hydrate barrier preventing gas contact with the aqueous liquid. With such a barrier, solubility data may be more appropriately considered as two-phase ($L_W$-H) equilibria rather than three-phase ($L_W$-H-V) equilibria.

## 4.3 Three-Phase ($L_W$-H-V) Equilibrium Calculation

The conditions of three-phase ($L_W$-H-V) equilibrium have attained the widest use in the history of hydrates. In particular, methods to calculate the temperature and pressure at which hydrates form from a given gas composition and free water have interested both researchers and practitioners. The discussion in Section 4.2.1 on gas solubility at the hydrate point reflects the paucity of data on the composition of the aqueous liquid phase. There are comparatively few data on the composition of the hydrate phase, perhaps due to difficulty of measurement. On hydrate formation, water is often occluded in the hydrate mass, and consequently the ratio of water:hydrocarbon is often inaccurate. Only recently have experimental techniques, such as spectroscopy, become accurate enough to determine the degree of filling of hydrate cavities with different types of molecules.

In this section two calculation techniques are discussed. While both techniques enable the user to calculate the pressure and temperature of hydrate formation from a gas, only the final technique allows the calculation of the hydrate composition. The calculations via the statistical thermodynamics method of the following chapter provide access to the hydrate composition and the fraction of each cavity filled by various molecule types.

The two calculation techniques may be regarded as successive approximations (increasing both in accuracy and in sophistication) to hydrate phase equilibria. The first method, in Section 4.3.1, based upon the parameter of gas gravity, is a simple graphical technique which provides a first-order accuracy estimate of hydrate formation. The second calculation technique of Section 4.3.2, based on distribution coefficients, sometimes called the $K_i$-value method, is both more accurate and slightly more complex than the gravity method. It is suitable for either hand calculation or use with the IBM-PC compatible floppy disk in the frontpapers.

Yet a third technique is based upon the statistical thermodynamics approach of van der Waals and Platteeuw (1959) and provides the best approximation to hydrate phase equilibria. It has the additional advantage of being extendable to all of the phase equilibrium regions discussed above. Unfortunately this third method is too lengthy for inclusion in this chapter without disruption of the concepts presented here. For ease of understanding, the discussion of the third method is deferred until Chapter 5; a second computer program on the floppy disk in the frontpapers provides access to that method.

### 4.3.1. The Gas Gravity Method

The simplest method of determining the temperature and pressure of a gas mixture three-phase ($L_W$-H-V) conditions is available through the gas gravity charts of Katz (1945). Gas gravity is defined as the molecular mass of the gas divided by that of air. In order to use this chart, first presented as Figure 1-2, the gas gravity is calculated and either the temperature or pressure of interest is specified. The second intensive variable (either pressure or temperature) at which hydrates will form is read directly from the chart. The following example from the original paper illustrates the use of the chart.

**Example 4-2:** Find the pressure at which a gas composed of 92.67 mol % methane, 5.29% ethane, 1.38% propane, 0.182% i-butane, 0.338% n-butane, and 0.14% pentane form hydrates from free water at a temperature of 283.2 K ($50^0$F).

**Solution:** The gas gravity is calculated as 0.603 by the below procedure.

| Component | Mol Fraction | Mol Wt. | Mols in Mixture |
|---|---|---|---|
| Methane | 0.9267 | 16.043 | 14.867 |
| Ethane | 0.0529 | 30.070 | 1.591 |
| Propane | 0.0138 | 44.097 | 0.609 |
| i-Butane | 0.00182 | 58.124 | 0.106 |
| n-Butane | 0.00338 | 58.124 | 0.196 |
| Pentane | 0.0014 | 72.151 | 0.101 |
| | 1.000 | | 17.470 |

$$\text{Gas Gravity} = \frac{\text{Mol Wt of Gas}}{\text{Mol Wt of Air}} = \frac{17.470}{28.966} = 0.603$$

At 283.2K , the hydrate pressure is read as 3.1 MPa.

---

The purpose of the original chart was to enable the determination of the hydrate limits to adiabatic expansion of a gas, as detailed in Section 4.3.1. The hydrate formation lines on Figure 1-2 appear to be fairly linear at low temperatures with a slope change above 288 K for each gravity; this slope change is intentional and has been ascribed to the change in hydrate structure (Katz, 1983).

All of the qualitative trends shown on the chart are correct and thus provide a valuable heuristic as a check for more sophisticated calculations. For example, the chart correctly indicates that the logarithmic pressure increases linearly with temperature over a short range. Over a wider temperature range the logarithmic pressure is more nearly linear with reciprocal absolute temperature. The chart also shows that gases with heavier components cause hydrates to form at lower pressures for a given temperature (or at higher temperatures for a given pressure).

The hydrate formation chart was generated from a limited amount of experimental data, and a more substantial amount of calculated data with the calculations done (and therefore the accuracy determined) via the $K_i$-value method of Section 4.3.2. The calculated accuracy limits per-

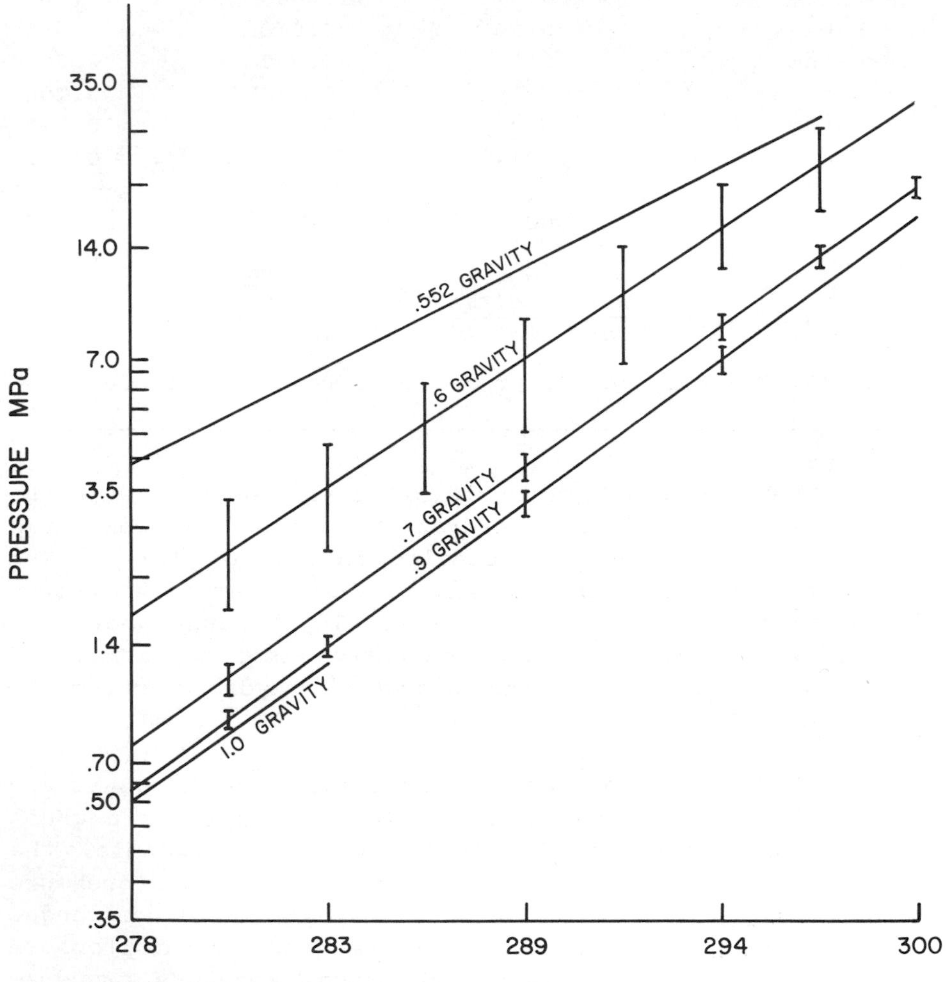


4-7 Gas Gravity Chart Error Bars
(Reproduced, with permission, from the Proc. 63rd Annual GPA Convention (Sloan 1984), by the Gas Processors Association.)

formed in this laboratory (Sloan (1984)) for the gas gravity chart are presented in Figure 4-7, using the statistical thermodynamic approach of Chapter 5. Figure 1-2 was generated for gas containing only hydrocarbons, and so should be used with caution for those gases with substantial amounts of non-combustibles. While this method is very simple, it should be considered as approximate. In the forty years since it was conceived, more hydrate data and more accurate computer methods have been determined. Nevertheless the gas gravity method has served the gas processing industry well, as an initial estimate.

#### 4.3.1.a. Hydrate Limits to Gas Expansion

The generation of the so-called Joule-Thomson charts, such as Figure 4-8, for hydrate limits to gas adiabatic expansion, was the original goal of construction of the hydrate gas gravity chart. A series of such charts for gas gravities between 0.55 and 1.0 are available in the original article by Katz (1945). Only Figures 4-8, 4-9, and 4-10 for gas gravities of 0.6, 0.7, and 0.8, respectively, are presented as those of the highest utility. The generation of the Joule-Thomson charts was made available through the initial enthalpy-entropy charts for natural gas by Brown (1942), with the assumption that any free water present did not affect the properties of the natural gas in the single-phase region.

The pressure and temperature of the gas normally decreased upon expansion along an isenthalpic curve until the intersection with the hydrate boundary of Figure 1-2 was encountered, which provided one point on a figure such as 4-8. Multiple points were determined to construct each figure. The charts enabled the user to estimate the limits to adiabatic expansion before the hydrate stability region was encountered. The below examples of the use of Figure 4-8 were also taken from Katz's original reference.

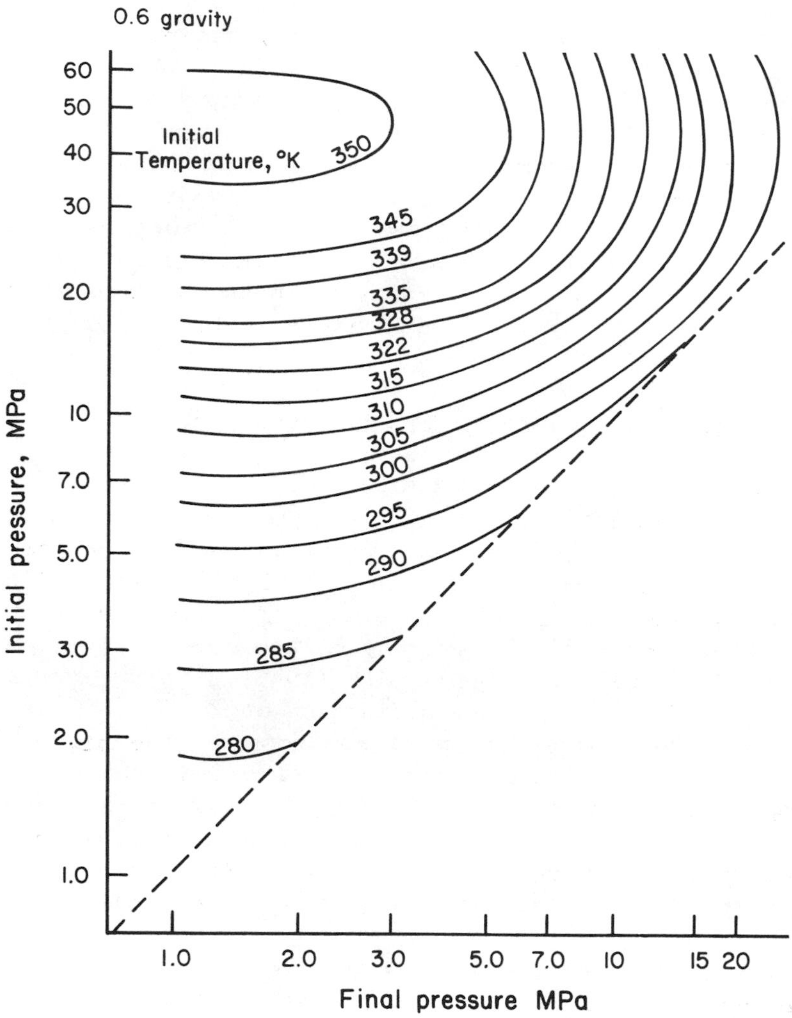


4-8 Joule-Thomson Limits to 0.6 Gravity Gas Expansion (Reproduced, with permission, from Transactions AIME (Katz, 1945), by the American Institute of Mining, Metallurgical, and Petroleum Engineers.)

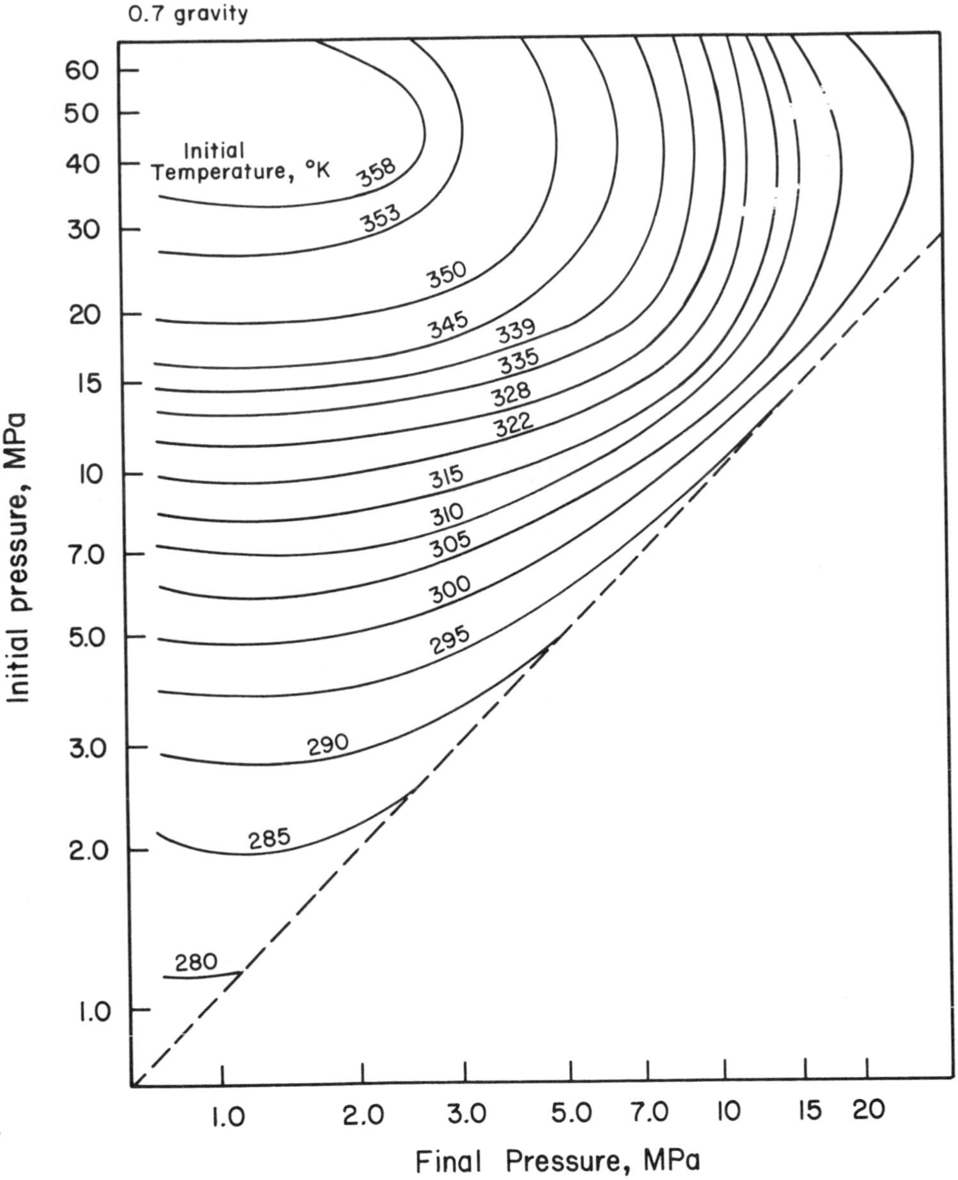


4-9 Joule-Thomson Limits to 0.7 Gravity Gas Expansion (Reproduced, with permission, from Transactions AIME (Katz, 1945), by the American Institute of Mining, Metallurgical, and Petroleum Engineers.)

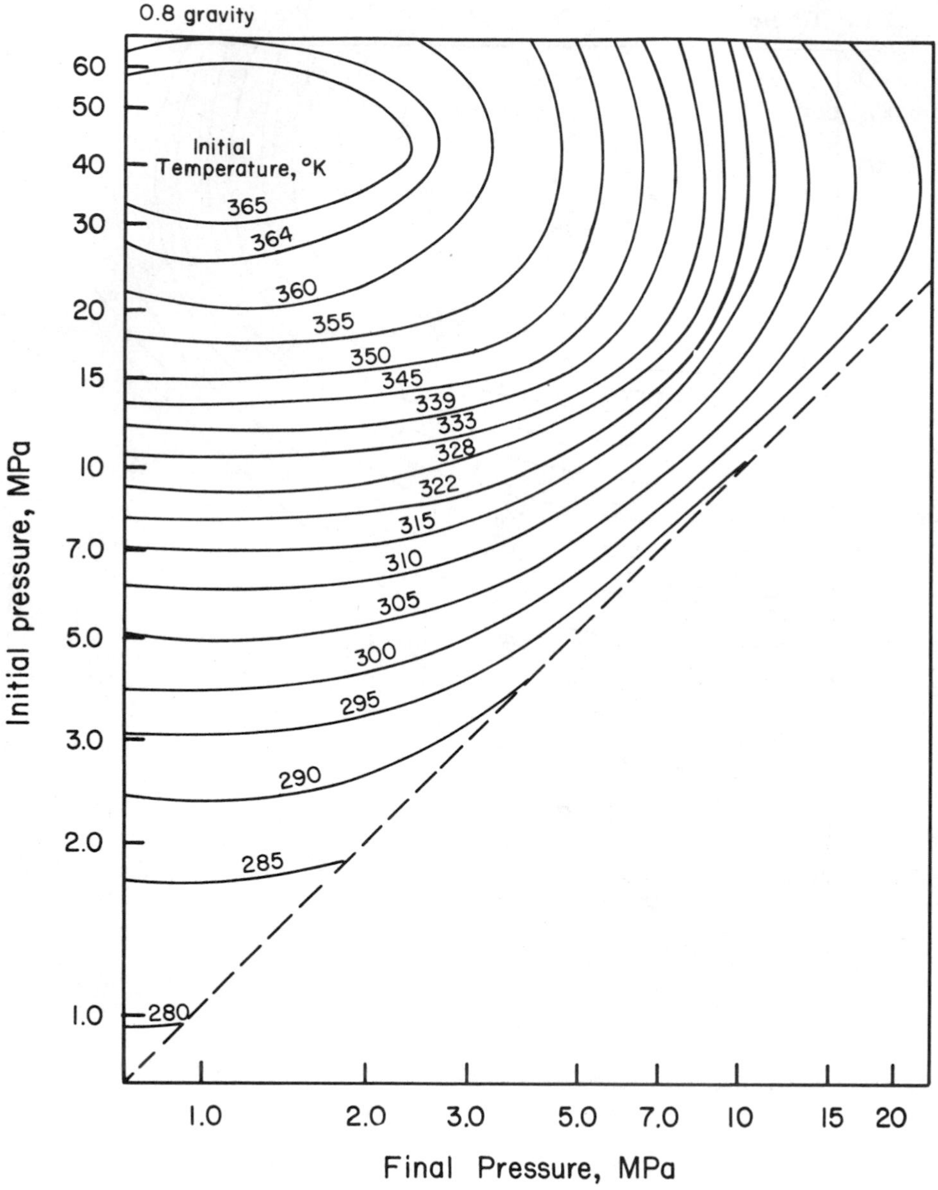


4-10 Joule-Thomson Limits to 0.8 Gravity Gas Expansion (Reproduced, with permission, from Transactions AIME (Katz, 1945), by the American Institute of Mining, Metallurgical, and Petroleum Engineers.)

**Example 4.3: Calculations of Hydrate Limits to Gas Expansion:**

1. To what pressure may a 0.6 gravity gas at 13.8 MPa (2000 psia) and 311K (100$^0$F) be expanded without danger of hydrate formation?

**Solution:** From Figure 4-8 read 7.24 MPa (1050 psia).

2. A 0.6 gravity gas is to be expanded from 10.34MPa (1500 psia) to 3.45 MPa (500 psia). What is the minimum temperature that will permit the expansion without danger of hydrates?

**Solution:** From Figure 4-8 the answer is read as 310 K (99$^0$F) or above.

---

Figures 4-8, 4-9, and 4-10 are provided for hydrate limits to isenthalpic Joule-Thomson expansions, such as occurs when a gas with entrained free water droplets flows through a valve. As an aside, a similar set of charts could in principle be determined for hydrate limits to isentropic expansions such as would occur when a gas flows through an ideal turboexpander of a modern gas processing plant. To date however, no such charts have been generated.

The inaccuracies listed in the previous section for the gas gravity chart are inherent in the expansion charts of Figures 4-8, 4-9, and 4-10 due to their method of derivation. Accuracy limits to these expansion curves have been determined by Loh et al. (1983) who found, for example, that the allowable 0.6 gravity gas expansion from 339 K and 24MPa was 2.8 MPa rather than the value of 4.8 MPa, given in Figure 4-8.

The work of Loh et al. (1983) was done using the same principles as those used to generate Figure 4-8. That is, from the initial temperature and pressure, an isenthalpic cooling curve, and its intersection with the hydrate three-phase locus, was determined. However the isenthalpic line was determined via the Soave-Redlich-

Kwong equation-of-state rather than the charts of Brown, and the statistical thermodynamic prediction method of Chapter 5 was substituted for the three-phase hydrate line prediction by the gas gravity chart of Katz. The interested reader should refer to Examples 8.1 and 8.2 for similar calculations of isenthalpic and isentropic expansions using a third computer program in the frontpapers, with discussion in Appendix A.

### 4.3.2 The Distribution Coefficient

The distribution coefficient method, sometimes called the $K_i$-value method, was conceived by Wilcox, Carson, and Katz (1941) and finalized by Carson and Katz (1942). The best methane, ethane, and propane charts are from the latter reference. Updated charts for carbon dioxide (Unruh and Katz (1949)), hydrogen sulfide (Noaker and Katz (1954)), nitrogen (Jhaveri and Robinson(1965)), isobutane (Wu et al.(1976)) and n-butane (Poettmann, 1984), as well as method revisions (Mann et al. (1989)) have been presented more recently.

Carson and Katz noted that their experimental hydrate composition changed at different temperatures and pressures in a manner indicative of a solid solution, rather than segregated macroscopic quantities of hydrocarbon within the hydrate. The concept of a solid solution enabled the notion of the mol fraction of a guest component in the solid phase mixture, on a water-free basis. Carson and Katz defined a vapor-solid distribution coefficient ($K_{vsi}$) for each component as:

$$K_{vsi} \equiv y_i/x_{si} \qquad (4.1)$$

where: $y_i$ = mol fraction of component i in the water-free vapor

$x_{si}$ = mol fraction of component i in the water-free solid.

The $K_{vsi}$-values for natural gas components are presented as a function of temperature and pressure in Figures 4-11 through 4-18. By viewing these charts one may

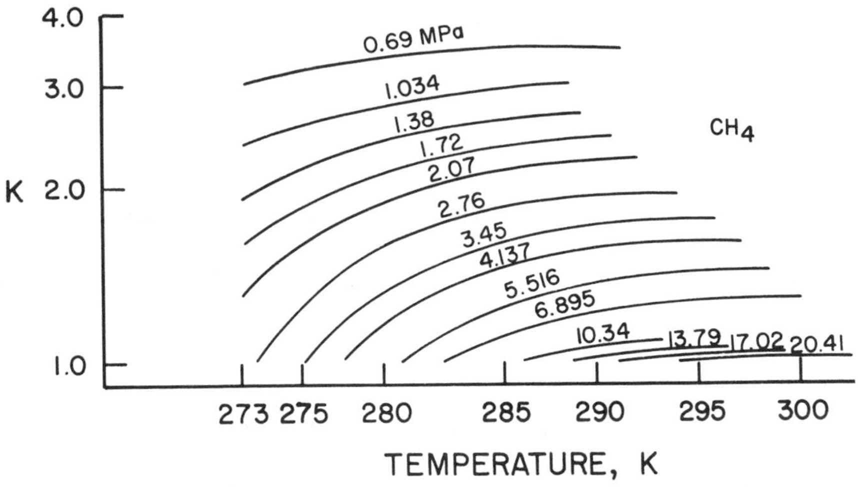


4-11 Methane $K_{vs}$ Chart (Reproduced, with permission, from <u>Transactions AIME</u> (Carson and Katz, 1942), by the American Institute of Mining, Metallurgical, and Petroleum Engineers.)

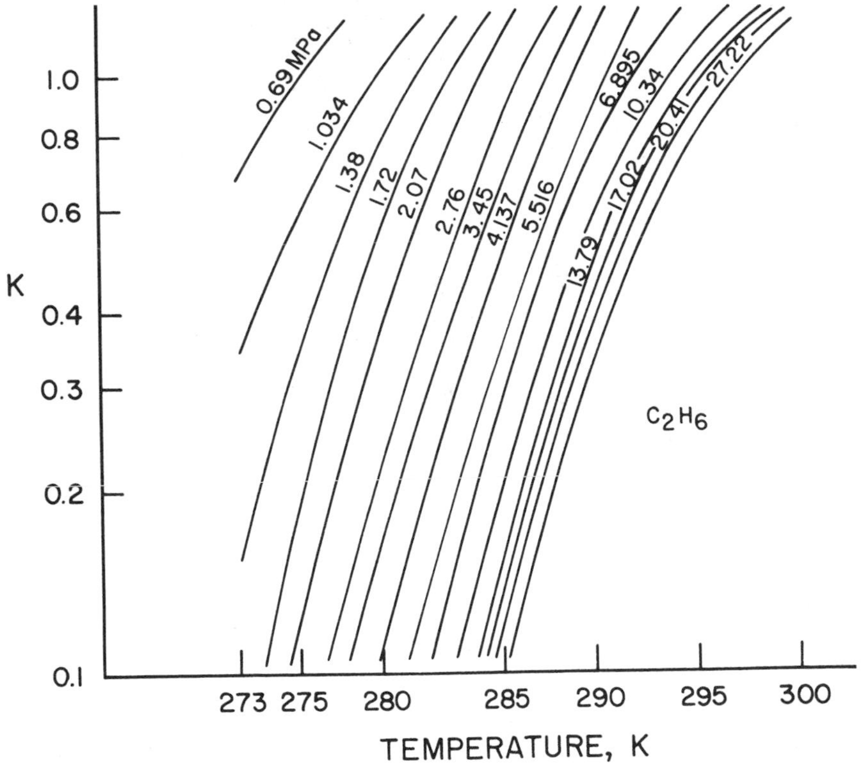


4-12 Ethane $K_{vs}$ Chart (Reproduced, with permission, from <u>Transactions AIME</u> (Carson and Katz, 1942), by the American Institute of Mining, Metallurgical, and Petroleum Engineers.)

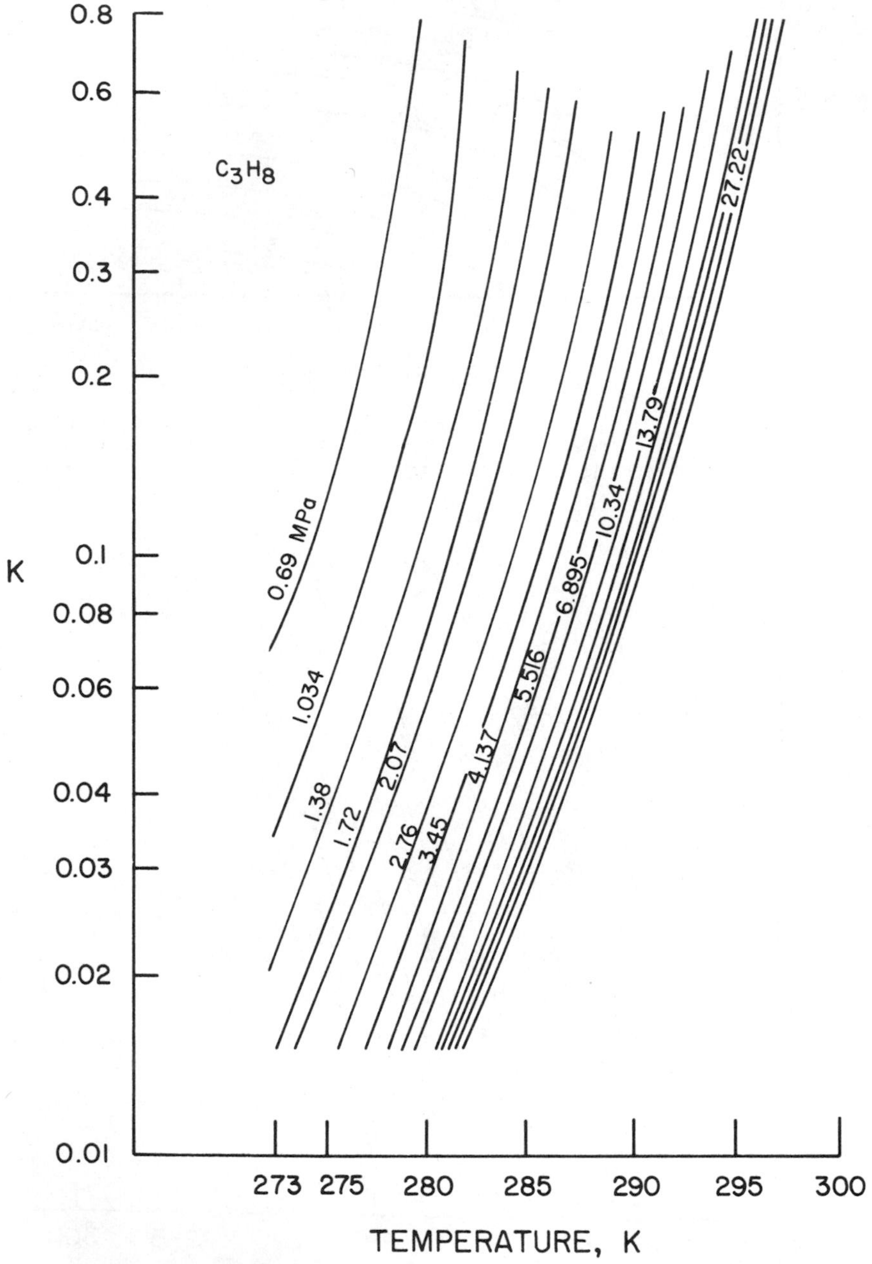


4-13 Propane $K_{vs}$ Chart (Reproduced, with permission, from Transactions AIME (Carson and Katz, 1942), by the American Institute of Mining, Metallurgical, and Petroleum Engineers.)

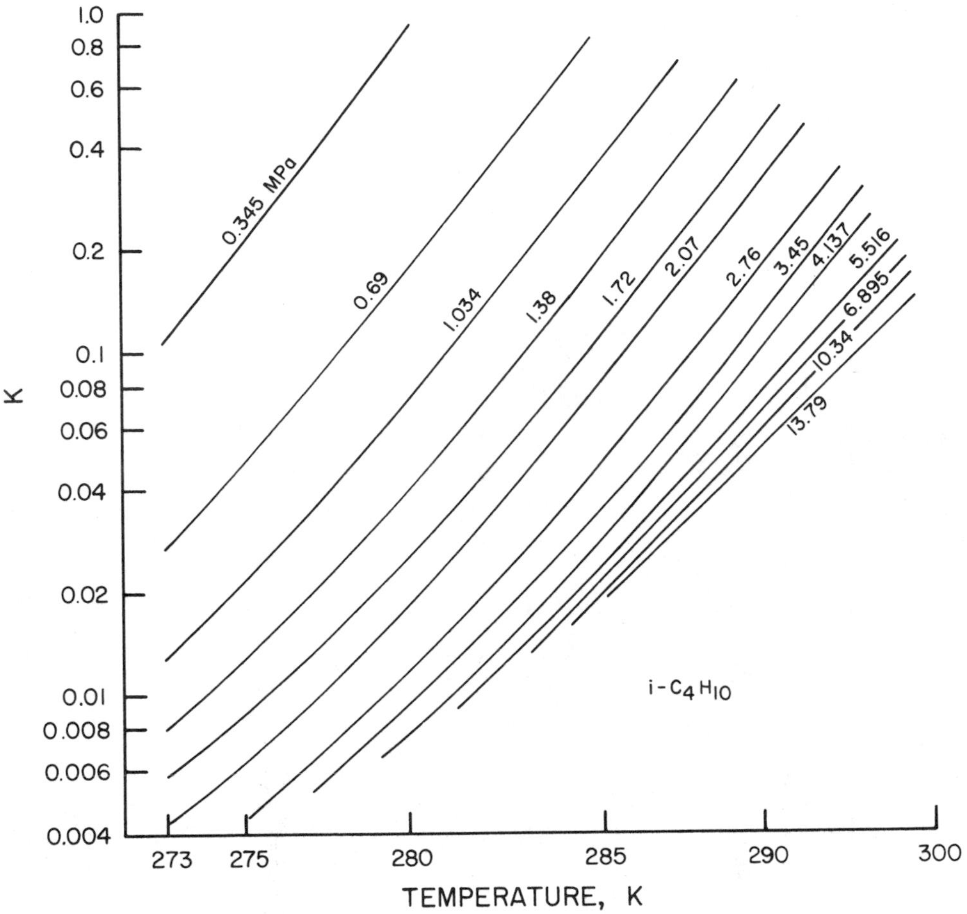


4-14 Iso-Butane $K_{vs}$ Chart
(Reproduced, with permission, from the Journal of Chemical Thermodynamics (Wu et al., 1976), by Academic Press, Ltd.)

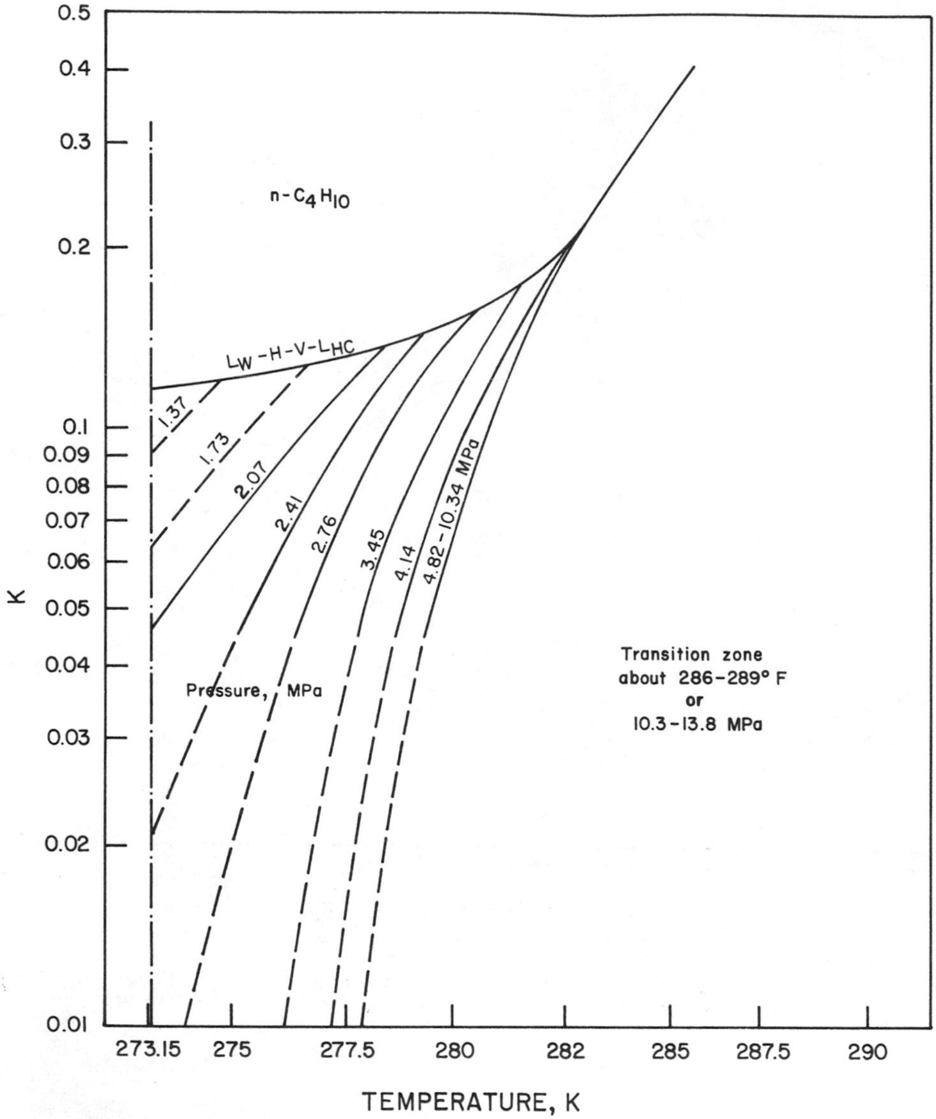


4-15 Normal-Butane $K_{vs}$ Chart
(Reproduced, with permission, from Hydrocarbon Processing (Poettmann, 1984), by Gulf Publishing Co.)

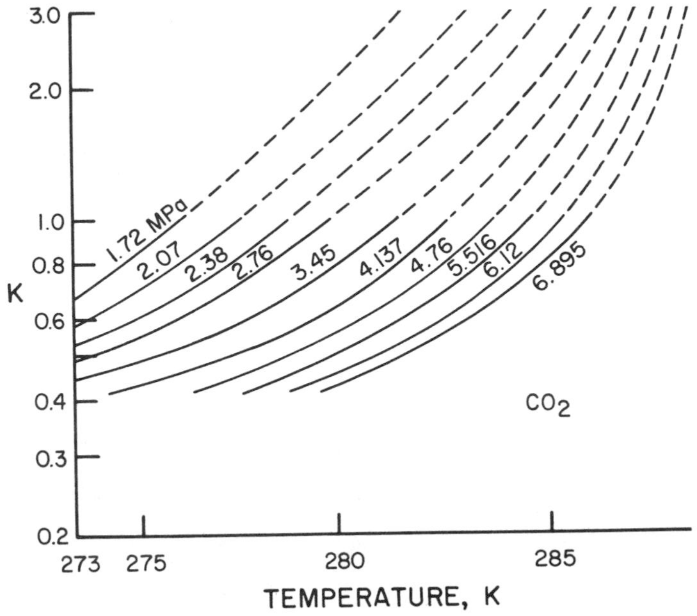


4-16 Carbon Dioxide $K_{vs}$ Chart (Reproduced, with permission, from Transactions AIME (Unruh and Katz, 1949), by the American Institute of Mining, Metallurgical, and Petroleum Eningeers)

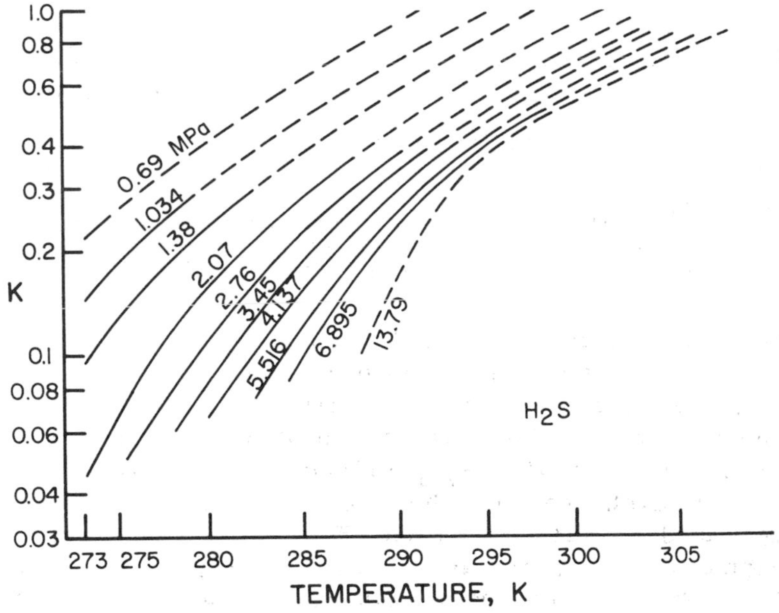


4-17 Hydrogen Sulfide $K_{vs}$ Chart
(Reproduced, with permission, from Transactions AIME (Noaker and Katz, 1954), by the American Institute of Mining, Metallurgical, and Petroleum Engineers.)

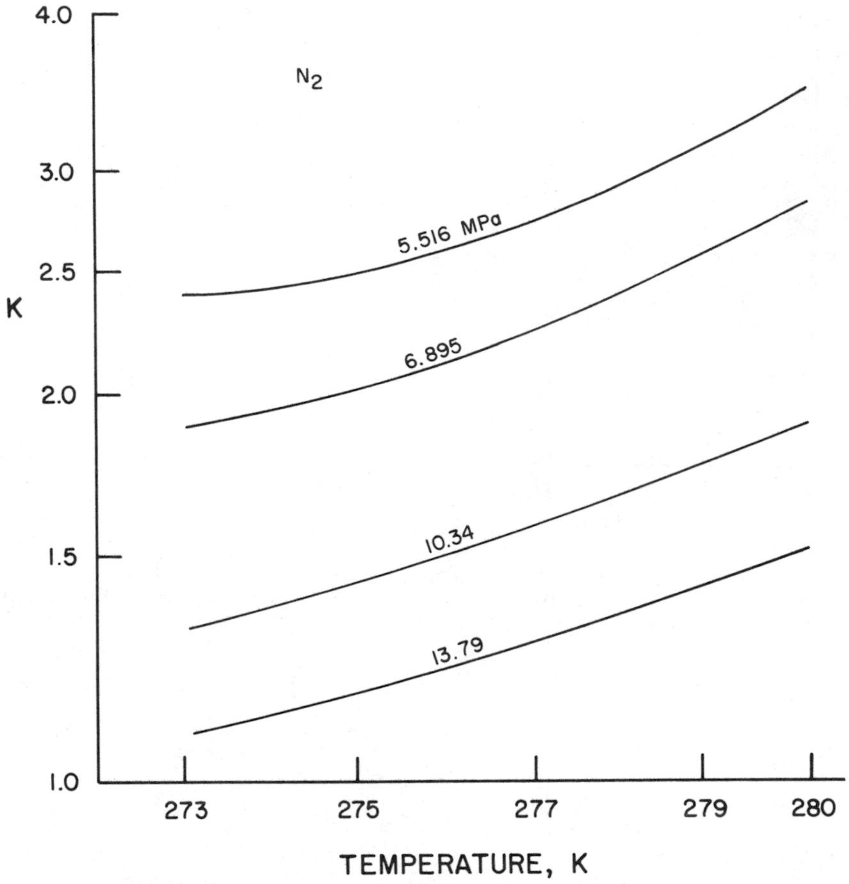


4-18 Nitrogen $K_{vs}$ Chart
(Reproduced, with permission, from <u>Canadian Journal of Chemical Engineering</u> (Jhaveri and Robinson, 1965), by the Canadian Society for Chemical Engineering.)

qualitatively determine in which phase a component will concentrate. For example components like nitrogen have $K_{vsi}$ values always greater than unity, so they concentrate in the vapor rather than the hydrate; those components like isobutane with $K_{vsi}$ values normally less than unity are concentrated in the hydrate phase. Equation (4.2) was used to fit all of the $K_{vsi}$ values in Figures 4-11 through 4-18. Table 4-2 presents the parameters of Equation (4.2), with an indication of the correlation coefficient.

**Table 4-2**
**Parameters and Accuracy of Prediction of Equation (4.2) for All Hydrate Components of Natural Gas**

| Compnt | Parameters: A | B | C | D | E |
|---|---|---|---|---|---|
| $CH_4$ | 1.63636 | 0.0 | 0.0 | 31.6621 | -49.3534 |
| $C_2H_6$ | 6.41934 | 0.0 | 0.0 | -290.283 | 2629.10 |
| $C_3H_8$ | -7.8499 | 0.0 | 0.0 | 47.056 | 0.0 |
| $i\text{-}C_4H_{10}$ | -2.17137 | 0.0 | 0.0 | 0.0 | 0.0 |
| $n\text{-}C_4H_{10}$ | -37.211 | 0.86564 | 0.0 | 732.20 | 0.0 |
| $N_2$ | 1.78857 | 0.0 | -0.001356 | -6.187 | 0.0 |
| $CO_2$ | 9.0242 | 0.0 | 0.0 | -207.033 | 0.0 |
| $H_2S$ | -4.7071 | 0.06192 | 0.0 | 82.627 | 0.0 |

| | F | G | H | I |
|---|---|---|---|---|
| $CH_4$ | -5.31E-6 | 0.0 | 0.0 | 0.128525 |
| $C_2H_6$ | 0.0 | 0.0 | -9.0E-8 | 0.129759 |
| $C_3H_8$ | -1.17E-6 | 7.145E-4 | 0.0 | 0.0 |
| $i\text{-}C_4H_{10}$ | 0.0 | 1.251E-3 | 1.0E-8 | 0.166097 |
| $n\text{-}C_4H_{10}$ | 0.0 | 0.0 | 9.37E-6 | -1.07657 |
| $N_2$ | 0.0 | 0.0 | 2.5E-7 | 0.0 |
| $CO_2$ | 4.66E-5 | -6.992E-3 | -2.89E-6 | -6.223E-3 |
| $H_2S$ | -7.39E-6 | 0.0 | 0.0 | 0.240869 |

| | J | K | L | M | N |
|---|---|---|---|---|---|
| $CH_4$ | -0.78338 | 0.0 | 0.0 | 0.0 | -5.3569 |
| $C_2H_6$ | -1.19703 | -8.46E4 | -71.0352 | 0.596404 | -4.7437 |
| $C_3H_8$ | 0.12348 | 1.669E4 | 0.0 | 0.23319 | 0.0 |
| $i\text{-}C_4H_{10}$ | -2.75945 | 0.0 | 0.0 | 0.0 | 0.0 |
| $n\text{-}C_4H_{10}$ | 0.0 | 0.0 | -66.221 | 0.0 | 0.0 |
| $N_2$ | 0.0 | 0.0 | 0.0 | 0.0 | 0.0 |
| $CO_2$ | 0.0 | 0.0 | 0.0 | 0.27098 | 0.0 |
| $H_2S$ | -0.64405 | 0.0 | 0.0 | 0.0 | -12.704 |

**Table 4-2 (continued)**

| Compnt | Parameters O | P | Q | R | Correl Coeff |
|---|---|---|---|---|---|
| $CH_4$ | 0.0 | -2.3E-7 | -2.0E-8 | 0.0 | 0.999 |
| $C_2H_6$ | 7.82E4 | 0.0 | 0.0 | 0.0 | 0.998 |
| $C_3H_8$ | -4.48E4 | 5.5E-6 | 0.0 | 0.0 | 0.998 |
| $i\text{-}C_4H_{10}$ | -8.84E2 | 0.0 | -5.4E-7 | -1.0E-8 | 0.999 |
| $n\text{-}C_4H_{10}$ | 9.17E5 | 0.0 | 4.98E-6 | -1.26E-6 | 0.996 |
| $N_2$ | 5.87E5 | 0.0 | 1.0E-8 | 1.1E-7 | 0.999 |
| $CO_2$ | 0.0 | 8.82E-5 | 2.55E-6 | 0.0 | 0.996 |
| $H_2S$ | 0.0 | -1.3E-6 | 0.0 | 0.0 | 0.999 |

**Accuracy of Predictions Using HYDK Program with Equation (4.2)**

| Prediction of Hydrate | Temperature(K) | Pressure(kPa) |
|---|---|---|
| Number of Data Predicted | 559 | 583 |
| Percentage Convergence, % | 71.1 | 51.3 |
| Absolute Average Deviation,% | 8.13 | 17.2 |

$$\ell n\, K_{vsi} = A + B*T + C*\Pi + D*T^{-1} + E*\Pi^{-1} + F*\Pi*T + G*T^2 + H*\Pi^2 + I*\Pi*T^{-1} + J*\ell n(\Pi*T^{-1}) + K*(\Pi^{-2}) + L*T*\Pi^{-1} + M*T^2*\Pi^{-1} + N*\Pi*T^{-2} + O*T*\Pi^{-3} + P*T^3 + Q*\Pi^3*T^{-2} + R*T^4 \quad (4.2)$$

where $\Pi$ = pressure, psia, T = temperature, $^{0}F$

The $K_{vsi}$-value charts or equations are used to determine the temperature or pressure of three-phase ($L_W$-H-V) hydrate formation. The condition for initial hydrate formation from free water and gas is calculated from an equation analogous to the dew point in vapor-liquid equilibria, at the following condition:

$$\sum_{i=1}^{n} \frac{y_i}{K_{vsi}} = 1.0 \quad (4.3)$$

At the three-phase pressure at a given temperature and gas phase composition, the sum of the mol fraction of each component in the vapor phase divided by the $K_i$ value of that component must equal unity. In order to have hydrates present with a gas mixture, it is always necessary to have at least one $K_i$ value greater than unity and at least one $K_i$ value less than unity. The pressure is changed in an iterative manner, and other $K_i$ values are determined until the above sum equals one at the point of hydrate formation (or dissociation). A similar technique is followed to determine the three-phase temperature at a given pressure. The technique is illustrated with the following example from Carson and Katz for the calculation of the pressure for hydrate formation.

**Example 4.4:** Determine the pressure of hydrate formation at 283.2 K (50°F) from a gas with the below composition.

**Solution:**

| Component | Mol Fraction | $K_i$ at 2.07MPa (300 psia) | $y_i/K_i$ | $K_i$ at 2.41MPa (350 psia) | $y_i/K_i$ |
|---|---|---|---|---|---|
| Methane | 0.784 | 2.04 | 0.384 | 1.90 | 0.412 |
| Ethane | 0.060 | 0.79 | 0.076 | 0.63 | 0.0953 |
| Propane | 0.036 | 0.113 | 0.318 | 0.09 | 0.400 |
| i-butane | 0.005 | 0.046 | 0.108 | 0.034 | 0.1471 |
| n-butane | 0.019 | $\infty$ | 0.0 | $\infty$ | 0.0 |
| nitrogen | 0.094 | $\infty$ | 0.0 | $\infty$ | 0.0 |
| carbon dioxide | 0.002 | 3.0 | 0.0007 | 2.3 | 0.0009 |
| SUM | 1.000 | | 0.8874 | | 1.0553 |

Interpolating linearly, the chart values yield $\sum y_i/K_i$ = 1.0 at 2.3 MPa (333 psia). The experimental value of hydrate formation at 283.2 K (50° F) is 2.24 MPa (325 psia)

A computer program by Al-Ubaidi (1988) for performing such a calculation has been compiled in the FORTRAN language, and is provided within the 5 1/4 inch floppy disk in the frontpapers. To access the program, an IBM-PC compatible machine with one floppy disk drive and 256K of RAM is needed. After the computer is "booted" simply mount the disk and type "HYDK (Return)". Follow the instructions for the input of data. A Newton-Raphson convergence scheme is used for the iterations in the calculation. Müller-Bongartz (1989) tested the accuracy of predictions from Equation (4.2) against the ternary and multicomponent data in Chapter 6 with the results given in Table 4-2. From these comparisons, it can be seen that the polynomial fit of Equation (4.2) is not entirely satisfactory, but it will often serve as an acceptable estimate, which may be refined through use of the charts themselves or with the method given in Chapter 5.

The accuracy of the $K_{vsi}$-value method is impressive, considering the fact that the method preceded the knowledge of the crystal structure. Carson and Katz (1942) labelled their charts as tentative, yet the original methane, ethane, and propane charts are still in use. The $K_{vsi}$ chart of methane was constructed from three data points at 4.14 MPa (600 psia), while the curves at other pressures were based on two data points and drawn symmetrical to the curve at 4.14 MPa.

The $K_{vsi}$ charts for components other than methane were derived from binary experimental data, with the $K_{vsi}$ values for the second component based on that for methane. That is, at an experimental hydrate formation temperature, pressure, and binary gas composition, the values of $y_i$ were fixed and the methane value of $K_{vsi}$ was determined by the original methane chart. The $K_{vsi}$ value of the second component was then calculated in order to satisfy Equation (4.3). With this calculation method, one might expect the $K_{vsi}$ charts for other components to be less accurate than that for methane, because any inaccuracy in the methane chart is incorporated in succeeding charts.

In the discussion appendix of the original paper by Carson and Katz, Hammerschmidt indicated that, while the method was acceptable for gases of "normal" natural gas composition, an unacceptable deviation was obtained for a gas rich in ethane, propane, and butanes. It should also be noted that more work is required to revise the $K_{vsi}$ value charts for two components, namely carbon dioxide and nitrogen. In three-phase hydrate formation from binary mixtures of carbon dioxide and propane, Robinson and Mehta (1971) determined that the $K_{vsi}$ method for carbon dioxide gave unsatisfactory results. The API Data Book shows the $K_{vsi}$ values for nitrogen to be only a function of pressure, without regard for temperature; Daubert (1987) indicated that data were insufficient for temperature dependence over a wide range of conditions.

Figures 4-19 and 4-20 present $K_{vsi}$ values of this method together with $K_{vsi}$ values obtained from the more accurate statistical thermodynamics method in Chapter 5, for three natural gases studied by Deaton and Frost (1945). In the figures, the $K_{vsi}$ values are presented as functions of temperature, with gas compositions as parameters. While both methods predict the three-phase condition of the gases acceptably, it is important to note that there are some substantial differences in the $K_{vsi}$ values by each method.

In particular, the methane $K_{vsi}$ values of Carson and Katz appear more accurate than those of the other components; this fact rationalizes the longevity of the charts because methane is normally the major component of a natural gas. The differences in the $K_{vsi}$ values shown in Figures 4-19 and 4-20 suggest that the charts be used with caution, particularly for gases with significant amounts of heavy components, or non-combustible components.

A second limitation to the $K_{vsi}$ value charts occurs in the limited range of temperatures and pressures. The charts are only applicable for temperatures above the ice

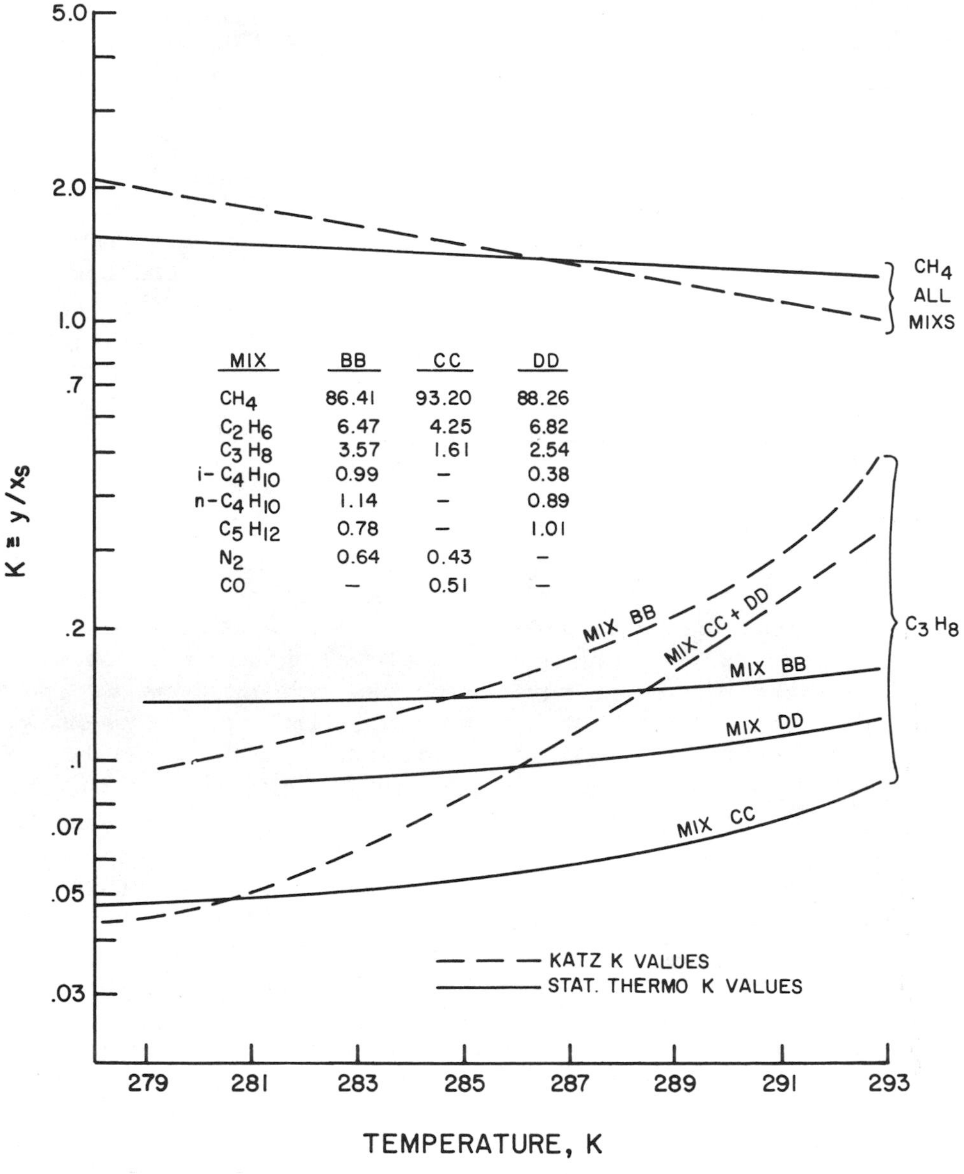


4-19 $K_{vs}$ Values for Methane and Propane
from Chapter 4 Katz Charts (solid lines)
and Chapter 5 Statistical Method (dashed lines)
(Reproduced, with permission, from the Proc. 63rd Annual GPA Convention (Sloan 1984), by the Gas Processors Association.)

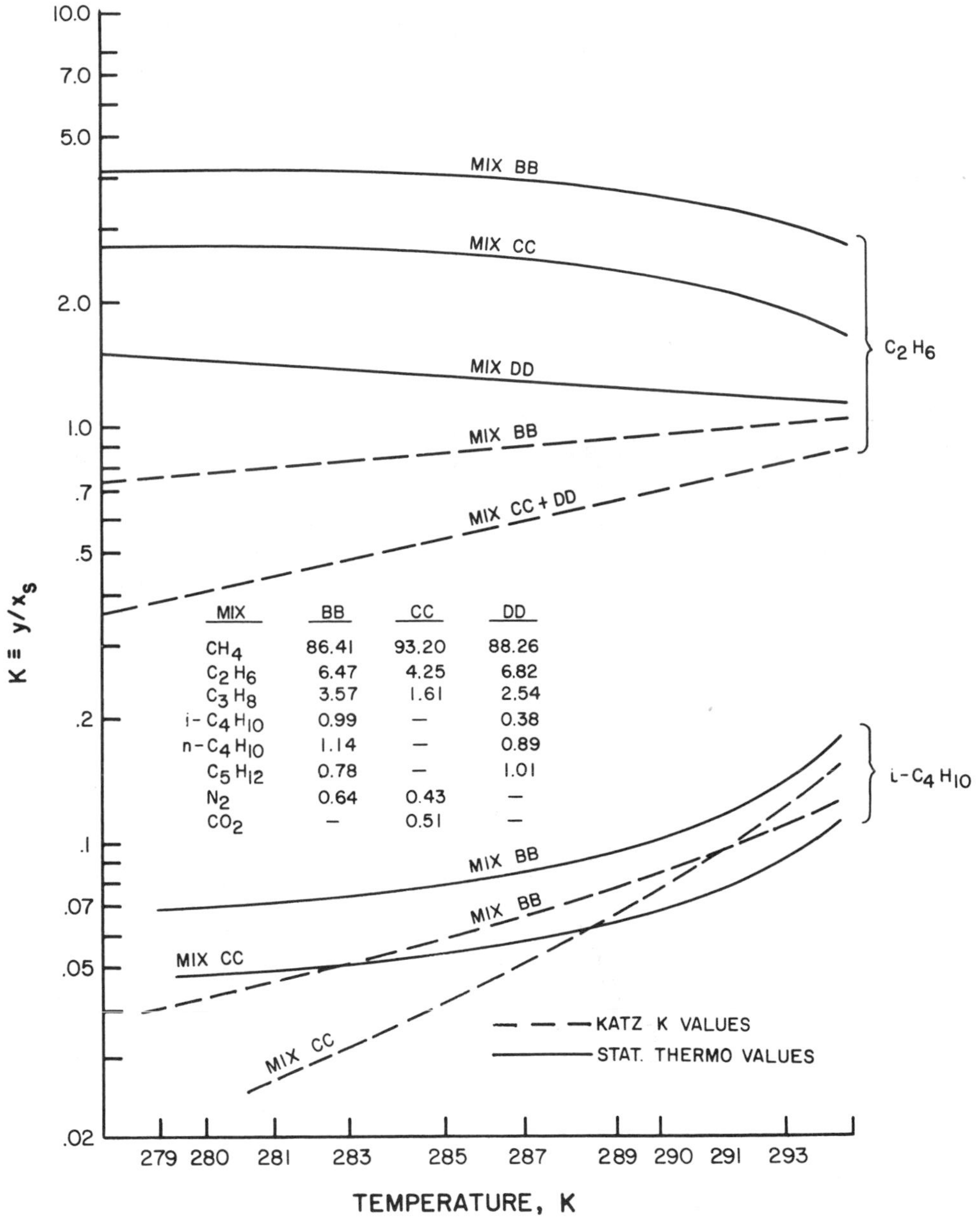


4-20 $K_{vs}$ Values for Ethane and Iso-Butane from Chapter 4 Katz Charts (solid lines) and Chapter 5 Statistical Method (dashed lines) (Reproduced, with permission, from the Proc. 63rd Annual GPA Convention (Sloan 1984), by the Gas Processors Association.)

**Table 4-3 Accuracy and Applicability of $K_{vsi}$ Values Compared Against The Statistical Thermodynamic Method of Chapter 5**

| | |
|---|---|
| Total Number of Data Points for 20 Natural Gases | 141 |
| Data Points Not Predictable Using the $K_{vsi}$ Charts | 40(28%) |
| Average Error[1] in Those Data Which were Predicted | |
| Using the $K_{vsi}$ Charts | 12.3% |
| Using the Statistical Thermodynamics Method | 9.6% |
| Average Error in All Data Points Predicted | |
| Using the Statistical Thermodynamic Method | 8.8% |

1. Average Error means average absolute error in pressure for a given temperature

point. Table 4-3 presents the results of a comparison of the experimental three-phase data on hydrates with the predictions of the $K_{vsi}$ charts and the predictions from the statistical thermodynamics method in Chapter 5. In addition to the accuracy comparisons, it should be noted that 28% of the three-phase data could not be predicted via the $K_{vsi}$ charts, principally due to chart temperature range limitations.

It should be noted that the use of the $K_{vsi}$ charts implies that both the gas phase and the hydrate phase can be represented as ideal solutions. This means that the $K_{vsi}$ of a given component is independent of the other components present, with no interaction between molecules. While the ideal solution model is acceptable for hydrocarbons in the hydrate phase (perhaps due to a shielding effect by the host water cages) the ideal solution assumption is not as accurate for the gas phase.

Work by Mann, et al. (1989) indicated that gas gravity may be a viable way of including gas non-idealities as a composition variable. That work might be considered as a combination of the gas gravity and the $K_{vsi}$ value charts via the calculations of statistical thermodynamic method of the following chapter. The calculations of Mann et al., which are amenable for use on a programmable calculator, indicate that better $K_{vsi}$ value charts may be obtained through a set of charts for each hydrate structure.

The $K_{vsi}$-value method was conceived prior to the determination of the hydrate crystal structure. As such, it is a fine representation of the intuitive insight which characterizes Katz's work. The method provided a substantial advance in hydrate prediction beyond the gas gravity method. The $K_{vsi}$-value method was the first predictive method, and it was used as the basis for the calculations in the gravity method, so it is logical that the $K_{vsi}$-value method should be more accurate.

To a first approximation, a more rapid method exists to estimate the pressure and temperature three-phase ($L_W$-H-V) loci of a simple hydrate. For the five simple natural gas hydrate formers which have both middle and upper quadruple points, $Q_1$ and $Q_2$, the three-phase loci may be approximated by semilogarithmic interpolation (logarithm pressure versus reciprocal absolute temperature) of the values in Table 4-4. The location of both quadruple points, using the methods of the present section, is discussed in the following section.

**Table 4-4: Natural Gas Component Quadruple Points**

| Component | T,P at $Q_1$<br>K, MPa | T,P at $Q_2$<br>K, MPa |
|---|---|---|
| Methane | 272.9, 2.563 | No $Q_2$ |
| Ethane | 273.1, 0.530 | 287.8, 3.39 |
| Propane | 273.1, 0.172 | 278.8, 0.556 |
| iso-Butane | 273.1, 0.113 | 275.0, 0.167 |
| n-Butane | does not form hydrates as single hydrocarbon | |
| Carbon Dioxide | 273.1, 1.256 | 283.0, 4.499 |
| Nitrogen | 271.9,14.338 | No $Q_2$ |
| Hydrogen Sulfide | 272.8, 0.093 | 302.7, 2.239 |

## 4.4 The Quadruple Points $Q_1$ and $Q_2$ and Equilibria of Three Condensed Phases ($L_W$-H-$L_{HC}$)

Both the gas gravity method and the $K_i$-value method enable the estimation of three-phase ($L_W$-H-V) equilibria between the upper and middle quadruple points ($Q_2$ and $Q_1$). For immiscible hydrate formers of natural gas components,the quadruple points of simple hydrate formers (excluding $Q_2$ for methane and nitrogen) are reproduced in Table 4-4 from Table 1-2. The upper quadruple point marks the lowest pressure of the three condensed phase ($L_W$-H-$L_{HC}$) equilibria.

### 4.4.1 The Location of the Quadruple Points

The middle quadruple point $Q_1$ (I-$L_W$-H-V) invariably is located at the intersection of the three-phase ($L_W$-H-V) pressure-temperature locus with the ice point (273.15 K) within a degree. The intersection temperature closely approximates the ice point because the solubility of immiscible hydrate formers in water is too small to significantly change the freezing point of water. Therefore the methods of the previous section may be used for a rapid estimation of the pressure and temperature of $Q_1$. The approximate location of the upper quadruple point $Q_2$ ($L_W$-H-V-$L_{HC}$) is only slightly more difficult to determine.

The quadruple point $Q_2$ is located by the intersection of the $L_W$-H-V line with the $L_W$-V-$L_{HC}$ line. The locus of the $L_W$-V-$L_{HC}$ line is normally very close to the vapor pressure of the hydrocarbon, terminating in a three-phase ($L_W$-V-$L_{HC}$) critical point which approximates the two-phase ($V_{HC}$-$L_{HC}$) critical point of the pure hydrocarbon. As a consequence, the intersection of the pure hydrocarbon vapor pressure with the $L_W$-H-V line, determined as in Section 4.3 by either the gas gravity or $K_i$-value method, normally provides a good estimate of pressure and temperature of the upper quadruple point $Q_2$.

The three-phase ($L_W$-V-$L_{HC}$) pressure-temperature locus approximates the vapor pressure ($V_{HC}$-$L_{HC}$) locus for the pure component due to two effects, both of which are caused by the hydrogen bond phenomena described in Chapter 2. First the hydrogen bonds cause almost complete immiscibility between the hydrocarbon liquid and the aqueous liquid, so that the total pressure may be closely approximated by the sum of the vapor pressures of the pure hydrocarbon and water. Secondly, the hydrogen bonds cause such an self-attraction of the water molecules that the water vapor pressure is only a small fraction of the hydrocarbon vapor pressure at any temperature. Because each immiscible liquid phase exerts its own vapor pressure, the vapor pressures are additive. Because the vapor pressure of water is relatively small, the pure hydrocarbon vapor pressure is a very good approximation of the three-phase ($L_W$-V-$L_{HC}$) locus.

### 4.4.2. Condensed Three-Phase Equilibria

Katz (1972) first noted that hydrates could form from heavy liquids such as crude oils which have dissolved gases suitable for hydrate formation. He suggested that the point of hydrate formation from water and a liquid hydrocarbon phase (no gas present) could be predicted using the vapor-hydrate distribution coefficient $K_{vsi}$ of Equation (4.1) together with the more common vapor-liquid distribution coefficient $K_{v\ell i}$ ($\equiv y_i/x_{\ell i}$). In this case Equation (4.3) should be modified to be:

$$\sum_{i=1}^{n} \frac{y_i}{K_{vsi}} = \sum_{i=1}^{n} \frac{x_{\ell i}\, K_{v\ell i}}{K_{vsi}} = 1.0 \qquad (4.3a)$$

The substitution of $y_i = x_{\ell i}\, K_{v\ell i}$ in the numerator of Equation (4.3a) suggests that this equation applies at the bubble point, or the quadruple point ($L_W$-H-V-$L_{HC}$) which marks the lowest pressure of a three-phase ($L_W$-H-$L_{HC}$) region. Katz noted that Scauzillo (1956) had measured systems which did not appear to conform to the above equation. As noted in Section 7.1.3, later measurements by Verma (1974) and Holder (1976) appear to confirm

Katz's analysis. Further discussion of hydrate formation from condensed systems is provided in Section 8.1.5.

For condensed three-phase ($L_W$-H-$L_{HC}$) hydrate equilibria at pressures above the upper quadruple point, the pressure changes extremely rapidly with only a small change in temperature. This is because all three phases are relatively incompressible and highly immiscible, so that only a small thermal expansion is needed to cause a large pressure change. As a consequence, the accurate determination of a P-T locus is a stringent test for even the most accurate equation-of-state to predict both the density and the solubility of the liquid hydrocarbon phase and the aqueous phase. As a first approximation, the incompressible three-phase ($L_W$-H-$L_{HC}$) condition can be acceptably estimated by a vertical line on the pressure-temperature plot. Figure 1-1 shows the condensed phase P-T plot above $Q_2$ to be almost vertical for simple hydrate formers of natural gas components.

The fact that the three-phase ($L_W$-H-$L_{HC}$) equilibria can be approximated by a line of infinite slope on a P-T diagram led nineteenth century investigators to suggest that the upper quadruple point $Q_2$ represented the maximum temperature of hydrate formation. By the same reasoning, without an upper quadruple point methane and nitrogen hydrates are considered to have no upper temperature of formation. The pure component critical temperatures of methane and nitrogen (190.6K and 126.2K, respectively) negate any possible intersection of their vapor pressures with the three-phase $L_W$-H-V loci which only exist at temperatures above 273.15 K.

However, the condensed three-phase P-T locus is not perfectly vertical. Ng and Robinson (1977) experimentally measured the $L_W$-H-$L_{HC}$ equilibria for a number of structure II hydrate mixtures and suggested that a better estimation of the slope dP/dT might be obtained through the Clapeyron equation:

$$\frac{dP}{dT} = \frac{\Delta H}{T\Delta V} \tag{4.4}$$

where $\Delta H$ and $\Delta V$ represent the enthalpy and volume respectively accompanying the process of conversion of liquid water and liquid hydrocarbon into the hydrate unit cell. The value of $\Delta H$ was found to be essentially constant at 65,400 ± 2,100 J/mole for many gas mixtures. Therefore, to a very good approximation, the volume change $[\Delta V \equiv V_H - (V_{LW} + V_{LHC})]$ at the quadruple point $Q_2$ determines the slope of the condensed phase equilibria. Of the 21 gas mixtures studied by Ng and Robinson in the $L_W$-H-$L_{HC}$ region, the value for dP/dT ranged between 3.4 MPa/K and 66.3 MPa/K, with an average value of 10.16 MPa/K; therefore the large value of $\Delta H$ causes dP/dT to be very high in all cases.

For the upper temperature for hydrate formation, Makogon (1981) suggested a better criteria than the location of $Q_2$ is the P-T condition at which the density of the combined hydrocarbon and water is equal to that of the hydrate. He assumed complete liquid immiscibility and used the inverted Clapeyron relation:

$$\frac{d\ \ell nT}{dP} = \frac{\Delta V}{\Delta H} \qquad (4.5)$$

where: $\Delta H_H$ = heat of hydrate formation from liquid water and liquid hydrocarbon[1]

$\Delta V$ = the molar volume of the hydrate less that of the hydrocarbon and liquid water $(\equiv V_H - V_{LHC} - V_{LW})$.

Since the value of $\Delta H_H$ remains constant over a large range of pressures, the maximum in T (or its logarithm) is determined by the point at which the molar volume change is zero. The volume comparison must be made between the pure liquid hydrocarbon, liquid water, and hydrate, since the hydrocarbon must exist as liquid at

---

1. The translation indicates hydrocarbon vapor, but the condensed hydrocabon phase is clearly required by the pressure and temperature conditions

pressures between the vapor pressure and the critical pressure. However, since the liquid density of the most simple hydrate formers is always substantially greater than that in the hydrate phase, a value of $\Delta V = 0$ is never realized. Maxima in hydrate formation temperatures above $Q_2$ have been calculated, but they have yet to be measured.

### 4.5 Effect of Inhibitors on Hydrate Formation

The means of hydrate prevention and dissociation are discussed in detail in Chapter 8. In the present section we consider the lowering of the three-phase ($L_W$-H-V) temperature or the increase of the $L_W$-H-V pressure via an inhibitor.

By the Phase Rule, a second intensive variable is needed (in addition to either temperature or pressure) to specify the three-phase binary system with the addition of an inhibitor. Typically the concentration of the inhibitor in the free water phase is specified. Substances which have considerable solubility in the aqueous phase, such as alcohols, glycols, and salts, normally act as inhibitors to hydrate formation. The mechanism for inhibition of formation stems from the increased competition for the water molecules by the dissolved inhibitor molecule or ion.

As a first approximation, the temperature depression for hydrate inhibition might be considered to be similar to the depression of the freezing point of ice by an equivalent mass fraction of the inhibitor. Nielsen and Bucklin (1983) provided a derivation to indicate that the hydrate depression temperature will always be less than the ice depression temperature by the factor [(heat of fusion of ice)/ (heat of hydrate dissociation)], which has a numerical value between 0.6 and 0.7.

Figure 4-21 shows the correlation of experimental data of Hammerschmidt (1935) with five inhibitors. The

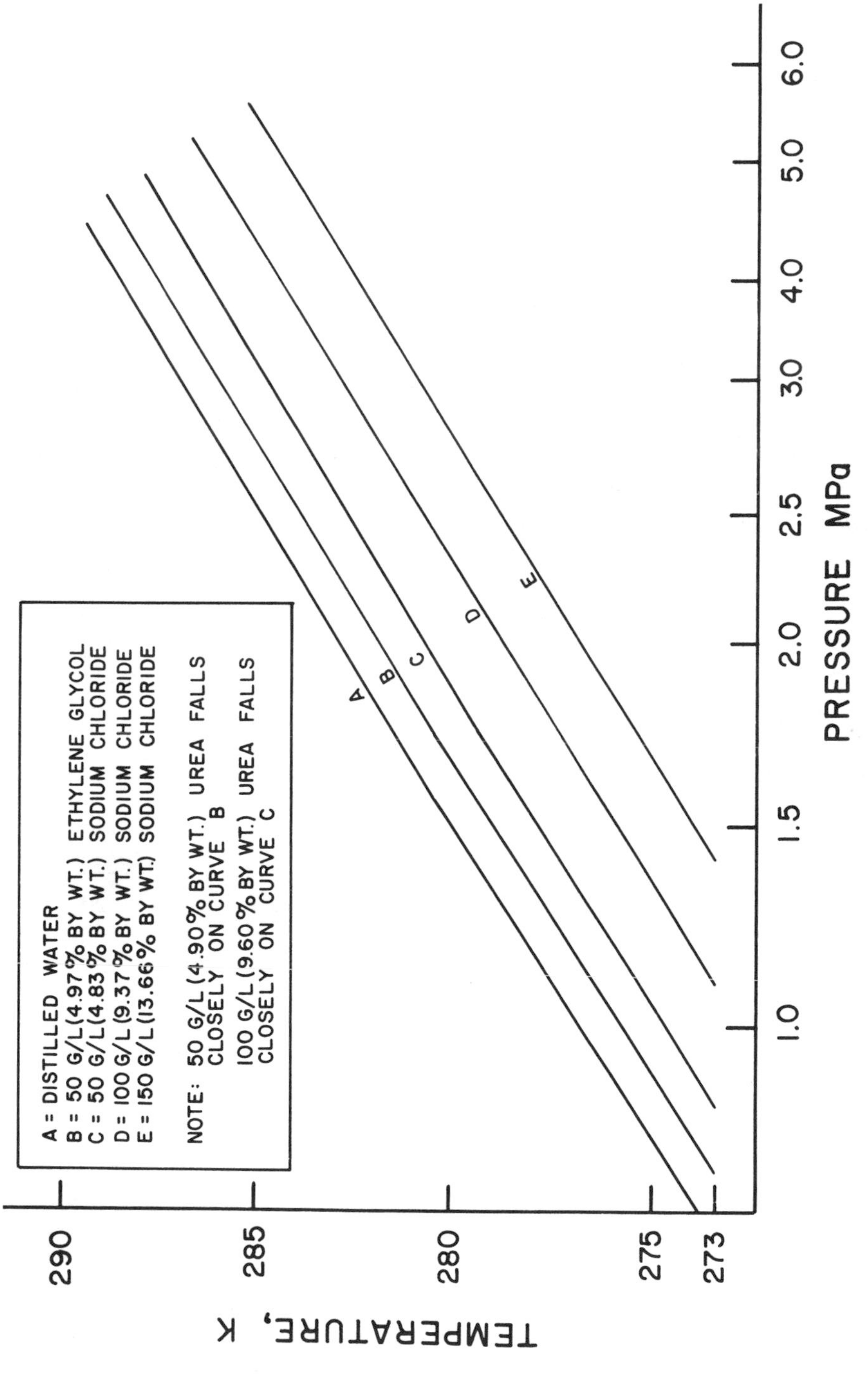


4-21 Hydrate Equilibrium Curves with Various Inhibitors (Reproduced courtesy of the United States Bureau of Mines, (Deaton and Frost, 1946))

striking feature of Figure 4-21 is the degree to which all of the solid experimental lines are parallel, for the inhibition effect of both alcohols and salts relative to pure water. The parallel solid lines provide some indication of the molecular nature of the inhibition. Normally a phase transformation is considered relative to Gibbs Free Energy defined as:

$$\Delta G \equiv \Delta H - T\Delta S \qquad (4.6)$$

The two components of $\Delta G$ are 1)an energetic part $\Delta H$ and 2) a structure part $\Delta S$, for the equilibrium $(V + L_W \rightleftharpoons H)$ at constant temperature and pressure. The Gibbs Free Energy must be increased by the inhibitor in order for a lower temperature to be required for hydrate formation. Using the Clapeyron equation with the data of Figure 4-21 to relate the slope $(d\ell nP/dT)$ to the enthalpy of formation $\Delta H$, one can determine that the value of $\Delta H$ is relatively constant. Therefore it appears that the energetic effects are not appreciably affected by the inhibitors. In order to increase the Gibbs Free Energy, the primary effect of the inhibitor is on the structure of the water phase. The inhibitor must encourage non-randomness (or structure other than hydrate-like clusters) in the water structure, in order to be effective.

Several types of inhibitors have been tried, but the glycols and alcohols have proved to be the most successful. As an example, ammonia was initially determined to be twice as effective an inhibitor as methanol; over long times however, ammonia reacts with carbon dioxide and water form ammonium carbonate, bicarbonate and carbamate through the reactions:

$$2\ NH_3 + H_2O + CO_2 \rightarrow (NH_4)_2CO_3$$

$$NH_3 + H_2O + CO_2 \rightarrow NH_4\ HCO_3$$

$$CO_2 + 2\ NH_3 \rightarrow NH_4\ CO_2\ NH_2$$

The solid ammonium carbonate and carbamate caused more problems than the hydrates (Townsend and Reid, p. 100 (1978)). The natural gas industry has opted for methanols and glycols which may be injected into pipelines and processes without undesirable side reactions.

### 4.5.1 Hydrate Inhibition Via Alcohols and Glycols

The alcohols (in the homologous series ending with butanol) all hydrogen bond to water with their hydroxyl group. However, as summarized in Chapter 2, a substantial body of work (reviewed by Franks (1973) and Ben Naim (1978)) indicates that the hydrocarbon end of the alcohol molecule causes a clustering effect on water molecules similar to that of hydrate formers. Alcohols therefore have two effects on water which compete with dissolved apolar molecules for clusters: the hydroxyl group hydrogen bonds the water molecules, and the hydrocarbon end of the alcohol tends to organize the water into solvent sheaths in direct competition with the dissolved apolar molecules.

Makogon (1981, p.134) and Berecz and Balla-Achs (1983,p. 102) indicated that methanol can increase the temperature of hydrate formation at concentrations less than 5 mass per cent (presumably due to the clustering effect), while higher concentrations inhibit formation. Nakayama and Hashimito (1980) also suggested that several of the alcohols could form hydrates; yet further study by Nakayama with Davidson et al. (1981) provided another interpretation of the original methanol data, without resorting to hydrate formation. Further studies by Svartas (1988) proved that small methanol amounts do not increase hydrate thermodynamic stability. However, small amounts of methanol may cause a slight reduction in the amount of hydrate metastability on formation; this hypothesis may be used to explain the erroneous interpretation from the three works mentioned at the first of this paragraph.

Of alcohols, methanol has been the most popular inhibitor, due to its cost and its effectiveness. Katz et al. (1959,p.218) indicated that the inhibition ability of alcohols decrease with volatility, i.e. methanol > ethanol > isopropanol. Typically methanol is vaporized into the gas stream of a transmission line, then dissolves in any free water present, preventing hydrate formation. Makogon (1981, p. 133) noted that, in 1972 the Soviet gas industry used 0.3 kg of methanol for every 1000 cubic meters of gas extracted. Stange et al. (1989) indicated that North Sea methanol usage may surpass the ratio given by Makogon by an order of magnitude. Nielsen and Bucklin (1983) present calculations to indicate that methanol is less expensive than drying with either alumina or molecular sieves. Nevertheless, the use of methanol in the North Sea has become so expensive that methanol recovery and return lines have become economical.

The glycols (ethylene glycol(EG), diethylene glycol(DEG), and triethylene glycol(TEG)) provide more hydrogen bonding opportunity with water through one more hydroxyl group than alcohols, as well as through oxygen atoms in the case of the larger glycols. The glycols generally have higher molecular weights which inhibit volatility. Thus they may be recovered and recycled more readily than alcohols, but are not as frequently injected into transmission lines in practice. An illustration of ethylene glycol injection is given in Example 8.2.

Through a comprehensive set of experimental studies, Ng and Robinson (1983) determined that methanol inhibited hydrate formation more than an equivalent mass fraction of glycol in the aqueous liquid. The preference for methanol versus glycol may also be determined by economic considerations (Nelson, 1973).

Almost all of the techniques for hydrate inhibition with methanol enable the determination of the methanol concentration in the aqueous liquid in equilibrium with hydrates at a given temperature and pressure. With the technique for the determination of the amount of methanol in the liquid phase, the user is left with the additional step of estimating the amount of methanol to be injected in the vapor. This problem was addressed first by Jacoby (1953) and then by Nielsen and Bucklin (1983), who presented a revised methanol injection technique. The amount of injected methanol represents a combination of three factors: (1) methanol lost to the vapor phase, (2) methanol dissolved in the aqueous phase, and (3) methanol dissolved in any liquid hydrocarbon phase. Example 8.3 illustrates the calculation and shows that substantial methanol can be dissolved (and lost) in both the liquid and vapor hydrocarbon phase.

To approximate the hydrate depression temperature for several inhibitors in the aqueous liquid, the natural gas industry has used the original Hammerschmidt (1939) expression for four decades:

$$\Delta T = \frac{2,335\ W}{100M - MW} \tag{4.7}$$

where:

$\Delta T$ = hydrate depression, $^0F$,
M = molecular weight of the alcohol or glycol, and
W = weight per cent of the inhibitor in the liquid.

Equation (4.7) was based on more than 100 experimental determinations of equilibrium temperature lowering in a given natural gas - water system in the concentration range 5 - 25 wt%. The equation was used to correlate data for alcohols and ammonia inhibitors. Hammerschmidt (1939) provided for a modification of the molecular weight M when salts were used as inhibitors. Unfortunately, no information on the gas composition or on the individual experimental data were provided. The assumption is normally made that the gases used by Hammerschmidt were methane-rich.

Pieroen (1955) and Nielsen and Bucklin (1983) presented derivations to show the theoretical validity of the Hammerschmidt equation. The latter work suggested that the equation applies only to typical natural gases, and to methanol concentrations less than 0.20 mol fraction (typically for system operation at temperatures above 250K). Due to a cancellation of errors, the equation (without modification) is applicable for aqueous ethylene glycol concentrations to about 0.40 mol fraction (typically for system operation to 233K).

A comparison of Hammerschmidt's equation, as well as the prediction by the freezing point depression of water for methanol inhibition is summarized in Table 4-5.

Nielsen and Bucklin (1983) presented an improved version of the Hammerschmidt equation which is accurate over a wider range, i.e. to concentrations as large as 0.8 mol fraction. They suggested that Equation (4.8) may be effectively used to design methanol injection systems operating as low as 165 K:

$$\Delta T = -129.6 \ \ell n \ (1 - x_{MeOH}) \tag{4.8}$$

where $\Delta T$ is in $^0F$. Makogon (1981, p 134) indicated that the inhibition effect is a function (albeit much smaller) of pressure as well as that of temperature.

### 4.5.2 Hydrate Inhibition Via Salts

The action of salts as inhibitors is somewhat different than that of alcohols or glycols. The salt ionizes in solution and interacts with the dipoles of the

**Table 4-5**
**Comparison of Two Simple Prediction Methods for Hydrate Inhibition by Methanol**

| Simple Component | Wt% MeOH | Number of Hydrate Data Pts | Avg % Error in Temperature by Hammerschmidt Equation (4.7) | Avg % Error in Temperature by Freezing Point Depression |
|---|---|---|---|---|
| Methane | 10 | 4 | 3.98 | 4.21 |
| Methane | 20 | 3 | 7.56 | 15.3 |
| Ethane | 10 | 5 | 2.03 | 8.97 |
| Ethane | 20 | 2 | 1.26 | 23.5 |
| Propane | 5 | 3 | 1.62 | 3.13 |
| Carbon Dioxide | 10 | 3 | 7.63 | 3.53 |

water molecules with a much stronger bond than the van der Waals forces which cause clustering around the apolar solute molecule. This clustering also causes a decrease in the solubility of potential hydrate guest molecules in the water (a phenomenon known as "salting-out") as a secondary effect. Both of these effects combine to require substantially more subcooling to overcome the structural changes and cause hydrates to form.

Makogon (1981 p. 124) indicates that salt inhibition is approximately a direct function of charge and an inverse function of ion radius. The best inhibitors have cations with a maximum charge and a minimum radius. Table 4-6 (from Makogon) ranks, in decreasing effectiveness, the various suitable anions according to these characteristics, together with their chloride salts. Higher values of salting-out capacity $A_p$ are indicative of better inhibitors. Table 4-6 also provides the salting-out capacity of the chloride salt with each anion.

Anions of the salt should have similar requirements on charge and ion radius, but are additionally restricted by factors of solubility and cost. Of the anions, the chlorides have found the most utility, followed by the nitrates and the sulfates. Considerations of cost and effectiveness have caused calcium chloride to predominate as a salt inhibitor, followed in usage frequency by sodi-

**Table 4-6**
**Salting-Out Capacity $A_p$ of Cation**

| Element | Charge | Ion Radius Å | $A_p$ | $A_{MeCl_n}$ |
|---|---|---|---|---|
| Be | +2 | 0.34 | 0.65 | 0.073 |
| Al | +3 | 0.57 | 0.20 | 0.041 |
| Mg | +2 | 0.78 | 0.11 | 0.028 |
| Ca | +2 | 1.06 | 0.05 | 0.018 |
| Na | +1 | 0.98 | 0.04 | 0.0157 |
| K | +1 | 1.33 | 0.02 | 0.010 |

um chloride. Frequently special precautions must be taken because aqueous solutions of calcium chloride are much more corrosive than those of sodium chloride. Figure 4-22 from Katz et al. (1959) shows a comparison of the effectiveness of $CaCl_2$ and NaCl relative to methanol as an inhibitor.

The original Hammerschmidt Equation (4.7) has been modified for the prediction of salt inhibition. For example, Makogon (1981) suggested that the constant be changed to 2320 for sodium chloride. It is interesting to note that no simple technique has been determined for predicting the effect of mixtures of salts, such as might be found in sea water. Section 5.3.1 presents a method by Englezos and Bishnoi (1988a) for predicting the inhibiting effect of salt mixtures when the statistical thermodynamic method is used.

### 4.6 Two-Phase Equilibria of Hydrates with Fluid Hydrocarbons

Hydrates may also exist in equilibrium with only a fluid hydrocarbon phase, either vapor or liquid, when there is no aqueous phase present. Two-phase (H-V or H-$L_{HC}$) regions are shown on every diagram of Figure 4-2. By the Phase Rule, in the three-phase regions, as discussed in Sections 4.3 and 4.4, only one intensive variable is needed to specify a binary system. However, two variables are needed to specify a two-phase binary system; typically water concentration in the hydrocarbon fluid is determined as the second variable at a specified temperature or pressure. The determination of the equilibrium water concentration enables the engineer to maintain the hydrocarbon fluid in the single-phase region, without hydrate solids for fouling or flow obstruction.

Two common misconceptions exist concerning the presence of water to form hydrates in pipelines, both of which are illustrated via the T-x phase equilibrium diagrams in Figure 4-2. The first and most common misconcep-

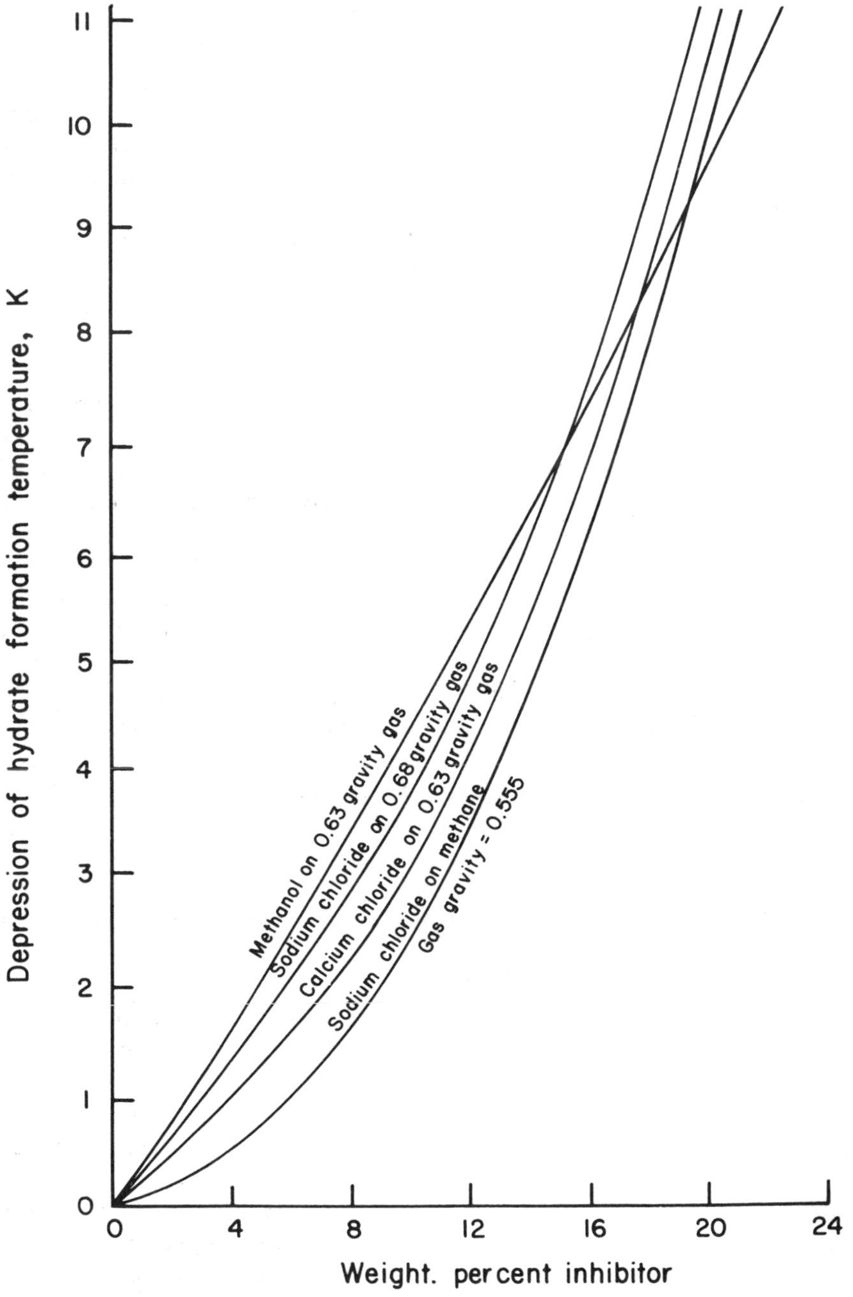


4-22 Hydrate Temperature Depression by Various Inhibitors (Reproduced, with permission, from the Handbook of Natural Gas Engineering (Katz et al., 1959), by McGraw-Hill Bk. Co.)

tion is that a free water phase is absolutely necessary for the formation of hydrates. The upper three-phase ($L_W$-H-V) line temperature marks the condition of hydrate formation from free water and gas. Below that temperature and to the right of the hydrate line however, are two-phase regions in which hydrates are in equilibrium only with hydrocarbon vapor or liquid containing a small (<1000 ppm) amount of water. From a strict thermodynamic standpoint then, a vapor or liquid with sufficient water can form hydrates at the H-V or the H-$L_P$ boundaries when no free water phase is present. The question of the accumulation of water molecules into a hydrate phase is a question of kinetics, dependent upon the time necessary for hydrate nuclei to attain a critical size. This time may be in excess of that available for laboratory study, but may occur in processes which operate over extended periods of days, months, or years.

In addition there have been multiple studies (Sloan et al., 1976; Cady, 1983a,b; Devlin, 1987; Kobayashi et al., 1987) which demonstrate that hydrate growth can easily occur from a hydrocarbon fluid phase if a hydrate nucleus is either already present, at adsorbed sites on a wall, or on a third surface. Therefore, from a practical standpoint one should require that the hydrocarbon fluid be maintained in the thermodynamic single phase region if hydrate growth is to be prevented.

The second misconception about two-phase hydrates concerns the dew point of the hydrocarbon phase. Bucklin et al. (1985) correctly point out that the extrapolated points from the vapor-liquid water region at higher temperatures give meta-stable dew points. This effect is indicated in the Gas Processors Association Handbook (1981, Figure 15-14). There is an additional metastability to be considered, however. It is also incorrect to determine the dew point, assuming that ice is the condensed water phase; such equilibrium occurs very infrequently as seen in only a few sub-diagrams (Figure 4-2,g-k), and then only over a very narrow range of temperatures and pressures. Much more frequently hydrate is the condensed water phase at low temperature, or high pressure. Details of the more common dew point (with hydrates) are given in Section 4.6.1, while the less common dew point (with ice) is described in Section 4.6.2. Section 8.1.1 discusses the processing and production implications of these equilibrium.

### 4.6.1 Water Content of Vapor in Equilibrium with Hydrate

The water content of the vapor phase in (H-V) equilibria is very small (typically less than 0.001 mol fraction) and therefore difficult to measure accurately. As a consequence, in the history of gas processing, semilogarithmic straight lines (gas water content versus reciprocal absolute temperature) from the $L_W$-V region were extrapolated into the H-V region with only limited justification. A typical chart for water content from this period is presented in Figure 4-23. In Figure 4-23 the water content chart at temperatures above the hydrate stability conditions is based primarily on the data of Olds et al. (1942) while the data of Skinner (1948) are the basis for extrapolations to temperatures below the hydrate formation point. However, below the initial hydrate formation conditions Figure 4-23 represents metastable values, as observed (in gas field data) by Records and Seely (1951). Kobayashi and Katz (1955), indicated that such concentration extrapolations across hydrate phase boundaries yield severe errors.

Laboratory confirmation that the water content of gas in equilibrium with hydrate should be much lower than the extrapolated values has been verified by Sloan et al. (1976) for methane hydrates, and by Song and Kobayashi (1982) for methane-propane hydrates. A typical replacement chart is shown in Figure 4-24. In this figure the high temperature $L_W$-V region is separated from the low temperature H-V region by a line representing the three-phase ($L_W$-H-V) boundary.

The isobaric data in the vapor-hydrate region of Figure 4-24 follow semilogarithmic straight lines when plotted against reciprocal absolute temperature, but

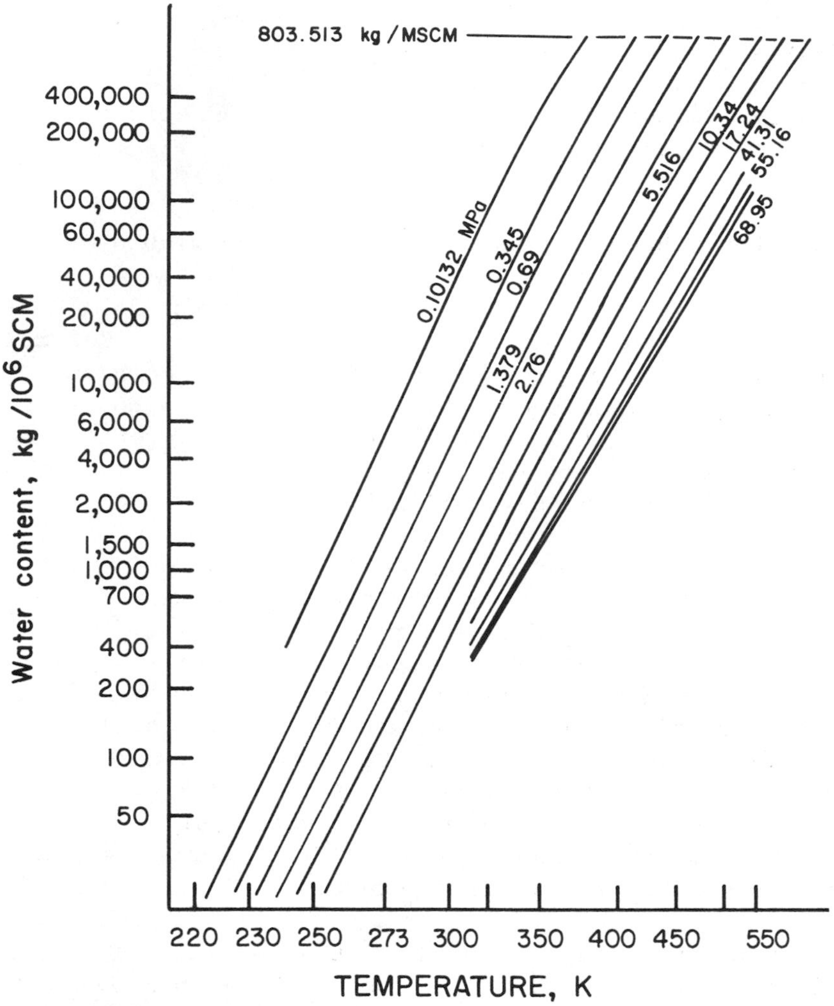


4-23 Metastable Water Content of Gas (Reproduced, with permission, from Petroleum Engineering Handbook (Kobayashi et al., 1987), by the Society of Petroleum Engineers.)

these lines have slopes different from the straight lines in the $L_W$-V region. In addition, the three-phase ($L_W$-H-V) point at which the slope change occurs is a function of gas composition. The change in the slope of an isobar from the $L_W$-V region occurs at different temperatures for differing compositions. With the above complexities, a comprehensive water content chart (or series of charts)

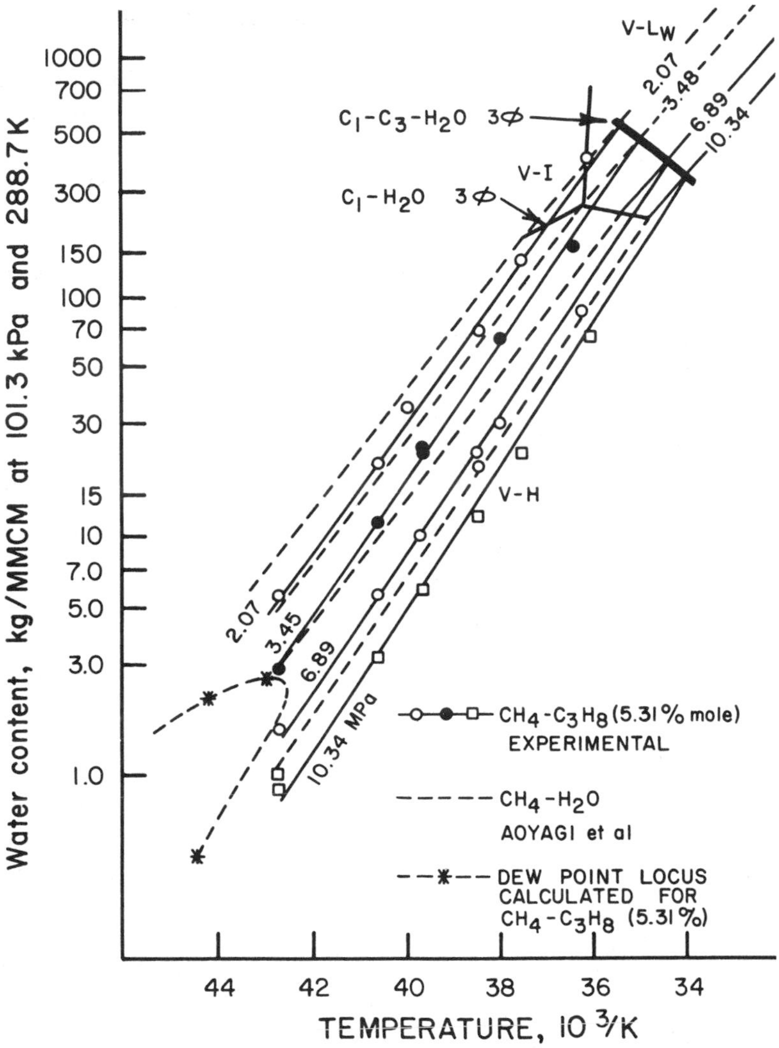


4-24 Equilibrium Water Content Chart for Methane (Dashed) and for Methane+5.31%Propane (Solid)
(Reproduced, with permission, from Industrial and Engineering Chemistry Fundamentals (Song and Kobayashi, 1982), by the American Chemical Society.)

for gases of differing compositions would be cumbersome. Instead, a mathematical method for determining the water content of gases in the V-H region is used. The algorithm in Table 4-7 below is corrected slightly from that presented recently by Kobayashi, Song, and Sloan (1987).

## Table 4-7
## Algorithm to Calculate the Water Content of Methane-Rich Vapor in Equilibrium with Hydrates

In order to calculate the water content of vapor in equilibrium with hydrates at a given temperature and pressure, the six steps below should be followed:

1. Calculate the metastable water content $W_{MS}$ at the temperature and pressure of interest. This may be done via Figure 4-24 or by a regression of the data that of figure provided by Equation (4.9) for pressures between 200 and 2,000 psia and for temperatures between -40 and 120$^{o}$F:

$$W_{MS} = \exp\left[ C_1 + C_2/T + C_3 (\ell n\ P) + C_4/T^2 + C_5(\ell n\ P)/T + C_6(\ell n P)^2 \right] \quad (4.9)$$

where:

$W_{MS}$ = metastable water content, $lb_m/10^6 SCF$,
T = temperature, $^{o}R$,
P = pressure, psia, and
$C_i$ = constants whose values are given in Table 4-8.

2. Calculate the three-phase ($L_W$-H-V) temperature at the given pressure and water free composition, using either the gas gravity or the $K_i$-Value method of Section 4.3. Obtain a temperature difference $\Delta T$ by subtracting the temperature of interest from the calculated three-phase temperature.

3. Calculate the displacement from the metastable water content ($\Delta W$) at the above $\Delta T$ and pressure of interest using both Equation (4.10a) for methane and Equation (4.10b) for a 94.69 mol% methane/5.31 mol% propane mixture.

$$\Delta W = \exp[C_1 + C_2(\ell n\ P) + C_3(\ell n\ \Delta T) + C_4(\ell n\ P)^2 + C_5(\ell n\ P)(\ell n\ \Delta T) + C_6(\ell n\ \Delta T)^2 + C_7(\ell n\ P)^3 + C_8(\ell n\ P)^2(\ell n\ \Delta T) + C_9(\ell n\ P)(\ell n\ \Delta T)^2 + C_{10}(\ell n\ \Delta T)^3 + C_{11}(\ell n\ P)^4 + C_{12}(\ell n\ P)^3(\ell n\ \Delta T) + C_{13}(\ell n\ P)^2(\ell n\ \Delta T)^2 + C_{14}(\ell n\ P)(\ell n\ \Delta T)^3 + C_{15}(\ell n\ \Delta T)^4] \quad (4.10)$$

In Equation (4.10) the constants for (4.10a) are in Table 4-8 for methane, with a regression of the methane data in the pressure range between 500 and 1,500 psia and the temperature range of -28 to 26 $^{o}$F. The constants for the mixture (Equation 4.10b) were generated in the pressure range of 500 to 1,500 psia, and in the temperature range of -38 to 40 $^{o}$F.

4. Calculate the $\Delta W$ for the gas composition of interest by a linear interpolation between the $\Delta W$ for methane (gravity 0.552) and the $\Delta W$ for the mixture containing 5.31 % propane (gravity 0.603), using gravity as an interpolation parameter.

5. Calculate the equilibrium water content by subtracting the $\Delta W$ value obtained in Step 4 from the metastable water value obtained in Step 1.

6. Consider the range of the data used to determine the regression constants of Equations (4.9) and (4.10) to determine whether the answer obtained in Step 5 is within the bounds of the correlation.

**Table 4-8**
**Regression Constants in Equations (4.9) and (4.10a,b)**

| | Equation(4.9) | Equation(4.10a) | Equation(4.10b) |
|---|---|---|---|
| $C_1$ | 2.8910758E01 | -1.605505E03 | 2.59097E03 |
| $C_2$ | -9.6681464E03 | 8.181485E02 | -1.51351E03 |
| $C_3$ | -1.6633582E00 | 9.289352E02 | -1.16506E02 |
| $C_4$ | -1.3082354E05 | -1.578381E02 | 3.26066E02 |
| $C_5$ | 2.0353234E02 | -3.899544E02 | 6.65280E01 |
| $C_6$ | 3.8508508E-2 | -2.009926E01 | -1.17697E01 |
| $C_7$ | | 1.368723E01 | -3.05990E01 |
| $C_8$ | | 5.500387E01 | -1.20352E01 |
| $C_9$ | | 4.088990E00 | 2.94244E00 |
| $C_{10}$ | | 1.517650E00 | 7.83747E-1 |
| $C_{11}$ | | -4.524342E-1 | 1.04913E00 |
| $C_{12}$ | | -2.590273E00 | 7.23943E-1 |
| $C_{13}$ | | -2.465990E-1 | -2.79450E-1 |
| $C_{14}$ | | -7.543630E-2 | 7.08799E-2 |
| $C_{15}$ | | 1.034443E-1 | -1.24938E-1 |

**Example 4.5: Water Content of Vapor in the Hydrate-Vapor Region**

Determine the water content of a gas whose gravity is 0.575 in equilibrium with hydrate at 1,000 psia and 8.4 $^{0}$F.

**Solution:**

1. The metastable water content of the gas is calculated from Equation (4.9) at 1,000 psia and 8.4 $^{0}$F as 2.745 $lb_m/10^6$SCF, where 1 SCF is defined as a cubic foot of gas at 101.328 KPa (1 atm) and 288.7 K (60$^{0}$F).
2. The three-phase temperature at 1,000 psia is calculated by the methods of Section 4.3 to be 55.01 $^{0}$F. The $\Delta T$ value is 46.61$^{0}$F (= 55.01 - 8.4).
3. The displacement from the metastable conditions $\Delta W$ is calculated for methane to be 0.746 $lb_m/10^6$ SCF from Equation (4.10a). The displacement from the metastable condition $\Delta W$ is calculated for the mixture containing 5.31% propane as 1.80 $lb_m/10^6$SCF from Equation (4.10b).
4. An interpolation between the values of $\Delta W$ obtained in Step 3 (based upon gas gravity) is done to determine the displacement $\Delta W$ for a gas with a gravity of 0.575. The resulting displacement is 1.223 $lb_m/10^6$ SCF.
5. The displacement in Step 4 is subtracted from the metastable value in Step 1 to obtain a water concentration of 1.522 $lb_m/10^6$ SCF. The 0.575 gravity gas must be dried to less than 1.522 $lb_m/10^6$ SCF to prevent hydrate formation at 1,000 psia and 8.4$^{0}$F.
6. A check of the conditions of regression of Equations (4.9) and (4.10a,b) indicates that the correlations should apply for the conditions of the example. Unfortunately, the experimental data are not available at present to verify the calculation beyond the bounds measured for methane and the methane-propane mixture.

---

It is worthwhile to emphasize that all of the data presented in the section thus far have been for methane-rich gases. For heavier gases, or for non-combustible gases, there are almost no data in the V-H two-phase region; the sole exception is the water content study of Song and Kobayashi (1983a,b) for carbon dioxide.

### 4.6.2 Water Content of Vapor in Equilibrium with Ice

A third two-phase region (marked V-I) exists on water concentration diagrams such as Figure 4-24. This region is normally of minor importance relative to the hydrate-vapor region. On the propane-water phase diagrams of Figure 4-2, the V-I region can only be observed on a few diagrams 4-2(g-k) and then only over a small temperature range. For natural gases, at pressures below approximately 3.5 MPa, the isobaric water lines will intersect the ice-liquid water-hydrate ($I-L_W-H$) boundary rather than the ($L_W-H-V$) three-phase boundary. The intersection occurs at approximately 273 K as the temperature is lowered along an isobar from the $L_W-V$ region. At temperatures below the ($I-L_W-H$) intersection the isobars extend into the ice-vapor (I-V) two-phase region. Figure 4-25 shows that the thermodynamic isobars (solid lines) in this region also deviate from the metastable semilogarithmic extrapolations from the vapor-liquid water region (dashed lines). The thermodynamic isobars (solid lines) of Figure 4-25 were not measured, but calculated from enhanced ice vapor pressure. In this case the solid phase is pure and is thus more amenable to thermodynamic calculations. The calculation, provided by Equation (4.11), should be very accurate and provides another illustration of the inaccuracies of metastable extrapolations:

$$y_W \phi_W P = P_i^o \phi_i^o \exp\left[(v_i\{P-P_i^o\})/RT\right] \qquad (4.11)$$

where: $y_W$ = water mol fraction in the vapor,
$\phi_W$ = fugacity coefficient of water in the vapor (obtained from an equation-of-state),
$P_i^o$ = vapor pressure of pure ice,
$\phi_i^o$ = correction for ice vapor pressure non-ideality ($\approx 1.0$)

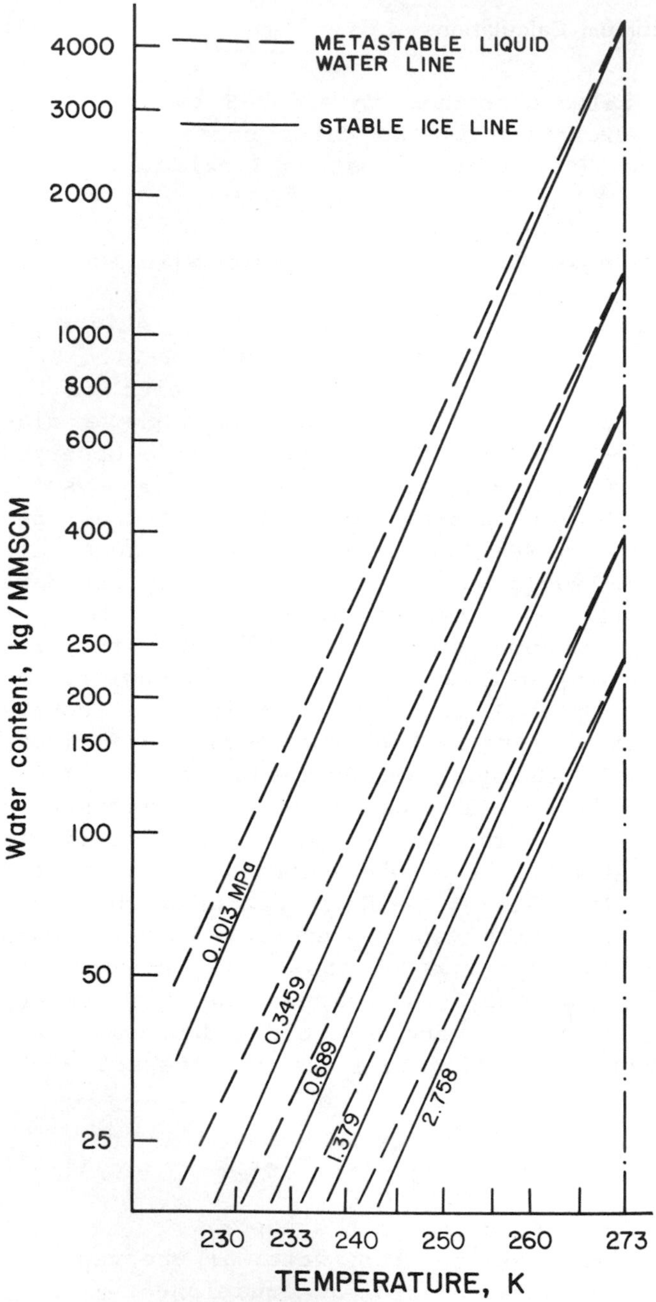


4-25 Equilibrium (Solid) and Metastable (Dashed) Water-Content Chart for Methane Gas in Equilibrium with Ice (Reproduced, with permission, from Petroleum Engineering Handbook (Kobayashi et al., 1987), by the Society of Petroleum Engineers.)

$v_i$ = molar volume of ice, and
R = universal gas constant.

At still lower temperatures along an isobar in the I-V region, the I-H-V boundary intersection occurs. As for the $L_W$-H-V boundary, the temperature and pressure (and therefore the water concentration) of the I-H-V intersection varies according to the hydrocarbon concentration of the vapor, which may be calculated using the $K_{vsi}$ method of Mann et al. 1989). Below the I-H-V boundary the concentration of water in the vapor for H-V equilibria may be calculated with the procedure of Table 4-7.

### 4.6.3 Water Content of Liquid Hydrocarbon

In the H-$L_{HC}$ two-phase region there is a severe paucity of both data and a simple calculation scheme. The only data available are from this laboratory (Sloan et al.(1986, 1987)). The latter work provides a prediction scheme which is discussed in the following chapter using the statistical thermodynamics method. There is no simple method for inclusion here.

Because there are so few data for water content of the fluid hydrocarbon in either of the two-phase (H-V or H-$L_{HC}$) regions, their accuracy cannot be determined. These data are very difficult to obtain due to the low concentrations (typically $<$ 100 ppm mol). The inaccuracies in normal experimental data in other phase regions are frequently greater than the absolute values of the water content in the H-$L_{HC}$ region.

With low concentrations of water a substantial amount of time may be required before the water molecules can agglomerate into a hydrate structure. Experimental time to acquire each data point is normally on the order of days or weeks, rather than hours. Appreciable metastability is observed, and long times are required for the formation of critical hydrate nuclei. Nevertheless, the long time involved should not be taken as an indication that hydrates are not thermodynamically stable in the two-phase region. The phase diagram analyses presented in Section 4.1 clearly indicate the thermodynamic validity of this region.

## 4.7 Hydrate Number and Enthalpy from Phase Equilibria

Historically, two periods occurred for the determination of the number of hydrate water molecules per guest molecule. In the first century (1810 - 1900) after their artificial fabrication, the hydrate number was determined directly. That is, the amounts of hydrated water and guest molecules were each experimentally measured via various methods. The experimental difficulties stemmed from two facts: (1) the water phase could not be completely converted to hydrate without some occlusion and (2) the reproducible measurement of the inclusion of guest molecules was hindered by hydrate metastability. As a result, the hydrate numbers differed widely for each substance, with a general reduction in the ratio of water molecules per guest molecule as the methods became refined with time. After an extensive review of experiments of the period Villard (1897) proposed "Villard's Rule" to summarize the work of that first century of hydrate research:

> "The dissociable (hydrate) compounds, that form through the unification of water with different gases and that are only stable in the solid form, all crystallize reguularly and have the same constitution that can be expressed by the formula M + 6 $H_2O$, where M designates a molecule of the respective gas."

After 1900, however, the direct determination of hydrate number was abandoned in favor of the second, indirect method. This method was originally proposed by de Forcrand (1902) relying on his experience in other fields. The indirect method is still in use today, and is based on calculation of the enthalpies of formation of hydrate from gas and water, and from gas and ice. With this more accurate method many exceptions were found to Villard's Rule. The historical summary provided in Chapter 1 indicates that, while the number of hydrated water molecules was commonly thought to be an integer, frequently that integer was determined to differ from 6, particularly after de Forcrand had proposed his method.

The method considers the equilibrium of gas and n moles of liquid water (or ice) with hydrates on either side of the ice point:

$$\text{Gas} + n(\text{Liquid Water}) \rightleftharpoons \text{Hydrate} \qquad \Delta H_1 \tag{4.12}$$

$$\text{Gas} + n(\text{Ice}) \rightleftharpoons \text{Hydrate} \qquad \Delta H_2 \tag{4.13}$$

When Equation (4.13) for the $L_W$-H-V equilibria is set equal to Equation (4.12) for I-H-V equilibria at the middle quadruple point $Q_1$ (approximately 273 K), they may be subtracted, with the result of the number of moles of liquid water converted to ice:

$$n(\text{Liquid Water}) \rightleftharpoons n(\text{Ice}) \qquad \Delta H_3 \tag{4.14}$$

where $\Delta H_3 = \Delta H_1 - \Delta H_2$

Because the enthalpy of fusion, $\Delta H_f$, of water is well known, $\Delta H_3$, the difference in the $\Delta H$ values of Equations (4.12) and (4.13), may be divided by $\Delta H_f$ to obtain n, the number of moles of water (or ice) converted to hydrates.

This method has been found to be much more accurate than the direct method during the current century. One reason for its accuracy is related to the determination of $\Delta H_1$ and $\Delta H_2$ from three-phase ($L_W$-H-V or I-H-V) equilibrium measurements of pressure and temperature via the Clapeyron equation:

$$\frac{dP}{dT} = \frac{\Delta H}{T\Delta V} \tag{4.15}$$

where $\Delta H$ may be taken as the enthalpy change in either

Equation (4.12) or (4.13), $\Delta V$ is the corresponding volume change, and P and T are the phase equilibrium points along the appropriate three-phase line. This equation represents the primary relationship of pressure-temperature phase equilibria data to enthalpic data.

**Example 4.6 Determining Hydrate Number from Pressure-Temperature Data for $L_W$-H-V and I-H-V.**

Sortland and Robinson (1964) measured the formation conditions of sulfur hexafluoride hydrates from 264K to 297K. Using the Clapeyron equation with their data they determined values of $\Delta H_1$ = -29,570 cal/gmol, and $\Delta H_2$ = -5,140 cal/gmol. When these two values are subtracted and the result is divided by the molar heat of fusion of water (1435.3 cal/gmol) a value of 17.02 gmol $H_2O$ per gmol $SF_6$ is obtained in the hydrate. This value is significantly different from that of Villard's Rule, and indicates that the $SF_6$ molecules essentially fill all of the large cavities in structure II hydrate. If each of the large cavities in structure II were filled the ratio would be exactly 17 (=136 water molecules/8 cavities).

---

In theory, with this indirect method the problems in the direct method (metastability and occlusion) are avoided because the P-T measurements are at equilibrium and they are not dependent on the amounts of each phase present. The question about the validity of the method centers on the validity of the Clapeyron equation to the three-phase hydrate equilibria, as discussed in the following section.

### 4.7.1 The Clapeyron Equation and Hydrate Equilibria

In the most common thermodynamic case, the Clapeyron equation is used with pure components to obtain the heat of vaporization from pure component two-phase (vapor pressure) data. Thus, in hydrate equilibria, it seems unusual to apply it to binary systems of three-phase ($L_W$-H-V or I-H-V) equilibria to obtain the heats of dissociation in Equations (4.12) and (4.13). As van der Waals and Platteeuw (1959b) point out however, the application of the Clapeyron equation is thermodynamically correct, as long as the system is univariant, as is the case for simple hydrates.

In Equations (4.12) and (4.13) if the volume of hydrate approximates that of water (or ice) in the hydrate formation reaction, then to a good approximation, $\Delta V \approx V_g$ (= zRT/P, where z is compressibility). The substitution of this expression for $\Delta V$ leads to a more useable form of the Clausius-Clapeyron equation:

$$\frac{d\ \ell n\ P}{d\ (1/T)} = -\ \frac{\Delta H}{zR} \qquad (4.16)$$

Semilogarithmic plots of formation pressure versus reciprocal absolute temperature yield straight lines, over limited temperature ranges, for hydrate formation from either liquid water or ice. From Equation (4.16) such linear plots either indicate (a) constant values of the three factors: (1) heat of formation $\Delta H$, (2) compressibility factor, z, (3) stoichiometry ratios of water to guest, or (b) cancellation of curvilinear behavior in these three factors.

The most recent confirmation of the validity of the Clausius-Clapeyron equation for hydrates was by Handa (1986b,c), who measured the heat of dissociation (via calorimetry) of the normal paraffins which form simple hydrates. Table 4-9 shows the literature values for hydrate numbers, all obtained using de Forcrand's method of enthalpy differences around the ice point. However Handa's values for the enthalpy differences were determined calorimetrically, while the other values listed were determined using phase equilibrium data and the Clausius-Clapeyron equation. The agreement appears to be very good for simple hydrates. Verification of the Clausius-Clapeyron equation for hydrates of gas mixtures has not been accomplished.

**Table 4-9**
**Hydrate Number ($M \cdot nH_2O$) for Simple Hydrates of Natural Gas Components from Handa (1986b,c)**

| Component | n | Reference |
|---|---|---|
| Methane | 6.00 | Handa (1986b,c) |
| | 5.77 | Glew (1962) |
| | 7.00 | Roberts et al.(1941) |
| | 7.18 | Frost and Deaton (1946) |
| | 6.00 | Galloway et al. (1970) |
| | 7.4 | de Roo et al. (1983) |
| | 6.3 | de Roo et al. (1983) |
| Ethane | 7.67 | Handa (1986b,c) |
| | 7.00 | Roberts et al.(1941) |
| | 8.25 | Frost and Deaton (1946) |
| | 8.24 | Galloway et al. (1970) |
| Propane | 17.0 | Handa (1986b,c) |
| | 5.7 | Miller and Strong (1946) |
| | 17.95 | Frost and Deaton (1946) |
| | 18.0 | Knox et al. (1961) |
| | 19.7 | Ceccotti (1966) |
| | 17.0 | Cady (1983) |
| Isobutane | 17.0 | Handa (1986c) |
| | 17.1 | Uchida and Hyano (1964) |
| | 17.5 | Rouher and Barduhn (1969) |

Roberts, Brownscombe, and Howe (1940) and Barrer and Edge (1967) present similar, detailed derivations to consider the use of the Clapeyron equation for hydrate binary and multicomponent systems. The reader is referred to the latter article for the precise meaning of dP/dT and the details of the derivation. Barrer and Stuart (1957) and Barrer (1959) point out that the problem in the use of the Clapeyron equation evolves from the non-stoichiometric nature of the hydrate phase. Fortunately that problem is not substantial in the case of hydrate equilibria, because the non-stoichiometry does not change significantly over small temperature ranges. At the ice point, where the hydrate number is usually calculated, the non-stoichiometry is essentially identical for each three-phase system at an infinitesimal departure on either side of the quadruple point.

### 4.7.2 Hydrate Numbers by the Miller and Strong Method

After de Forcrand's Clapeyron method, a third method for the determination of hydrate number, proposed by Miller and Strong (1946), was determined to be applicable when simple hydrates were formed from a solution with an inhibitor, such as a salt. They proposed that a thermodynamic equilibrium constant K be written for the physical reaction of Equation (4.12) to produce one mol of guest M, and n moles of water from one mole of hydrate. Writing the equilibrium constant K as multiple of the activity of each product over the activity of the reactant, each raised to its stoichiometric coefficient, one obtains:

$$M\bullet(H_2O)_n \rightleftharpoons M + nH_2O \tag{4.12a}$$

where $M\bullet(H_2O)_n \equiv$ hydrate (H)

$$K = \frac{(a_M)(a_W)^n}{a_H} \quad . \tag{4.17}$$

As is the normal practice, K, the thermodynamic reaction equilibrium constant is assumed to be only a function of temperature. In Equation (4.17) $a_M$, the activity of the guest in the vapor phase, is taken as the fugacity of the pure component divided by that at the standard state, normally taken as 1 atm. The fugacity of the pure vapor is a function of temperature and pressure, and may be determined through the use of a fugacity coefficient such as that of Appendix A. The method also assumes that $a_H$, the activity of the hydrate, is essentially constant at a given temperature regardless of the other phases present.

The activity of water $a_W$ in Equation (4.17) is normally taken as unity disregarding the solubility of the gas. At a given temperature, if an inhibitor such as a salt is present, the activity of the water decreases and the activity of the gas must increase in order to maintain a constant product $K \cdot a_H$ at that temperature. Thus writing a second equation for the formation of a hydrate from an inhibited liquid we get:

$$a_M \cdot a_W = K \cdot a_H \qquad \text{without the inhibitor} \tag{4.17a}$$

and

$$a'_M \cdot a'_W = K \cdot a_H \qquad \text{with the inhibitor} \tag{4.18}$$

Subtracting (4.18) from (4.17a), with the right sides constant, one may replace the activity of the guest M with its fugacity to obtain:

$$f_M \cdot a_W{}^n = f'_M \cdot a'_W{}^n \tag{4.19}$$

By dropping the subscripts and remembering that the (′) denotes the presence of an inhibitor, Equation (4.19) may be rearranged to obtain the hydrate number n as:

$$n = \frac{\ell n(f/f')}{\ell n(a'/a)} \tag{4.20}$$

In Equation (4.20) the activity of pure water a is taken as unity and the activity of the water with the inhibitor a′ is taken as the product of the water concentration $x_W$ and the activity coefficient $\gamma$. The water concentration is known and the activity coefficient $\gamma$ is easily obtained from colligative properties for the inhibitor, such as the freezing point depression. For instance the activity of water in aqueous sodium chloride solutions may be obtained from Robinson and Stokes (1959, pp. 476) or from any of several handbooks of chemistry and physics.

With the above data, Equation (4.20) indicates that the hydrate number n may be obtained from a measurement of the increase of the hydrate pressure with an inhibitor present at a given temperature. The fugacity may be calculated for a pure component from any of a number of thermodynamic methods given a temperature and pressure. Only in the case of an ideal gas (very low pressure or very high temperature) may the fugacities be replaced with the pressure itself.

It should be noted that this method contains several key assumptions, as follows:

1. that the equilibrium constant K does not change when the inhibitor is added to the aqueous fluid, but that it is only a function of temperature,
2. that the activity of the hydrate phase is constant at a given temperature regardless of the other phases present,
3. that the vapor presence of any component other than the hydrate former may be neglected, including any water or inhibitor present in the liquid, and
4. that the aqueous phase without inhibitor is pure water.

Exceptions may be found to the above assumptions and consequently the method might be expected to be more limited than the de Forcrand method presented at the first of this section. On the other hand, the de Forcrand method requires more data. Rouher and Barduhn (1969) indicate better results are achieved with the Miller and Strong method when NaCl solutions are in the range between 5 and 15 wt%. Most recently Patil (1987) determined the hydrate number of simple propane hydrates to be 18.95 by the de Forcrand method; using the Miller and Strong method he obtained hydrate numbers of 19.29, 19.95, and 19.89 for NaCl solutions of 3, 5, and 10 wt%, respectively. Wilms and van Haute (1973) presented the mathematically correct version of the Miller and Strong method, which eliminates some of the above assumptions, together with the statistical thermodynamic method of the following chapter. Wilms and van Haute indicated that the Miller and Strong method is a special case of their more rigorous method.

### 4.8 Summary and Relationship to Chapters which Follow

The object of this chapter is to provide the reader with a qualitative understanding of hydrate phase equilibria. Such an understanding implies a historical overview, which also provides successive approximations to hydrate phase equilibria in terms of accuracy. In the following chapter, the most accurate method available is discussed for the determination of hydrate equilibria - that of statistical thermodynamics. The consideration of this method ties the macroscopic phase equilibria, such as has been discussed qualitatively in the present chapter, to the microscopic structure discussed in Chapter 2. The bridging of the microscopic and macroscopic phenomena is satisfying both from a theoretical and from a pragmatic standpoint.

The concepts of this chapter are illustrated in the experimental methods and data compiled in Chapter 6. Applications of the concepts of this chapter are also found in the final two chapters. Hydrates in the earth provide natural examples of phase equilibrium as detailed in Chapter 7. Applications to artificially fabricated hydrates and their problems in gas production and processing are found in Chapter 8.

# 5

# A Statistical Thermodynamics Approach to Hydrate Phase Equilibria

The object of Chapter 4 was to provide an overview of phase equilibria, which is more easily obtained through phase diagrams and the approximate, historical methods available. With Chapter 4 as background the subject of this chapter is the phase equilibrium calculation method which is both the most accurate and the most comprehensive available.

The statistical thermodynamic method discussed here provides a bridge between the molecular crystal structures of Chapter 2 and the macroscopic thermodynamic properties of Chapter 4. It also provides a comprehensive means of correlation and prediction of all of the hydrate equilibrium regions of the phase diagram, without separate prediction schemes for two-phase regions, three-phase regions, inhibition, etc. as in Chapter 4. However, for a qualitative understanding of trends and an approximation of the results of the various prediction schemes of the present chapter, the reader is referred to the previous chapter.

Section 5.1 presents the fundamental method as the heart of the chapter. In this section is presented the statistical thermodynamics approach to hydrate phase equilibria. The basic statistical thermodynamic equations are developed, and the relationships to measurable, macroscopic hydrate properties are given. The parameters for the method are shown to be determined from simple hydrate equilibrium measurements. A calculation algorithm is given for multicomponent three-phase ($L_W$-H-V) equilibria, for comparison with those methods previously discussed in Section 4.3. Finally, recent modifications to the method are presented along with an assessment of the accuracy of the method.

The remaining sections of the chapter consider the extension of the method to other phase equilibrium regions. Section 5.2 discusses the method of determining condensed phase equilibria ($L_W$-H-V-$L_{HC}$ and $L_W$-H-$L_{HC}$) with the method, in analogy to those equilibria detailed in Section 4.4. Section 5.3 discusses the prediction of the inhibitor effect on hydrate equilibria as first discussed in Section 4.5. Finally Section 5.4 presents the extension of the method to two-phase (H-V or H-$L_{HC}$) equilibria, analogous to that discussed in Section 4.6.

At the onset, it should be noted that the application portions presented here have been the subject of two recent articles. The reviews by Holder et al. (1988), and Munck et al. (1988) provide slightly different perspectives on the material given in this chapter after Section 5.1.4. The reader is sure to gain additional insights from these other vantage points.

Since the method is too involved for hand calculation, an IBM PC-compatible computer program is provided on the floppy disk in the frontpapers of the book. The use of the program is discussed in a preliminary way in each of the sections of the chapter, beginning in Section 5.1.5. A detailed description of the program, together with input data and example illustrations, is provided in Appendix B.

While the method of the present chapter may appear detailed and comprehensive, the reader is cautioned that the calculation is limited by the available data. The accuracy of the prediction method is invariably less than the accuracy of good experimental data. For each region of phase equilibrium prediction, the limitations on both the accuracy and data availability are discussed. In formulating designs based on the methods of this chapter, the user should carefully consider the data available to support the calculations. The methods presented may be useful for interpolations between available data sets. The reader is urged to use the method with caution for extrapolations beyond the range of the available data; that may warrant further experiments.

## 5.1 Statistical Thermodynamics of Hydrate Equilibria

After the determination of the hydrate crystal structures in the early 1950's, it was possible to generate theories for equilibria of macroscopic properties based upon microscopic properties. With the knowledge of distinguishable cavities containing at most one guest particle per cavity came the necessity to describe the distribution of the guest particles via statistics (i.e. "How many ways can one distribute M indistinguishable particles in L distinguishable boxes with at most one particle per box?"). The resulting improvement in theoretical ability led to a more accurate calculation method. Currently, the calculation of hydrate equilibria is a leading exemplar of the industrial use of statistical thermodynamics on a routine basis.

The initial work in the area was done by Barrer and Stuart (1957), with a more accurate method by van der Waals and Platteeuw (1959), who are considered the founders of the method which is widely used. In the present section we follow the latter development, constructed as the following Sections:

The first two of the above sections are a simplification and slight expansion of the derivation from the review article by van der Waals and Platteeuw (1959). They were written, assuming that the reader has a background in statistical thermodynamics on the level of an introductory text, such as that of Hill (1960) or McQuarrie (1976), as a minimum. The reader who does not have an interest in statistical thermodynamics may choose to obtain a pragmatic overview by reviewing the basic assumptions listed in 5.1.1 and 5.1.4 before progressing to the final equations and the calculation algorithm in 5.1.6.

For the reader's convenience, Table 5-1 gives the nomenclature used in Sections 5.1.1 and 5.1.2 as well as a parenthetical listing of the equations in which each first appears.

### 5.1.1 Grand Canonical Partition Function for Water

To develop the model, four fundamental assumptions based upon structure were made and are stated as follows:

1) The host molecules' contribution to the free energy is independent of the occupation of the cavity. This assumption also implies that encased molecules do not distort the cavity.
2) Each cavity can contain at most one guest molecule, which cannot diffuse from the cavity.

**Table 5-1 Nomenclature for Subsections 5.1.1 and 5.1.2**

$A^{MT}$ = Helmholtz free energy of empty host lattice (5.1)
$f_k$ = fugacity of a molecule of type k (5.22)
$N_{J_i}$ = number of solute (guest) molecules of type J within a type i cavity (5.2)
$N_W$ = number of host (water) molecules (5.2)
Q = canonical partition function for host lattice (5.4)
$q_{J_i}$ = partition function of a J molecule in a type i cavity (5.3)
$y_{k_i}$ = probability of finding a molecule of type k in a cavity of type i (5.17)
$\lambda_J$ = absolute chemical activity of guest molecule J (5.5b)
$\nu_i$ = number of type i cavities per water molecule (5.2) (for structure I $\nu_1 = 1/23$, $\nu_2 = 3/23$ for structure II $\nu_1 = 2/17$, $\nu_2 = 1/17$)
$\mu_i$ = chemical potential of component i (5.5a)
$\Xi$ = grand canonical partition function for the guest-host ensemble (5.5)

Subscripts and Superscripts

H = hydrate
i = type of cavity (for each structure i=1,2)
I = ice
J = type of guest molecule ( $1 \leq J \leq M$)
M = total number of possible guest components in the mixture
MT = property of the empty hydrate crystal
o = standard state

3) There are no interactions of the solute molecules, i.e. the energy of an encased guest molecule is independent of the number and types of other solute molecules.
4) No quantum effects are needed; classical statistics are valid.

As a convenient starting point for the model, the grand canonical partition function is developed from the canonical partition function, to incorporate the above assumptions. The canonical partition function is written as the product of three factors, as follows:

1) the exponential of the empty lattice Helmholtz free energy divided by kT, where k is Boltzmann's constant,

$$\exp(-A^{MT}/kT) \tag{5.1}$$

2) the number of ways to distribute the indistinguishable guest molecules of type J in distinguishable cavities of type i, with a upper limit of 1 guest per cavity.

If we choose only one species of guest ($N_{J_i}=N_{1_i}$) and define L as the number of boxes, then we obtain the mathematical permutation formula for the number of ways $N_{1_i}$ indistinguishable objects can be placed in L distinguishable boxes with at most one object per box

$$\frac{L!}{(L - N_{1_i})!N_{1_i}!} \tag{5.2a}$$

In our case however, we have $\nu_i N_W$ distinguishable boxes of type i and we wish to distribute $N_{J_i}$ indistinguishable objects, with no more than one object per box, so Equation (5.2a) is modified to obtain the second factor in the canonical partition function as:

$$\frac{(\nu_i N_W)!}{(\nu_i N_W - \sum_J N_{J_i})! \prod_J N_{J_i}!} \tag{5.2}$$

3) The product of all individual particle partition functions, $q_{J_i}$, raised to the number of J particles in a type i cavity, $N_{J_i}$.

For the third factor, the analogy is to an ideal gas mixture of $N_1$ molecules of type 1 and $N_2$ molecules of type 2, so that the canonical partition function for the ideal gas mixture is

$$Q(N_1, N_2, V, T) = \frac{q_1^{N_1}\, q_2^{N_2}}{N_1!\, N_2!} \tag{5.3a}$$

where the factorial product in the denominator accounts for the indistinguishability of the molecules. In the clathrate, however, the molecules are distinguished by the cage they are in, eliminating any analogy with $N_1!N_2!$ in the denominator. Thus the third product term in the canonical partition function becomes:

$$\prod_J q_{J_i}^{N_{Ji}} \tag{5.3}$$

Multiplying all three factors (5.1), (5.2), and (5.3) together over the type i cavities van der Waals and Platteeuw obtained the canonical partition function as:

$$Q = \exp\left(\frac{-A^{MT}}{kT}\right) \prod_i \left[\frac{(\nu_i N_W)!}{(\nu_i N_W - \sum_J N_{J_i})!\, \prod_J N_{J_i}!} \prod_J q_{J_i}^{N_{Ji}}\right] \tag{5.4}$$

In order to obtain the grand canonical partition function $\Xi$ from the canonical partition function Q, we use the standard transformation

$$\Xi = \sum_N Q\, e^{\mu N/kT} \tag{5.5a}$$

and since the chemical potential $\mu$ is related to the absolute activity $\lambda$ by

$$\mu = kT \,\ell n\, \lambda \quad \text{or} \quad \lambda = e^{\mu/kT} \tag{5.5b}$$

then we can multiply (5.4) by

$$\lambda_A^{N_{A_1}} \lambda_A^{N_{A_2}} \cdots \lambda_B^{N_{B_1}} \lambda_B^{N_{B_2}} \cdots \lambda_M^{N_{M_1}} = \prod_i \prod_J \lambda_J^{N_{J_i}}$$

and sum over all values of $N_{J_i}$, to obtain the grand canonical partition function

$$\Xi = \exp\left(\frac{-A^{MT}}{kT}\right) \sum_{N_{J_i}} \prod_i \left\{ \frac{(\nu_i N_W)!}{(\nu_i N_W - \sum_J N_{Ji})! \prod_J N_{J_i}!} \prod_J q_{Ji}^{N_{Ji}} \lambda_J^{N_{Ji}} \right\} \tag{5.5}$$

Consider the summation term in Equation (5.5) for one type of cavity (i = 1) and for two types of guests (J = A, B), e.g mixtures of $C_3H_8$+i-$C_4H_{10}$ in $5^{12}6^4$ cavities:

$$\sum_{N_A} \sum_{N_B} \frac{(\nu_1 N_W)!}{(\nu_1 N_W - N_A - N_B)!\, N_A!\, N_B!} q_A^{N_A} q_B^{N_B} \lambda_A^{N_A} \lambda_B^{N_B} (1)^{(\nu_1 N_W - N_A - N_B)} \tag{5.5a}$$

with the final unity factor as the partition function for the empty cavity. Note also that the empty cavity has

$$\mu^{MT} = 0, \qquad \therefore \quad \lambda^{MT} = 1.$$

By the mathematical multinomial theorem, we know that

$$(x_1 + x_2 + \cdots + x_m)^N = \sum_{N=\sum_1^m n_i} \frac{N!}{n_1!\, n_2! \cdots n_m!} x_1^{n_1} x_2^{n_2} \cdots x_m^{n_m} \tag{5.6}$$

When we consider the analogy between the right hand side of Equation (5.5a) and (5.6) we obtain, from (5.5a)

$$(1 + q_A\lambda_A + q_B\lambda_B)^{\nu_1 N_W} \tag{5.5b}$$

So we have a product of terms, such as in Equation (5.5b), one for each cavity type i. Equation (5.5) then becomes simplified to its final form

$$\Xi = \exp\left(\frac{-A^{MT}}{kT}\right) \prod_i \left(1 + \sum_J q_{J_i} \lambda_J\right)^{\nu_i N_W} \tag{5.7}$$

### 5.1.2 Deriving the Chemical Potential of Water

Note that, while Equation (5.7) is the grand partition function with respect to the solute (guest) molecule, it appears to be the ordinary partition function with respect to the solvent (host-water) because $\lambda^{MT} = 1$, so that we have

$$\Xi^{combined} = Q^{host}\,\Xi^{guest}$$

$$\therefore \quad kT\,\ell n\,\Xi^{combined} = kT\,\ell n\,Q^{host} + kT\,\ell n\,\Xi^{guest} \tag{5.8}$$

Now, using the letter "h" to denote the host, and "g" to denote the guest property, each of the partition functions in Equation (5.8) can be related to their thermodynamic potentials in the usual way (see McQuarrie, 1976) as

$$d\,(kT\,\ell n\,Q^h) = -dA^h = S^h dT + PdV^h - \mu_W^h\,dN_W \tag{5.9}$$

and

$$d\,(kT\,\ell n\,\Xi^g) = d(PV^g) = S^g dT + PdV^g + \sum N_J d\mu_J \tag{5.10}$$

Since entropy and volume are extensive properties, they can be combined,

$$S = S^g + S^h \qquad \text{and} \qquad V = V^g + V^h$$

so that when we add Equations (5.9) and (5.10) we get a revised form of Equation (5.8) as

$$d(kT\,\ell \eta\,\Xi^{combined}) = SdT + PdV + \sum_J N_J d\mu_J - \mu_W dN_W \tag{5.11}$$

By dropping the superscript "combined", taking the leftmost derivative, and using $d\mu_J = kT\,d\,\ell n\,\lambda_J$ we get:

$$kT\,d\ell n\,\Xi = (-k\ell n\,\Xi + S)dT + PdV + \sum_J kTN_J\,d\ell n\,\lambda_J - \mu_W^H dN_W \tag{5.12}$$

With the development of Equation (5.12) for the relationship between the partition function and the macroscopic properties, all of the macroscopic thermodynamic properties may be derived from Equation (5.7). For example, differentiating $\ell n\ \Xi$ with respect to the absolute activity $\lambda$ of k, provides the sum of guest molecules "k" over all the cavities i

$$N_k = \sum_i N_{k_i} = \lambda_k \left( \frac{\partial \ell n\ \Xi}{\partial\ \lambda_k} \right)_{T,V,N_W,\lambda_{J \neq k}} \tag{5.13}$$

Equation (5.13) may be applied to the logarithm of Equation (5.7), which is:

$$\ell n\ \Xi = -\frac{A^{MT}(T,V,N_W)}{RT} + \sum_i \nu_i N_W\ \ell n\ \left( 1 + \sum_J q_{J_i} \lambda_J \right) \tag{5.14}$$

to yield the total number of guest molecules $N_k$ as

$$N_k = \sum_i \left( \frac{\nu_i N_W q_{k_i} \lambda_k}{(1 + \sum_J q_{J_i} \lambda_J)} \right) \tag{5.15}$$

Since $N_k$ must be a linear, homogeneous function of $\nu_i$, the number of cavities of different types, it follows from Equation (5.15) that

$$N_{k_i} = \frac{\nu_i N_W q_{k_i} \lambda_k}{(1 + \sum_J q_{J_i} \lambda_J)} \tag{5.16}$$

Equation (5.16) may be used to determine the simple probability ($y_{k_i}$) of finding a molecule of type k in a cavity of type i. This value may be obtained by dividing the number of molecules of k in such a cavity by the total number of cavities of type i, $\nu_i N_W$

$$\boxed{y_{k_i} = \frac{N_{k_i}}{\nu_i N_W} = \frac{q_{k_i} \lambda_k}{(1 + \sum_J q_{J_i} \lambda_J)}} \tag{5.17}$$

The chemical potential of the solvent $\mu_W^H$ may also be obtained from Equation (5.12) as

$$\frac{\mu_W^H}{kT} = -\left(\frac{\partial \ell n\ \Xi}{\partial N_W}\right)_{T,V,\lambda_J}$$

so that, from (5.14)

$$\boxed{\frac{\mu_W^H}{kT} = \frac{\mu_W^{MT}}{kT} - \sum_i \nu_i\ \ell n\ (1 + \sum_J q_{J_i}\lambda_J)} \qquad (5.18)$$

Equations (5.17) and (5.18) are important in that they enable the determination of the hydrate composition and the chemical potential of the hydrated water as a function of variables $(T,V,N_W,\lambda_1,\cdots\lambda_M)$. Equation (5.17) may be simplified somewhat by finding expressions for the absolute activity $\lambda_k$ and the individual particle partition function $q_{k_i}$ in terms of an experimentally measured or fitted parameters. To achieve such a simplification, we first consider the chemical potential of an ideal gas and its relation to the particle partition function.

For an ideal gas, the canonical partition function Q may be written as:

$$Q = \frac{1}{N!}\ q^N$$

and the ideal gas chemical potential $\mu$ is calculated by:

$$\mu = -kT\left(\frac{\partial \ell nQ}{\partial N}\right)_{T,V} = -kT\ \ell n\ \frac{q}{N}$$

where separability may be assumed for the individual particle partition function q into a translational part and a second part containing the internal modes of energy, i.e. $q = q_{trans}q_{int}$ with:

$$\frac{q_{trans}}{N} = \left(\frac{2\pi mkT}{h^2}\right)^{3/2}\frac{V}{N}$$

where the square root of the quantity in parentheses is called the mean thermal De Broglie wavelength. For an ideal gas V/N = kT/P, therefore

$$\mu = -\,kT\,\ell n\left[\left(\frac{2\pi mkT}{h^2}\right)^{3/2} kT\right] - kT\,\ell n\, q_{int} + kT\,\ell n\, P \qquad (5.19)$$

and since the chemical potential is normally defined in reference to a standard chemical potential $\mu^0$ as:

$$\mu = \mu^0(T) + kT\,\ell n\, P \qquad (5.20)$$

the identity can be made between the standard chemical potential with the first two terms on the right of Equation (5.19) as:

$$\mu^0 = -kT\,\ell n\left[\left(\frac{2\pi mkT}{h^2}\right)^{3/2} kT\right] - kT\,\ell n\, q_{int}$$

Without the standard chemical potential, Equation (5.19) becomes in terms of absolute chemical potential ($\mu = kT\,\ell n\,\lambda$),

$$\mu = kT\,\ell n \frac{P}{kT\left(\frac{2\pi mkT}{h^2}\right)^{3/2} q_{int}}$$

or for the absolute activity $\lambda$

$$\lambda = \frac{P}{kT\left(\frac{2\pi mkT}{h^2}\right)^{3/2} q_{int}}$$

The absolute activity and the individual particle partition function may be taken into account through a constant $C_{k_i}$, defined as

$$C_{k_i} \equiv \frac{q_{k_i}\lambda_k}{P_k} = \frac{q_{k_i}}{kT\left(\frac{2\pi mkT}{h^2}\right)^{3/2} q_{int}} \qquad (5.21)$$

Note that Equation (5.21) contains in the denominator the internal portion of the particle partition function and the ideal gas contribution, so that the division indicated accounts for the non-ideal gas effect. When Equation (5.21) is put into Equation (5.17), $y_{k_i}$, the fractional filling of cavity i by a type k molecule, is obtained:

$$y_{k_i} = \frac{C_{k_i} P_k}{1 + \sum_J C_{J_i} P_J} \tag{5.22}$$

For an ideal gas Equation (5.22) may be considered as the elementary probability of cavity i occupation by molecule k. This equation is one of the most useful equations in the method of hydrate prediction. It may be recognized as the Langmuir isotherm. If the equation were written for one guest component k, it would contain the Langmuir constant $C_{k_i}$ as the only unknown. Equation (5.21) shows that the Langmuir constant is a direct function of the particle partition function within the cavity $q_{k_i}$; in particular $C_{k_i}$ contains the non-ideal gas translation term.

When the fluid in equilibrium with the hydrate is a non-ideal gas, the pressure of component k in Equation (5.22) is replaced with its fugacity, $f_k$. Thus corrected, the equation finds many uses in the calculation of hydrate properties. The equation relies on the fitting of the Langmuir constant $C_{k_i}$ to experimental hydrate conditions. The method of relating the Langmuir constant to experimental conditions is given in Section 5.1.4.

Equation (5.22) enables the calculation of the chemical potential of water in the hydrate as a function of the fractional occupation in the cavities. Equation (5.18) derives the chemical potential of water in terms of that for the empty hydrate, as well as the product of the individual cavity partition function and the absolute activity:

$$\frac{\mu_W^H}{kT} = \frac{\mu_W^{MT}}{kT} - \sum_i \nu_i \; \ell n \; ( \; 1 + \sum_J q_{J_i} \lambda_J \; ) \qquad (5.18)$$

Since the final product is $q_{k_i}\lambda_k = C_{k_i}P_k$ by Equation (5.21) then

$$\frac{\mu_W^H}{kT} = \frac{\mu_W^{MT}}{kT} - \sum_i \nu_i \; \ell n \; ( \; 1 + \sum_k C_{k_i} P_k \; ) \qquad (5.18a)$$

Now consider the logarithmic term in Equation (5.18a). It may be simplified through the use of Equation (5.22), which relates $y_{k_i}$, the fractional occupation of a cavity of type i by a molecule of type k, to the Langmuir constant

$$\ell n \; ( \; 1 - \sum_k y_{k_i} \; ) = \ell n \left( 1 - \sum_k \frac{C_{k_i} P_k}{1 + \sum_k C_{k_i} P_k} \right)$$

or

$$\ell n \; ( \; 1 - \sum_k y_{k_i} \; ) = \ell n \left( \frac{1}{1 + \sum_k C_{k_i} P_k} \right)$$

Substitution of the above equation into (5.18a) yields:

$$\boxed{\mu_W^H = \mu_W^{MT} + kT \sum_i \nu_i \; \ell n \; ( \; 1 - \sum_k y_{k_i} )} \qquad (5.23)$$

Equation (5.23) may be used with Equation (5.22) to determine the chemical potential of water in hydrate $\mu_W^H$, which is one of the major contributions of the model. The combination of these two equations is of vital importance to phase equilibrium calculations, since the method equates the chemical potential of a component in different phases, at constant temperature and pressure.

Equation (5.23) shows that the fractional filling of cavities reduces the chemical potential of the water in the hydrate, thereby making the hydrate more thermodynamically stable. The higher the percentage filling of the cavities, the lower the value of $\mu_W^H$ and the more stable the hydrate becomes, since logarithms of small fractions $(1-\Sigma y_{ki})$ can be large negative numbers.

If the value of $y_{k_i}$ were uncharacteristically small in Equation (5.23), the final term could be replaced by $-kT\Sigma\Sigma\nu_i y_{k_i}$. The resulting equation is van't Hoff's law:

$$\mu_W^H = \mu_W^{MT} - kT \sum_i \sum_k \nu_i \, y_{k_i} \tag{5.23a}$$

which is used in other applications to model ideal, dilute, solid solutions (Lewis and Randall, 1923, p. 238). Thus the above limiting form of Equation (5.23) lends mathematical credence to the hydrate solid solution model, used here and in Katz's distribution coefficient method of Section 4.3.2. Equation (5.23a) clearly shows that hydrates are stabilized by lowering the chemical potential by an increase in $y_{k_i}$, the cavity occupation.

**Example 5.1: Determination of Hydrate Non-Stoichiometry**

Cady (1983) provided the illustration below of how Equations (5.22) and (5.23) may be used to determine the hydrate number for seven simple hydrates of structure I.

From Chapter 2, recall that structure I has 46 water molecules in the basic crystal with 6 large ($L \equiv 5^{12}6^2$) cavities and 2 small ($S \equiv 5^{12}$) cavities. For ideal hydrates, if all of the cavities were filled the hydrate number (water molecules per guest molecule) would be $n = 46/8$. For simple hydrates, the hydrate number is related to the fractional filling of the large and small cavities, $y_L$ and $y_S$, respectively as:

$$n = \frac{46}{6y_L + 2y_S} = \frac{23}{3y_L + y_S} \qquad \text{(E5.1-1)}$$

where Equation (5.22) applied to each size cavity gives

$$y_L = \frac{C_L P}{1 + C_L P} \quad \text{(E5.1-2a)} \quad \text{and} \quad y_S = \frac{C_S P}{1 + C_S P} \qquad \text{(E5.1-2b)}$$

Now if the formation pressure is considered for a simple hydrate at 273.15 K, we may obtain from Equation (5.23), with Davidson's (1973) suggestion that $\mu_W^H - \mu_W^{MT} = -1108$ J/gmol

$$-0.4885 = \frac{3}{23}\,\ell n(1-y_L) + \frac{1}{23}\,\ell n(1-y_S) \qquad \text{(E5.1-3)}$$

Cady suggested that a value of n may be estimated at a pressure corresponding to 273.15 K for a hydrated guest, using De Forcrand's method with the Clapeyron equation (Section 4.7). Equations (E5.1-1) and (E5.1-3) may then be solved simultaneously for $y_L$ and $y_S$. In turn, $y_L$ and $y_S$ may be substituted into Equations (E5.1-2) to calculate $C_L$ and $C_S$. Since the values of $C_L$ and $C_S$ are constant at constant temperature, they may be used to determine the hydrate numbers at pressures higher than the equilibrium value, using Equations (E5.1-1) and (E5.1-2).

Using the above method, Cady compared the calculated hydrate numbers to his experimental values, with results as shown in Figure 5-1. The fit of the equations to the experiment seems to be remarkably good, with the exception of Davy's original hydrate, chlorine.

---

### 5.1.3 The Langmuir Adsorption Analogy

In single component Langmuir adsorption, one finds many analogies to the process of guest encapsulation in the hydrate cavity, which provide a physical interpreta-

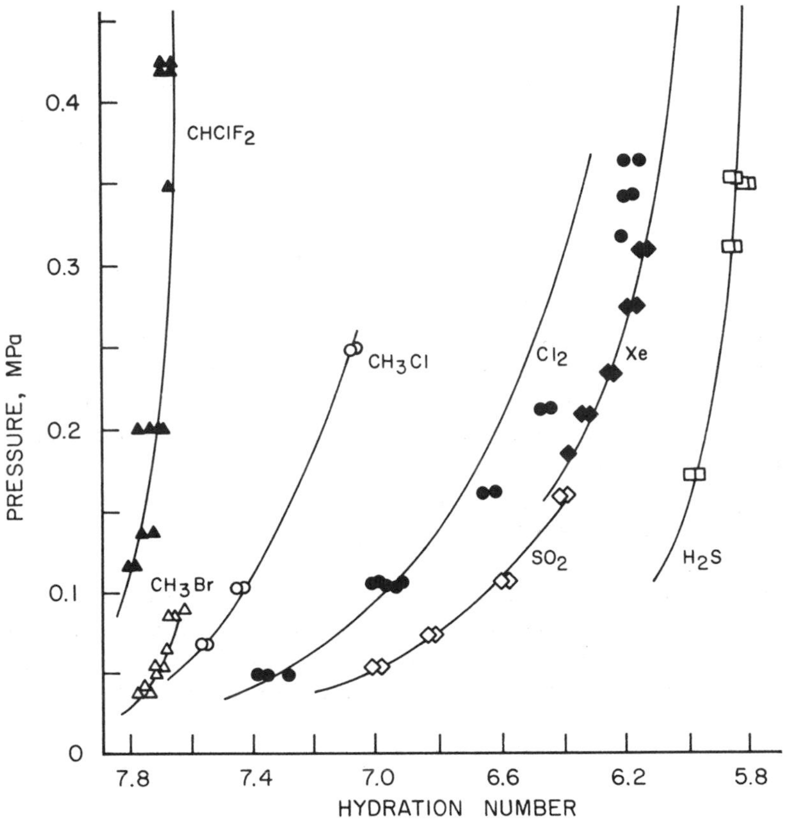


5-1 Change of Hydration Number with Pressure
(Reproduced, with permission, from the Journal of Physical Chemistry (Cady, 1983a), by the American Chemical Society.)

tion of the guest containment. In the below assumptions of the single component Langmuir adsorption isotherm, the analogies are readily apparent through a replacement of the words "adsorption or desorption" with "enclathration or declathration", and the word "sites" by "cavities":

1. The adsorption of gas molecules occurs at discrete sites on the surface.
2. The energy of adsorption on the surface is independent of the presence of other adsorbed molecules.

3. The maximum amount of adsorption corresponds to a monolayer, or one molecule per site.
4. The adsorption is localized and occurs by collision of gas phase molecules with vacant sites.
5. The desorption rate depends only on the amount of material on the surface.

To illustrate the analogy more clearly, it is necessary to consider the derivation of the Langmuir adsorption isotherm. We can incorporate the above assumptions into an equilibrium expression which equates the rate of adsorption $r_{ads}$ to that of desorption $r_{des}$ of gas molecules of type J. The desorption rate is directly proportional to the fraction of monolayer sites occupied $\theta_J$, and is expressed as

$$r_{des} = \varkappa' \theta_J$$

where $\varkappa'$ is taken as the proportionality constant, sometimes called the desorption rate constant. The rate of adsorption is proportional to the product of the gas pressure $P_k$ and the number of unoccupied sites $(1-\theta_J)$ in the equation

$$r_{ads} = \varkappa\ P_J(1-\theta_J)$$

At equilibrium, when the above rates of adsorption and desorption are equated, an expression is obtained for the fraction of sites occupied $\theta_J$, which appears identical to Equation (5.22) for simple hydrates of one component:

$$\theta_J = \frac{\varkappa P_J}{\varkappa' + \varkappa P_J} = \frac{KP_J}{1 + KP_J} \qquad (5.22a)$$

where the equilibrium Langmuir adsorption constant ($K \equiv \varkappa/\varkappa'$) is analogous to the Langmuir hydrate constant $C_{J_i}$ in Equation (5.22). Similarly, the fraction of the adsorbed monolayer $\theta_J$ is analogous to the fractional occupation of the cavities of type i, $y_{k_i}$. Finally the Langmuir constants for both adsorption and enclathration are only functions of temperature for each molecule type retained at the individual site (or cavity).

### 5.1.4 Relating the Langmuir Constant to Cell Potential Parameters

With the adsorption analogy in mind, our next purpose is to relate the Langmuir constants $C_{k_i}$ to experimental variables, thereby providing another link between these mathematics and the physical world. In order to consider the Langmuir constant, it is first necessary to determine the individual particle energy potential within the cavity. We must first make two assumptions, which are in addition to the four assumptions made for the hydrate in Section 5.1.1:

5) The internal motion partition function of the guest molecules is the same as that of an ideal gas. That is, the rotational, vibrational, nuclear and electronic energies are not affected by enclathration, as supported by spectroscopic results summarized by Davidson (1971) and Davidson and Ripmeester (1984).

6) The potential energy of a guest molecule at a distance r from the cavity center is given by the spherically symmetrical potential $\mathcal{W}(r)$ proposed by Lennard-Jones and Devonshire (1937, 1938).

Assumptions (5) and (6) are more restrictive than (1) through (4), in that they apply more to monatomic or spherical molecules than for oblate or polar molecules. The less accurate predictions of the model for certain gases may be related to inaccuracies of these assumptions.

In this model the interactions of the guest with the nearest neighbor $z_i$ water molecules of a spherical cage are summed in a pairwise manner. The model obtains a function $\mathcal{W}(r)$ describing the resulting field, averaged over all orientations of the molecules within the cavity. The fundamental intermolecular potential between a water molecule of the cavity wall and a solute molecule may be described by a number of pair potentials. The original work used the Lennard-Jones 6-12 pair potential. McKoy and Sinanoğlu (1963) suggested that the Kihara (1951) core potential was better for both larger and non-spherical molecules; this is the potential commonly used, with parameters fitted to experimental hydrate dissociation data. However, it should be pointed out that the equations presented below are for a spherical core.

The pair potential energy $\Phi$ between the guest molecule and any water molecule is related to the force F each exerts on the other by $F = -\partial\Phi/\partial\mathcal{R}$, where $\mathcal{R}$ is the molecular center distance between the two. The potential, which itself is a function of separation distance, is unique to every molecular type and is given by:

$$\Phi(\mathcal{R}) = \infty \qquad \text{for } \mathcal{R} \le 2a \qquad (5.24a)$$

$$\Phi(\mathcal{R}) = 4\varepsilon\left\{\left(\frac{\sigma}{\mathcal{R}-2a}\right)^{12} - \left(\frac{\sigma}{\mathcal{R}-2a}\right)^{6}\right\} \qquad \text{for } \mathcal{R} > 2a \qquad (5.24b)$$

where $\sigma$ = distance between cores at zero potential energy ($\Phi$=0), where attraction and repulsion balance,
$a$ = radius of the spherical core, and
$\varepsilon$ = maximum attractive potential (at $\mathcal{R} = \sqrt[6]{2}\,\sigma$).

The Lennard-Jones Devonshire theory (as summarized by Fowler and Guggenheim, 1952, pp. 336 ff.) averaged the pair potentials of (5.24a,b) between the solute and each water, for the $z_i$ molecules in the surface of the spherical cavity to obtain a cell potential $\mathcal{W}(r)$ of:

$$\mathcal{W}(r) = 2z\varepsilon\left[\frac{\sigma^{12}}{R^{11}r}\left(\delta^{10} + \frac{a}{R}\delta^{11}\right) - \frac{\sigma^{6}}{R^{5}r}\left(\delta^{4} + \frac{a}{R}\delta^{5}\right)\right] \qquad (5.25a)$$

where:

$$\delta^{N} = \frac{1}{N}\left[\left(1 - \frac{r}{R} - \frac{a}{R}\right)^{-N} - \left(1 + \frac{r}{R} - \frac{a}{R}\right)^{-N}\right] \qquad (5.25b)$$

where N = 4, 5, 10 or 11, indicated in Equation (5.25a),
z = the coordination number of the cavity
R = the free cavity radius[1], and
r = distance of the guest molecule from the cavity center.

---

[1] Values given in Table 2-1, minus 1.45Å for the free cavity radius of water to obtain R

The averaging of the potential function over all the angles of interaction with the wall enables the potential of Equation (5.25) to be expressed solely in terms of distance r from the cavity center for a given guest molecule. It should be noted that the parameters $\varepsilon$, a, and $\sigma$ are unique to every guest molecule, but they do not change in the different cavity types. On the other hand, the parameters z and R have been uniquely determined for each type cavity by X-ray diffraction data (see Chapter 2) and do not change as a function of guest molecules.

Following van der Waals and Platteeuw (1959, pp. 26ff.) the individual particle partition function is related to the product of three factors: (a) the cube of the De Broglie wavelength, (b) the internal partition function, and (c) the configurational triple integral, as:

$$q_{ki} = \left(\frac{2\pi mkT}{h^2}\right)^{3/2} q_{int} \int_0^{2\pi} \int_0^{\pi} \int_0^{R} \exp\left(-\frac{\omega(r)}{kT}\right) r^2 \sin\theta \, dr \, d\theta \, d\phi \tag{5.26}$$

The cavities are assumed to be spherically symmetric, which enables the elimination of the two angular portions of the triple integral, resulting in $4\pi$. The substitution of the resulting equation in Equation (5.21) yields the final expression for the Langmuir constant in terms of the particle potential within the cavity.

$$C_{k_i} = \frac{4\pi}{kT} \int_0^{R} \exp\left(-\frac{\omega(r)}{kT}\right) r^2 \, dr \tag{5.27}$$

The evaluation of the Langmuir constant may then be determined from a minimum of experimentally fitted Kihara parameters via an integration over the cavity radius. Equation (5.27) shows the Langmuir constant to be only a function of temperature for a given component within a given cavity. The experimentally fitted hydrate guest Kihara parameters in the cavity potential $\omega(r)$ of Equation (5.25) are not congruent with those found from second virial coefficients or viscosity data for several reasons, two of which are listed here. First, the Kihara potential itself does not adequately fit pure water virials over a wide range of temperature and pressure, and thus will not be adequate for water-hydrocarbon mixtures. Secondly, with the spherical Lennard-Jones Devonshire theory the pointwise potential of water molecules is "smeared" to yield an averaged spherical shell potential, which causes the water parameters to become indistinct. As a result, the Kihara parameters for the guest within the cavity are fitted to hydrate formation properties for each component.

A typical potential $\mathcal{W}(r)$ is shown in Figure 5-2. Note that the potential is more negative (high attrac-

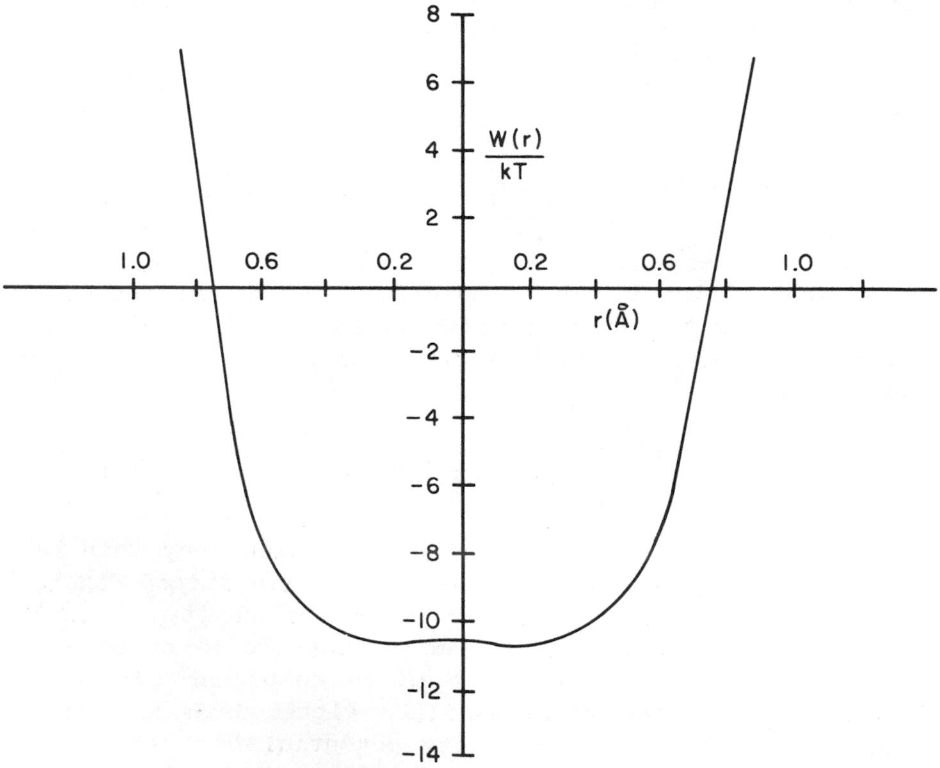


5-2 Typical Spherically Symmetrical Cavity Potential Function Between Guest and Cell. (Reproduced, with permission, from the Journal of Chemical Physics (McKoy and Sinanaoglu, 1963), by the American Institute of Physics.)

tion) in the center of a cell, or at some distance from the cell wall, with high repulsion (positive values) at the cell wall. As the guest molecule approaches one wall of the cavity, it is both repulsed by that wall and attracted by the opposite wall, causing it to exist in the center as much as possible. Equation (5.27) then shows that the Langmuir constants get their dominant contributions from the interior of the cells. This is a partial explanation of the fact that Assumption (6) for "smeared water molecules" works so well as an approximation for the cavity shell. This approximation is examined further in Section 5.1.7.

The value of fitting the Langmuir constants to simple hydrate formation data is in the prediction of mixture hydrate formation. In principle, if formation data for the simple hydrates can be adequately fitted, then the hydrate formation of mixtures of those guest components can be predicted with no adjustable parameters. Since there are only seven simple hydrate formers of natural gas, but an infinite variety of mixtures, such an advantage represents a tremendous saving of time and effort.

### 5.1.5 Fitting Three-Phase Simple Hydrate Formation

For the convenience of the reader, Table 5-2 contains the equations of the former sections which are used in this discussion. The fit of Kihara potential parameters for simple hydrates is accomplished having a value for $\Delta\mu_W'(\equiv \mu_W^{MT} - \mu_W^H)$, the difference of the chemical potential of water in Equation (5.23) in the empty and in the filled hydrate at each hydrate formation point. Combining Equation (5.23) and (5.22) with a value for the vapor fugacity of the guest molecule enables the determination of $C_{k_i}$. With multiple values of $C_{k_i}$ (one at each experimentally measured three-phase condition for the hydrate), an optimum set of Kihara potential parameters ($\varepsilon$, $\sigma$, and a) may be obtained through Equations (5.24), (5.25) and (5.27) for each simple hydrate. Normally the "a" values are set equal to those obtained from virial coefficient data, while the values of "$\varepsilon$" and "$\sigma$" are op-

**Table 5-2**
**Summary of the Equations of Subsection 5.1.4**
**For Use in Fitting Simple Hydrate Formers**

$$y_{k_i} = \frac{C_{k_i} f_k}{1 + \sum_J C_{J_i} f_J} \tag{5.22}$$

$$\mu_W^H = \mu_W^{MT} + kT \sum_i \nu_i \, \ell n \left( 1 - \sum_k y_{k_i} \right) \tag{5.23}$$

$$\Phi(\mathcal{R}) = \infty \qquad \text{for } \mathcal{R} \le 2a \tag{5.24a}$$

$$\Phi(\mathcal{R}) = 4\varepsilon \left\{ \left( \frac{\sigma}{\mathcal{R}-2a} \right)^{12} - \left( \frac{\sigma}{\mathcal{R}-2a} \right)^{6} \right\} \qquad \text{for } \mathcal{R} > 2a \tag{5.24b}$$

$$\mathcal{W}(r) = 2z\varepsilon \left[ \frac{\sigma^{1}}{R^{11} r} \left( \delta^{10} + \frac{a}{R} \delta^{11} \right) - \frac{\sigma^{6}}{R^{5} r} \left( \delta^{4} + \frac{a}{R} \delta^{5} \right) \right] \tag{5.25a}$$

where:

$$\delta^N = \frac{1}{N} \left[ \left( 1 - \frac{r}{R} - \frac{a}{R} \right)^{-N} - \left( 1 + \frac{r}{R} - \frac{a}{R} \right)^{-N} \right] \tag{5.25b}$$

$$C_{k_i} = \frac{4\pi}{kT} \int_0^R \exp\left( - \frac{\mathcal{W}(r)}{kT} \right) r^2 \, dr \tag{5.27}$$

timized for the set of $C_{ki}$ fit the the hydrate experimental data. However, the entire method is predicated on the ability to determine the value for $\Delta\mu_W'$, defined as the difference between the chemical potential of water in the empty and in the filled hydrate lattice.

With the specification of the above prescription for determining the Kihara potential parameters, the remainder of this section concerns the determination of $\Delta\mu_W'$.

In order to determine $\Delta\mu_W'$, the normal phase equilibrium condition is followed by equating the chemical potential of water in the hydrate to that in the condensed water phase (either free water $\mu_W^L$ or ice $\mu_W^\alpha$).

$$\mu_W^H = \mu_W^L \quad (5.28a) \qquad \text{or} \qquad \mu_W^H = \mu_W^\alpha \quad (5.28b)$$

If the condensed water phase is assumed to be pure, then the appropriate chemical potential may be substituted for $\mu_W^H$ in Equation (5.23). The problem of determining $\Delta\mu_W'$ becomes the problem of determining $\Delta\mu_W$, defined as the chemical potential difference between the condensed water phase and the empty hydrate. As indicated below, the value of $\Delta\mu_W$ may be equated to macroscopic quantities calculated by integration of thermodynamic relations.

Saito, Marshall, and Kobayashi (1964) determined a method of obtaining $\Delta\mu_W$, so that potential parameters may be obtained as in the above paragraph. Since $\Delta\mu_W$ is a function of temperature and pressure only, changes are given by:

$$d\left(\frac{\Delta\mu_W}{RT}\right) = -\left(\frac{\Delta h_W}{RT^2}\right) dT + \left(\frac{\Delta v_W}{RT}\right) dP \qquad (5.29)$$

where $\Delta h_W$ and $\Delta v_W$ are the enthalpy and volume difference between ice or water and the empty hydrate. The fact that three-phase equilibrium for a binary system is univariant in either pressure or temperature enabled Saito et al. to provide a change of variable so that the pressure derivative could be replaced by the product of the slope of the experimentally determined equilibrium P-T curve times the temperature derivative, as:

$$dP = \left(\frac{dP}{dT}\right) dT$$

The combination of the above two equations yields an expression from classical thermodynamics for finding $\Delta\mu_W$ along the three-phase line, from some reference temperature $T_o$:

$$\left(\frac{\Delta\mu_W}{RT}\right) - \left(\frac{\Delta\mu_W}{RT}\right)_{T_o} = -\int_{T_o}^{T} \left(\frac{\Delta h_W}{RT^2}\right) dT + \int_{T_o}^{T} \left(\frac{\Delta v_W}{RT}\right)\left(\frac{dP}{dT}\right) dT \qquad (5.30)$$

The above equation applies for a pure condensed water phase, such as ice or liquid water without a solute. If the condensed water phase were not pure, the above equation should be modified to include a final term for the activity of water $\gamma_W x_W$ (or $x_W$ for very dilute solutions). In order to maintain the quantity $\Delta\mu_W$ as the difference between pure phases, the relation ($\mu_W = \mu_W^{pure} + RT \ln \gamma_W x_W$) is used to obtain an additional term on the right of Equation (5.30), so that:

$$\left(\frac{\Delta\mu_W}{RT}\right) - \left(\frac{\Delta\mu_W}{RT}\right)_{To} = -\int_{T_o}^{T} \left(\frac{\Delta h_W}{RT^2}\right) dT + \int_{T_o}^{T} \left(\frac{\Delta v_W}{RT}\right)\left(\frac{dP}{dT}\right) dT - \ln\gamma_W x_W \quad (5.31)$$

By using parameters for the reference values of $\Delta\mu_W$, $\Delta h_W$ and $\Delta v_W$ in Equation (5.31), Saito et al. determined potential parameters for methane, argon and nitrogen, via the method outlined in the first paragraph of this section. As an aside, argon and nitrogen have since been determined to be structure II simple hydrates, while Saito et al. fitted them as structure I hydrates.

Parrish and Prausnitz (1972) used a variation of Equation (5.31), which included a reference gas, for which the derivative dP/dT was obtained. A second pressure correction ($\Delta v_W[P-P_{ref}]/RT$) was needed to translate from the reference hydrate pressure to the pressure of interest. A third pressure term corrected the empty hydrate chemical potential from zero pressure to the dissociation pressure of the reference hydrate at $T_o$. Holder, et al. (1980) simplified the method considerably by direct integration of Equation (5.29) to obtain:

$$\frac{\Delta\mu_W}{RT} = \frac{\Delta\mu_W^o}{RT_o} - \int_{T_o}^{T} \frac{\Delta h_W}{RT^2} dT + \int_{0}^{P} \frac{\Delta v_W}{RT} dP - \ln \gamma_W x_W \quad (5.32)$$

Equation (5.32) has two advantages over previous equations used to determine $\Delta\mu_W$: (a) direct integration with pressure and temperature eliminates the need for the dP/dT term in (5.31) and the other two pressure corrections, and (b) the integration of $\Delta h_W$ with temperature can be done at low pressure so that only the correction in the third term on the right of (5.32) is necessary.

The concentration of the water $x_W$ is very close to unity, due to the low aqueous solubility of hydrate-forming components. The solubility of guest molecules k in the free water phase is included in the final term of Equation (5.32) using the corrected Henry's law constant expression of Krichevsky and Kasarnovsky (1935):

$$x_W = 1 - x_k = 1 - \frac{f_k}{H_{kW} \exp\left(\frac{P\bar{V}}{RT}\right)}. \tag{5.33}$$

The Henry's law constant $H_{kW}$ is fitted to solubility data as a function of temperature in the form

$$\ell n\ H_{kW} = H_{kW}^{(0)} - H_{kW}^{(1)}/T + H_{kW}^{(2)} \times \ell nT + H_{kW}^{(3)} \times T. \tag{5.34}$$

The parameters of Henry's constant equations such as (5.34) for solutes in water or in aqueous solutions of methanol are provided by Nguyen (1985), Anderson and Prausnitz (1986), or Holder et al.(1988). In Equation (5.32) the activity coefficients for the water phase $\gamma_W$ may normally be taken as unity without appreciable error, unless inhibitors are present.

Table 5-3 presents the other experimental parameters needed with Equation (5.32) to obtain the values of $\Delta\mu_W$ at values along the three-phase curve. The parameters included in this table represent only one self-consistent set of many parameters available (see for example the parameters of Table 5-5) which appear to be in evolutionary stages. Discussions of the determination of such parameters are found in the work of Dharmawardhana et al. (1980, 1981) and of Holder et al. (1984). In the table the values of $\Delta\mu_W^0$ and $\Delta v_W$ (taken as a constant $\Delta v_W^0$) may be used directly in Equation (5.32). The value for $\Delta h_W^0$ in the table should be corrected for temperature as:

$$\Delta h_W = \Delta h_W^o + \int_{T_{ref}}^{T} \Delta C_{p_W}\, dT \qquad (5.35a)$$

and $$\Delta C_{p_W} = \Delta C_{p_W}^{\ o} + a(T-273.15) \qquad (5.35b)$$

The determination of the above parameters enable the calculation of $\Delta\mu_W$, and consequently the calculation of the Kihara potential parameters from experimental three-phase hydrate formation data, as indicated in the initial portion of this section. Table 5-3 provides Kihara parameters for all of the natural gas hydrate formers, fitted with the fugacities of Equation (5.22) calculated from the Soave (1972) modification of the Redlich-Kwong equation-of-state. In Section 5.1.8 figures will be presented to illustrate the accuracy to which the simple hydrate three-phase data are fit with the method. With the fit of the simple hydrate parameters, the extension to the prediction of mixture hydrate formation data is straightforward.

### 5.1.6 Prediction of Mixture Hydrate Formation

Parrish and Prausnitz (1972) were the first to extend the statistical method to multicomponent systems. In the method for mixtures shown below, no new equations or parameters are introduced. The method for mixtures uses the previous fit of the pure component parameters along with the basic Equations (5.22), (5.23), and (5.32), listed in Table 5-4.

Calculation Algorithm for Three-Phase Equilibria:

The easiest method to comprehend is the three-phase ($L_W$-H-V or I-H-V) pressure calculation at a given temperature. The other three-phase calculations require variations of the scheme described below. In order to predict the three-phase ($L_W$-H-V) hydrate formation pressure at a given temperature the algorithm (with a flow diagram pre-

**Table 5-3**
**Fitted Parameters for Hydrate Formation**

| | Structure I | Structure II |
|---|---|---|
| $\Delta\mu_W^0$ (J/mol) | 1297 | 937 |
| $\Delta h_W^0$ (J/mol) | 1389 | 1025 |
| $\Delta v_W^0$ (cc/mol) | 3.0 | 3.4 |
| $\Delta v^{\alpha-L}$ (cc/mol) | 1.598 | |
| $\Delta C_p^{\alpha-L}$ (J/mol-K) | $= 38.12 - 0.0336\times(T-273.15)$ | |

| Component | $\varepsilon/k$ K | $\sigma$ Å | a Å |
|---|---|---|---|
| Methane | 154.54 | 3.1650 | .3834 |
| Ethane | 176.40 | 3.2641 | .5651 |
| Propane | 203.31 | 3.3093 | .6502 |
| i-Butane | 225.16 | 3.0822 | .8706 |
| n-Butane | 209.00 | 2.9125 | .9379 |
| $H_2S$ | 204.85 | 3.1530 | .3600 |
| $N_2$ | 125.15 | 3.0124 | .3526 |
| $CO_2$ | 168.77 | 2.9818 | .6805 |

| | $H_{kW}^{(0)}$ | $H_{kW}^{(1)}$ | $H_{kW}^{(2)}$ | $H_{kW}^{(3)}$ |
|---|---|---|---|---|
| Methane | -365.183 | 18,016.7 | 49.7554 | -0.000285 |
| Ethane | -533.392 | 26.565.0 | 74.6240 | -0.004573 |
| Propane | -628.866 | 31,638.4 | 88.0808 | 0.0 |
| i-Butane | 190.982 | -4,913.0 | -34.5102 | 0.0 |
| n-Butane | -639.209 | 32,785.7 | 89.1483 | 0.0 |
| $H_2S$ | -297.158 | 16,347.7 | 40.2024 | 0.002571 |
| $N_2$ | -327.850 | 16,757.6 | 42.8400 | 0.016765 |
| $CO_2$ | -317.658 | 17,371.2 | 43.0607 | -0.002191 |

Note $H_{kW}$ is in atmospheres and T in Kelvins.

sented in Figure 5-3) is followed for the calculation:

1. Using the simple hydrate Kihara parameters for each component in the gas, calculate the Langmuir constant for each component in each cavity of both hydrate structures, using a numerical integration of Equation (5.27), together with Equation (5.25).

2. Estimate a pressure for the three-phase condition at the temperature of interest. Convergence should occur over a fairly wide range of initial estimates for the pressure.

3. Calculate the gas phase fugacity of each component at the given temperature and estimated pressure, with an

**Table 5-4**
**Summary of Equations[1] Used for the Hydrate Prediction of Mixtures**

$$y_{k_i} = \frac{C_{k_i} f_k}{1 + \sum_J C_{J_i} f_J} \quad (5.22)$$

$$\mu_W^H = \mu_W^{MT} + kT \sum_i \nu_i \, \ell n \left( 1 - \sum_k y_{k_i} \right) \quad (5.23)$$

$$\frac{\Delta\mu_W}{RT} = \frac{\Delta\mu_W^0}{RT_0} - \int_{T_0}^{T} \frac{\Delta h_W}{RT^2} dT + \int_0^P \frac{\Delta v_W}{RT} dP - \ell n \, \gamma_W x_W \quad (5.32)$$

1. Note that equations in molecular units may be translated into mol units through use of the conversion factor $k = R/N_A$, where $N_A$ is Avogadro's number.

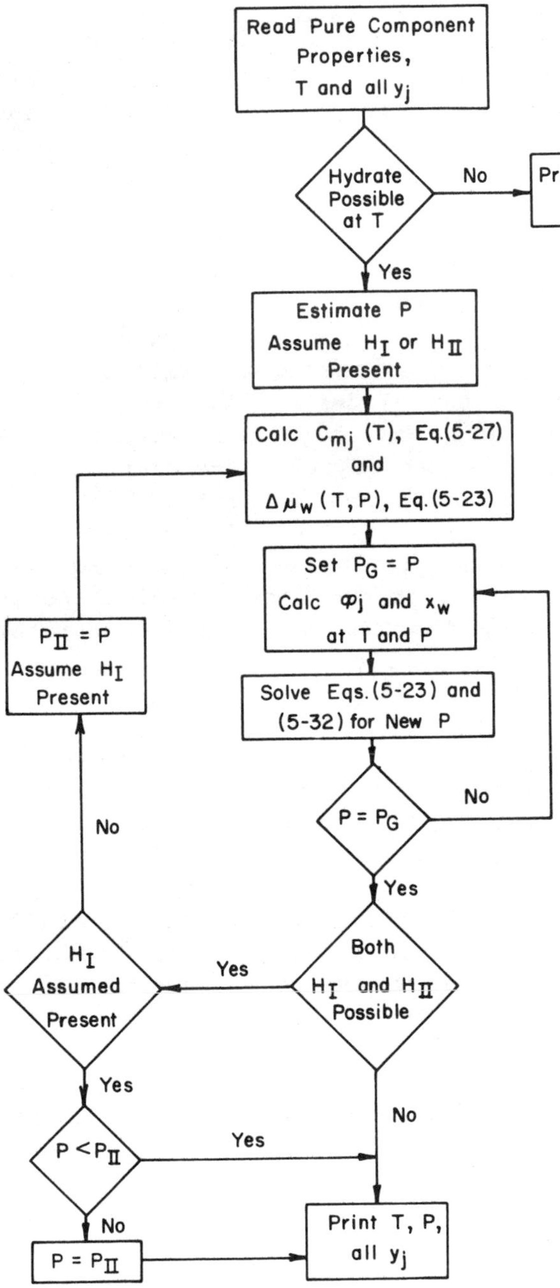


5-3 Flow Algorithm for Computer Code Construction of CSMHYD (Modified, with permission, from <u>Industrial and Engineering Chemistry Process Design and Development</u> (Parrish and Prausnitz, 1972), by the American Chemical Society.)

equation-of-state such as the Soave (1972) modification of the Redlich-Kwong equation, or the Peng-Robinson (1976) equation-of-state. The equations for the calculation of fugacity are discussed in Appendix A, because their inclusion here would cause an unwieldy digression.

4. Without natural gas components which form structure II as simple hydrates (nitrogen, propane, and iso-butane) only structure I hydrates can form, as discussed in Chapter 2. For most natural gases however, propane and heavier components are present, so structure II should be initially assumed to form. This assumption will be checked in the final step of the algorithm, but it initially requires the calculation of only the structure II Langmuir constants and the use of structure II properties such as $\Delta\mu_W^0$, $\Delta h_W^0$, $\Delta v_W^0$, and $\Delta C_p$ from Table 5-3.

5. Calculate the fractional cavity filling via Equation (5.22) and the first value for the water chemical potential difference using Equation (5.23) at the temperature and the estimated pressure in Step 2.

6. Determine the mol fraction of water in the free water phase using Equation (5.33). If necessary, calculate the activity coefficient for water mol fractions which differ significantly from unity. If the temperature is less than 273.15 K, the free water phase is assumed to be ice.

7. Integrate the quantities in Equation (5.32), including those in Equation (5.35) to obtain the second value for the chemical potential difference, for comparison with the first value obtained in Step 5.

8. If the values for the chemical potential differences in Steps 5 and 7 compare favorably, then the correct three-phase pressure has been calculated for the assumed hydrate structure. Otherwise iteration is required until the values compare within an acceptable tolerance.

9. After the above calculation for the formation of structure II hydrate, Steps 4 through 8 should be repeated assuming structure I to be the hydrate crystal formed. The crystal structure with the lower formation pressure

will be favored thermodynamically. In the unusual case of equality of the pressures of both structures, the two crystal structures may coexist, along with the vapor and condensed water phases.

---

While the calculation in the above algorithm appears straightforward in principle, several of the steps require the use of a digital computer. In particular, the numerical integration in Step 1 to obtain the Langmuir constants, the calculation of component fugacities in Step 3, and the correction scheme in Step 8 are especially difficult to perform by hand. The procedure requires one iteration for each structure which negates the use of hand calculation. As a consequence, this method is only performed via a digital computer. Only within the last few years did the capability of personal computers become sufficient to perform such calculations efficiently.

A computer program suitable for performing such a calculation has been compiled in the FORTRAN language, and is provided within the 5 $^1/_4$ inch floppy disk in the frontpapers. To access the program, an IBM-PC compatible machine with one floppy disk drive and 512K of RAM is needed. An 8087 numeric data coprocessor chip, while not a necessity, will speed the calculation by a factor of approximately three. After the computer is operational in the DOS system, simply mount the floppy disk in the computer and type "CSMHYD" (RETURN). Follow the instructions for the input of data. For a three-phase prediction of pressure at a given temperature, a type "1" calculation (or type "2" for the determination of three-phase temperature at a given pressure) is required. As a guide to the user, detailed examples of input data and program output are provided in Appendix B.

A Newton-Raphson convergence scheme is included within a secant convergence scheme to expedite and stabilize the iterative aspects of the calculation. This numerical method provides convergence for a wide range of initial estimates for the calculated variable (either pressure or temperature). If convergence difficulty occurs, the $K_i$-Value program discussed in Section 4.3.3 might be used to check the initial estimate for the calculated variable. As an alternative, the gas gravity method (Section 4.3.3) might be used for an estimate.

**Example 5.2: Hydrate azeotropes for ternary systems (two guest components + water)**

In vapor-liquid equilibria it is common to encounter azeotropes, in which case the vapor and liquid compositions are identical. The prediction of such vapor-liquid equilibrium azeotropes is normally considered a stringent test of prediction schemes (Holmes and van Winkle, 1970). There is a direct analogy to such azeotropes in vapor-hydrate equilibria in which the vapor and hydrate compositions are equal. Hydrate azeotropes were first recognized and measured by Platteeuw and van der Waals (1959c) for the system hydrogen sulfide + propane at 270 K. Other measurements of hydrate azeotropes have been made for $CH_4$+$C_2H_4$ by Snell, Otto, and Robinson (1961), $CH_4$+$C_3H_8$ by Verma (1974) and $CH_4$+$C_2H_6$ by Holder and Grigoriou (1980) and by Thakore and Holder (1987).

While vapor-liquid azeotropes are caused by non-ideality in the liquid phase, vapor-hydrate azeotropes are caused by molecules which optimally fit the cavity diameters. Based on the van der Waals and Platteeuw model of this chapter, Holder and Manganiello (1982) specified analytical conditions for the molecules and compositions of vapor-hydrate azeotropes which they determined to occur only for ternary systems. The primary requisite for an azeotrope is that guest molecules be chosen so that the value of $\sigma_i^*/d_i \approx 0.44$, where $d_i$ represents the cavity diameter (see Table 2-1), and $\sigma_i^*$ values were obtained from Parrish and Prausnitz (1972). This optimum azeotrope value is similar to (but more stringent than) the hydrate formation ratio requirement of guest:cavity diameter ratio discussed at length in Chapter 2.

Predictions of three of these binary azeotropes are given using the current theory in Figures 5-4 through 5-6. Because the two cavity dimensions of structure II differ substantially, two different sizes of guest components will cause a marked pressure decrease at the azeotrope. Figures 5-4 and 5-5 demonstrate this for structure II azeotropes for mixtures of large molecules with hydrogen sulfide. Hydrogen sulfide appears to be an optimal size ($\sigma_i^*/d_i = 0.444$) for the small cavity of structure II and provides azeotropes with large molecules which are characterized by shallow y-P curves. Hydrates formed by such ternary mixtures with hydrogen sulfide are stable over a wide range of $H_2S$ composition, due to azeotropes.

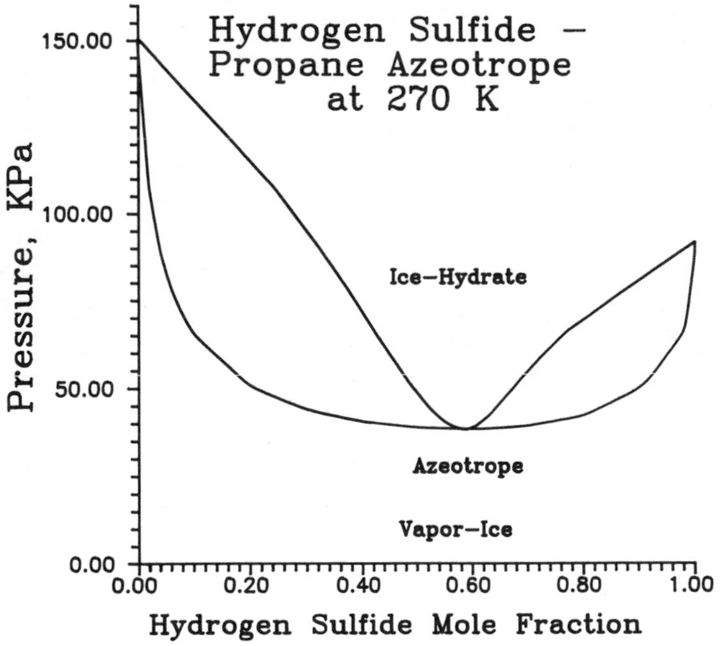


5-4 Predicted Azeotrope for Mixtures of Propane+Hydrogen Sulfide

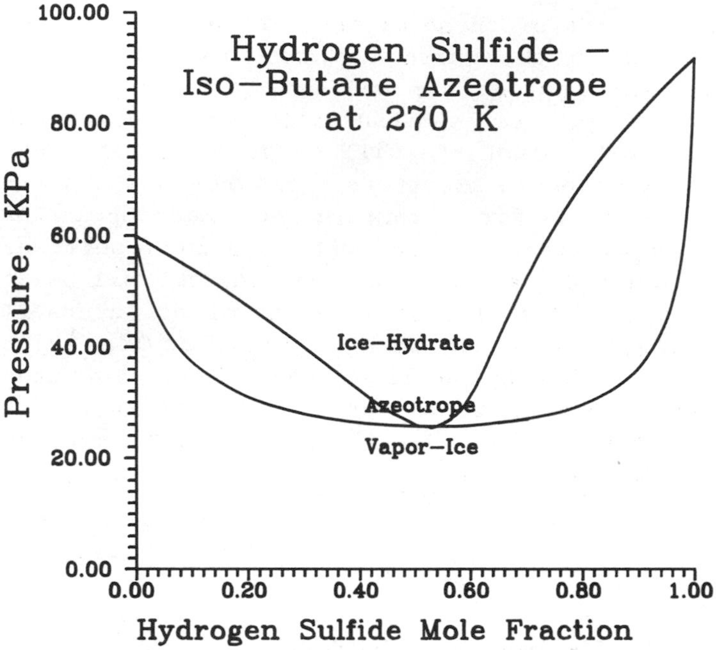


5-5Predicted Azeotrope for Mixtures of I-Butane+Hydrogen Sulfide

The shapes of the curves of Figures 5-4 and 5-5 provide an explanation of the effects empirically observed by early investigators, such as de Forcrand (1883), who used $H_2S$ extensively as an aid in forming hydrates with more difficult substances. Von Stackelberg and Müller (1954) suggested a separate "double hydrate" classification for hydrates of gas mixtures containing $H_2S$. Platteeuw and van der Waals (1959c) later showed that the unusual behavior was caused by azeotrope formation predicted by their theory, without need for a distinctive classification of $H_2S$ hydrates.

In contrast, the azeotropes for hydrates of methane with ethane or methane with propane, shown in Figure 5-6, are less well-defined than those of hydrogen sulfide. While the optimal size of the

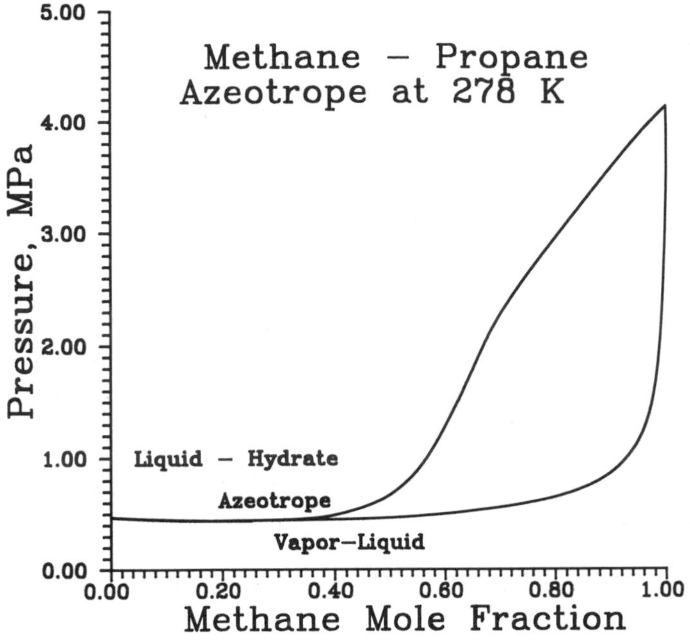


5-6 Predicted Azeotrope for Mixtures of Methane+Propane

molecule seems to represent the primary criteria for azeotrope formation, temperature appears to be a second-order effect on the appearance of an azeotrope. For "weak" azeotropes such as those with methane, temperature plays a vital role in determining azeotropic formation. For example Holder and Grigoriou (1980) indicated that azeotropes of 10%$CH_4$-90%$C_2H_6$ will form only above 281 K.

A final important point is to be seen in Figures 5-4 through 5-6. The right lobes of the curves in each figure demonstrate that, while only a small amount of the heavy component may be present in the vapor, that component is highly concentrated the hydrate phase at a given pressure. This means that, in a constant volume apparatus, hydrates will act to "denude" the gas phase of its heavy components, leaving the gas enriched in the light components. The left lobes of Figures 5-4 and 5-5 indicate that hydrogen sulfide, due to its unique size, is also "denuded" from the gas by inclusion in the smaller cavities of structure II.

---

### 5.1.7 Modifications of the Method

With the fundamental principles of the model given above, modifications have been made with three goals:
1) to increase the accuracy of prediction for mixtures,
2) to improve upon the six assumptions made in the statistical thermodynamics development of van der Waals and Platteeuw, and
3) to broaden the applicability of the theory to other phase equilibrium regions.

Initially the extensions for increasing the accuracy and improving the assumptions of the basic method will be discussed. Both the empirical corrections of Ng and Robinson and the theoretical improvements of Holder and co-workers should enhance the accuracy of the method appreciably beyond the more ordinary theory such as that found in the computer program accompanying this text. Section 5.1.8 discusses the accuracy which may be expected from the best current methods. Extensions of the method to other equilibria regions are the subject of Sections 5.2,5.3, and 5.4.

#### 5.1.7.a. Increasing the Accuracy of the Predictions

After the Parrish and Prausnitz method of the previous section was determined, several investigators have determined values of two of the thermodynamic parameters in Equations (5.32) and (5.35a), namely $\Delta\mu_W^o$ and $\Delta h_W^o$. An increase in accuracy of thermodynamic parameters would enable more accurate Kihara parameters from the fitting method (Sections 5.1.4 and 5.1.5). The excellent x-ray diffraction work of von Stackelberg and Müller (1954) has enabled the values of $\Delta v_W^o$ to remain unchallenged. Table 5-5 lists some of the values of $\Delta\mu_W^o$ and $\Delta h_W^o$ which have

**Table 5-5 Differences in Thermodynamic Properties Between Ice and the Empty Hydrate Lattice at 273K and 101.328 KPa**

| Structure I $\Delta\mu_W^0$ J/mol | Structure I $\Delta h_W^0$ J/mol | Structure II $\Delta\mu_W^0$ J/mol | Structure II $\Delta h_W^0$ J/mol | Reference |
|---|---|---|---|---|
| 699 | 0 | 820 | 0 | van der Waals and Platteeuw(1959) |
| | | 366-537 | | Barrer and Ruzicka (1962a) |
| | | 883 | | Sortland and Robinson (1964) |
| 1255.2 | 753 | 795 | 837 | Child (1964) |
| 1264 | 1150 | | 808 | Parrish and Prausnitz (1972) |
| 1155 | 381 | | 0 | Holder and Hand (1982) |
| 1297 | 1389 | 937 | 1025 | Dharmawardhana, Parrish and Sloan(1980) |
| 1299 | 1861 | | | Holder, Malekar, and Sloan(1984) |
| 1120 | 931 | 1714 | 1400 | John, Papadopoulos, and Holder (1985) |
| 1297 | | | | Davidson, Handa, and Ripmeester (1986) |
| 1287 | 931 | 1068 | 764 | Tse et al. (1986) |

been determined using various calculation and experimental techniques. In constructing such a table, care must be taken that thermodynamic parameters derived from data with simple hydrate formers, such as Ar and Kr (recently determined to be structure II hydrates - see Chapter 2), are not included as structure I properties. The variation of the values in Table 5-5 indicate that the parameters are very difficult to determine; however, some of the discrepancies have been resolved in the last decade.

Nuclear magnetic resonance (NMR) measurements of xenon occupation values by Davidson et al.(1986) seem to confirm the earlier macroscopic measurements of Dharmawardhana et al.(1980) as $\Delta\mu_W^0$ = 1297 J/mol for structure I. Recently, Holder et al. (1988) suggested that the value of $\Delta\mu_W^0$ should vary according to the type of guest molecule, due to a possible distortion of the hydrate cavities; this idea has not been evaluated at the present time. The thermodynamic values of Handa and Tse (1986) in Table 5-5 represent the only parameters determined for simple hydrates of molecules (xenon for structure I and krypton for structure II) which fill both the large and the small cavities in each structure.

The version of the model used in current flowsheet simulation programs, such as PROCESS or HYSIM is based upon the modification by Ng and Robinson (1976). They suggested an empirical modification to one of the equations in the method of Parrish and Prausnitz to increase the accuracy of the fit obtained from simple hydrates, as extended to mixtures of natural gases. Ng and Robinson suggested that, when Equation (5.23) is used for mixtures it should be modified by the correction factor:

$$1 + 3(\alpha-1)Y_1^2 - 2(\alpha-1)Y_1^3 \qquad (5.36)$$

where $\alpha$ is a constant of the order of unity for a binary pair and $Y_1$ is the mol fraction of the more volatile hydrate former in the gas, on a water-free basis. The constant is obtained through a fit of binary experimental data, with the other constants for $C_{ij}$ being determined, as before, from data for simple hydrates. When equilibria for multicomponent mixtures were calculated, the value of $\Delta\mu_W$ in Equation (5.23) should be changed to:

$$\Delta\mu_W = kT \left[ \prod_k \{1 + 3(\alpha_k-1)Y_k^2 - 2(\alpha_k-1)Y_k^3\}\right]\left[\sum_i \nu_i \ \ell n(1-\sum_k Y_{k_i})\right] \qquad (5.37)$$

where $\alpha_k$ is the interaction between the least volatile and each of the other more volatile hydrate forming molecule k and $Y_k$ is the mol fraction of the component k. An inspection of the interaction parameters in the original work indicates that all of the values for $\alpha_k$ are within 4% of unity except that of carbon dioxide-propane. In this modification, all of the other equations and parameters remain as previously determined.

The reasons for the inclusion of the $\alpha$ correction parameter were of pragmatic, rather than theoretical, origin. Ng and Robinson's work indicated that the correction is most important for binary systems with substantial size differences (e.g. methane+isobutane) or type differences (e.g. carbon dioxide+propane), thereby implying some interaction between guest molecules. Davidson and Ripmeester (1978) noted that the entire correction factor of Equation (5.36) is greatest at $Y_k = 1$, which suggests that the departure of $\alpha$ from unity measures experimental inconsistencies between dissociation pressures measured for simple hydrates and hydrates of two-component gases. Holder et al. (1988) suggested that the use of $\alpha$ implies that the quantity $\Delta\mu_W^0$ in Equation (5.32) is a function of hydrate component composition.

While the Ng-Robinson modification represents the model of most widespread current usage, it should be applied with some prudence. For example, Holder and Manganiello (1982) indicated that the use Equation (5.37) led Ng and Robinson to determine a "$\sigma$" value which would cause structure I to form as a simple hydrate of isobutane (a well-known structure II simple hydrate). Such a consideration suggests that the model should be more inaccurate with high fractions of isobutane. Avlonites (1988) recently indicated that the accuracy achieved through the use of Equation (5.37) could be easily obtained through the use of Equation (5.23), without the use of any fitting constants.

### 5.1.7.b. Correcting Assumptions of the Statistical Model

John and Holder (1981,1982a,b,1985) considered the assumptions (listed in Sections 5.1.1, and 5.1.4) of van der Waals and Platteeuw in the formulation of the statistical thermodynamics model. In particular, they were concerned about the assumptions of sphericity of each cavity, and the contribution of water molecules in other cavities to the potential of each central cavity. Their work, as summarized by Holder, et al.(1988), made two corrections as factors of the Langmuir constant:

$$C = C^{*}Q^{*} \tag{5.38}$$

where $C^{*}$ accounted for the change of the potential by water molecules in other cavities and $Q^{*}$ accounted for the non-sphericity of the cavities and molecules. In order to calculate $C^{*}$, the potential $\omega(r)$ of (5.25) was modified to include contributions from the cavity of interest $\omega_1(r)$, and the second and third shells of water molecules around the cavity, $\omega_2(r)$ and $\omega_3(r)$, respectively, at progressively larger intervals of distance as:

$$C^{*} = \frac{4\pi}{kT}\int_0^R \exp\left(-\frac{\omega_1(r) + \omega_2(r) + \omega_3(r)}{kT}\right) r^2\,dr \tag{5.39}$$

While the second and third shell of water molecules surrounding a cavity can contribute substantially to the Langmuir Constant due to the exponential nature of the potential, both $\omega_2$ and $\omega_3$ give an approximately constant contribution to $\omega$; this is probably because the distances to the outer shells are appreciable compared to the radius of the inner shell. The contributions do not however, give constant additive contributions to the Langmuir constants, due to the integration in Equation (5.39).

The second term in Equation (5.38), $Q^{*}$, was fitted

as a function of Kihara parameters, ($\varepsilon$, $\sigma$, and a) in a microscopic corresponding states equation as:

$$Q^* = \exp\left(-a_0\left[\omega\left(\frac{\sigma}{R-a}\right)\left(\frac{\varepsilon}{kT_0}\right)\right]^n\right) \tag{5.40}$$

where the fitting constants "$a_0$" and "n" were empirical parameters which depended upon the cavity type, but were independent of the species enclathrated. While the values of "$a_0$" and "n" in Equation (5.40) lack physical interpretation, it was intended that the parameters "$\varepsilon$", "$\sigma$", and "a" were to be very similar to the Kihara parameters obtained from second virial coefficient or viscosity data. However, some small differences in the Kihara parameters and in the acentric factor, $\omega$, remained between those hydrate parameters and the parameters obtained from other thermodynamic and transport property data. Since Langmuir constants are extremely sensitive to small variations in Kihara parameters, those values obtained from other than hydrate data should be used with extreme caution. However, the method itself holds promise as a considerable refinement in the theory.

John and Holder (1985) extended their theory in a singular use of the rod-shaped Kihara potential for linear molecules. Those results seem to indicate that the rotational motion of linear molecules is inhibited and that they can penetrate the hydrate cavity in certain configurations. However, the measurement of diffusion coefficients for linear molecules in the lattice, has yet to be accomplished.

### 5.1.8 Accuracy of the Method

The accuracy of the method is obtained by comparison against experimental data for three-phase ($L_W$-H-V) data which are available in the literature. Comparisons of the prediction methods do not yield significant differences for the simple hydrate formers such as those shown in Figures 5-7, and 5-8. One would expect the prediction

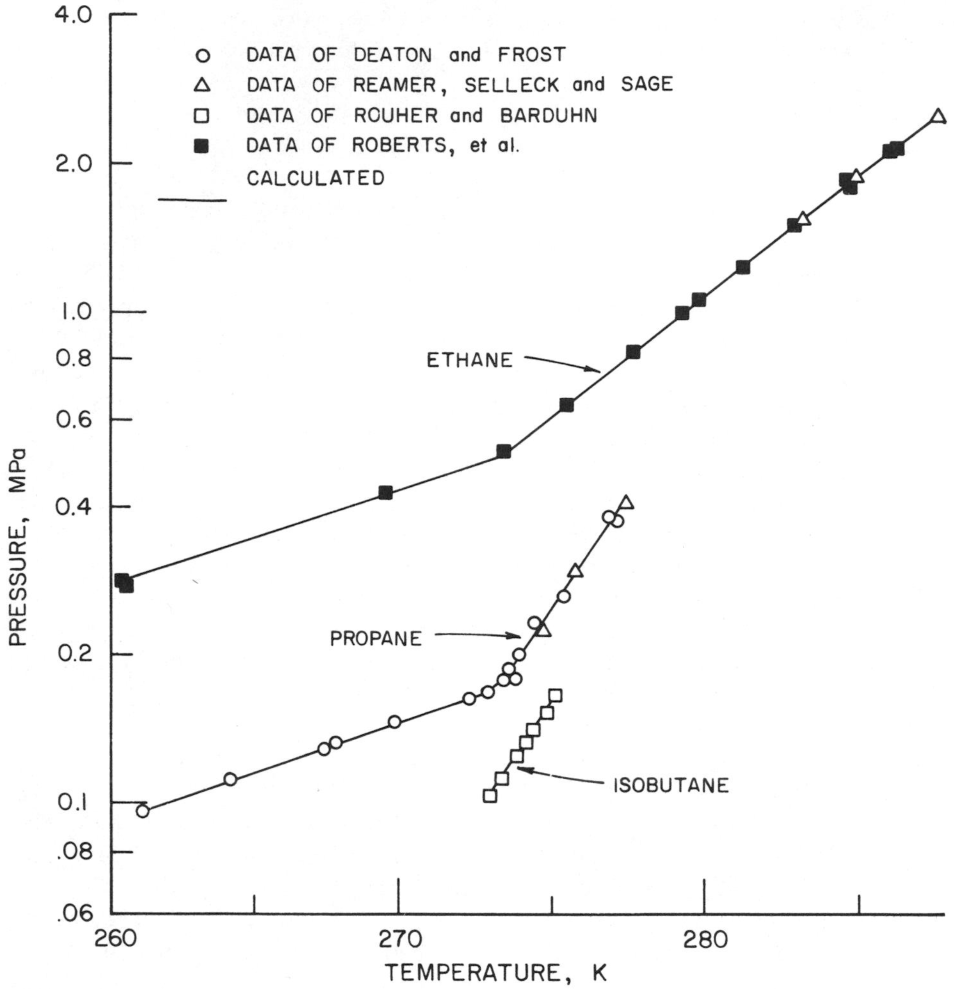


5-7 Measured Versus CSMHYD Fit of Simple Hydrate Three-Phase Conditions

schemes to be comparable in accuracy, primarily because the fits of Kihara parameters are normally heavily weighted by the simple hydrate formation data.

More stringent tests of the prediction schemes can be obtained by comparison to mixture data such as in Fig-

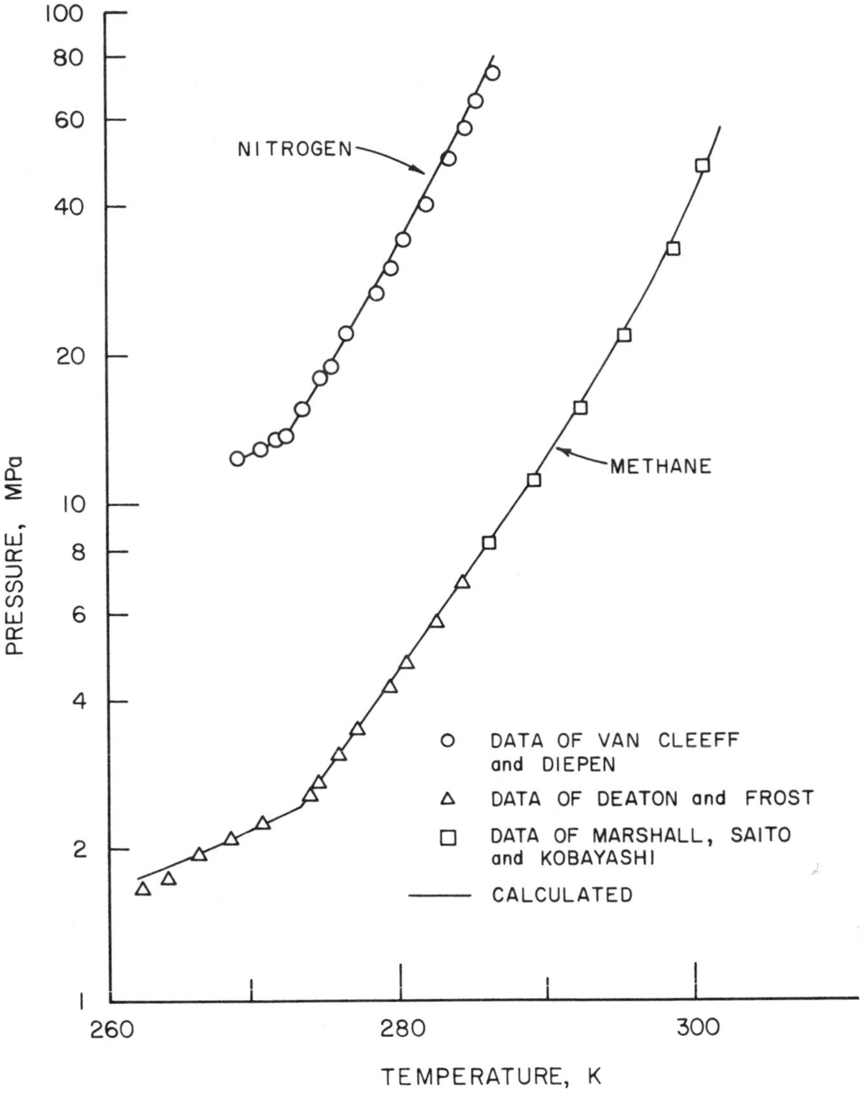


5-8 Measured Versus CSMHYD Fit of Simple Hydrate Three-Phase Conditions ure 5-9. In Section 2.1.3 the size of the molecules were considered as a determining factor in hydrate structure. Of particular interest was the fact that the addition of a small amount of propane to methane would cause the hydrate structure to change from I to II. Comparison of the pure methane and 1% propane three-phase lines in Figure 5-9 demonstrates that such a structure change can effectively halve the pressure required for hydrate formation at a given temperature. Hydrate structure changes, coexistence curves of both structures (Holder and Hand, 1982), and hydrate azeotrope predictions (Example 5.2) are some of the most stringent tests of program accuracy.

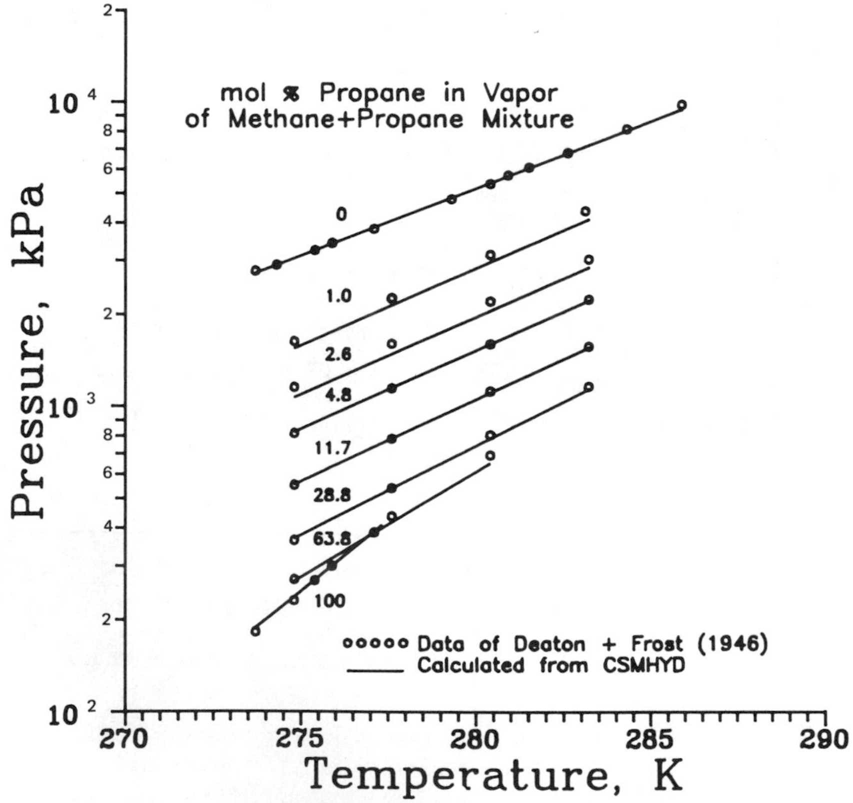


5-9 CSMHYD Predicted and Measured Three-Phase Hydrate Conditions for Mixtures of Methane+Propane

The ultimate test of prediction accuracy is comparison with experimental data which were obtained under similar conditions required for application.

For mixed hydrates the average absolute deviation of pressure for the method included in the disk in the frontpapers is 9%. A recent thesis by Avlonitis (1988) compared several prediction methods, including those presented here. Currently one may expect an absolute average pressure prediction deviation which is not less than 7%.

Mixed hydrate prediction comparisons are not normally made below the first quadruple point $Q_1$, because in that three-phase (I-V-H) region, the formation of a second solid phase (ice) may be reason enough to maintain a higher temperature. Solid phases are normally avoided due to their inhibition of gas flow, heat transfer, etc.

## 5.2 Condensed Three-Phase and Quadruple Calculations

Quadruple point $Q_1$ (I-$L_W$-H-V) is always determined by the intersection of the three-phase ($L_W$-H-V) line of the previous section with the ice point. On a pressure-temperature plot (such as Figure 4-1), when the $L_W$-H-V line intersects the three fluid phase ($L_W$-V-$L_{HC}$) line, the upper quadruple point $Q_2$ ($L_W$-H-V-$L_{HC}$) results. Because, as discussed in Section 4.4.1, the liquid water phase has little effect on the three-phase ($L_W$-V-$L_{HC}$) line, that locus may be approximated as the vapor pressure line for the hydrocarbon. If the hydrocarbon contains more than one hydrate former, the vapor pressure line is replaced by the phase envelope of the hydrocarbon, and the upper quadruple point ($Q_2$) becomes a line extending between the dew point and the bubble point of the hydrocarbon, as discussed in Section 4.1.3.

Modern studies in condensed three-phase equilibria, by Verma (1974) and by Ng and Robinson (1976) estimated the upper quadruple point (or line), by determining the intersection temperature and pressure of the hydrocarbon phase envelope with the three-phase ($L_W$-H-V) line. For the case of such an upper quadruple line (see line $\overline{KC}$ of Figure 4-5) the line is bounded by a vapor (dew point K) and a liquid (bubble point C) at the same hydrocarbon composition. The lower point of the line is determined by the intersection of the $L_W$-H-V line with the dew point portion of the constant composition phase envelope. The upper point of the upper quadruple line is determined by the intersection of another $L_W$-H-V line with the bubble point of the phase envelope; the vapor composition used in the $L_W$-H-V calculation is that which is in equilibrium with the liquid at its bubble point. The fundamentals of this calculation are discussed in Section 4.1.3.

For three-phase ($L_W$-H-$L_{HC}$) equilibria at pressures above the quadruple point on a pressure-temperature plot, Section 4.4.2 suggests that the slope is very steep and may be accurately approximated for either hydrate structure by the Clapeyron relation. In order to achieve a better estimation of condensed three-phase equilibria, Equations (5.22),(5.23), and (5.32) may be used just as before, except that the vapor fugacity of the hydrate former will be replaced by that of the liquid in Equation (5.22). Unfortunately, the determination of liquid phase densities and fugacities represent the properties of the greatest challenge to an equation-of-state, although Trebble and Bishnoi (1987) recently have indicated some success with a four parameter cubic equation-of-state.

The experimental data collected to date are insufficient to warrant correction from a straight line in the $L_W$-H-$L_{HC}$ region. The line should emanate from the upper quadruple point, with a slope determined by the Clapeyron equation with constant values of $\Delta H$ for each hydrate structure, and values as in Section 4.4.2. Most of the more sophisticated prediction schemes are limited in pressure, terminating at pressures less than 40 MPa. Minor exceptions to the straight line estimation are observed for simple hydrates of carbon dioxide (Song and Kobayashi, 1987) and those of hydrogen sulfide (Selleck et al., 1952). Since industrial practice seldom concerns simple hydrates of these non-combustible components, the straight line estimation should be satisfactory for ($L_W$-H-$L_{HC}$) conditions near the quadruple point or line.

### 5.3 Prediction of the Effect of Hydrate Inhibition

Hydrate inhibitors may be incorporated directly into the equations for the statistical thermodynamic model, in a straightforward way. The inhibitors fall into two broad categories: (1) those compounds, such as alcohols and glycols (initially discussed in Section 4.5.1), which have appreciable vapor pressure, and (2) those salt compounds (initially discussed in Section 4.5.2) which have very low vapor pressure.

Makogon (1981, p. 180) indicates that on a weight basis, calcium chloride is more effective than methanol which, on a weight basis, was determined to be more effective than glycol by Ng, Chen, and Robinson (1985). The relatively high vapor pressures of the alcohols, however, pose both advantages and disadvantages for their use. First, the high vapor pressure of methanol allows it to be injected directly into the hydrocarbon vapor phase, so that methanol flows to the point of any free water, where liquid saturation and hydrate inhibition occurs. The use of the solid calcium chloride does not afford this transportation advantage. On the other hand, the high vapor pressures of methanol causes substantial losses in the gas phase, where recovery of the methanol is difficult. In places where high inhibition is required such as the North Sea, it becomes necessary to have two pipelines: one to carry the hydrocarbon/methanol mixture from the offshore platform to the land, and a smaller methanol return pipeline.

#### 5.3.1 Predicting Hydrate Inhibition With Salts

Because salts have a very low vapor pressure, and because they are entirely excluded from the hydrate phase, the inhibition effect of salts may be determined by treating solely the liquid phase. The fundamental model for this treatment was initially proposed for a single inhibitor by Menten, Parrish, and Sloan (1981).

The effect of single salts may be calculated **a priori**, using data from the freezing point depression of the salt on ice, with data such as are commonly found in handbooks of chemistry and physics. When no inhibitors are present, the activity coefficient of the water phase may normally be taken as unity, due to the small solubility of gas components in the water. However, the presence of the inhibitor dramatically affects the activity coefficient of the water, which one may calculate through colligative properties, such as freezing point depression or vapor pressure lowering. The activity co-

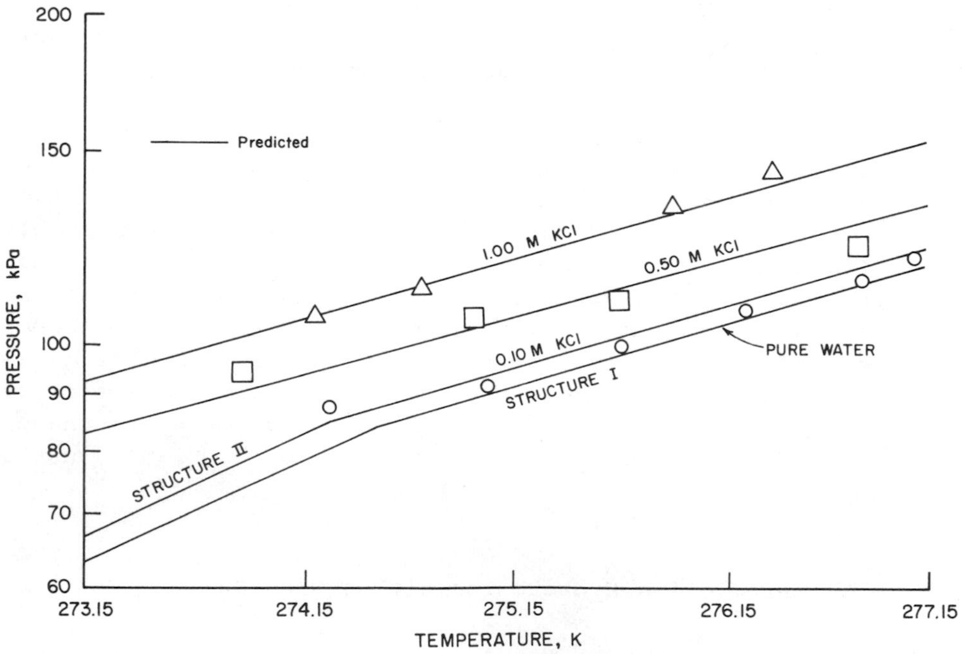


5-10 Measured and Predicted Inhibition of Cyclopropane Hydrates with Potassium Chloride
(Reproduced, with permission, from Industrial and Engineering Chemistry Process Design and Development (Menten et al., 1981), by the American Chemical Society.)

efficient thus obtained may be used directly in Equation (5.32) with the water concentration:

$$\frac{\Delta\mu_W}{RT} = \frac{\Delta\mu_W^o}{RT_o} - \int_{T_o}^{T} \frac{\Delta h_W}{RT^2} dT + \int_0^P \frac{\Delta v_W}{RT} dP - \ell n\, \gamma_W x_W \qquad (5.32)$$

Equation (5.32) is then used together with Equations (5.22) and (5.23) in the manner just as before, with very acceptable results, as indicated in Figures 5-10 and 5-11 for inhibition of cyclopropane by potassium chloride and calcium chloride, respectively. With very small amounts of effective inhibitors the water concentration $x_W$ remains close to unity, but the activity coefficients have a predominant effect resulting in substantial changes for $\Delta\mu_W$.

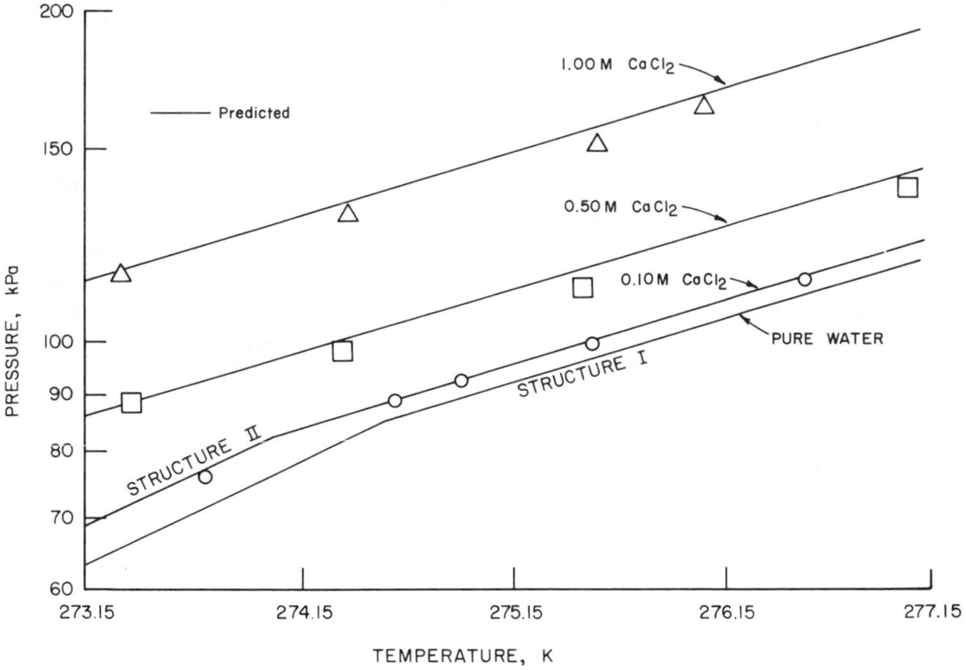


5-11 Measured and Predicted Inhibition of Cyclopropane Hydrates with Calcium Chloride
(Reproduced, with permission, from Industrial and Engineering Chemical Process Design and Development (Menten et al., 1981), by the American Chemical Society.)

Very recently, a new model has been proposed to extend the above treatment to systems of several electrolytes, such as seawater (Englezos and Bishnoi, 1988a). They followed the treatment of Patwardhan and Kumar (1986) to obtain the activity of water $a_W$ with mixed electrolytes (to replace $\gamma_W x_W$ in Equation (5.32)) by the relation:

$$\ell n \, a_W = \sum_{1}^{c} \left( \frac{m_k}{m_k^o} \right) \ell n \, a_{W,k}^o \tag{5.41}$$

where $m_k$ is the molality of electrolyte k in the mixed solution, $m_k^o$ is the molality of a solution containing only electrolyte k, which has the same ionic strength as that of the mixed solution and $a_{W,k}^o$ is the activity of water in the single salt solution. The method appears to be very promising, but experimental data on mixed electrolyte hydrate equilibria have yet to be published.

### 5.3.2 Prediction of Inhibition by Methanol

As discussed in Section 4.5.1, almost all of the techniques for hydrate inhibition enable the determination of the inhibitor concentration in the aqueous liquid in equilibrium with hydrates at a given temperature and pressure. The user is left with the additional step of estimating the amount of alcohol/glycol to be injected in the vapor. Of the alcohol/glycol inhibitors, the most volatile and the most frequently used is methanol. For that reason the prediction of methanol inhibition is presented here, as the most useful case and, at the same time, the worst case for inhibitor loss by vaporization and/or solubility in any hydrocarbon fluid which may be present.

In treating the equilibrium of hydrates with inhibitors which vaporize, the calculation technique should rigorously consider not only the primary presence of inhibitors in the aqueous phase but also in the hydrocarbon phase. However, of the published methods derived from that of van der Waals and Platteeuw (1959), only the method of Anderson and Prausnitz (1986) considers the presence of methanol in the hydrocarbon phase. Most of the methods consider only the methanol in the aqueous liquid and neglect its presence in the vapor or the liquid hydrocarbon phase, where its concentration may be smaller, (Menten et al., 1981). When methanol is used to inhibit hydrate formation from the liquid hydrocarbon and aqueous liquid regions, the solubility of methanol in the liquid hydrocarbon is frequently neglected in the hydrate prediction schemes (Makogon, 1988). A small collection of solubility measurements by Noda et al.(1975) and by Ng and Robinson (1983, 1985) of methanol solubility in hydrocarbon phases confirm the low values of methanol solubility; however the large flows of the hydrocarbon phases may cause total methanol losses to be large. In Chapter 8, Examples 8.2 and 8.3 provide the current state-of-the-art calculation methods for inhibitor injection, which account for losses to non-aqueous phases.

All of the methods for prediction of hydrate inhibition with methanol require the activity coefficient in Equation (5.32). Usually the activity coefficient parameters are determined from vapor-liquid equilibria, or liquid-liquid equilibria such as that of Noda et al. (1975). The various models of activity coefficients which have been used are the Margules equation (Holder, et al., 1988), the Wilson equation (Menten et al., 1981), the Parameters from Group Contributions (PFGC) method (Wagner, et al., 1985), and the UNIQUAC method (Anderson and Prausnitz, 1986).

Figures 5-12 to 5-15 show methanol inhibition predictions for methane and hydrogen sulfide by various methods, including the computer program with this book. In each figure data points taken from those of Chapter 6 are plotted for three-phase ($L_W$-V-H) inhibition together with correlations for the uninhibited hydrate formation, the freezing point depression of water from the uninhibited hydrate formation, and the Hammerschmidt Equation (4.7) as presented and discussed in Chapter 4. A comparison with the computer inhibition method (CSMHYD) of this

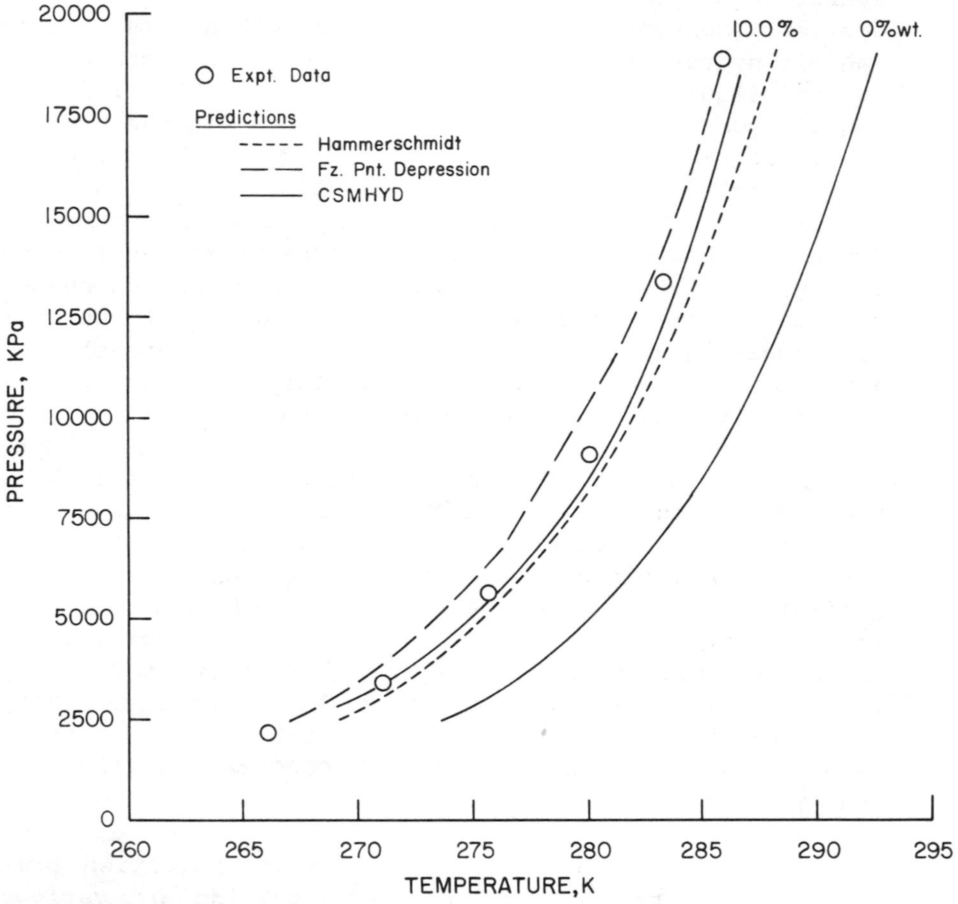


5-12 Measured and Predicted Inhibition of Simple Methane Hydrates with 10 wt% Methanol

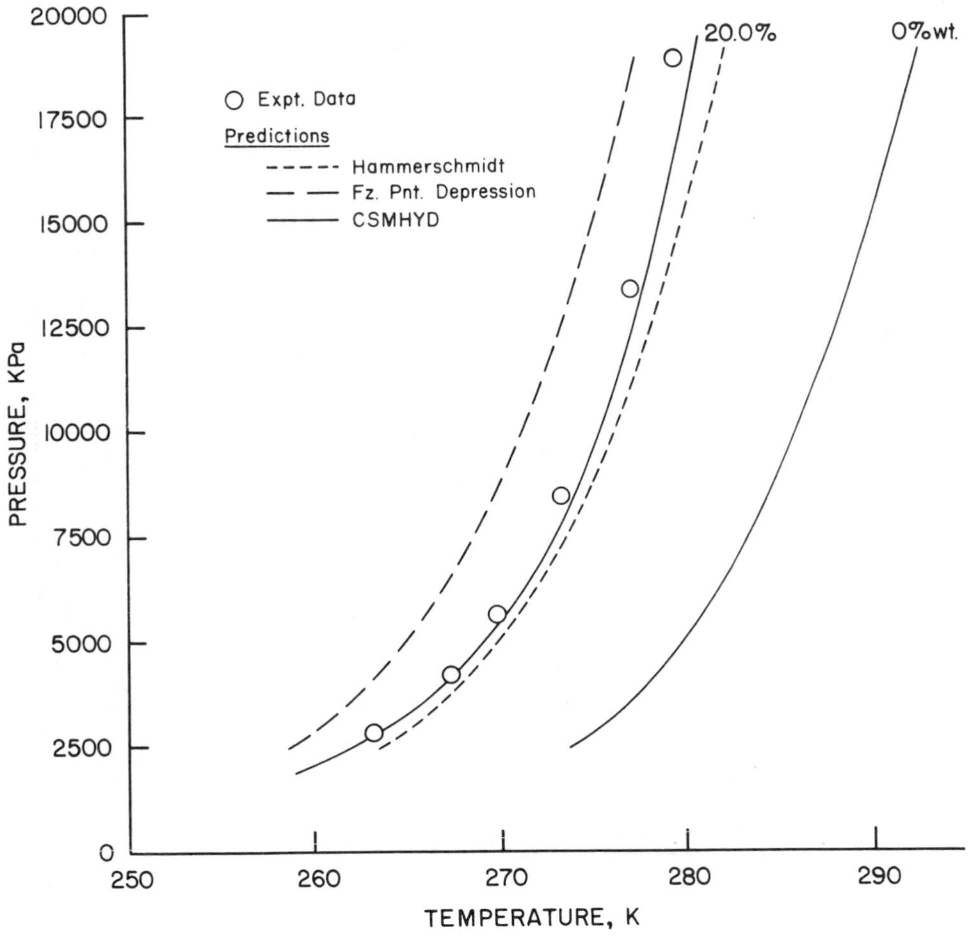


5-13 Measured and Predicted Inhibition of Simple Methane Hydrates with 20 wt% Methanol

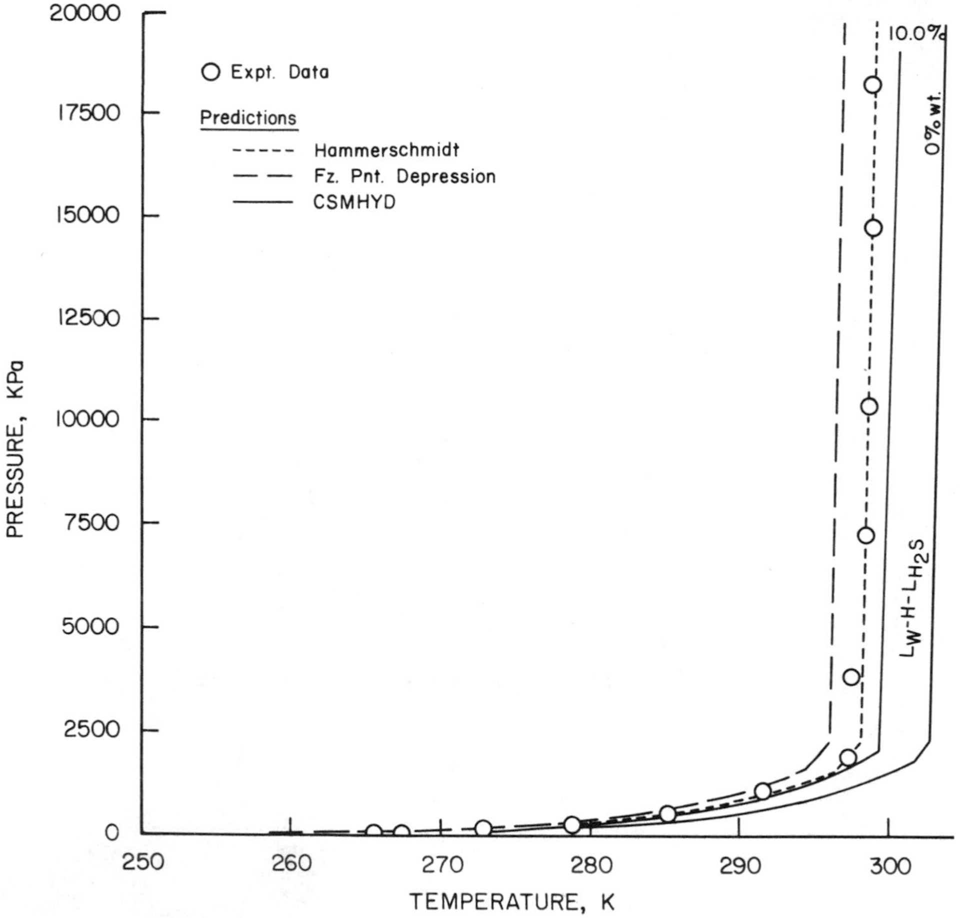


5-14 Measured and Predicted Inhibition of Simple Hydrogen Sulfide Hydrates with 10 wt% Methanol

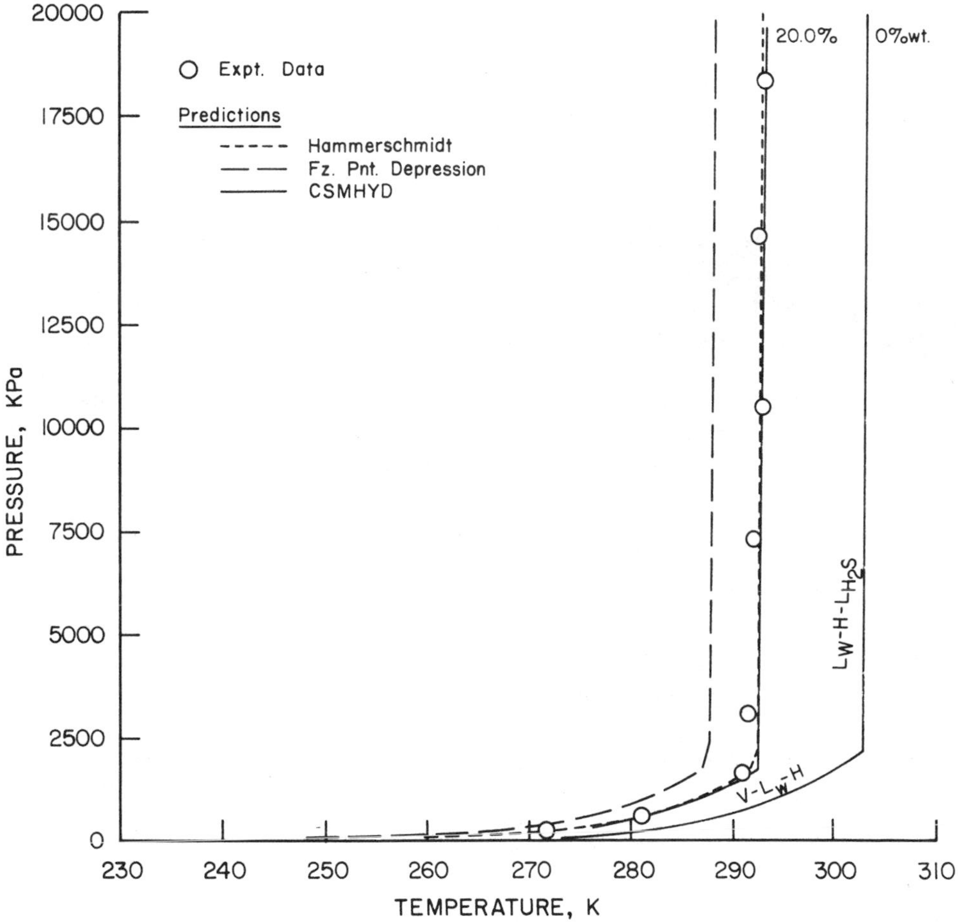


5-15 Measured and Predicted Inhibition of Simple Hydrogen Sulfide Hydrates with 20 wt% Methanol chapter, which uses Equation (5.32), shows the computer method to have substantial advantages.

## 5.4 Two-Phase (Hydrate-Fluid Hydrocarbon) Predictions

When free water is absent, the fluid hydrocarbon phase may still contain enough water to form a hydrate phase, usually over a prolonged period of time. The two-phase equilibrium regions were initially shown on every diagram of Figure 4-2 and discussed in Section 4.6. In each case, the water content of the hydrocarbon phase is less than that in the three-phase condition. In other words, the three-phase water content in the hydrocarbon phase represents an upper limit to the water content of the hydrocarbon phase in two-phase equilibria.

The measurements of hydrocarbon water content are the most difficult experimentally for two reasons:(1) low water concentrations in the hydrocarbon phase require exacting measurement systems, and (2) long experimental times are required for water molecules to form critical hydrate nuclei for further growth. For these reasons there is an extreme paucity of data for hydrate-hydrocarbon equilibria, against which to compare such prediction methods. Most of the two-phase data have evolved from the laboratory of Kobayashi (Sloan, et al.,1976, Aoyagi et al., 1978,1979, Song and Kobayashi, 1982,1987), with a few systems studied in this laboratory (Bourrie et al.,1986, Sloan et al.,1987). The inaccuracy of water composition prediction is estimated at ± 10 ppm when it is applied to normal hydrocarbon components, but the inaccuracy may be much larger for carbon dioxide rich fluids in equilibrium with hydrates (Song and Kobayashi, 1987). In the next section the prediction method for hydrate-vapor equilibria is presented, followed by the modification for the hydrate-liquid equilibria.

### 5.4.1 Hydrate-Hydrocarbon Vapor Equilibria

For two-phase, hydrate-hydrocarbon vapor equilibria to exist, the temperature must be equal to or below the three-phase temperature at a given pressure; the two-phase pressure must be equal to or above that of the three-phase pressure at a given temperature. The concept of tying the two-phase prediction to the statistical thermodynamics method at the three-phase boundary was first determined by Sloan, Khoury, and Kobayashi (1976). Since that time several other hydrate-vapor prediction methods have been presented by Ng and Robinson (1980), Anderson and Prausnitz (1986), Song and Kobayashi (1982), Holder, Zetts, and Pradhan (1988), and Munck, Skjold-Jørgensen, and Rasmussen (1988).

The presentation below follows the extension of the initial development by Sloan, Khoury, and Kobayashi (1976). They related the fugacity of water $f_W^H$ to the chemical potential difference of water in the filled and empty hydrate by the following equation:

$$f_W^H = f_W^{MT} \exp\left(\frac{\mu_W^H - \mu_W^{MT}}{RT}\right) \tag{5.42}$$

where $f_W^{MT}$ is the fugacity of water in the hypothetical empty hydrate phase, and $(\mu_W^H - \mu_W^{MT})$ represents the connection to the previous Equations (5.23) and (5.22), reproduced below.

$$y_{k_i} = \frac{C_{k_i} f_k}{1 + \sum_J C_{J_i} f_J} \tag{5.22}$$

$$\mu_W^H = \mu_W^{MT} + kT \sum_i \nu_i \, \ell n\left(1 - \sum_k y_{k_i}\right) \tag{5.23}$$

In Equations (5.22) and (5.23) only the fugacity of the vapor guest components should be changed to accommodate the two-phase calculation. The Langmuir constants are only functions of temperature for given components and thus will remain at their three-phase values for equivalent temperatures.

The empty hydrate fugacity $f_W^{MT}$ in Equation (5.42) was related to a hypothetical empty hydrate vapor pressure $P_W^{MT}$ by the common phase equilibrium equation to correct the vapor pressure:

$$f_W^{MT} = P_W^{MT} \phi_W^{MT} \exp \int_{P_W^{MT}}^{P} \frac{v_W^{MT}}{RT} dP \qquad (5.43)$$

where the fugacity correction $\phi_W^{MT}$ is unity and the exponential term, sometimes called the Poynting correction factor, is set equal to unity at pressures below 3.5 MPa.

The concept in Equation (5.43) of an universal empty hydrate vapor pressure for each structure, prompted Dharmawardhana (1980) to calculate the $P_W^{MT}$ from a number of simple hydrate three-phase (I-H-V) equilibria, which represent the concentration limit to the method. By equating the fugacity of water in the hydrate phase (through Equations (5.42) and (5.43)) to that of pure ice at the three-phase line, he obtained the equation:

$$P_{ice}^{vap} \phi_{ice}^{vap} \exp \int_{P^{vap}}^{P} \frac{v_{ice} dP}{RT} = P_W^{MT} \phi_W^{MT} \exp \int_{P_W^{MT}}^{P} \frac{v_W^{MT} dP}{RT} \exp\left(\frac{\Delta\mu_W^H}{RT}\right) \qquad (5.44)$$

On the left of Equation (5.44) all of the properties were obtained from the vapor pressure of ice while $\Delta\mu_W^H$, the primary property on the right, was obtained from Equations (5.22) and (5.23) for the three-phase (I-H-V) equilibria. The values of $P_W^{MT}$ were determined at a number of temperatures for a number of different components. The results are plotted in Figure 5-16 for structure I and Figure 5-17 for structure II simple hydrates, implying the universality of a single temperature function of $P_W^{MT}$ for each structure. The equations for the two empty hydrate structures vapor pressures are:

$$\ell n\ P_W^{MT} = 17.440 - 6{,}003.9/T \text{ for Structure I} \qquad (5.45a)$$

and

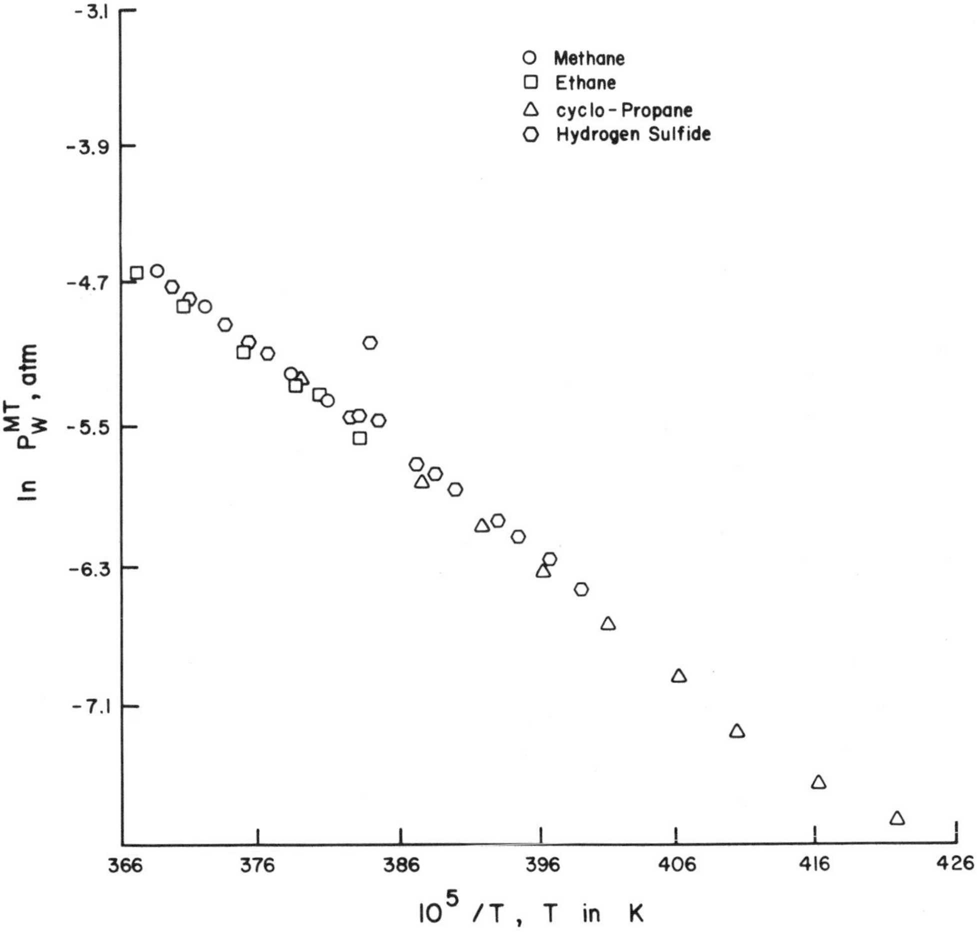


5-16 Empty Hydrate Hypothetical Vapor Pressures for Structure I Derived from Simple Hydrate Three-Phase Data (Reproduced, with permission, from Industrial Engineering Chemistry Research (Sloan et al., 1987), by the American Chemical Society.)

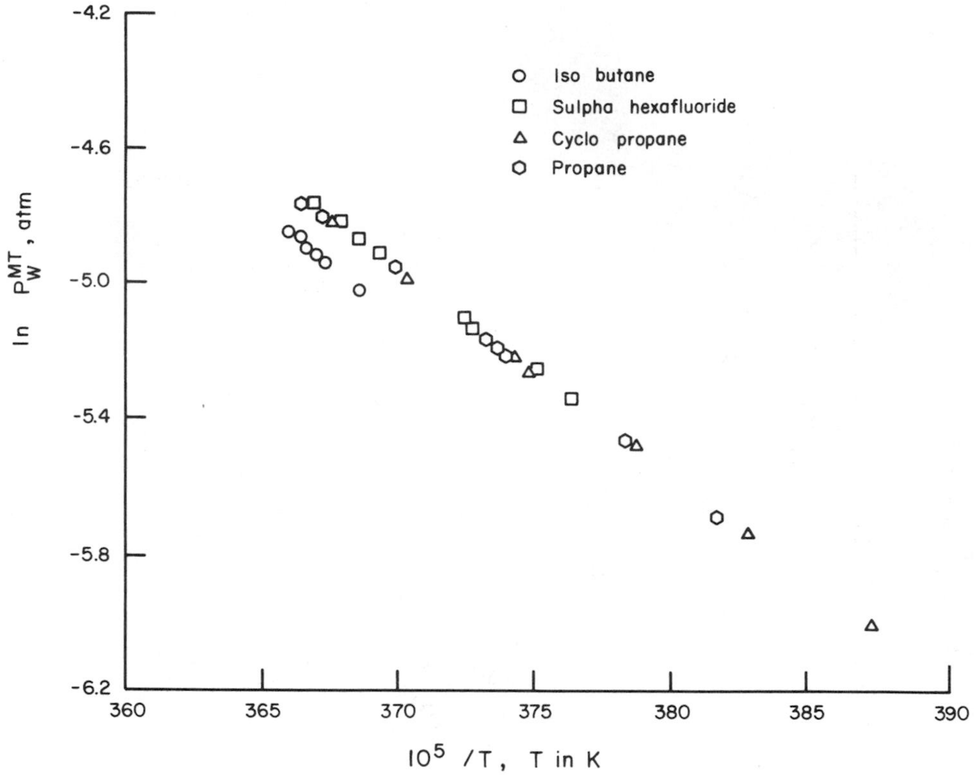


5-17 Empty Hydrate Hypothetical Vapor Pressures for Structure II Derived from Simple Hydrate Three-Phase Data (Reproduced, with permission, from <u>Industrial Engineering Chemistry Research</u> (Sloan et al., 1987), by the American Chemical Society.)

$$\ell n\ P_W^{MT} = 17.332 - 6{,}017.6/T \text{ for Structure II} \qquad (5.45b)$$

with $P_W^{MT}$ in atmospheres and T in Kelvins. The above equations, which were derived from three-phase (I-H-V) data may then be used, together with Equations (5.42) and (5.43) to determine the fugacity of water dissolved in the hydrocarbon in the two-phase vapor-hydrate region.

The fugacity of water in the hydrocarbon vapor $f_W^G$ is calculated by the relation:

$$f_W^G = y_W \, \phi_W^G \, P \tag{5.46a}$$

where $\phi_W^G$ is calculated from an equation-of-state via the normal method, as specified in Appendix A. By equating Equations (5.42) and (5.46a) the concentration of water in the hydrocarbon vapor in equilibrium with the hydrate is determined at a specified temperature and pressure.

$$y_W = \frac{f_W^H}{\phi_W^G P} = \frac{1}{\phi_W^G P} f_W^{MT} \exp\left[\frac{\mu_W^H - \mu_W^{MT}}{RT}\right] \tag{5.47a}$$

As indicated in Appendix A, the calculation of the fugacity coefficient $\phi_W^G$ on the right of Equation (5.47a) requires the concentration of water in the vapor phase $y_W$, so that an iterative calculation is required. The calculation of $\phi_W^G$, using Equations given in Appendix A, requires abnormally high values of the interaction parameter $\delta_{ij}$ (ca. 0.5) (Peng and Robinson, 1976) to account for the hydrocarbon-water interaction. Temperature dependence might be incorporated into $\delta_{ij}$ in order to fit the two-phase data. However, as stated by Anderson and Prausnitz (1986), when the $\delta_{ij}$ value has an additional pressure dependence, thermodynamic inconsistency results. Anderson and Prausnitz (1986) determined thermodynamic inconsistencies in the other methods of two-phase prediction (except theirs and the one presented here), which may inhibit the ability to predict data beyond those to which they were fit.

The computer program on the disk in the frontpapers provides an estimation of the water content of the hydrocarbon phase in equilibrium with hydrates. The algorithm for the calculation is prescribed in Table 5-6, for those who may wish to construct their own program. Figure 5-18 shows typical vapor phase water concentration data and predictions from the method in the current section.

## Table 5-6
## Hydrate-Fluid Hydrocarbon Prediction Algorithm

1. Determine which hydrate structure will form by performing a three-phase ($L_W$-H-V or I-H-V) calculation at the temperature of interest and noting the hydrate structure present.

2. Ensure that hydrate will form in the two-phase region for the conditions of interest. This condition requires that the temperature be lower and/or the pressure be greater than the corresponding value at the three-phase condition.

3. Select the hydrocarbon fluid phase (vapor or liquid) by a dew point/bubble point calculation for the water-free hydrocarbon at the temperature and pressure of interest.

4. Calculate the Langmuir constants in Equation (5.22) using Equations (5.25) and (5.27) with the Kihara parameters of the hydrate forming components.

5. Calculate the fugacity of water in the empty hydrate $f_W^{MT}$ via Equations (5.45a or b) for the appropriate hydrate structure.

6. Estimate the mol fraction of water in the hydrocarbon fluid phase. Usually this may be taken as a value less than 0.001.

7. Calculate the fugacity of the components in the hydrocarbon fluid phase, using the method specified in Appendix A, with a reliable equation-of-state.

8. Calculate $y_{ki}$, the fractional filling of cavity i by a type k molecule, using Equation (5.22).

9. Calculate the chemical potential difference of water between the empty and filled hydrate, using Equation (5.23).

**Table 5-6 (continued)**
**Hydrate-Fluid Hydrocarbon Prediction Algorithm**

10. Calculate the fugacity of water in the hydrate $f_W^H$ with Equation (5.42).

11. Calculate an improved estimate of water in the hydrocarbon fluid phase, with either Equation (5.47a) or (5.47b).

12. If the improved estimate of the water concentration compares with the preceding estimate to within an acceptable tolerance, terminate the calculation. Otherwise use the improved estimate of the water concentration for the next iteration in Step 6.

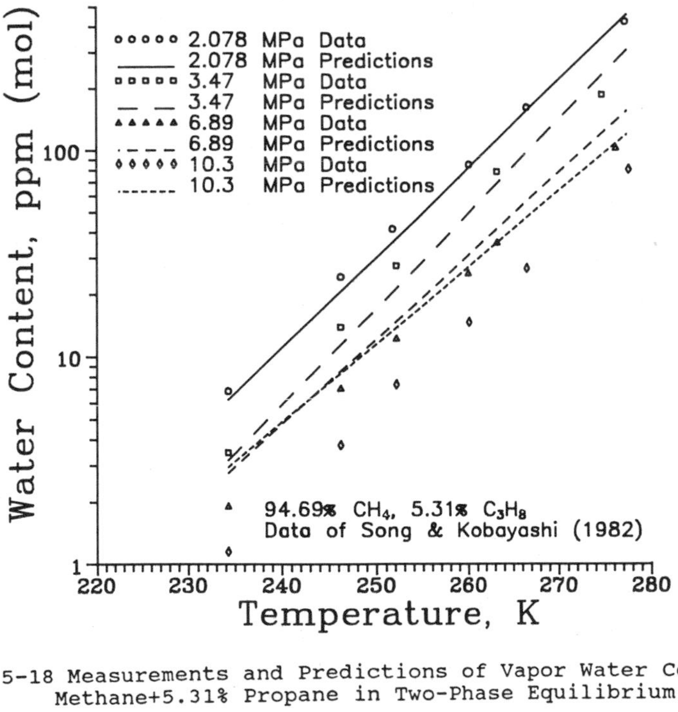


5-18 Measurements and Predictions of Vapor Water Content of Methane+5.31% Propane in Two-Phase Equilibrium with Hydrates

### 5.4.2 Hydrate-Hydrocarbon Liquid Equilibria

The extension of the above method to the hydrate-liquid hydrocarbon equilibria was proposed by Sloan et al.(1987). All of the above equations are valid, with the replacement of $y_W$ with $x_W$ in Equations (5.46) and (5.47) so that they become:

$$f_W^L = x_W \; \phi_W^L \; P \tag{5.46b}$$

where $\phi_W^L$ for the liquid phase is calculated from an equation-of-state via the normal method, as specified in Appendix A. In order to combine Equation (5.46b) with (5.42) to obtain the concentration of water in the liquid hydrocarbon, one other change must be made. The fugacity of the hydrocarbon component in the liquid phase must be used with Equations (5.22) and (5.23) in order to get the $\Delta\mu_W^H$ to use in (5.42). Combining (5.46b) and (5.42) the final expression is obtained as :

$$x_W = \frac{1}{\phi_W^L P} \; f_W^{MT} \; \exp\left[\frac{\mu_W^H - \mu_W^{MT}}{RT}\right] \tag{5.47b}$$

The algorithm prescription in Table 5-6 represents that used to construct the computer program in the disk in the frontpapers. Figure 5-19 shows a typical comparison of data with predictions of the water content of liquid hydrocarbon in equilibrium with hydrate.

## 5.5 Chapter Summary; Relationship to Following Chapters

The statistical thermodynamic method presented in this chapter represents the state-of-the-art for the prediction of the following types of phase equilibria:

(a) compressible three-phase ($L_W$-H-V or I-H-V),
(b) incompressible three-phase ($L_W$-H-$L_{HC}$),
(c) inhibition of equilibria in (a) or (b),
(d) quadruple points/lines($L_W$-H-V-$L_{HC}$ or I-$L_W$-H-V), and
(e) two-phase (H-V or H-$L_{HC}$) equilibria.

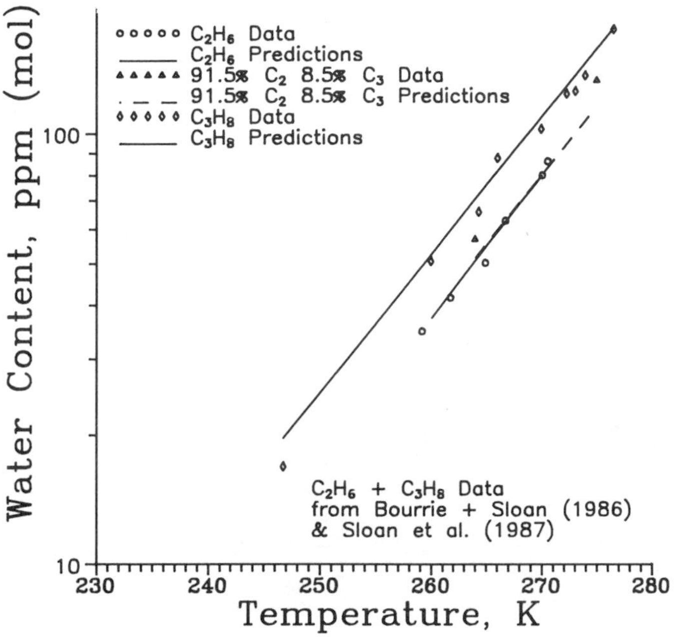


5-19 Measurements and Predictions of Liquid Water Content of Ethane and Propane in Two-Phase Equilibrium with Hydrates

The following chapter presents the experimental methods and the data upon which the present chapter is based. Such data will form the foundation for future modifications of theory in hydrate phase equilibria.

Chapter 7 then considers the formation of hydrates in geology, such as in the permafrost and deep oceans of the earth. In such situations geologic time mitigates the necessity for kinetic formation effects and allows the use of thermodynamic conditions, such as those in the three-phase portions of the present chapter, for identification, exploration, and recovery.

Chapter 8 presents problems of natural gas production and processing that are related to hydrates. Because a standard kinetic treatment method is some years away, the thermodynamic methods of the present chapter represent the foundation for treatment of such problems. With the thermodynamics for pristine systems relatively well-defined, hydrate research seems to trend toward more unusual problems, e.g systems with high concentrations of condensible, non-combustibles, and inhibitors, or systems with third surfaces.

# 6

# Experimental Methods and Measurements of Hydrate Properties

Chapters 4 and 5 were concerned with the fitting and prediction of hydrate thermodynamic data. Those two chapters indicate how hydrate theoretical developments have dramatically changed over their history, particularly due to advances in knowledge of molecular structure, statistical thermodynamics, and computing power. Yet the powerful tools provided by all of these predictive methods are only as good as the measurements upon which they are based.

While theoretical methods have changed appreciably, the experimental methods and data for macroscopic properties have remained relatively stable. For example, the current, static experimental apparatus for the measurement of hydrate formation conditions from water has not changed appreciably over the last half-century from that of Deaton and Frost (1937). In part, this is due to the fact that the simplest apparatus is both the most elegant and reliable simulation of hydrate formation in industrial systems. In Section 6.1 experimental apparati for determination of hydrate thermodynamic and transport properties are briefly reviewed.

Thermodynamic property data for hydrate formation will be the basis for future theoretical developments because, for the most part, they have been painstakingly obtained at several laboratories. Usually a period of several months (or even years) is required to construct the experimental apparatus and, due to long metastable periods, it is not uncommon to obtain only one pressure-temperature data point per twenty-four hours of experimental effort. Calorimetric data are even more difficult to obtain. Perhaps because there are such efforts associated with the experiments, the accumulation of hydrate thermodynamic and transport data is small enough to summarize in one place. The data bank of hydrate properties is presented in Section 6.2 for comparison with future prediction methods, or with the methods presented in this monograph.

This chapter deals only with macroscopic thermodynamic and transport hydrate properties of natural gas components and water, with and without soluble inhibitors. Kinetic data and apparati have been previously presented in Chapter 3, while the effects of insoluble substances (such as porous media, clays, or muds) are presented in Chapter 8.

## 6.1 Experimental Apparati and Methods

The experimental apparati for hydrate phase equilibria underwent considerable evolution over the first century after their discovery. During the last half-century the methods of determining macroscopic equilibria have not changed substantially. Instead experimentalists have concentrated upon spectroscopic methods for microscopic properties and unusual macroscopic conditions, such as high concentrations of inhibitors, different phases in equilibrium, and unusual temperatures, pressures, or gas compositions. The first portion of this section deals with the evolution of the current apparati for the measurement of phase equilibria.

The last fifty years have witnessed considerable progress in measurement methods for calorimetric and transport properties. The last portion of the section deals with the methods for measurement of those macroscopic properties dealing with the transmission, storage, and phase change due to heating and cooling.

### 6.1.1 Measurement Methods of Hydrate Phase Equilibria

During the first century after their discovery, hydrates were regarded as a scientific curiosity. Researchers worked primarily either with gases which were highly soluble, or under conditions which enabled formation of hydrates at low pressures. With the notable exception of Roozeboom, very early workers were concerned primarily with the establishment that hydrates of certain compounds did indeed exist, and secondly the determination of the number of water molecules associated with each gas molecule. Those research objectives, combined with the properties of hydrate formers, necessitated only the use of hand-blown glass apparati. While such apparati are of historical interest, they are currently considered to be too costly and too fragile to simulate industrial or **in situ** hydrate conditions.

Beginning in the 1880's, de Forcrand and his former collaborator Villard began one of the most perseverant research endeavors in the history of hydrates, over a forty-five year period. Via an ingenious glass apparatus, de Forcrand and Villard (1888a) were able to exclude most of the mother-liquor (water) on the formation of $H_2S$

hydrates, so that the hydrate number (gas•water) for hydrogen sulfide was reduced from the previous value of $H_2S \cdot 12H_2O$ to $H_2S \cdot 7H_2O$. As indicated in Chapter 1, Villard was the first to determine hydrates of methane, and ethane (1888), and propane (1890), but he was not successful in the formation of nitrogen hydrates. In order to form methane and ethane hydrates, he replaced the glass container of Cailletet (which was not suitable for use at very high pressures) with a round metal bell jar, and formed hydrates of methane at 26.9 MPa and 293.4 K. A model of the Cailletet apparatus is in current use at Delft University.

### 6.1.1.a. Principles for Equilibrium Apparatus Development

While these early workers did not have apparati suitable for modern use, the experimentalists during the first century of hydrates did prove three important principles to guide the development of succeeding apparati and methods:

1) Vigorous agitation is necessary for complete hydrate transformation as determined by Villard (1888). With minor exceptions, all of the early results showed that refinement of experimental agitation caused a decrease in the number of water molecules in the hydrate.

Such agitation is necessary for three reasons:(a) to provide surface renewal and exposure of liquid water to the hydrate former, and (b) to prevent occlusion of the mother-liquor. Without agitation, Villard (1894) showed for example, that nitrous oxide hydrate formation was continuous for a period longer than fifteen days under a pressure of 2 MPa. It was also Villard, who determined that in previous research the ratio of water to guest molecules had been analyzed as greater than $6H_2O \cdot G$ (Villard's Rule) due to either occlusion of water within the hydrate mass, or due to the loss of the guest component. Later, Hammerschmidt (1934) added the third reason for vigorous mixing; some agitation perhaps in the form of flow fluctuations, pressure cycles, bubbling gas through water, etc. was necessary to initiate hydrate formation, in order to decrease the metastability.

2) Hydrate metastability is always present on formation. Wroblewski (1882) was one of the first researchers to form ($CO_2$) hydrates, using equilibrium pressures higher than atmospheric for hydrate formation via Joule-Thomson expansions. Cailletet (1887) later found that the pressure must be increased beyond the hydrate equilibrium value and that hydrate formation results in a "relaxation" of the metastable pressure. With slow heating or depressurization for hydrate dissociation however, metastability did not occur. The endpoint of hydrate dissociation was thus much more reproducible and was taken as an indication of the upper limit to formation metastability.

Half a century later, the work of Carson and Katz (1942) provided a second reason for considering the dissociation condition as the hydrate equilibrium point. Their work clearly showed the solid solution behavior of hydrates formed by gas mixtures. This result meant, for example, that the hydrate preferentially encapsulated propane from a methane+propane gas mixture, so that a closed gas volume changed in composition as more hydrates formed. Thus by the conclusion of hydrate formation, the gas mixture composition had changed substantially and required measurement. On the other hand, when the last hydrate crystal dissociated the original gas composition was regained, minus a very small amount to account for solubility in the liquid phase. Data for solubility of hydrocarbons in water are in Section 3.1.2.

3) A marked change in the experimental pressure validates gas hydrate formation or dissociation in isochoric, or constant volume apparati. All of the early workers noted a concentration of the gas as it was encapsulated in the hydrate, which led to a decrease in pressure. Conversely on dissociation with heating, visual observation of the disappearance of the last hydrate crystal was accompanied by a decrease in the slope of a pressure versus temperature trace. This observation provided a means of obtaining higher hydrate equilibrium pressures than could be

achieved in a glass apparatus, simply by measuring the intersection point of the cooling and heating isochores, given as point H in Figure 3-5.

The following two subsections consider the apparati used for the most prominent phase equilibria measurements, i.e. the three-phase (I-H-V and $L_W$-H-V) regions respectively. Similar apparati have been used to measure other three-phase ($L_W$-H-$L_{HC}$), four-phase ($L_W$-H-V-$L_{HC}$), and even five phase ($L_W$-$H_I$-$H_{II}$-V-$L_{HC}$) conditions. For two-phase (H-$L_{HC}$ and H-V) regions different apparati have been used as discussed in the third subsection

### 6.1.1.b. Apparati for Use Above the Ice Point

Over the last fifty years, modern hydrate phase equilibrium apparati have been developed with the guidance of the above three principles. As a consequence of reading Schroeder's 1927 hydrate monograph (Katz, personal communication 1988), Hammerschmidt (1934) constructed a flow apparatus for simulation of hydrate formation in pipelines. The original flow apparatus consisted of a thermostatted Pyrex tube wherein hydrates were visually observed. Note that since this was a flow system with substantial gas available, pressure drops and changes in gas compositions were not observed at hydrate equilibrium conditions. After the gas flow was stopped, visual confirmation of hydrate disappearance, as well as temperatures, and pressures were measured with slow heating. As previously noted, while numerous hydrate equilibria data were obtained using the above apparatus, Hammerschmidt reported a correlation rather than the data points.

In 1937 Deaton and Frost constructed a static hydrate equilibrium apparatus which was to be the prototype for many others. The essential features of the apparatus are cited below and shown in Figures 6-1a,b, with a chronological listing of the modifications provided in Table 6-1. Note that Table 6-1 represents only the initiation of the use of such apparatus in each laboratory, providing a chronology of modifications. Most of the apparati have

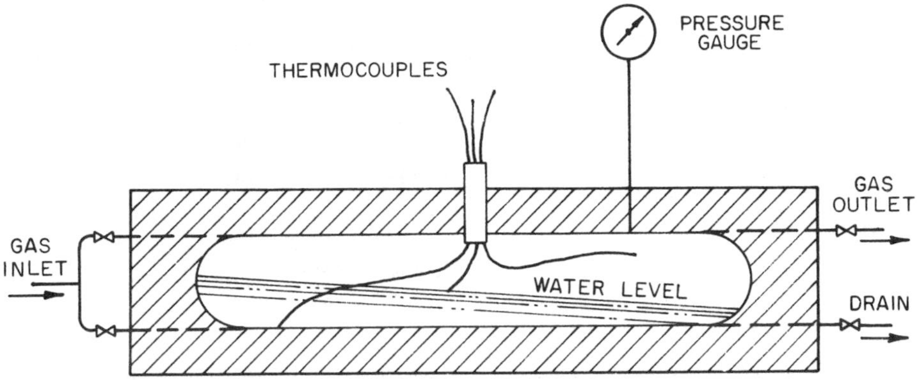


6-1a Detail of Deaton and Frost Hydrate Equilibrium Cell (Reproduced courtesy of U.S. Bureau of Mines, (Deaton and Frost, 1946))

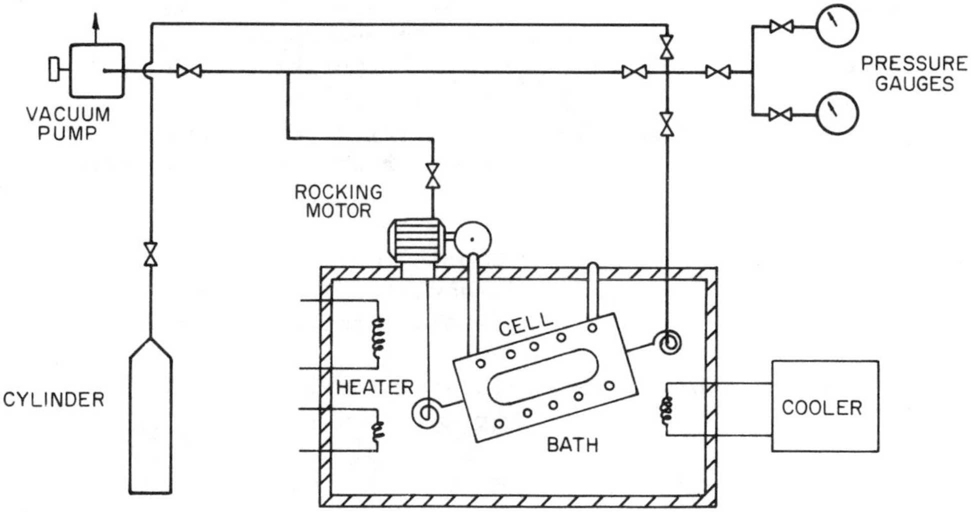


6-1b Typical Rocking Hydrate Equilibrium Apparatus

been used in each laboratory by many other investigators. For example, a similar apparatus of Carson and Katz (1942) was used for a number of years in the University of Michigan laboratory, most recently by Holder and Hand (1982).

The salient features of the apparatus are as follows:

**Table 6-1**
**Development of High Pressure Visual Hydrate Cell**

| Date | Investigator(s) | Modification |
|---|---|---|
| 1937 | Deaton & Frost | Basic Cell with options for gas flow above liquid, or sparged through liquid<br>Option for rocking cell<br>Cell in thermostatted bath |
| 1940 | Roberts, Brownscombe and Howe | Mercury(Hg) displacement of liquids; agitation |
| 1942 | Carson and Katz | Rocking hydrate cell;<br>Hg displacement of liquids;<br>hydrate decomposed for composition |
| 1952 | Reamer, Selleck, and Sage | Capillary sight glass; hydrocarbons and water over Hg; agitation |
| 1956 | Scauzillo | Cooling coils adjacent to sight glass; Hg displacement; rocking |
| 1960 | Otto and Robinson | Double window cell; rotated or agitated (after 1960) |
| 1961 | van Welie and Diepen | Hg pressurization;P>25 MPa; electromagnetic agitation |
| 1969 | Andryushchenko and Vasilchenko | 100 cc steel-coated organic glass cylinder; magnetic agitation |
| 1974 | Makogon | Bathyscaph for P≤20 MPa without agitation; filming capability. |

**Table 6-1 (Continued)**
**Development of High Pressure Visual Hydrate Cell**

| | | |
|---|---|---|
| 1974 1980 | Berecz and Balla-Achs | Single & double chamber multivibrator mixer |
| 1980 | Dharmawardhana, Parrish and Sloan | Bronze cell, plexiglas windows ultrasonic mixing |
| 1982b | John and Holder | Glass Windowed Cell without mixing used below the ice point |
| 1983a | Vysniauskas and Bishnoi | Cylinder with sight ports magnetically stirred |

• The heart of the apparatus consists of a sight glass for visual confirmation of hydrate formation and disappearance. Normally only a fraction of the cell is filled with water, the remainder being gas and hydrate.
• The cell is enclosed in a thermostatted bath but thermocouples are placed in the cell interior to measure any thermal lag between the cell and the bath.
• Since the pressure is usually too high to be measured via manometry, Bourdon tube gages or transducers, serve as sensors.
• Mixing at the gas-liquid interface is provided in a variety of ways, such as by rotation or rocking of the cell, by bubbling gas through the water in the cell, by ultrasonic agitation, or by the insertion of mechanical or magnetic agitators.

Operation of Standard Static Formation Apparatus. With the fundamental apparatus established, normal operation above 273 K, proceeds in one of three static modes:

• In isothermal operation the gas-liquid system is maintained at a pressure higher than the hydrate formation value. As hydrates are formed, encapsulation of the gas decreases the pressure to the three-phase ($L_W$-H-V) condition. The system pressure is attained through exchange of gas, aqueous liquid, or sometimes mercury with an external reservoir. After hydrate formation, the pressure is reduced gradually, and the visual observation of final hydrate crystal disappearance is taken as the equilibrium pressure for hydrate formation. At the three-phase condition, the pressure will remain constant until hydrates are depleted.

• In isobaric operation the system is maintained at a constant pressure, by the exchange of gas or liquid with an external reservoir. The temperature is decreased until the formation of hydrate occurs from the gas and liquid. After hydrate formation the temperature is slowly increased (maintaining constant pressure) until the last crystal of hydrate disappears. This point, taken as the equilibrium temperature of hydrate formation at constant pressure, may be determined by observation of hydrate dissociation or as a plateau of constant temperature as hydrates dissociate with heat input.

• Isochoric operation of the hydrate formation cell is illustrated by the pressure-temperature traces of Figure 3-5. The temperature of the closed cell is lowered from the vapor-liquid region, and isochoric cooling causes the pressure to decrease slightly. Hydrates form at the metastability limit, causing a marked pressure decrease, ending at the three-phase ($L_W$-H-V) pressure and temperature. The temperature is then slowly increased to dissociate the hydrates. On a pressure-temperature plot, the hydrate dissociation point is taken as the intersection of the hydrate dissociation trace with the initial isochoric cooling trace. This procedure is commonly used for high pressure hydrate formation, and provides an alternative to visual observation, which is the primary option in the previous two procedures.

In the 1940's a sight glass rupture resulted in the death of a hydrate experimentalist. Consequently there was a revival of interest in non-visual means of hydrate detection, especially at high pressures. The development of the van der Waals and Platteeuw statistical theory in 1959 also rekindled interest in high pressure experimen-

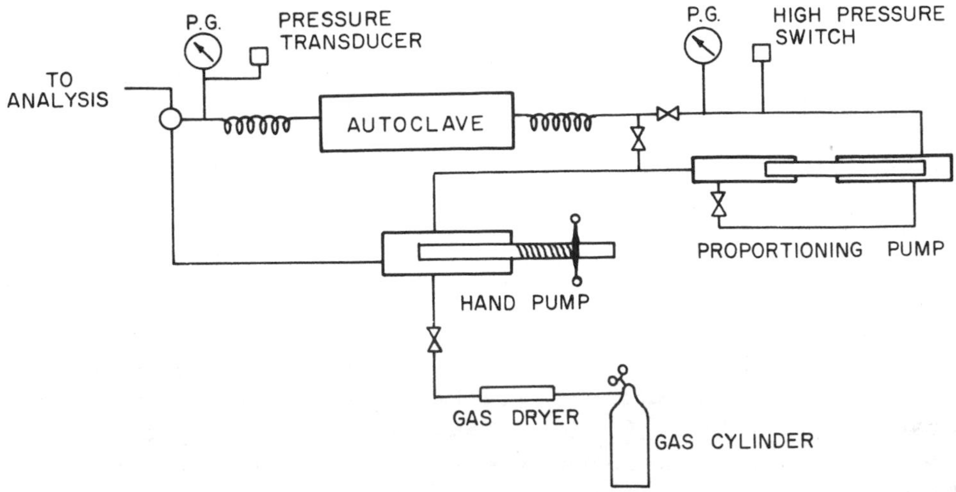


6-2 Kobayashi's Ball Mill Hydrate Apparatus for Three-Phase and for Two-Phase Hydrate Equilibria
(Reproduced, with permission, from (Aoyagi et al., 1980) by Gas Processors Association)

tal work to complement and test the theory. In the 1960's Kobayashi extended a series of studies he had begun with Katz at the University of Michigan; the new work represented high pressure non-visual experimental studies which have continued over thirty years at Rice University. Of particular note are the high pressure studies of Kobayashi with co-workers Saito (1964), Marshall and Saito (1964(2)) and the advances in theory made by Nagata and Kobayashi (1966a,b) and Galloway et al. (1970). The apparatus consisted of a high pressure stainless steel cylinder, which rotated about its axis. Inside the cylinder Galloway inserted stainless steel balls, suggested by Barrer and Edge (1967), to renew the surface area upon rotation, and to enable all of the water present to be converted to ice. Aoyagi et al. (1979) have modified the apparatus, as shown in Figure 6-2, so that gas may be recirculated through the cell for semi-batch operation.

### 6.1.1.c. Apparati for Use Below the Ice Point

Normally experimental conditions have been defined in large part by the needs of the natural gas industry, which in turn determined that experiments be done above the ice point. Below 273.15 K there is the danger of ice as a second solid phase (in addition to hydrates) to cause fouling of transmission or processing equipment. However, since the development of the statistical theory, there has been a need to fit the formation conditions of pure component hydrate formers with the objective of predicting mixtures, as suggested in the Chapter 5. Because most of the upper quadruple points of the hydrocarbon hydrate formers limit the temperature range of simple hydrate formers to a few degrees above the ice point, the region below the ice point was measured to provide more extensive data. However, as recorded in Section 6.2 substantially fewer data exist below the ice point than at higher temperatures.

Deaton and Frost (1946) suggested that the same apparatus be used for conditions below the ice point by first bubbling the gas through water above 273 K, to form a "honeycomb mass" of hydrate. Then free water was drained from the cell before the cell was cooled below the ice point. After the temperature was stabilized, gas was removed in small increments until a region of constant pressure was obtained, which indicated dissociation of the hydrate phase at that temperature and pressure. Deaton and Frost used this procedure only for equilibria of simple hydrates, since the hydrate mass of guest mixtures was not constrained to be of uniform composition, and consequently would have decomposed at different pressures.

For three-phase (I-H-V) equilibria the operating procedure was altered with an apparatus similar to the above, through the use of an aqueous solution of ethylene glycol by Platteeuw and van der Waals (1959) and van der Waals and Platteeuw (1959). Ethylene glycol was chosen because the molecule is too large to fit into the hydrate cage, and the vapor pressure is so low that it will not contribute significantly to the total pressure of the experiment. However, just as ethylene glycol depresses the freezing point of ice, it has been shown to be an effective hydrate inhibitor (see Section 4.5.1 and the inhibitor data of this chapter, Section 6.2.1). As a result, lower temperatures or higher pressures are required to form hydrates, and the results do not quantitatively represent the pure aqueous phase. Rather, the data obtained from the apparatus with ethylene glycol should be considered to provide important qualitative results, such as the illustration of azeotropic behavior in Section 5.1.6.

For glass-tube hydrate equilibria below the ice point, Barrer and co-workers (Rusicka, 1962a,b, and Edge, 1967) and later Falabella and Vanpee (1974,1975) used glass beads or stainless steel balls to provide surface renewal in a shaken glass tube. At the ice point Cady (1981, 1983a,b, 1985) was able to condense hydrates from mixtures of water and guest molecules in a visual glass apparatus. In other innovative experiments below the ice point Holder and co-workers (Godbole, 1982, Kamath, 1982, and John, 1982) performed low pressure measurements below 273.15 K using an electrobalance, a non-visual sampling cylinder, and a static sight glass, respectively.

In general, below the ice point more time is required to equilibrate the two solid phases, ice and hydrate. Byk and Fomina (1968) suggest that water molecule rearrangement is very difficult between the ice non-planar hexagonal structure and the hydrate planar pentagonal cage faces, unless an intermediate structure occurs. Since many molecules in each structure must have some mobility to allow the transition necessary, a liquid-like structure was hypothesized on a molecular scale.

### 6.1.1.d. Apparati for Two-Phase Equilibria

In two-phase (H-V or H-$L_{HC}$) equilibria, one less phase is present than in the more common three-phase measurements; therefore one more intensive variable (in addition to temperature and pressure) must be measured. Typically the water concentration of the hydrocarbon fluid phase is determined. The fluid phase contains very low concentrations of water, and a special means must be used to measure minute water concentrations. The development of a special chromatograph for measurement of such water concentrations was determined by Ertl et al. (1976) for use in the Rice University laboratory.

By far the most productive two-phase (H-V or H-$L_{HC}$) equilibrium apparati has been that used by Kobayashi and co-workers. The same apparatus has been used for two-phase systems such as methane+water (Sloan et al.,1976; Aoyagi et al., 1979), methane+propane+water (Song and Kobayashi, 1982), and carbon dioxide+water (Song and Kobayashi, 1987). The basic apparatus was described previously in Section 6.1.1; however it was used in a unique way for two-phase studies. With two-phase measurements, excess gas was used to convert all of the water to hydrate at a three-phase ($L_W$-H-V) line before the conditions were changed to temperature and pressures in the two-phase region. This requires very careful conditioning of the hydrate phase to prevent metastability and occlusion. Kobayashi and co-workers have equilibrated the hydrate phase by using the ball mill-type apparatus, to continually recondition the hydrate phase.

A second apparatus for the measurement of hydrate-liquid hydrocarbon equilibria was described by Sloan et al. (1986,1987). It was used to measure hydrate equilibria with ethane, propane, ethane+propane, and a quarternary liquid. The basic apparatus used a modified Clausius-Mossotti function to determine the presence of a hydrate phase by a change in slope of the dielectric constant with temperature. The method was described in detail in the theses of Johnson (1981) and Sparks (1983). Because a single equilibrium point required over seven measurements, each with over 24 hours of laboratory effort, the use of this method was considered to be too unwieldy to pursue in a long-range research program.

### 6.1.2 Methods for Measurement of Thermal Properties

The number of measurements for natural gas hydrate thermal properties is several orders of magnitude less than that for measurement of phase equilibrium properties. The experimental difficulties in thermal measurements center on the determination of the composition of the system prior to measurement. The difficulty of system composition determination is due to two factors. First, at temperatures above the ice point, high equilibrium pressures cause decomposition when the apparatus is loaded with preformed hydrate. Second, hydrate metastability and component occlusion causes extreme difficulty in completely converting all of the water to hydrate.

As a consequence of the above experimental difficulties, more experimentalists have chosen an easier route of measurements of the thermal properties of the cyclic ethers, e.g. ethylene oxide for structure I, or tetrahydrofuran for structure II, which are totally miscible with liquid water. Such compounds may be formed in the apparatus at the theoretical hydrate compositions, without problems of diffusion or occlusion. In such measurements, the concept was to obtain the hydrate crystal structure in the hope that differences in guest molecules would contribute in a minor way to the thermal properties. Table 6-2 is a listing of such measurements on cyclic ethers and other non-natural gas components.

#### 6.1.2.a. Heat Capacity and Heat of Dissociation

In a thorough review of calorimetric studies of clathrates and inclusion compounds, Parsonage and Staveley (1984) presented no direct calorimetric methods used for natural gas hydrate measurements. Instead, the heat of dissociation has been indirectly determined via the Clapeyron equation by differentiation of three-phase equilibrium pressure-temperature data. This technique is presented in detail in Section 4.7. However, as discussed by Barrer (1959), there is an inherent difficulty in

**Table 6-2. Measurements of Thermal Properties for Hydrates of Non-Natural Gas Components**

**Heat Capacity and Heat of Dissociation**

| Investigator(s)/date | Component | Range of Experiments |
|---|---|---|
| Ross and Andersson/1982 | Tetrahydrofuran | $C_v$:100-260K;P<1.5GPa |
| Leaist,Murray,Post and Davidson/1982 | Ethylene Oxide | $\Delta H$ and $C_p$: 120-260 K |
| | Tetrahydrofuran | $\Delta H$ and $C_p$: 120-260 K |
| Callanan and Sloan/1982 | Ethylene Oxide | $\Delta H$ and $C_p$: 240-270 K |
| | Tetrahydrofuran | $\Delta H$ and $C_p$: 240-270 K |
| | Cyclopropane | $C_p$: 240-265 K |
| Handa/1984 | Tetrahydrofuran | $\Delta H$ and $C_p$: 100-270 K |
| Handa/1985 | Trimethylene Oxide | $\Delta H$ and $C_p$: 85-270 K |
| Rueff and Sloan/1985 | Tetrahydrofuran | $\Delta H$ and $C_p$: 240-265 K |
| White and MacLean/1985 | Tetrahydrofuran | $C_p$: 17-261 K |
| Handa/1986a | Xenon | $\Delta H$:273K;$C_p$:150-230 K |
| Handa/1986b | Xenon and Krypton | $\Delta H$ and $C_p$: 85-270 K |

**Thermal Conductivity**

| Investigator(s)/date | Component | Range of Experiments |
|---|---|---|
| Cook and Laubitz/1981 | Ethylene Oxide<br>Tetrahydrofuran | at ambient freezing point |
| Ross, Andersson, and Bäckström/1981 | Tetrahydrofuran | 100-277 K; 0.1 GPa |
| Ross and Andersson/1982 | Tetrahydrofuran | 100-260 K; P≦1.5 GPa |

**Table 6-2 (continued)**
**Measurements of Thermal Properties of Hydrates of Non-Natural Gas Components**

**Thermal Conductivity**

| | | |
|---|---|---|
| Andersson and Ross/1983 | 1,3 Dioxolane<br>Cyclobutanone | 100-260K;$0.05 \le P \le 1.0$GPa<br>100-260K;$0.05 \le P \le 1.5$GPa |
| Ashworth, Johnson Lai/1985 | Tetrahydrofuran | 45-160 K |
| Ahmad and Phillips/1987 | 1,3 Dioxolane | T < 200 K; ambient pressure |
| Asher/1987 | Tetrahydrofuran | 273 K; ambient pressure |

the Clapeyron method, particularly when there is significant non-stoichiometry, as in the case for molecules which occupy the smaller cavities (see Example 5-1.) Additionally, while the Clapeyron equation often provides acceptable estimates of the heat of dissociation, no information about the hydrate heat capacity is directly determined by that equation.

Recently three experimental techniques have been used to measure the clathrate hydrate heat capacity and enthalpy of dissociation for natural gas components. These activities reflect a growing interest in the thermal properties of hydrates as related to recovery from **in situ** reservoirs, which is discussed further in Chapter 7.

At the Canadian National Research Council, Handa modified a Setaram, model BT Tian-Calvet calorimeter for high pressures. This device was used to measure the heat capacity and heat of dissociation between 85 and 270 K for methane, ethane, propane (Handa, 1986b), isobutane (Handa, 1986c), and the heat of dissociation of two natu-

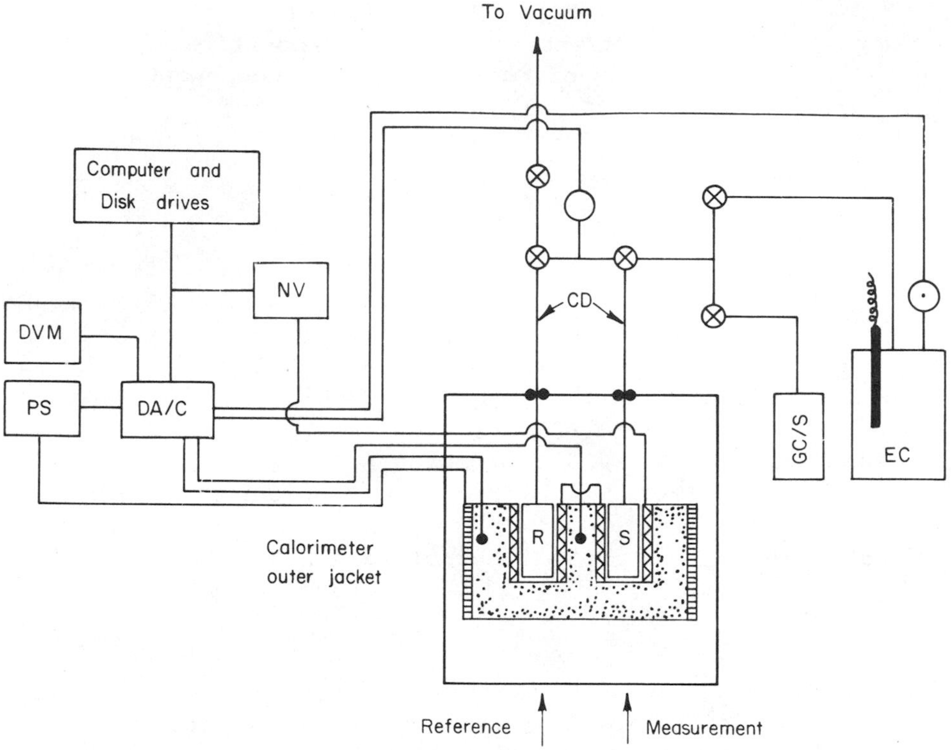


6-3 Schematic of Tian-Calvet Calorimeter at Canadian National Research Council (Reproduced, with permission, from (Handa, 1986c))

rally occurring hydrates (Handa, 1988). An abbreviated diagram of the calorimeter is provided in Figure 6-3.

Using the instrument in Figure 6-3, Handa measured the heat input to a hydrate sample in the sample container S, relative to helium at ambient temperature and 5 kPa in reference cell R. The hydrate was externally prepared from ice in a rolling-rod mill, before a 4 gram sample was loaded into the calorimeter at liquid nitrogen temperatures. For heat capacity measurements, a pressure greater than the hydrate dissociation pressure was maintained in the sample cell. By cycling the temperature around 273.15 K, Handa was able to determine the amount

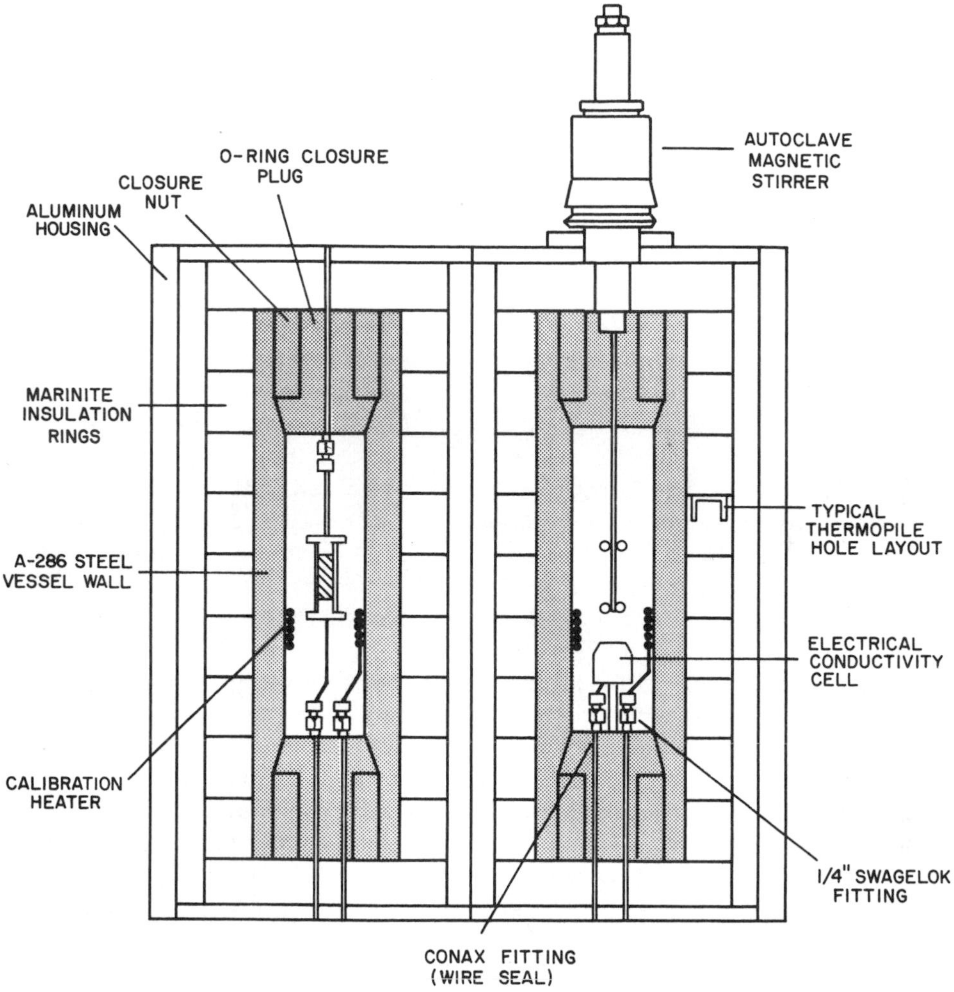


6-4 Schematic of Heat Flux Calorimeter at Rice University (Reproduced, with permission, from (Lievois, 1987))

of ice present in the sample, and to convert most of the ice to hydrate. For hydrate dissociation measurements, the amount of gas released from the hydrates was determined by PVT analysis. Handa estimated the precision to be ±1% for all measurements; the accuracies for $C_p$ >100 K and for $\Delta H$ were ±1%, while the accuracy for $C_p$ <100 K was ±1.5%.

In this laboratory, Rueff et al. (1988) used a Perkin-Elmer differential scanning calorimeter (DSC-2), with sample containers modified for high pressure, to obtain methane hydrate heat capacity (245-259 K) and heat of dissociation (285 K), which were accurate to within 20%. Rueff (1985) was able to analyze his data to account for the portion of the sample which was ice, in an extension of work done earlier (Rueff and Sloan, 1985) to measure thermal properties of hydrates in sediments.

At Rice University, Lievois (1987) developed the high pressure twin-cell heat flux calorimeter shown in Figure 6-4. In the right cell of the calorimeter, he made heat of dissociation measurements at 278.15 K, 4.36 MPa and at 283.15 K, 7.35 MPa. The left cell was intended for future use in the measurement of heat of dissociation of hydrates in cores. A magnetic agitator was used to promote hydrate formation in the right cell. Unlike Handa's apparatus, Lievois circumvented the experimental difficulty of complete conversion of the water phase by monitoring partial water phase conversion. During hydrate formation the amount of water encapsulated was determined through an electrical conductivity increase of a salt solution (salts are excluded by the hydrate phase). The amount of gas encapsulated was determined by PVT analysis. Hydrates were dissociated isothermally with a stated accuracy of ±2.6%.

### 6.1.2.b. Methods for Thermal Conductivity Measurements

Of the few thermal conductivity measurements which have been performed on natural gas hydrates, only two experimental methods have been determined: the transient method and the steady state method. Afanaseva and Groisman (1973) first measured hydrate thermal conductivity to be the same as ice within their stated accuracy (to within 10%).

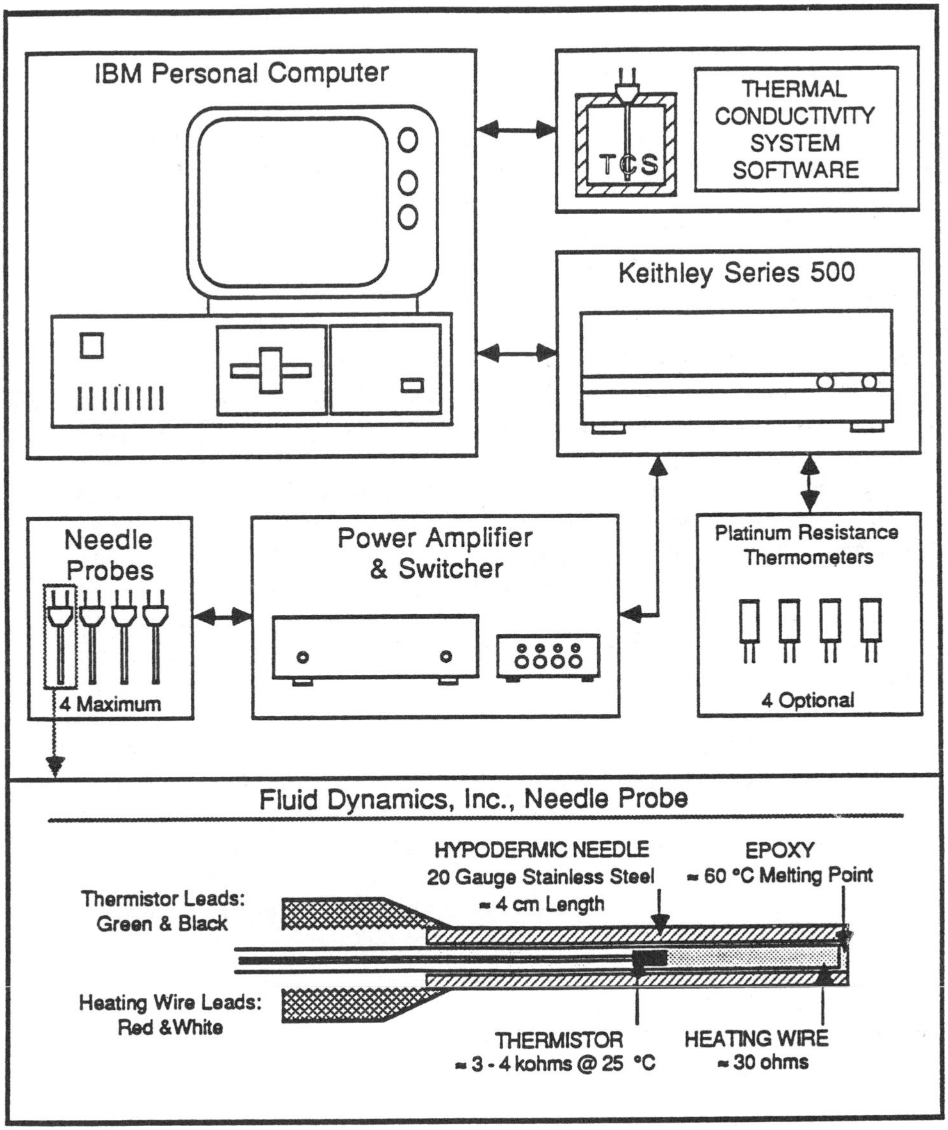


6-5 Schematic of Thermal Conductivity System
at Colorado School of Mines
(Asher, 1987)

Stoll and Bryan (1979) first used the transient needle probe method on propane hydrates which had been compacted after formation. That probe was modelled after the probe of von Herzen and Maxwell (1959) which had been used with marine sediments. A slightly modified version of the probe is shown in Figure 6-5 to be a stainless steel 20 gage hypodermic tube containing a full length heater wire, a mid-tube thermistor, and a solid filling the remainder of the annulus. When a step power change is input to the heater, the probe temperature varies with the thermal conductivity of the material surrounding the probe. With good media contact at the probe boundary, the logarithmic increase of temperature with time gives an inverse relationship with the thermal conductivity of the surrounding hydrate. In this laboratory the apparatus shown in Figure 6-5 was computerized and the mathematical model was refined and extended to short times by Asher (1987) for methane hydrates in sediments, as a possible well-logging tool. The accuracy of the thermal conductivity provided by short-time solution analysis is estimated to be within 8% for this transient method.

A conventional steady-state guarded hot-plate method for thermal conductivity measurement was used by Cook and Leaist (1983). Their apparatus, shown in Figure 6-6, was used to perform an exploratory measurement of methane hydrate with data scatter of 12%, which also approximates the inaccuracy of the method. In the apparatus, a sample of methane hydrate was made externally, pressed, and placed in the hot-plate cell at the point labelled "Sample Disc" in Figure 6-6. The lower sample heater (marked with crosses) had thermocouples (dots) contacting the top and the bottom of the sample to determine the temperature gradient.

Much more remains to be done in the area of thermal conductivity of hydrates. There seems to be a general consensus among such experimentalists that neither the types of guest molecules nor the types of hydrate crystal structure appear to have a significant effect on the thermal conductivity, which is anomalously low, relative

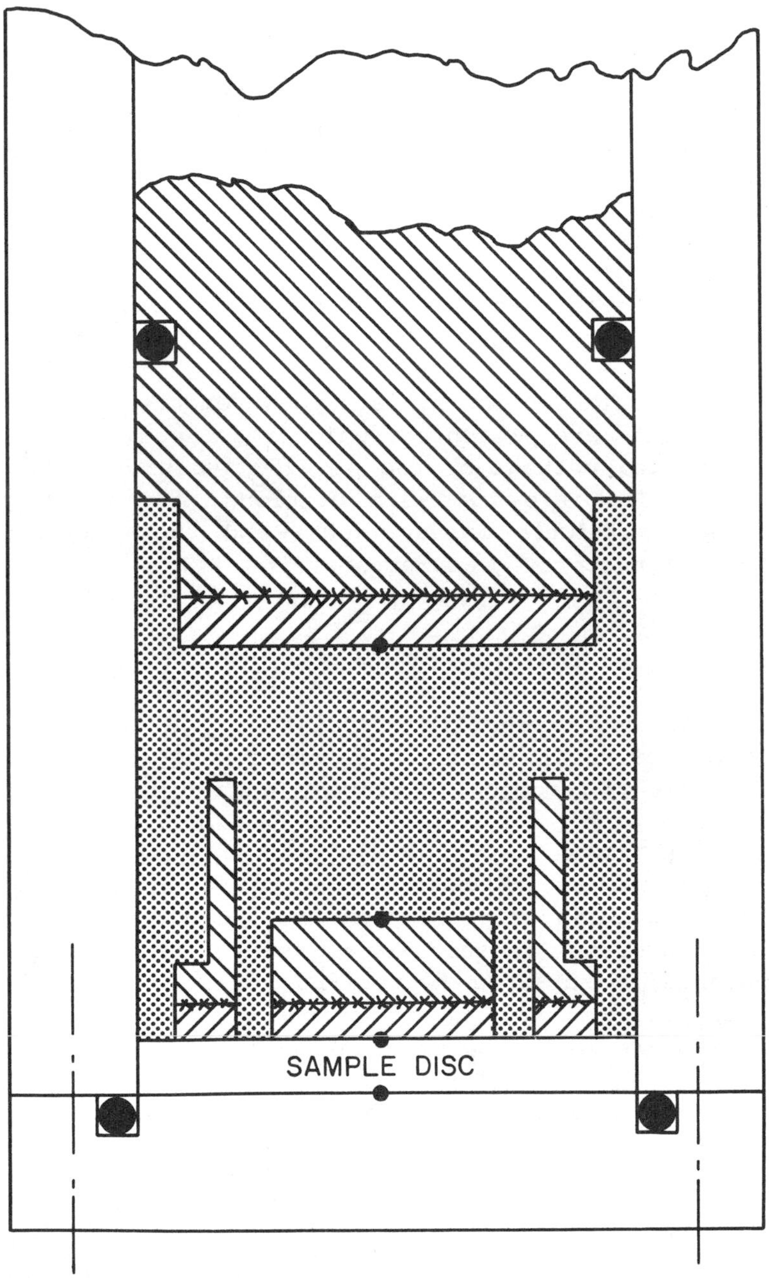


6-6 Steady State Thermal Conductivity System
at Canadian National Research Council
(Reproduced, with permission, from Geophys. Rsch. Lett.,(Cook and Leaist, 1983) by American Geophysical Union)

to the ice thermal conductivity. Indeed, as shown by Asher (1987) this substantial difference in thermal conductivities between ice and hydrate can be used to quantify the presence of hydrate in a sample. Three hypotheses for the relatively low hydrate thermal conductivity are given in Section 2.2.3.

## 6.2 Data for Natural Gas Hydrate Phase Equilibria and Thermal Properties

In this section hydrate measurements for natural gas components are presented, which result primarily from experimental methods of the previous section. The section is divided into two main portions: the first presents phase equilibria measurements, and the second presents thermal property measurements.

### 6.2.1 Phase Equilibria Data

The phase equilibria data for hydrates are divided into four sections for hydrates of single, binary, ternary, and multicomponent guests, with a final section for hydrates with inhibitors. In the nomenclature of this section, note that simple hydrates (single guest components) are actually binary mixtures in a thermodynamic sense, with water present as the second component; binary guest hydrates are ternary mixtures, etc. The guest components, water-free hydrocarbon phase compositions (except in the two-phase regions), and the equilibrium phases are indicated with each data set.

For ease of access and accuracy, only raw data dating from Hammerschmidt (1934) have been listed in chronological order. The correlation methods and smoothed data are relegated to Chapter 4. Table 6-3 is presented as a summary of the number of phase equilibria studies which have been performed for the simple and the binary hydrates of natural gas components over this period. The values along the matrix diagonal in the table represent the number of simple hydrate studies while the off-diago-

**Table 6-3**
**Number of Single and Binary Guest Hydrate Phase Equilibria Studies**

| | $CH_4$ | $C_2H_6$ | $C_3H_8$ | $i\text{-}C_4H_{10}$ | $n\text{-}C_4H_{10}$ | $N_2$ | $CO_2$ | $H_2S$ |
|---|---|---|---|---|---|---|---|---|
| $CH_4$ | 12 | 3 | 5 | 5 | 5 | 1 | 1 | 1 |
| $C_2H_6$ | | 9 | 2 | 0 | 0 | 0 | 0 | 0 |
| $C_3H_8$ | | | 10 | 2 | 2 | 1 | 1 | 0 |
| $i\text{-}C_4H_{10}$ | | | | 5 | 1 | 0 | 0 | 0 |
| $n\text{-}C_4H_{10}$ | | | | | 0 | 0 | 0 | 0 |
| $N_2$ | | | | | | 3 | 0 | 0 |
| $CO_2$ | | | | | | | 8 | 0 |
| $H_2S$ | | | | | | | | 2 |

nal values provide the number of binary studies. For example there have been 10 experimental studies of simple propane hydrates, but only two sets of measurements on the propane+isobutane binary hydrate. One conclusion of such a table is that there is a paucity of binary guest hydrate data for systems (a) which do not contain methane, and (b) for systems with the non-combustible components, i.e. nitrogen, carbon dioxide, and hydrogen sulfide. Data for ternary and multicomponent guest hydrates are even more sparse.

### 6.2.1.a. Equilibria of Simple Natural Gas Components

The accuracy of hydrate data has seldom been specified by the various experimentalists. Only a cursory effort has been made to exclude inaccurate data for simple hydrates. All of the three-phase data sets for simple hydrates were plotted (as logarithm pressure versus absolute temperature) to determine outliers.

In general the accuracy of the data were surprisingly good. For example, while Deaton and Frost (1946, p.

13) specified that their "pure" ethane contained 2.1% propane and 0.8% methane, effects of those impurities may have counterbalanced each other; those impurities were insufficient to cause the data to fall outside the line formed by other ethane data sets. On the other hand, the simple hydrate data of Hammerschmidt (1934) for propane and isobutane all appear to be outliers on such semilogarithmic plots, because they are at temperatures much too far above the upper quadruple ($Q_2$) points. The obvious outlying data were excluded from this work; less obvious outliers may be determined by inspection of the plots and subsequent numerical comparisons. The data, followed by the semilogarithmic plots for each simple hydrate, are listed chronologically by compound, in the following order: methane, ethane, propane, isobutane, carbon dioxide, nitrogen, and hydrogen sulfide.

While most of the simple hydrate data consist of the three-phase and quadruple point type, the available two-phase simple hydrate data are listed for methane, ethane, propane, and carbon dioxide. Plots of these data are not suitable for comparison between data sets and are therefore not provided.

Methane

Hydrate: Methane
Reference: Roberts, O.L. et al. (1940)

| T(K) | P(MPa) | Phases | T(K) | P(MPa) | Phases |
|---|---|---|---|---|---|
| 259.1 | 1.648 | I-H-V | 286.5 | 10.63 | $L_W$-H-V |
| 273.2 | 2.641 | $L_W$-I-H-V | 286.7 | 10.80 | $L_W$-H-V |
| 280.9 | 5.847 | $L_W$-H-V | | | |

Hydrate: Methane
Reference: Deaton, W.M., Frost, E.M., Jr. (1946)
Phases: I-H-V and $L_W$-H-V

| T(K) | P(MPa) | T(K) | P(MPa) |
|---|---|---|---|
| | I-H-V | | |
| 262.4 | 1.79 | 268.6 | 2.22 |
| 264.2 | 1.90 | 270.9 | 2.39 |
| 266.5 | 2.08 | | |
| | $L_W$-H-V | | |
| 273.7 | 2.77 | 280.4 | 5.35 |
| 274.3 | 2.90 | 280.9 | 5.71 |
| 275.4 | 3.24 | 281.5 | 6.06 |
| 275.9 | 3.42 | 282.6 | 6.77 |
| 275.9 | 3.43 | 284.3 | 8.12 |
| 277.1 | 3.81 | 285.9 | 9.78 |
| 279.3 | 4.77 | | |

Hydrate: Methane
Reference: Kobayashi, R., Katz, D.L. (1949)
Phases: $L_W$-H-V

| T(K) | P(MPa) | T(K) | P(MPa) |
|---|---|---|---|
| 295.72 | 33.992 | 301.00 | 64.812 |
| 295.89 | 35.302 | 302.00 | 77.499 |

Hydrate: Methane
Reference: McLeod, H.O., Campbell, J.M. (1961)
Phases: $L_W$-H-V

| T(K) | P(MPa) | T(K) | P(MPa) |
|---|---|---|---|
| 285.7 | 9.62 | 292.1 | 21.13 |
| 286.3 | 10.31 | 295.9 | 34.75 |
| 286.1 | 10.10 | 298.7 | 48.68 |

| | | | |
|---|---|---|---|
| 285.7 | 9.62 | 300.9 | 62.40 |
| 289.0 | 13.96 | 301.6 | 68.09 |

Compound: Methane
Reference: Marshall, D.R., Saito, S., Kobayashi, R. (1964)
Phases: $L_W$-H-V

| T(K) | P(MPa) | T(K) | P(MPa) |
|---|---|---|---|
| 290.2 | 15.9 | 306.7 | 110.8 |
| 290.5 | 15.9 | 310.3 | 152.7 |
| 295.2 | 30.0 | 312.7 | 187.3 |
| 295.1 | 29.9 | 313.7 | 206.3 |
| 295.8 | 33.8 | 314.2 | 223.9 |
| 298.1 | 44.3 | 315.1 | 237.5 |
| 298.2 | 43.8 | 316.8 | 271.7 |
| 300.2 | 56.9 | 318.3 | 319.7 |
| 301.6 | 65.4 | 319.6 | 367.8 |
| 301.6 | 65.4 | 320.1 | 397.0 |

Hydrate: Methane
Reference: Jhaveri, J., Robinson, D.B. (1965)
Phases: $L_W$-H-V

| T(K) | P(MPa) | T(K) | P(MPa) |
|---|---|---|---|
| 273.2 | 2.65 | 287.3 | 11.65 |
| 277.6 | 4.17 | 288.9 | 14.05 |
| 280.4 | 5.58 | 291.7 | 20.11 |
| 284.7 | 8.67 | 294.3 | 28.57 |

Hydrate: Methane
Reference: Galloway,T.J. et al. (1970)
Phases: $L_W$-H-V

| T(K) | P(MPa) | T(K) | P(MPa) |
|---|---|---|---|
| 283.2 | 7.10 | 288.7 | 13.11 |
| 283.2 | 7.12 | 288.7 | 13.11 |

Hydrate: Methane
Reference: Falabella, B.J. (1975)
Phases: I-H-V

| T(K) | P(kPa) | T(K) | P(kPa) |
|---|---|---|---|
| 148.8 | 5.3 | 178.2 | 42.0 |
| 159.9 | 12.1 | 191.3 | 90.1 |
| 168.8 | 21.1 | 193.2 | 101.3 |

Hydrate: Methane
Reference: Verma, V.K. (1974)
Phases: $L_W$-H-V

| T(K) | P(MPa) | T(K) | P(MPa) |
|---|---|---|---|
| 275.2 | 3.020 | 288.5 | 13.044 |
| 276.7 | 3.689 | 290.7 | 16.961 |
| 278.6 | 4.385 | 291.2 | 18.554 |
| 285.4 | 9.191 | | |

Hydrate: Methane
Reference: Aoyagi, K., et al.(1979)
Phases: V-H

| T K | P MPa | $H_2O$ ppm(mol) | T K | P MPa | $H_2O$ ppm(mol) |
|---|---|---|---|---|---|
| 240.0 | 3.447 | 12.30 | 260.0 | 3.447 | 78.24 |
| 240.0 | 6.895 | 5.60 | 260.0 | 6.895 | 39.56 |
| 240.0 | 10.343 | 2.72 | 260.0 | 10.343 | 24.23 |

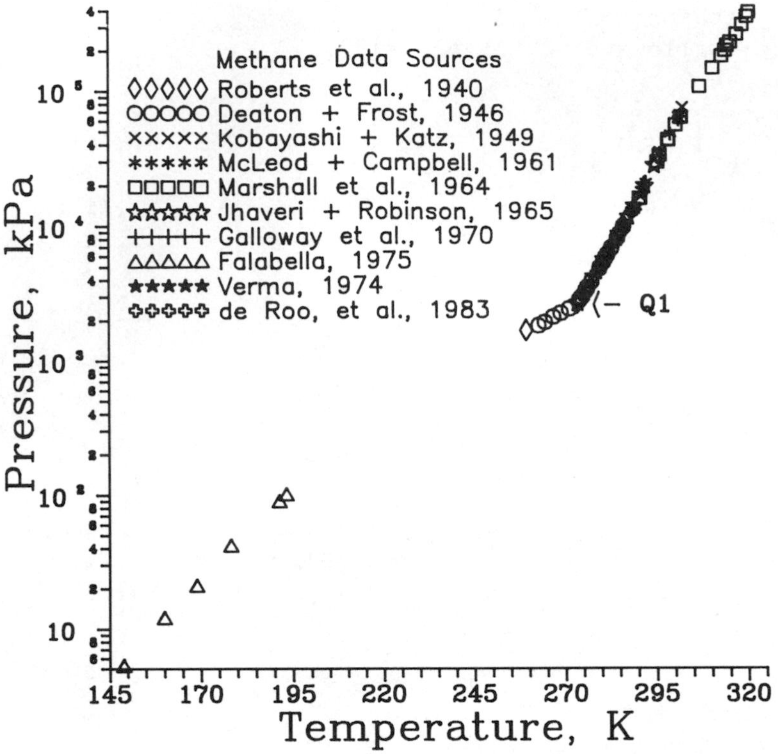


6-7 Three-Phase Data for Simple Hydrates of Methane

| | | | | | |
|---|---|---|---|---|---|
| 250.0 | 3.447 | 32.17 | 270.0 | 3.447 | 178.09 |
| 250.0 | 6.895 | 15.45 | 270.0 | 6.895 | 94.43 |
| 250.0 | 10.343 | 8.46 | 270.0 | 10.343 | 64.22 |

Hydrate: Methane
Reference: de Roo, J.L., et al. (1983)
Phases: $L_W$-H-V

| T(K) | P(MPa) | T(K) | P(MPa) |
|---|---|---|---|
| 273.30 | 2.69 | 282.80 | 7.04 |
| 275.41 | 3.34 | 283.98 | 8.05 |

| | | | |
|---|---|---|---|
| 275.95 | 3.34 | 284.98 | 9.04 |
| 279.49 | 5.04 | 285.98 | 10.04 |
| 281.29 | 6.04 | | |

Hydrate: Methane
Reference: Thakore, J.L., Holder, G.D. (1987)
Phases: $L_W$-H-V

| T(K) | P(MPa) | T(K) | P(MPa) |
|---|---|---|---|
| 275.35 | 2.87 | 278.15 | 4.50 |
| 276.15 | 3.37 | 279.15 | 4.90 |
| 277.15 | 3.90 | 281.15 | 6.10 |

## Ethane

Hydrate: Ethane
Reference: Roberts, O.L. et al. (1940)

| T(K) | P(kPa) | Phases | T(K) | P(kPa) | Phases |
|---|---|---|---|---|---|
| 260.8 | 294 | I-H-V | 285.8 | 2537 | $L_W$-H-V |
| 260.9 | 290 | I-H-V | 287.0 | 3054 | $L_W$-H-V |
| 269.3 | 441 | I-H-V | | | |
| | | | 287.7 | 4909 | $L_W$-H-$L_E$ |
| 273.4 | 545 | $L_W$-H-V | 287.8 | 3413 | $L_W$-H-$L_E$ |
| 275.4 | 669 | $L_W$-H-V | 287.8 | 4289 | $L_W$-H-$L_E$ |
| 277.6 | 876 | $L_W$-H-V | 288.1 | 3716 | $L_W$-H-$L_E$ |
| 279.1 | 1048 | $L_W$-H-V | 288.1 | 6840 | $L_W$-H-$L_E$ |
| 279.7 | 1131 | $L_W$-H-V | 288.2 | 4944 | $L_W$-H-$L_E$ |
| 281.1 | 1317 | $L_W$-H-V | 288.2 | 5082 | $L_W$-H-$L_E$ |
| 282.8 | 1641 | $L_W$-H-V | 288.3 | 4358 | $L_W$-H-$L_E$ |
| 284.4 | 2137 | $L_W$-H-V | 288.4 | 6840 | $L_W$-H-$L_E$ |
| 284.6 | 2055 | $L_W$-H-V | | | |

Hydrate: Ethane
Reference: Deaton, W.M., Frost, E.M., Jr. (1946)
Phases: I-H-V and $L_W$-H-V

| T(K) | P(kPa) | T(K) | P(kPa) |
|---|---|---|---|
| | I-H-V | | |
| 263.6 | 313 | 269.3 | 405 |
| 266.5 | 357 | 272.0 | 457 |
| | $L_W$-H-V | | |
| 273.7 | 510 | 280.4 | 1165 |
| 273.7 | 503 | 280.9 | 1255 |
| 274.8 | 579 | 281.5 | 1345 |
| 275.9 | 662 | 282.1 | 1448 |
| 277.6 | 814 | 282.6 | 1558 |
| 278.7 | 931 | 283.2 | 1689 |
| 278.7 | 931 | 284.3 | 1986 |
| 279.3 | 1007 | 285.4 | 2303 |
| 279.8 | 1083 | 285.4 | 2310 |
| 280.4 | 1165 | 286.5 | 2730 |

Hydrate: Ethane
Reference: Reamer, H.H. et al. (1952)
Phases: $L_W$-H-V

| T(K) | P(kPa) | T(K) | P(kPa) |
|---|---|---|---|
| 279.9 | 972 | 284.7 | 2129 |
| 282.8 | 1666 | 287.4 | 3298 |

Hydrate: Ethane
Reference: Galloway, T. J. et al. (1970)
Phases: $L_W$-H-V

| T(K) | P(kPa) | T(K) | P(kPa) |
|---|---|---|---|
| 277.6 | 814 | 282.5 | 1551 |
| 277.7 | 823 | | |

Hydrate: Ethane
Reference: Falabella, B.J. (1975)
Phases: I-H-V

| T(K) | P(kPa) | T(K) | P(kPa) |
|---|---|---|---|
| 200.8 | 8.3 | 240.4 | 98.1 |
| 215.7 | 22.1 | 240.8 | 101.3 |
| 230.2 | 56.4 | | |

Hydrate: Ethane
Reference: Holder, G.D., Grigoriou, G.C. (1980)
Phases: $L_W$-H-V

| T(K) | P(kPa) | T(K) | P(kPa) |
|---|---|---|---|
| 277.5 | 780 | 283.3 | 1660 |
| 278.1 | 840 | 284.5 | 2100 |
| 279.9 | 1040 | 286.5 | 2620 |
| 281.5 | 1380 | | |

Hydrate: Ethane
Reference: Holder,G.D., Hand,J.H. (1982)
Phases: $L_W$-H-V

| T(K) | P(kPa) | T(K) | P(kPa) |
|---|---|---|---|
| 278.8 | 950 | 286.0 | 2510 |
| 280.2 | 1140 | 286.5 | 2600 |
| 282.0 | 1450 | 288.2 | 3360 |
| 281.1 | 1280 | | |

Hydrate: Ethane
Reference: Ng,H.-J., Robinson, D.B. (1985)
Phases: $L_W$-H-$L_E$

| T(K) | P(MPa) | T(K) | P(MPa) |
|---|---|---|---|
| 288.0 | 3.33 | 288.5 | 6.99 |
| 288.1 | 3.84 | 289.2 | 10.39 |
| 288.2 | 5.00 | 289.7 | 13.95 |
| 288.4 | 6.06 | 290.6 | 20.34 |

Hydrate: Ethane
Reference: Sloan, E.D., et al. (1987)
Phases: H-$L_E$

Isobar at 3.45 MPa

| T(K) | Conc.(ppm)mol | T(K) | Conc.(ppm)mol |
|---|---|---|---|
| 259.1 | 34.8±2 | 266.7 | 62.7±4 |
| 261.7 | 41.6±3 | 270.0 | 80.0±5 |
| 264.9 | 50.1±3 | 270.5 | 86.2±5 |

Hydrate: Ethane
Reference: Avlonitis, D.A. (1988)
Phases: $L_W$-H-V

| T(K) | P(MPa) | T(K) | P(MPa) | T(K) | P(MPa) |
|---|---|---|---|---|---|
| 277.82 | 0.848 | 281.49 | 1.365 | 283.95 | 1.889 |
| 278.56 | 0.945 | 282.10 | 1.510 | 285.90 | 2.461 |
| 279.40 | 1.055 | 282.28 | 1.551 | 287.15 | 3.082 |
| 280.45 | 1.200 | | | | |

Propane

Hydrate: Propane
Reference: Wilcox, W.I., Carson, D.B., Katz, D.L. (1941)
Phases: $L_W$-H-$L_P$

| T(K) | P(kPa) | T(K) | P(kPa) |
|---|---|---|---|
| 278.9 | 807 | 279.2 | 2902 |
| 278.6 | 1296 | 278.8 | 4247 |
| 278.6 | 1758 | 278.9 | 6115 |
| 278.8 | 2034 | | |

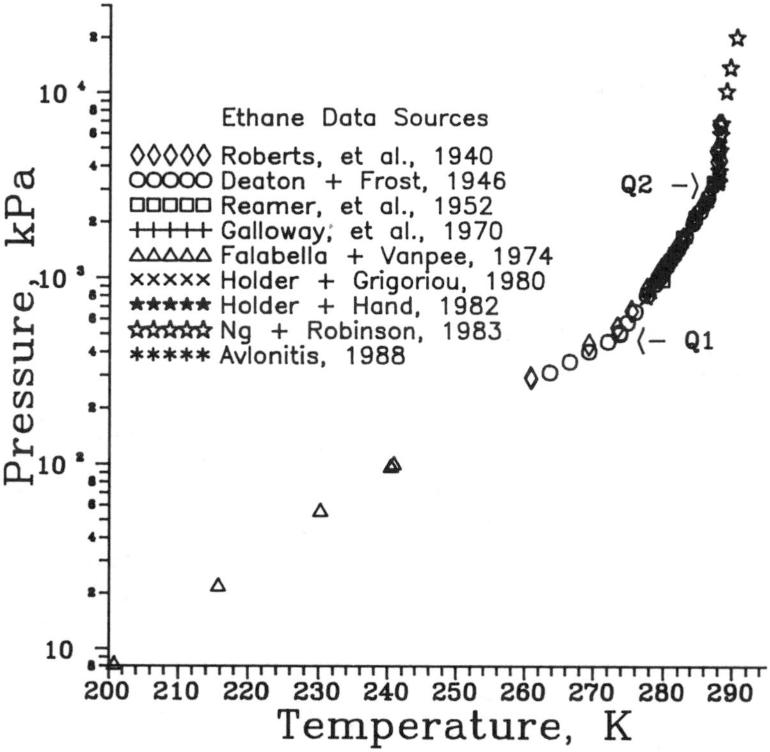


6-8 Three-Phase Data for Simple Hydrates of Ethane

Hydrate: Propane
Reference: Miller, B., Strong, E.R. (1945)
Phases: $L_W$-H-V

| T(K) | P(kPa) | T(K) | P(kPa) |
|---|---|---|---|
| 273.2 | 165 | 276.8 | 365 |
| 273.4 | 172 | 277.1 | 390 |
| 273.5 | 176 | 277.2 | 393 |

| | | | |
|---|---|---|---|
| 273.7 | 186 | 277.8 | 459 |
| 273.9 | 190 | 278.0 | 472 |
| 274.6 | 221 | 278.1 | 479 |
| 275.9 | 293 | 278.8 | 548 |
| 276.2 | 317 | 278.9 | 565* |
| 276.7 | 345 | | |

* $Q_2$ quadruple point ($L_W$-H-V-$L_P$)

Hydrate: Propane
Reference: Deaton, W.M., Frost, E.M., Jr. (1946)
Phases: I-H-V and $L_W$-H-V

| T(K) | P(kPa) | T(K) | P(kPa) |
|---|---|---|---|
| | I-H-V | | |
| 261.2 | 100 | 269.8 | 149 |
| 264.2 | 115 | 272.2 | 167 |
| 267.4 | 132 | 272.9 | 172 |
| 267.6 | 135 | | |
| | $L_W$-H-V | | |
| 273.7 | 183 | 275.4 | 270 |
| 273.7 | 183 | 275.9 | 301 |
| 274.8 | 232 | 277.1 | 386 |

Hydrate: Propane
Reference: Reamer, H.H., Selleck, F.T., Sage, B.H. (1952)
Phases: $L_W$-H-V and $L_W$-H-$L_P$

| T(K) | P(kPa) | T(K) | P(kPa) |
|---|---|---|---|
| | $L_W$-H-V | | |
| 274.3 | 241 | 277.2 | 414 |
| 275.7 | 305 | | |
| | $L_W$-H-$L_P$ | | |
| 278.6 | 684 | 278.8 | 2,046 |
| 278.7 | 1,477 | | |

Hydrate: Propane
Reference: Robinson, D.B., Mehta, B.R. (1971)
Phases: $L_W$-H-V

| T(K) | P(kPa) | T(K) | P(kPa) |
|---|---|---|---|
| 274.3 | 207 | 277.8 | 455 |
| 274.8 | 241 | 278.9 | 552* |
| 276.4 | 331 | | |

* $Q_2$ quadruple point ($L_W$-H-V-$L_P$)

Hydrate: Propane
Reference: Verma, V. K. (1974)
Phases: $L_W$-H-V and $L_W$-H-$L_P$

| T(K) | P(kPa) | T(K) | P(kPa) |
|---|---|---|---|
| | $L_W$-H-V | | |
| 273.9 | 188 | 276.7 | 361 |
| 274.6 | 219 | 277.4 | 425 |
| 275.1 | 250 | 278.0 | 512 |
| 275.7 | 288 | 278.4 | 562* |
| 276.2 | 322 | | |
| | $L_W$-H-$L_P$ | | |
| 278.42 | 3868 | 278.6 | 11,315 |
| 278.5 | 7026 | 278.5 | 16,762 |

* $L_W$-H-V-$L_P$ quadruple point.

Hydrate: Propane
Reference: Holder, G.D., Godbole, S.P. (1982)
Phases: I-H-V

| T(K) | P(kPa) | T(K) | P(kPa) |
|---|---|---|---|
| 247.9 | 48.2 | 258.2 | 81.1 |
| 251.4 | 58.3 | 260.8 | 90.5 |
| 251.6 | 58.3 | 260.9 | 94.5 |
| 255.4 | 69.6 | 262.1 | 99.4 |

Hydrate: Propane
Reference: Thakore, J.L., Holder, G.D. (1987)
Phases: $L_W$-H-V

| T(K) | P(kPa) | T(K) | P(kPa) |
|---|---|---|---|
| 274.15 | 217 | 277.15 | 450 |
| 275.15 | 248 | 278.15 | 510 |
| 276.15 | 310 | | |

Hydrate: Propane
Reference: Patil, S.L. (1987)
Phases: $L_W$-H-V

| T(K) | P(kPa) | T(K) | P(kPa) |
|---|---|---|---|
| 273.65 | 207 | 277.15 | 417 |
| 274.65 | 248 | 277.95 | 510 |
| 276.15 | 338 | | |

Hydrate: Propane
Reference: Sloan, E.D., et al. (1987)
Phases: H-$L_P$

Isobar at 0.772 MPa

| T(K) | Conc.(ppm)mol | T(K) | Conc.(ppm)mol |
|---|---|---|---|
| 246.7 | 16.9±2 | 272.2 | 123.9± 8 |
| 259.9 | 50.5±3 | 273.0 | 125.8± 8 |

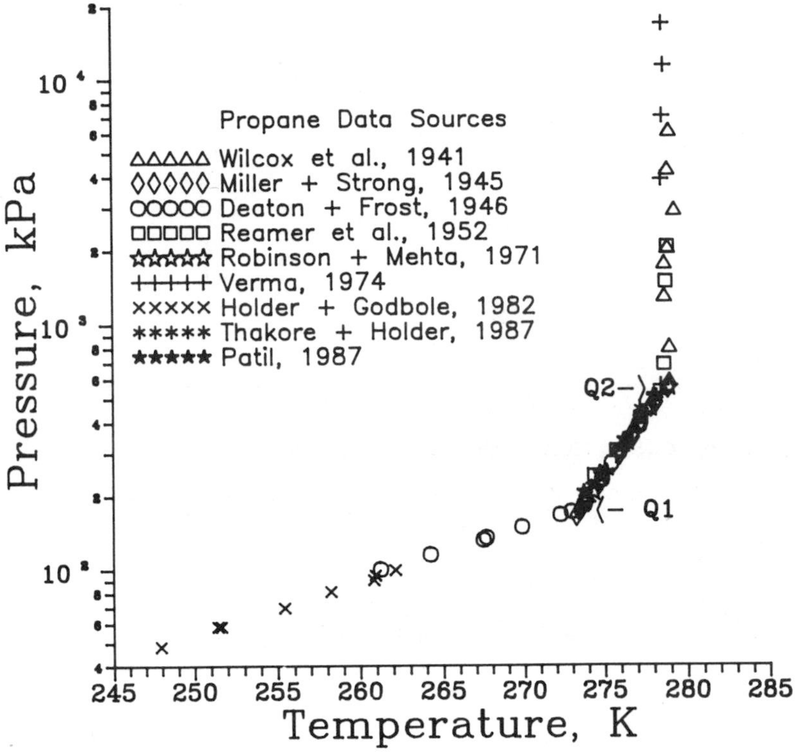


6-9 Three-Phase Data for Simple Hydrates of Propane

| | | | |
|---|---|---|---|
| 264.3 | 65.7±4 | 273.9 | 136.9± 8 |
| 266.0 | 87.6±5 | 276.5 | 176.1±11 |
| 269.9 | 102.5±6 | | |

Isobutane

Hydrate: Isobutane
Reference: Schneider, G.R., Farrar, J. (1968)
Phases: $L_W$-H-V and I-H-V

| T(K) | P(kPa) | T(K) | P(kPa) |
|---|---|---|---|
| | I-H-V | | |
| 273.1 | 109 | 272.3 | 105 |
| 273.1 | 109 | 272.2 | 103 |

| | | | |
|---|---|---|---|
| 272.8 | 109 | 271.2 | 95 |
| 272.8 | 102 | | |

$L_W$-H-V

| | | | |
|---|---|---|---|
| 275.1 | 167 | 273.9 | 130 |
| 275.0 | 165 | 273.6 | 124 |
| 274.9 | 163 | 273.4 | 117 |
| 274.4 | 141 | 273.2 | 109 |
| 274.2 | 137 | 273.2 | 110 |

Hydrate: Isobutane
Reference: Rouher, O.S., Barduhn, A.J. (1969)
Phases: $L_W$-H-V

| T(K) | P(kPa) | T(K) | P(kPa) |
|---|---|---|---|
| 273.18 | 115 | 274.07 | 137 |
| 273.34 | 118 | 274.15 | 140 |
| 273.48 | 122 | 274.20 | 140 |
| 273.50 | 122 | 274.30 | 143 |
| 273.57 | 123 | 274.40 | 147 |
| 273.58 | 124 | 274.55 | 151 |
| 273.68 | 126 | 274.56 | 151 |
| 273.76 | 129 | 274.65 | 157 |
| 273.86 | 132 | 274.82 | 160 |
| 273.95 | 135 | 274.96 | 164 |
| 273.96 | 134 | 275.00 | 168 |
| 273.98 | 135 | 275.05 | 169 |

Hydrate: Isobutane
Reference: Wu, B.-J., Robinson, D.B., Ng, H.-J. (1976)
Phases: $L_W$-H-$L_{IB}$

| T(K) | P(kPa) | T(K) | P(kPa) |
|---|---|---|---|
| 275.4 | 226 | 275.5 | 2,410 |
| 275.4 | 357 | 275.6 | 5,650 |
| 275.4 | 903 | 275.8 | 14,270 |

Hydrate: Isobutane
Reference: Holder, G.D., Godbole, S.P. (1982)
Phases: I-H-V

| T(K) | P(kPa) | T(K) | P(kPa) |
|---|---|---|---|
| 241.4 | 17.6 | 259.7 | 53.5 |
| 243.4 | 20.2 | 263.3 | 66.4 |
| 248.4 | 26.4 | 268.1 | 85.5 |
| 253.7 | 35.1 | 269.4 | 89.7 |
| 256.5 | 42.8 | 269.5 | 91.3 |

Hydrate: Isobutane
Reference: Thakore, J.L., Holder, G.D. (1987)
Phases: $L_W$-H-V

| T(K) | P(kPa) | T(K) | P(kPa) |
|---|---|---|---|
| 274.35 | 128 | 274.65 | 155 |

## Carbon Dioxide

Hydrate: Carbon Dioxide
Reference: Deaton, W.M., Frost, E.M., Jr. (1946)
Phases: $L_W$-H-V

| T(K) | P(kPa) | T(K) | P(kPa) |
|---|---|---|---|
| 273.7 | 1324 | 278.7 | 2427 |
| 273.7 | 1324 | 278.7 | 2413 |
| 274.3 | 1393 | 279.8 | 2758 |
| 274.3 | 1420 | 279.8 | 2786 |
| 274.3 | 1420 | 280.9 | 3213 |
| 275.4 | 1613 | 281.5 | 3530 |
| 276.5 | 1848 | 281.9 | 3709 |
| 277.6 | 2075 | 282.6 | 4130 |
| 277.6 | 2082 | 282.9 | 4323 |
| 277.6 | 2103 | | |

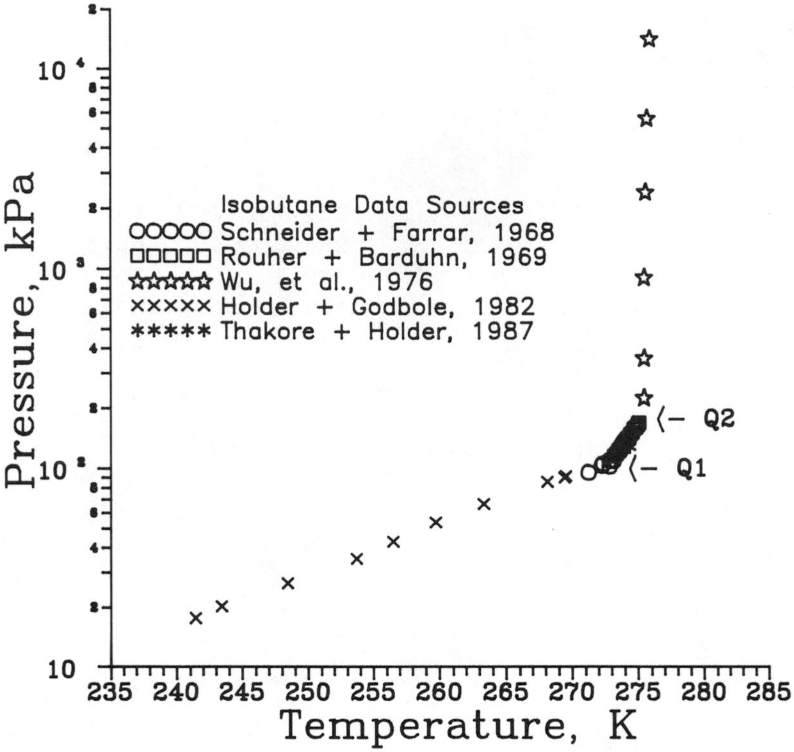


6-10 Three-Phase Data for Simple Hydrates of Iso-Butane

<u>Hydrate</u>: Carbon Dioxide
<u>Reference</u>: Unruh, C.H., Katz, D.L. (1949)
<u>Phases</u>: $L_W$-H-V

| T(K) | P(kPa) | T(K) | P(kPa) |
|---|---|---|---|
| 277.2 | 2041 | 281.9 | 3689 |
| 279.2 | 2586 | 283.1 | 4502* |
| 280.9 | 3227 | | |

* $Q_2$ quadruple point ($L_W$-H-V-$L_{CO_2}$)

<u>Hydrate</u>: Carbon Dioxide
<u>Reference</u>: Larson, S.D. (1955)
<u>Phases</u>: I-H-V and $L_W$-H-V

| T(K) | P(kPa) | T(K) | P(kPa) |
|---|---|---|---|
| **I-H-V** | | | |
| 256.8 | 545 | 270.7 | 1000 |
| 264.0 | 752 | 271.4 | 1027 |
| 267.4 | 869 | 271.7 | 1041 |
| 268.9 | 924 | 271.8 | 1048 |
| 270.0 | 972 | | |
| **$L_W$-H-V** | | | |
| 271.8 | 1048 | 277.8 | 2137 |
| 271.9 | 1048 | 278.0 | 2165 |
| 272.2 | 1089 | 278.6 | 2344 |
| 272.5 | 1110 | 278.8 | 2448 |
| 273.1 | 1200 | 279.1 | 2530 |
| 273.4 | 1234 | 279.2 | 2544 |
| 273.5 | 1241 | 279.8 | 2730 |
| 273.9 | 1317 | 280.1 | 2861 |
| 274.1 | 1351 | 280.2 | 2923 |
| 274.4 | 1386 | 280.5 | 3020 |
| 275.0 | 1510 | 280.8 | 3158 |
| 275.1 | 1496 | 281.1 | 3282 |
| 275.7 | 1634 | 281.5 | 3475 |
| 276.0 | 1682 | 281.9 | 3634 |
| 276.2 | 1717 | 282.0 | 3689 |
| 276.5 | 1806 | 282.3 | 3868 |
| 276.9 | 1889 | 283.1 | 4468 |
| 277.2 | 1951 | 283.2 | 4502 |
| **H-V-$L_{CO_2}$** | | | |
| 258.8 | 2337 | 272.1 | 3385 |
| 256.5 | 2179 | 273.1 | 3475 |

| | | | |
|---|---|---|---|
| 260.2 | 2434 | 274.4 | 3592 |
| 258.5 | 2310 | 275.1 | 3661 |
| 261.2 | 2503 | 276.3 | 3778 |
| 260.1 | 2420 | 277.0 | 3847 |
| 262.5 | 2599 | 278.9 | 4040 |
| 262.5 | 2599 | 279.1 | 4061 |
| 263.8 | 2696 | 281.5 | 4316 |
| 264.4 | 2744 | 281.4 | 4302 |
| 264.9 | 2779 | 283.1 | 4489 |
| 266.1 | 2875 | 282.8 | 4454 |
| 267.3 | 2972 | 283.5 | 4523 |
| 268.5 | 3068 | 283.7 | 4558 |
| 270.2 | 3220 | 284.5 | 4640 |
| 270.8 | 3268 | 285.0 | 4695 |

Hydrate: Carbon Dioxide
Reference: Miller, S.L., Smythe, W.D. (1970)
Phases: I-H-V

| T(K) | P(kPa) | T(K) | P(kPa) |
|---|---|---|---|
| 151.5 | 0.535 | 176.9 | 6.77 |
| 162.4 | 1.77 | 182.2 | 10.3 |
| 167.1 | 2.81 | 186.8 | 14.5 |
| 171.5 | 4.20 | 192.5 | 21.9 |

Hydrate: Carbon Dioxide
Reference: Robinson, D.B., Mehta, B.R. (1971)
Phases: $L_W$-H-V

| T(K) | P(kPa) | T(K) | P(kPa) |
|---|---|---|---|
| 273.9 | 1379 | 280.7 | 3130 |
| 275.2 | 1558 | 282.0 | 3840 |
| 276.1 | 1758 | 283.3 | 4468* |
| 278.9 | 2420 | | |

* $Q_2$ quadruple point ($L_W$-H-V-$L_{CO_2}$)

Hydrate: Carbon Dioxide
Reference: Falabella, B.J. (1975)
Phases: I-H-V

| T(K) | P(kPa) | T(K) | P(kPa) |
|---|---|---|---|
| 194.5 | 24.8 | 218.2 | 104.3 |
| 203.2 | 43.3 | 217.8 | 101.3 |
| 213.8 | 81.6 | | |

Hydrate: Carbon Dioxide
Reference: Ng, H.-J., Robinson, D.B. (1985)
Phases: $L_W$-H-V, $L_W$-H-$L_{CO_2}$

| T(K) | P(MPa) | T(K) | P(MPa) |
|---|---|---|---|
| | $L_W$-H-V | | |
| 279.59 | 2.74 | 282.79 | 4.36 |
| 282.10 | 4.01 | | |
| | $L_W$-H-$L_{CO_2}$ | | |
| 282.92 | 5.03 | 283.15 | 9.01 |
| 282.92 | 5.62 | 283.55 | 11.98 |
| 283.05 | 6.47 | 283.93 | 14.36 |

Hydrate: Carbon Dioxide
Reference: Song, K.Y., Kobayashi, R. (1987)

| T(K) | P(kPa) | Mole Fraction ( x $10^3$ ) | Phases |
|---|---|---|---|
| 251.8 | 690 | 0.1800 | V-H |
| 254.2 | 690 | 0.2190 | V-H |
| 265.2 | 690 | 0.5570 | V-I |
| 294.3 | 690 | 4.3276 | V-$L_W$ |
| 255.2 | 1380 | 0.1142 | V-H |

| | | | |
|---|---|---|---|
| 258.0 | 1380 | 0.1471 | V-H |
| 262.2 | 1380 | 0.2201 | V-H |
| 271.2 | 1380 | 0.4885 | V-H |
| 275.2 | 1380 | 0.6836 | V-$L_W$ |
| 273.2 | 2070 | 0.2775 | V-H |
| 275.7 | 2070 | 0.4368 | V-H |
| 288.7 | 2070 | 1.0656 | V-$L_W$ |
| 268.8 | 2070 | 0.2321 | V-H |
| 260.7 | 2070 | 0.1194 | V-H |
| 257.2 | 2070 | 0.0890 | V-H |
| 252.7 | 2070 | 0.2013 | $L_{CO2}$-H |
| 245.2 | 2070 | 0.1361 | $L_{CO2}$-H |
| 255.4 | 3450 | 0.2616 | $L_{CO2}$-H |
| 260.2 | 3450 | 0.3222 | $L_{CO2}$-H |
| 269.7 | 3450 | 0.4585 | $L_{CO2}$-H |
| 274.2 | 3450 | 0.2410 | V-H |
| 278.7 | 3450 | 0.3794 | V-H |
| 285.2 | 3450 | 0.6030 | V-$L_W$ |
| 293.2 | 3450 | 1.0010 | V-$L_W$ |
| 255.4 | 4830 | 0.3313 | $L_{CO2}$-H |
| 263.2 | 4830 | 0.4705 | $L_{CO2}$-H |
| 269.7 | 4830 | 0.5402 | $L_{CO2}$-H |
| 276.2 | 4830 | 0.7182 | $L_{CO2}$-H |
| 290.2 | 4830 | 0.8229 | V-$L_W$ |
| 298.2 | 4830 | 1.2787 | V-$L_W$ |
| 288.7 | 5240 | 0.6400 | V-$L_W$ |
| 288.7 | 5240 | 1.1200 | $L_{CO2}$-$L_W$ |
| 293.4 | 5790 | 0.8999 | V-$L_W$ |
| 293.4 | 5790 | 1.5000 | $L_{CO2}$-$L_W$ |
| 257.2 | 6210 | 0.5170 | $L_{CO2}$-H |
| 263.7 | 6210 | 0.6647 | $L_{CO2}$-H |
| 280.2 | 6210 | 1.0960 | $L_{CO2}$-H |
| 299.8 | 6690 | 1.2700 | V-$L_W$ |
| 299.8 | 6690 | 1.9541 | $L_{CO2}$-$L_W$ |
| 302.7 | 7170 | 1.4981 | V-$L_W$ |
| 302.7 | 7170 | 2.1940 | $L_{CO2}$-$L_W$ |
| 304.2 | 7390 | 2.1079 | 3-$\phi$ critical endpoint |
| 256.2 | 8280 | 1.0890 | $L_{CO2}$-H |
| 265.9 | 8280 | 1.5741 | $L_{CO2}$-H |
| 270.2 | 8280 | 1.8695 | $L_{CO2}$-H |
| 286.9 | 8280 | 2.7852 | $L_{CO2}$-$L_W$ |
| 298.2 | 8280 | 3.0152 | $L_{CO2}$-$L_W$ |

| | | | |
|---|---|---|---|
| 256.2 | 10340 | 1.2738 | $L_{CO2}$-H |
| 264.2 | 10340 | 1.6509 | $L_{CO2}$-H |
| 276.2 | 10340 | 2.4687 | $L_{CO2}$-H |
| 298.2 | 10340 | 3.3739 | $L_{CO2}$-$L_W$ |
| 255.4 | 13790 | 1.5091 | $L_{CO2}$-H |
| 260.1 | 13790 | 1.8057 | $L_{CO2}$-H |
| 267.7 | 13790 | 2.2043 | $L_{CO2}$-H |
| 275.9 | 13790 | 2.7441 | $L_{CO2}$-H |
| 286.3 | 13790 | 3.3627 | $L_{CO2}$-$L_W$ |

## Nitrogen

Hydrate: Nitrogen
Reference: van Cleeff, A., Diepen, G.A.M. (1960)
Phases: $L_W$-H-V

| T(K) | P(MPa) | T(K) | P(MPa) |
|---|---|---|---|
| 272.05 | 14.48 | 278.65 | 28.27 |
| 272.55 | 15.30 | 279.15 | 29.89 |
| 272.85 | 15.91 | 279.25 | 30.30 |
| 272.95 | 15.91 | 280.25 | 33.94 |
| 273.15 | 16.01 | 281.25 | 37.49 |
| 273.15 | 16.31 | 281.65 | 38.61 |
| 273.35 | 16.62 | 282.25 | 41.44 |
| 273.95 | 17.53 | 283.25 | 45.90 |
| 274.15 | 17.73 | 284.25 | 50.66 |
| 274.85 | 19.15 | 284.65 | 52.29 |
| 274.85 | 19.25 | 285.25 | 55.43 |
| 275.25 | 19.66 | 286.25 | 61.40 |
| 275.55 | 20.67 | 287.25 | 67.79 |
| 275.85 | 21.58 | 287.75 | 71.23 |
| 276.25 | 22.39 | 288.35 | 74.58 |
| 276.55 | 23.10 | 289.25 | 81.47 |
| 277.25 | 24.83 | 290.25 | 89.37 |
| 278.25 | 27.36 | 290.65 | 92.21 |
| 278.25 | 27.97 | 291.05 | 95.86 |

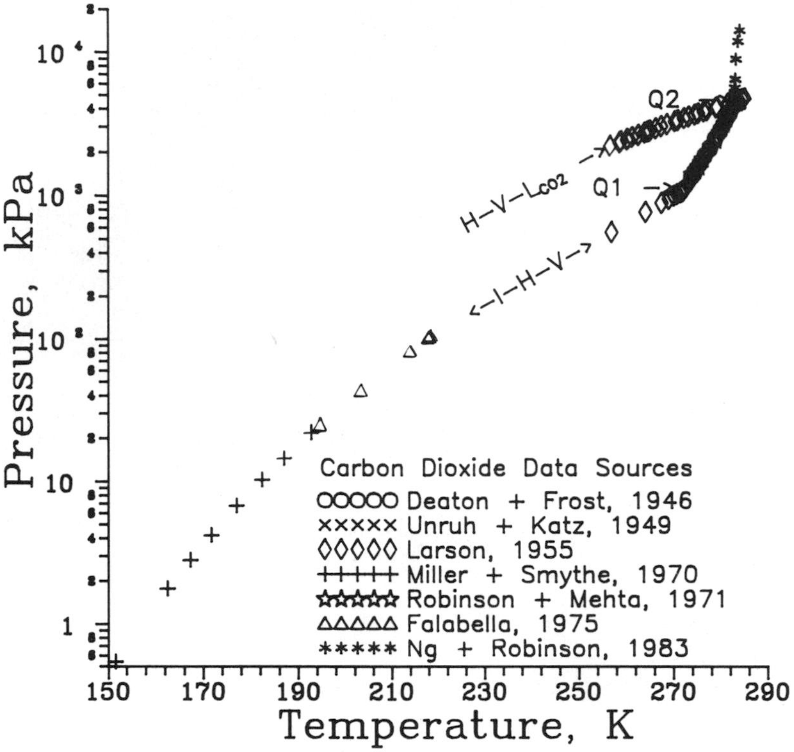


6-11 Three-Phase Data for Simple Hydrates of Carbon Dioxide

Hydrate: Nitrogen
Reference: Marshall, D.R., Saito, S, Kobayashi, R. (1964)
Phases: $L_W$-H-V

| T(K) | P(MPa) | T(K) | P(MPa) |
|---|---|---|---|
| 277.61 | 24.93 | 297.72 | 169.27 |
| 281.22 | 36.82 | 298.83 | 192.37 |
| 286.72 | 63.71 | 299.72 | 207.78 |
| 291.55 | 101.98 | 300.56 | 219.60 |
| 293.00 | 115.49 | 302.56 | 268.32 |
| 294.28 | 128.80 | 304.72 | 317.65 |
| 296.56 | 153.48 | 305.50 | 328.89 |

Hydrate: Nitrogen
Reference: Jhaveri, J., Robinson, D.B. (1965)
Phases: $L_W$-H-V

| T(K) | P(MPa) | T(K) | P(MPa) |
|---|---|---|---|
| 273.2 | 16.27 | 277.4 | 25.20 |
| 273.7 | 17.13 | 278.6 | 28.61 |
| 274.9 | 19.13 | 279.3 | 30.27 |
| 276.5 | 23.69 | 281.1 | 35.16 |

## Hydrogen Sulfide

Hydrate: Hydrogen Sulfide
Reference: Bond, D.C., Russell, N.B. (1949)
Phases: $L_W$-H-V

| T(K) | P(kPa) | T(K) | P(kPa) |
|---|---|---|---|
| 283.2 | 310 | 299.7 | 1496 |
| 291.2 | 710 | 302.7 | 2241 |

Hydrate: Hydrogen Sulfide
Reference: Selleck, F.T., Carmichael, L.T., and Sage, B.H. (1952)
Phases: $L_W$-H-V, I-H-V, $L_{H_2S}$-H-V, $L_W$-$L_{H_2S}$-H

| T(K) | P(kPa) | T(K) | P(kPa) |
|---|---|---|---|
| | $L_W$-H-V | | |
| 272.8 | 93 $Q_1$ | 295.7 | 1034 |
| 272.8 | 93 | 298.5 | 1379 |
| 277.6 | 157 | 299.8 | 1596 |
| 283.2 | 280 | 300.5 | 1724 |
| 285.2 | 345 | 302.1 | 2068 |
| 288.7 | 499 | 302.7 | 2239 $Q_2$ |

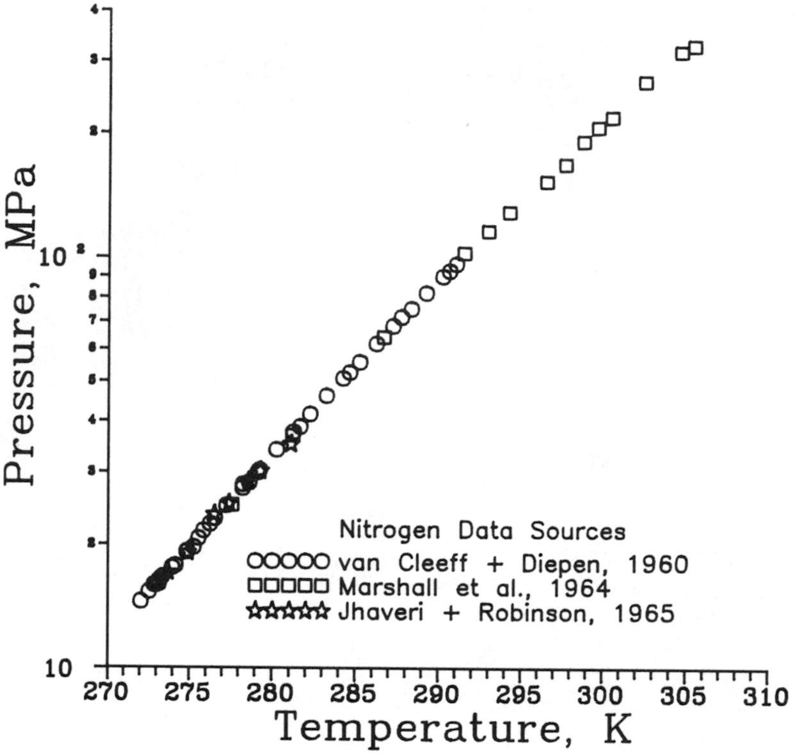


6-12 Three-Phase Data for Simple Hydrates of Nitrogen

| | | | |
|---|---|---|---|
| 291.8 | 689 | 302.7 | 2239 $Q_2$ |
| 294.3 | 890 | | |

I-H-V

| | | | |
|---|---|---|---|
| 250.5 | 34 | 265.3 | 69 |
| 255.4 | 44 | 266.5 | 72 |
| 258.2 | 50 | 269.3 | 81 |
| 260.9 | 57 | 272.1 | 90 |
| 263.7 | 64 | 272.8 | 93 $Q_1$ |

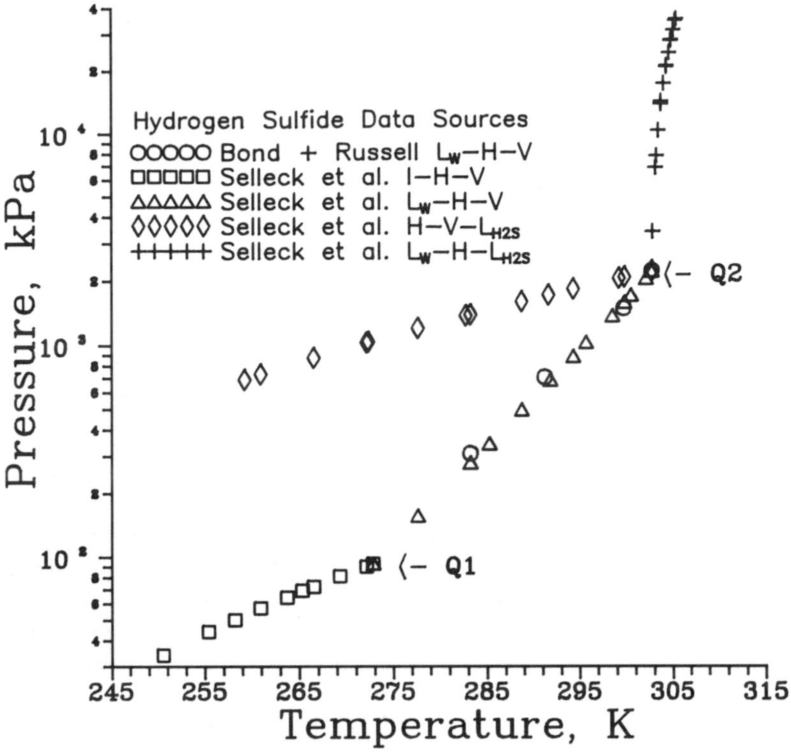


6-13 Three-Phase Data for Simple Hydrates of Hydrogen Sulfide

$\underline{H\text{-}V\text{-}L_{H_2S}}$

| | | | |
|---|---|---|---|
| 259.2 | 689 | 288.7 | 1605 |
| 260.9 | 731 | 291.6 | 1724 |
| 266.5 | 870 | 294.3 | 1839 |
| 272.1 | 1027 | 299.2 | 2068 |
| 272.3 | 1034 | 299.8 | 2097 |
| 277.6 | 1202 | 302.7 | 2239 $Q_2$ |
| 282.7 | 1379 | 302.7 | 2239 $Q_2$ |
| 283.2 | 1393 | | |

$\underline{L_W\text{-}H\text{-}L_{H_2S}}$

| | | | |
|---|---|---|---|
| 302.7 | 2239 $Q_2$ | 304.3 | 20685 |

| | | | |
|---|---|---|---|
| 302.7 | 2239 $Q_2$ | 304.3 | 20954 |
| 302.8 | 3447 | 304.6 | 24132 |
| 303.1 | 6895 | 304.8 | 27580 |
| 303.2 | 7826 | 304.8 | 27842 |
| 303.4 | 10342 | 305.1 | 31027 |
| 303.7 | 13790 | 305.3 | 34475 |
| 303.7 | 14190 | 305.4 | 35068 |
| 304.0 | 17237 | | |

$Q_1$ = middle quadruple point ($I$-$L_W$-$H$-$V$)
$Q_2$ = upper quadruple point ($L_W$-$H$-$V$-$L_{H_2S}$)

### 6.2.1.b. Equilibria of Binary Guest Mixtures

The phase equilibria data for binary guest mixtures are listed under the lighter component. For example, under the heading of binary guest mixtures of methane will be found data for methane+ethane, methane+propane, methane+isobutane, methane+n-butane, methane+nitrogen, methane+carbon dioxide, and methane+hydrogen sulfide. The heading for binary guest mixtures of ethane will thus contain no methane, only mixtures of ethane and heavier (or non-combustible) components. As an indication of consistency the data tabulations for each binary pair are followed by a semilogarithmic plot, where there are more than one data set.

As a second means to examine binary (and higher) hydrate data, one should determine the consistency of the simple gas hydrates which compose the mixed hydrate. For example the data for binary hydrates of methane and carbon dioxide by Berecz and Balla-Achs (1983, pp. 221ff.) show interesting new retrograde phenomena. However a nitrogen impurity in their carbon dioxide (op.cit., p.185) may have caused a systematic deviation from the other $CO_2$ data sets shown in Figure 6-11, and their methane purity was only 98%, providing a simple methane hydrate data deviation (op. cit., p.221). Because the simple hydrate data appeared quantitatively deviant they were excluded, and the resulting binary hydrate data are also suspect.

Gas impurities may have caused the systematic deviation of the Berecz and Balla-Achs $CH_4+CO_2$ hydrate data from those of Unruh and Katz (1954) and for those reasons the former data are excluded. A third data set for $CH_4+CO_2$ hydrates is needed to confirm this exclusion.

## Binary Guest Mixtures Containing Methane and Heavier (or Non-Combustible) Compounds

Hydrate: Methane+Ethane
Reference: Deaton W.M. and Frost, E.M. (1946)
Phases: $L_W$-H-V

| %$CH_4$ | T(K) | P(kPa) | %$CH_4$ | T(K) | P(kPa) |
|---|---|---|---|---|---|
| 56.4 | 274.8 | 945 | 95.0 | 283.2 | 4771 |
| 56.4 | 277.6 | 1289 | 97.1 | 274.8 | 2158 |
| 56.4 | 280.4 | 1758 | 97.1 | 277.6 | 2958 |
| 56.4 | 283.2 | 2434 | 97.1 | 280.4 | 4034 |
| 90.4 | 274.8 | 1524 | 97.8 | 274.8 | 2365 |
| 90.4 | 277.6 | 2096 | 97.8 | 277.6 | 3227 |
| 90.4 | 280.4 | 2889 | 97.8 | 280.4 | 4413 |
| 90.4 | 283.2 | 3965 | 97.8 | 282.6 | 5668 |
| 95.0 | 274.8 | 1841 | 97.8 | 283.2 | 6088 |
| 95.0 | 274.8 | 1841 | 98.8 | 274.8 | 2861 |
| 95.0 | 277.6 | 2530 | 98.8 | 277.6 | 3806 |
| 95.0 | 280.4 | 3447 | 98.8 | 280.4 | 5088 |

Hydrate: Methane+Ethane
Reference: McLeod, H.O., Campbell, J.M. (1961)
Phases: $L_W$-H-V

| %$CH_4$ | T(K) | P(MPa) | %$CH_4$ | T(K) | P(MPa) |
|---|---|---|---|---|---|
| 94.6 | 302.0 | 68.43 | 80.9 | 304.1 | 68.57 |
| 94.6 | 301.2 | 62.23 | 80.9 | 303.1 | 61.95 |
| 94.6 | 299.1 | 48.23 | 80.9 | 301.3 | 48.64 |
| 94.6 | 296.6 | 34.44 | 80.9 | 299.0 | 35.61 |

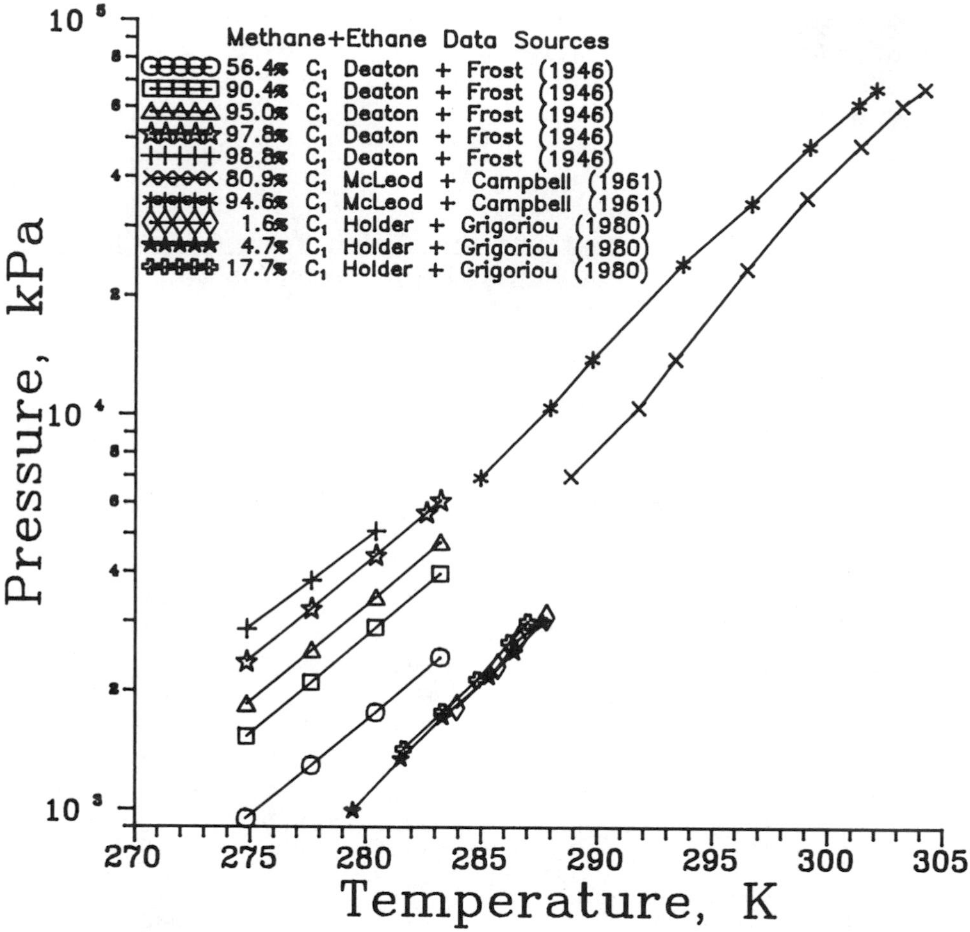


6-14 Methane+Ethane Mixture ($L_W$-H-V) Data

| | | | | | |
|---|---|---|---|---|---|
| 94.6 | 293.6 | 24.24 | 80.9 | 296.4 | 23.48 |
| 94.6 | 289.7 | 13.89 | 80.9 | 293.3 | 13.89 |
| 94.6 | 287.9 | 10.45 | 80.9 | 291.7 | 10.45 |
| 94.6 | 284.9 | 6.93 | 80.9 | 288.8 | 7.00 |

Hydrate: Methane+Ethane
Reference: Holder, G.D., Grigoriou, G.C. (1980)
Phases: $L_W$-H-V

| $\%CH_4$ | T(K) | P(kPa) | $\%CH_4$ | T(K) | P(kPa) |
|---|---|---|---|---|---|
| 1.6 | 283.9 | 1810 | 4.7 | 286.4 | 2510 |
| 1.6 | 285.7 | 2310 | 4.7 | 287.6 | 2990 |

| | | | | | |
|---|---|---|---|---|---|
| 1.6 | 286.6 | 2710 | 17.7 | 281.6 | 1420 |
| 1.6 | 287.8 | 3080 | 17.7 | 283.3 | 1770 |
| 4.7 | 279.4 | 990 | 17.7 | 284.8 | 2140 |
| 4.7 | 281.5 | 1340 | 17.7 | 286.2 | 2660 |
| 4.7 | 283.3 | 1710 | 17.7 | 287.0 | 3000 |
| 4.7 | 285.3 | 2170 | | | |

Hydrate: Methane+Propane
Reference: Deaton, W.M., Frost, E.M., Jr. (1946)
Phases: $L_W$-H-V

| %$CH_4$ | T(K) | P(kPa) | %$CH_4$ | T(K) | P(kPa) |
|---|---|---|---|---|---|
| 36.2 | 274.8 | 272 | 95.2 | 277.6 | 1138 |
| 36.2 | 277.6 | 436 | 95.2 | 280.4 | 1586 |
| 36.2 | 280.4 | 687 | 95.2 | 283.2 | 2227 |
| 71.2 | 274.8 | 365 | 97.4 | 274.8 | 1151 |
| 71.2 | 277.6 | 538 | 97.4 | 277.6 | 1593 |
| 71.2 | 280.4 | 800 | 97.4 | 280.4 | 2193 |
| 71.2 | 280.4 | 800 | 97.4 | 283.2 | 3013 |
| 71.2 | 283.2 | 1151 | 99.0 | 274.8 | 1627 |
| 88.3 | 274.8 | 552 | 99.0 | 277.6 | 2247 |
| 88.3 | 277.6 | 779 | 99.0 | 277.6 | 2255 |
| 88.3 | 280.4 | 1110 | 99.0 | 280.4 | 3123 |
| 88.3 | 283.2 | 1558 | 99.0 | 283.1 | 4358 |
| 95.2 | 274.8 | 814 | | | |

Hydrate: Methane+Propane
Reference: McLeod, H.O., Campbell, J.M. (1961)
Phases: $L_W$-H-V

| %$CH_4$ | T(K) | P(MPa) | %$CH_4$ | T(K) | P(MPa) |
|---|---|---|---|---|---|
| 96.5 | 290.5 | 6.93 | 96.5 | 290.7 | 6.93 |
| 96.5 | 303.7 | 62.47 | 94.5 | 293.1 | 7.41 |
| 96.5 | 304.4 | 68.98 | 94.5 | 292.8 | 7.41 |

| | | | | | |
|---|---|---|---|---|---|
| 96.5 | 299.1 | 34.51 | 94.5 | 300.6 | 34.58 |
| 96.5 | 296.6 | 20.86 | 94.5 | 302.7 | 48.37 |
| 96.5 | 301.6 | 48.37 | 94.5 | 304.9 | 62.23 |
| 96.5 | 303.7 | 62.23 | 94.5 | 298.5 | 23.62 |
| 96.5 | 294.5 | 13.89 | 94.5 | 296.2 | 13.89 |
| 96.5 | 293.3 | 10.45 | | | |

<u>Hydrate</u>: Methane+Propane
<u>Reference</u>: Verma, V.K., et al. (1974)
<u>Phases</u>: $L_W$-H-V and $L_W$-H-V-$L_{HC}$

<u>$L_W$-H-V</u>

| T(K) | P(MPa) | | T(K) | P(MPa) |
|---|---|---|---|---|
| | | 23.75% $CH_4$ | | |
| 274.88 | 0.263 | | 279.11 | 0.560 |
| 276.45 | 0.350 | | 280.23 | 0.689 |
| 277.74 | 0.443 | | 281.35 | 0.830* |
| | | 37.1% $CH_4$ | | |
| 274.45 | 0.270 | | 278.65 | 0.536 |
| 275.91 | 0.343 | | 280.16 | 0.691 |
| 277.08 | 0.419 | | 282.31 | 0.945 |

<u>$L_W$-H-V-$L_{HC}$</u>

| T(K) | P(MPa) | | T(K) | P(MPa) |
|---|---|---|---|---|
| | | 0.72% $CH_4$ | | |
| 279.62 | 0.662* | | 280.04 | 10.27 |
| 279.80 | 3.032 | | 280.21 | 15.25 |
| 279.91 | 6.996 | | | |
| | | 0.92% $CH_4$ | | |
| 280.39 | 0.739* | | 280.71 | 6.989 |
| 280.49 | 1.834 | | 280.91 | 11.00 |
| 280.59 | 3.166 | | 281.11 | 15.39 |

2.20% $CH_4$

| | | | |
|---|---|---|---|
| 282.22 | 0.949* | 282.57 | 3.170 |
| 282.35 | 1.756 | 282.88 | 6.003 |

4.46% $CH_4$

| | | | |
|---|---|---|---|
| 284.92 | 1.342* | 285.69 | 6.590 |
| 285.09 | 2.060 | 286.25 | 10.140 |
| 285.30 | 3.611 | 286.67 | 14.326 |

7.80% $CH_4$

| | | | |
|---|---|---|---|
| 287.96 | 2.032* | 288.72 | 7.362 |
| 288.01 | 2.811 | 289.27 | 11.496 |
| 288.27 | 4.804 | 289.84 | 15.339 |

13.70 % $CH_4$

| | | | |
|---|---|---|---|
| 290.18 | 2.708* | 291.56 | 9.375 |
| 290.37 | 3.363 | 291.66 | 10.389 |
| 290.66 | 5.507 | 292.10 | 13.133 |
| 290.91 | 6.362 | 292.73 | 16.939 |

25.70 % $CH_4$

| | | | |
|---|---|---|---|
| 294.18 | 4.962* | 295.62 | 10.816 |
| 294.24 | 5.259 | 296.39 | 13.795 |
| 294.85 | 7.810 | 297.15 | 17.063 |

48.23% $CH_4$

| | | | |
|---|---|---|---|
| 296.95 | 8.102* | 298.66 | 13.988 |
| 297.14 | 8.431 | 299.67 | 17.491 |
| 297.93 | 11.078 | | |

56.35% $CH_4$

| | | | |
|---|---|---|---|
| 297.58 | 9.161* | 299.69 | 15.381 |
| 298.09 | 10.430 | 300.22 | 17.408 |
| 298.98 | 13.181 | | |

59.40% $CH_4$

| | | | |
|---|---|---|---|
| 297.65 | 9.375* | 299.40 | 14.526 |
| 298.25 | 10.706 | 300.2 | 17.291 |
| 298.94 | 12.623 | | |

65.10% $CH_4$

| | | | |
|---|---|---|---|
| 297.67 | 9.561* | 298.93 | 13.891 |

| | | | |
|---|---|---|---|
| 298.02 | 10.600 | 300.12 | 17.125 |

* = $Q_2$ quadruple point ($L_W$-H-V-$L_{HC}$)

Hydrate: Methane+Propane
Reference: Song, K.Y., Kobayashi, R. (1982)
Phases: V-H

5.31 mol % propane in methane

| T(K) | P(MPa) | Mol Fraction ($\times 10^6$) | T(K) | P(MPa) | Mol Fraction ($\times 10^6$) |
|---|---|---|---|---|---|
| 234.2 | 2.07 | 6.86 | 246.2 | 6.89 | 7.03 |
| 246.2 | 2.07 | 24.28 | 252.1 | 6.89 | 12.25 |
| 251.7 | 2.07 | 41.54 | 260.0 | 6.89 | 25.42 |
| 260.1 | 2.07 | 85.20 | 263.2 | 6.89 | 35.78 |
| 266.5 | 2.07 | 161.99 | 276.2 | 6.89 | 103.70 |
| 277.2 | 2.07 | 427.28 | 234.2 | 10.34 | 1.15 |
| 234.2 | 3.45 | 3.47 | 246.2 | 10.34 | 3.75 |
| 246.2 | 3.45 | 13.85 | 252.1 | 10.34 | 7.33 |
| 252.1 | 3.45 | 27.50 | 260.1 | 10.34 | 14.67 |
| 263.2 | 3.45 | 78.76 | 266.5 | 10.34 | 26.75 |
| 274.7 | 3.45 | 187.89 | 277.6 | 10.34 | 81.15 |
| 234.2 | 6.89 | 1.92 | | | |

Hydrate: Methane+Propane
Reference: Thakore, J.L., Holder, G.D. (1987)
Phases: $L_W$-H-V

T = 275.15K

| %$CH_4$ | P(kPa) | %$CH_4$ | P(kPa) | %$CH_4$ | P(kPa) |
|---|---|---|---|---|---|
| 1.000 | 3370 | 0.420 | 279 | 0.054 | 245 |
| 0.903 | 672 | 0.366 | 279 | 0.046 | 245 |
| 0.765 | 424 | 0.352 | 269 | 0.037 | 245 |
| 0.727 | 393 | 0.190 | 245* | 0.021 | 245 |

| | | | | | |
|---|---|---|---|---|---|
| 0.700 | 365 | 0.083 | 245 | 0.000 | 278 |
| 0.516 | 303 | 0.081 | 245 | | |

T = 278.15K

| | | | | | |
|---|---|---|---|---|---|
| 1.000 | 4495 | 0.530 | 496 | 0.394 | 458 |
| 0.956 | 1306 | 0.510 | 489 | 0.390 | 458* |
| 0.947 | 1144 | 0.502 | 479 | 0.030 | 479 |
| 0.894 | 848 | 0.468 | 475 | 0.026 | 480 |
| 0.768 | 630 | 0.412 | 479 | 0.000 | 509 |

* = Azeotrope Point

Hydrate: Methane+Isobutane
Reference: Deaton, W.M., Frost, E.M., Jr. (1946)
Phases: $L_W$-H-V

| %$CH_4$ | T(K) | P(kPa) | %$CH_4$ | T(K) | P(kPa) |
|---|---|---|---|---|---|
| 98.9 | 274.8 | 1324 | 98.9 | 277.6 | 1841 |

Hydrate: Methane+Isobutane
Reference: McLeod, H.O., Campbell, J.M. (1961)
Phases: $L_W$-H-V

| %$CH_4$ | T(K) | P(MPa) | %$CH_4$ | T(K) | P(MPa) |
|---|---|---|---|---|---|
| 98.6 | 300.0 | 47.68 | 95.4 | 297.1 | 13.96 |
| 98.6 | 297.8 | 34.51 | 95.4 | 298.2 | 23.27 |
| 98.6 | 297.6 | 33.61 | 95.4 | 300.5 | 34.58 |
| 98.6 | 295.2 | 21.06 | 95.4 | 302.6 | 48.37 |
| 98.6 | 299.9 | 49.13 | 95.4 | 305.0 | 63.33 |
| 98.6 | 288.6 | 6.79 | 95.4 | 303.1 | 49.06 |
| 98.6 | 302.1 | 62.23 | 95.4 | 298.3 | 23.82 |
| 95.4 | 294.3 | 6.72 | 95.4 | 296.7 | 13.96 |
| 95.4 | 293.8 | 6.72 | 95.4 | 295.3 | 10.58 |
| 95.4 | 296.5 | 13.89 | 95.4 | 294.6 | 7.69 |

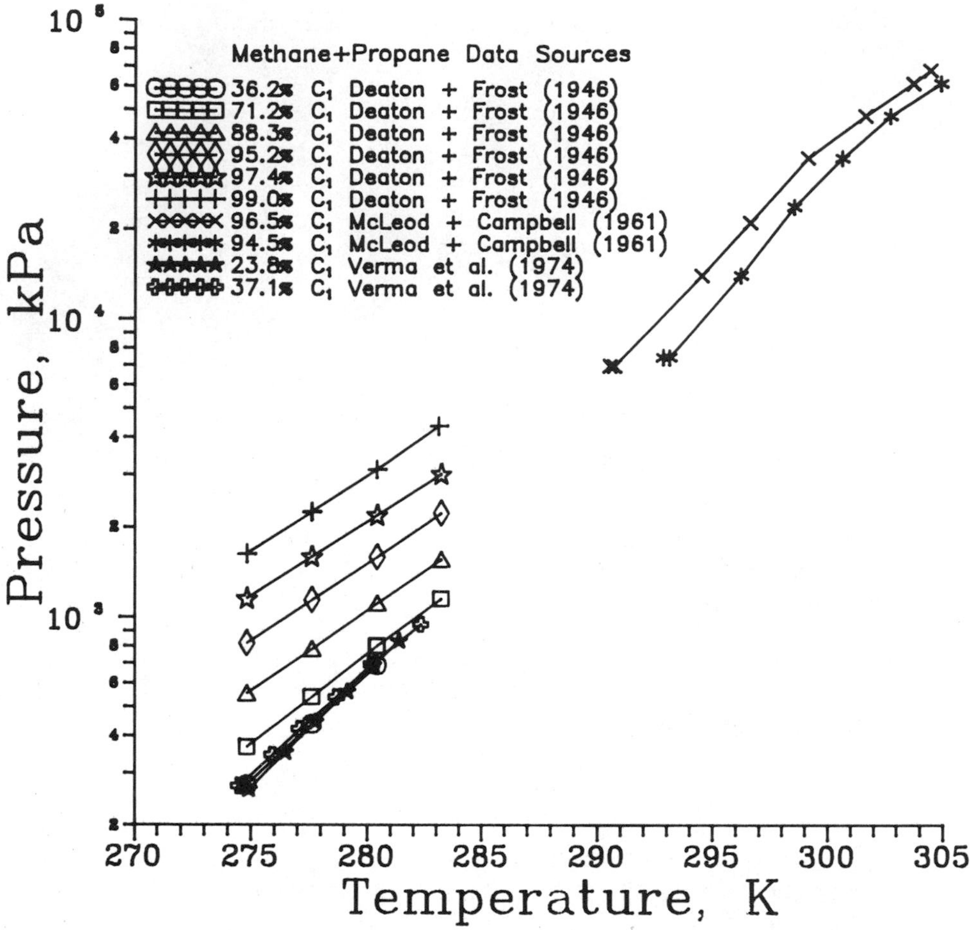


6-15a Methane+Propane Mixture ($L_W$-H-V) Data

Hydrate: Methane+Isobutane
Reference: Wu, B.-J. et al. (1976)
Phases: $L_W$-H-V and $L_W$-H-V-$L_{iC_4}$

| %i-$C_4H_{10}$ | T(K) | P(kPa) | %i-$C_4H_{10}$ | T(K) | P(kPa) |
|---|---|---|---|---|---|
| | | $L_W$-H-V | | | |
| 0.23 | 275.2 | 2080 | 2.50 | 277.8 | 1080 |
| 0.23 | 279.7 | 3440 | 2.50 | 279.8 | 1390 |

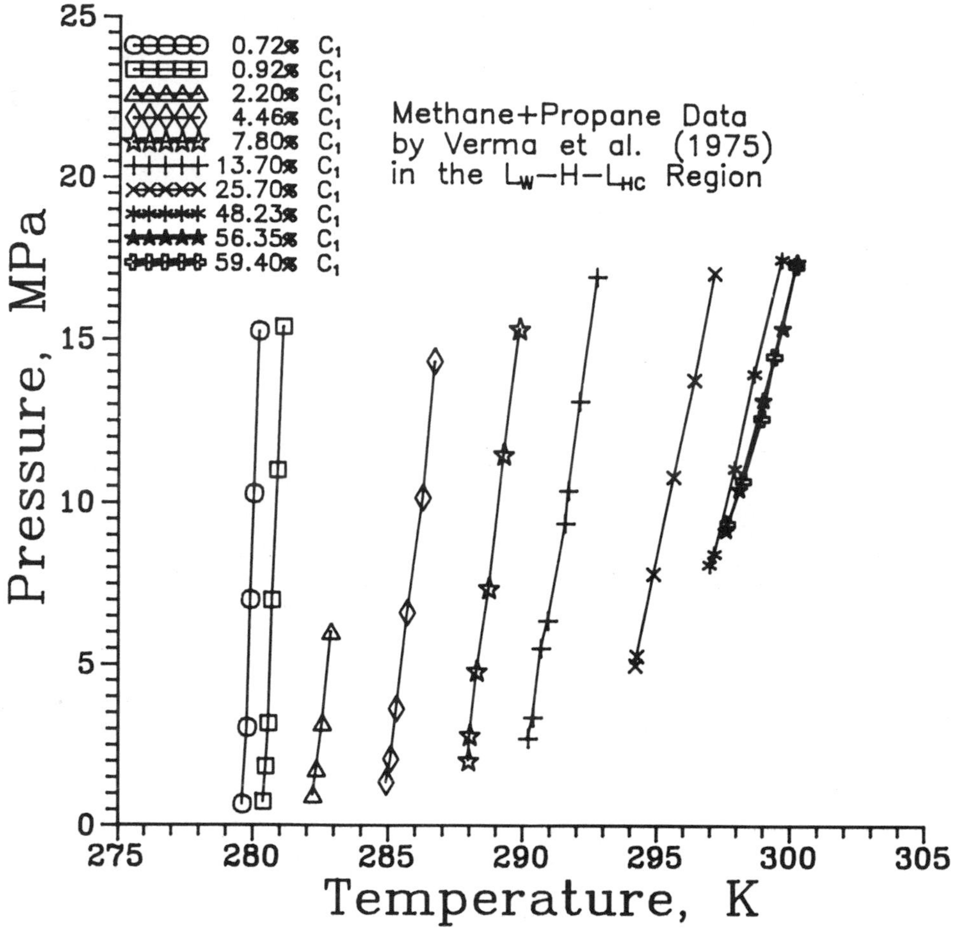


6-15b Methane+Propane Mixture ($L_W$-H-$L_P$) Data

| | | | | | |
|---|---|---|---|---|---|
| 0.27 | 284.6 | 6010 | 2.50 | 283.3 | 2150 |
| 0.36 | 285.4 | 6190 | 2.50 | 285.2 | 2740 |
| 0.37 | 288.0 | 9690 | 2.50 | 287.2 | 3450 |
| 0.40 | 276.2 | 1810 | 2.50 | 289.3 | 4560 |
| 0.43 | 285.3 | 5590 | 2.50 | 293.6 | 10070 |
| 0.45 | 282.0 | 3500 | 6.00 | 274.8 | 505 |
| 0.46 | 280.9 | 3150 | 6.00 | 280.4 | 1010 |
| 0.54 | 286.2 | 5480 | 6.00 | 284.5 | 1690 |
| 0.82 | 275.4 | 1270 | 6.00 | 288.5 | 2820 |
| 0.82 | 280.0 | 2190 | 15.20 | 274.0 | 304 |

| | | | | | |
|---|---|---|---|---|---|
| 0.82 | 283.5 | 3340 | 15.20 | 278.9 | 564 |
| 0.82 | 287.4 | 5900 | 15.20 | 283.4 | 1030 |
| 0.82 | 290.9 | 10040 | 15.20 | 288.9 | 2030 |
| 1.20 | 274.4 | 950 | 28.60 | 273.9 | 208 |
| 1.20 | 277.7 | 1390 | 28.60 | 277.2 | 356 |
| 1.20 | 279.9 | 1800 | 28.60 | 279.2 | 477 |
| 1.20 | 283.2 | 2700 | 28.60 | 280.8 | 602 |
| 1.20 | 284.9 | 3470 | 28.60 | 282.7 | 786 |
| 1.20 | 287.5 | 4880 | 63.60 | 273.8 | 159 |
| 1.20 | 290.0 | 6950 | 63.60 | 275.5 | 221 |
| 2.50 | 274.4 | 703 | 63.60 | 276.9 | 284 |

$L_W$-H-V-$L_{iC_4}$ (%i-$C_4H_{10}$ = vapor phase)

| | | | | | |
|---|---|---|---|---|---|
| 65.1 | 277.0 | 254 | 12.2 | 293.2 | 3990 |
| 44.5 | 279.6 | 427 | 12.0 | 293.3 | 4100 |
| 31.3 | 282.3 | 703 | 11.2 | 294.8 | 5560 |
| 23.8 | 284.7 | 1030 | 11.0 | 295.3 | 5760 |
| 18.2 | 287.5 | 1540 | 15.0 | 298.0 | 9990 |
| 13.8 | 290.8 | 2700 | 19.8 | 298.6 | 11570 |
| 13.5 | 291.5 | 2970 | | | |

Mixture: Methane+Isobutane
Reference: Ng, H.-J., Robinson, D.B. (1976a)
Phases: $L_W$-H-$L_{iC_4}$

| $\%CH_4$ | T(K) | P(kPa) | $\%CH_4$ | T(K) | P(kPa) |
|---|---|---|---|---|---|
| 0.1 | 275.4 | 179 | 15.2 | 288.9 | 1931 |
| 0.1 | 275.4 | 226 | 15.2 | 289.0 | 2441 |
| 0.1 | 275.4 | 357 | 15.2 | 289.4 | 3785 |
| 0.1 | 275.4 | 903 | 15.2 | 289.9 | 7129 |
| 0.1 | 275.6 | 2406 | 15.2 | 290.5 | 10577 |
| 0.1 | 275.7 | 5653 | 15.2 | 291.2 | 14073 |
| 0.1 | 275.8 | 14251 | 28.4 | 292.9 | 3792 |
| 4.3 | 282.2 | 682 | 28.4 | 293.0 | 4268 |
| 4.3 | 282.3 | 1048 | 28.4 | 293.7 | 6957 |
| 4.3 | 282.7 | 2179 | 28.4 | 294.8 | 10439 |
| 4.3 | 282.9 | 4474 | 28.4 | 295.0 | 13866 |

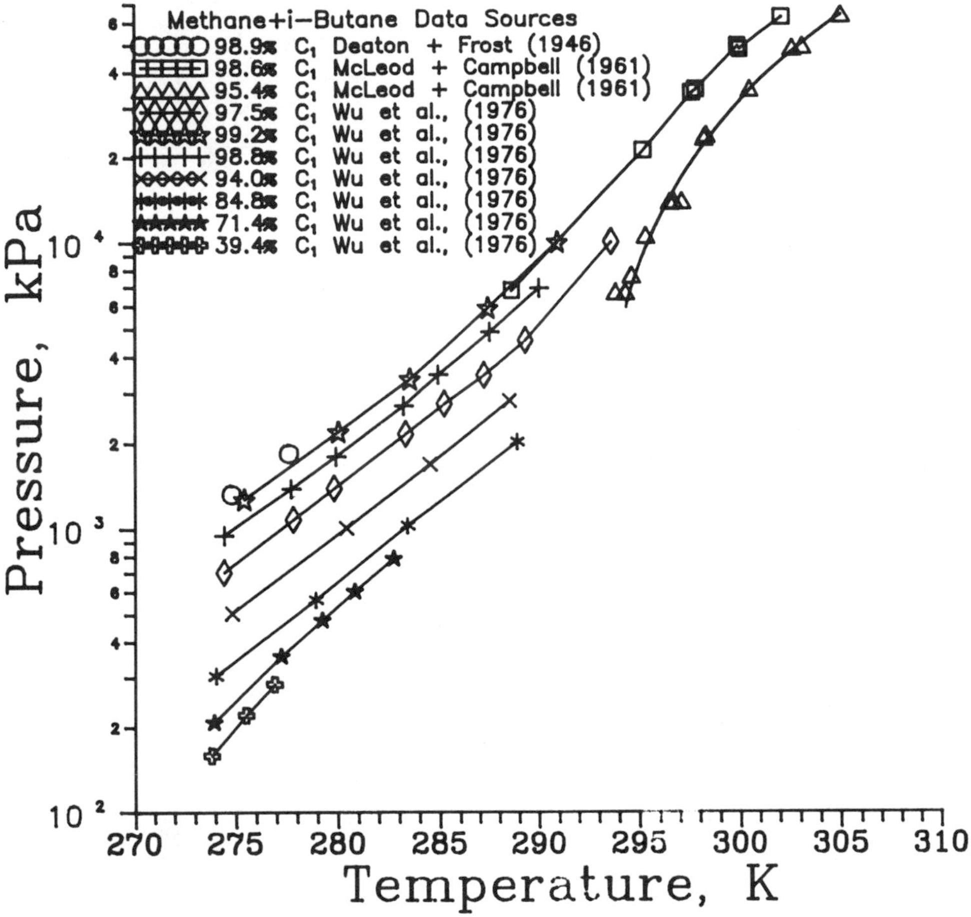


6-16a Methane+Iso-Butane Mixture ($L_W$-H-V) Data

| | | | | | |
|---|---|---|---|---|---|
| 4.3 | 283.3 | 8233 | 42.5 | 295.9 | 6619 |
| 4.3 | 283.9 | 14024 | 42.5 | 296.4 | 8784 |
| 8.7 | 286.4 | 1338 | 42.5 | 297.2 | 11321 |
| 8.7 | 286.5 | 1744 | 42.5 | 297.9 | 14093 |
| 8.7 | 286.8 | 3958 | 64.7 | 298.1 | 10204 |
| 8.7 | 287.2 | 6902 | 64.7 | 298.4 | 11232 |
| 8.7 | 287.8 | 11197 | 64.7 | 298.7 | 12528 |
| 8.7 | 288.4 | 14231 | 64.7 | 299.3 | 14548 |

<u>Hydrate</u>: Methane+Isobutane
<u>Reference</u>: Thakore, J.L., Holder, G.D. (1987)
<u>Phases</u>: $L_W$-H-V

| %$CH_4$ | P(kPa) | %$CH_4$ | P(kPa) | %$CH_4$ | P(kPa) |
|---|---|---|---|---|---|
| | | T = 274.35 K | | | |
| 1.000 | 3099 | 0.313 | 156 | 0.073 | 131 |
| 0.949 | 841 | 0.172 | 136 | 0.066 | 129 |
| 0.919 | 461 | 0.150 | 134 | 0.056 | 129 |
| 0.792 | 268 | 0.124 | 133 | 0.051 | 129 |
| 0.725 | 234 | 0.091 | 132 | 0.048 | 127 |
| 0.632 | 210 | 0.086 | 131 | 0.036 | 129 |
| 0.500 | 180 | 0.076 | 131 | 0.000 | 128 |

<u>Hydrate</u>: Methane+n-Butane
<u>Reference</u>: Deaton, W.M., Frost, E.M., Jr. (1946)
<u>Phases</u>: $L_W$-H-V

| %$CH_4$ | T(K) | P(kPa) | %$CH_4$ | T(K) | P(kPa) |
|---|---|---|---|---|---|
| 97.4 | 274.8 | 2048 | 97.5 | 274.8 | 2165 |
| 97.4 | 277.6 | 2875 | 99.2 | 274.8 | 3075 |
| 97.4 | 280.4 | 4061 | 99.2 | 277.6 | 4075 |

<u>Hydrate</u>: Methane+n-Butane
<u>Reference</u>: McLeod, H.O., Campbell, J.M. (1961)
<u>Phases</u>: $L_W$-H-V

| %$CH_4$ | T(K) | P(MPa) | %$CH_4$ | T(K) | P(MPa) |
|---|---|---|---|---|---|
| 97.4 | 285.0 | 7.69 | 94.7 | 295.1 | 34.16 |
| 97.4 | 287.7 | 12.45 | 94.7 | 297.9 | 48.23 |
| 97.4 | 295.7 | 34.58 | 94.7 | 300.1 | 61.61 |
| 97.4 | 301.1 | 65.95 | 94.7 | 301.1 | 68.43 |
| 97.4 | 286.3 | 10.45 | 94.7 | 290.3 | 17.96 |
| 97.4 | 285.7 | 9.07 | 94.7 | 288.8 | 13.89 |

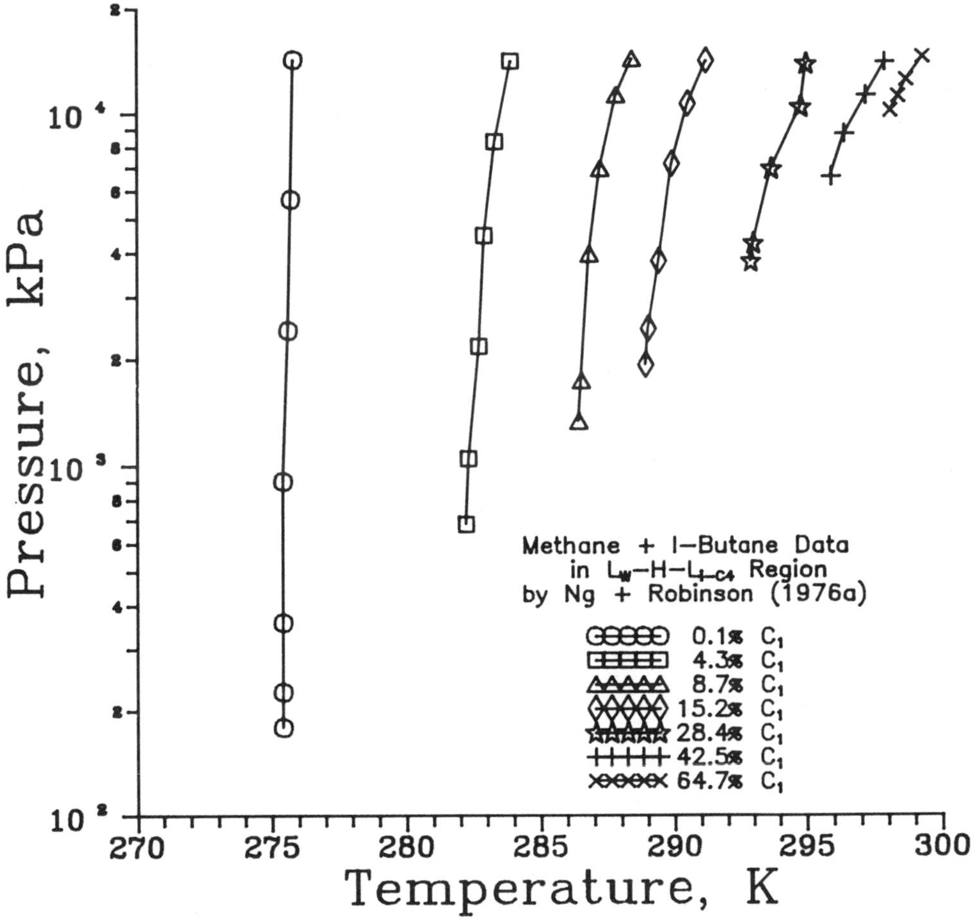


6-16b Methane+Iso-Butane Mixture ($L_W$-H-$L_{IB}$) Data

| | | | | | |
|---|---|---|---|---|---|
| 97.4 | 282.5 | 5.76 | 94.7 | 287.6 | 10.65 |
| 94.7 | 287.5 | 10.65 | 94.7 | 286.6 | 8.69 |
| 94.7 | 292.4 | 23.89 | 94.7 | 285.3 | 7.00 |

Hydrate: Methane+n-Butane
Reference: Ng, H.-J., Robinson, D.B. (1976b)
Phases: $L_W$-H-V

| %$CH_4$ | T(K) | P(MPa) | %$CH_4$ | T(K) | P(MPa) |
|---|---|---|---|---|---|
| 1.64 | 276.04 | 2.48 | 2.48 | 284.75 | 7.47 |
| 1.64 | 279.44 | 3.82 | 2.48 | 286.40 | 10.40 |
| 1.64 | 283.42 | 6.65 | 3.91 | 276.91 | 2.15 |
| 1.64 | 286.05 | 10.08 | 3.91 | 279.74 | 3.14 |
| 1.64 | 287.42 | 12.06 | 3.91 | 283.12 | 5.09 |
| 1.64 | 288.52 | 13.72 | 3.91 | 285.95 | 8.16 |
| 2.48 | 276.35 | 2.30 | 3.91 | 287.55 | 11.05 |
| 2.48 | 279.72 | 3.59 | 5.82 | 277.99 | 2.05 |
| 2.48 | 282.35 | 5.13 | 5.82 | 281.43 | 3.29 |

Hydrate: Methane+n-Butane
Reference: Ng, H.-J., Robinson, D.B. (1977)
Phases: $L_W$-H-$L_{n-C_4}$

%$CH_4$ in $L_{n-C_4}$

| %$CH_4$ | T(K) | P(MPa) | %$CH_4$ | T(K) | P(MPa) |
|---|---|---|---|---|---|
| 8.7 | 275.04 | 1.24* | 21.8 | 282.70 | 9.05 |
| 8.7 | 275.55 | 3.41 | 21.8 | 283.40 | 12.13 |
| 8.7 | 275.69 | 6.01 | 21.8 | 283.31 | 12.34 |
| 8.7 | 276.26 | 10.06 | 42.4 | 285.85 | 6.62* |
| 8.7 | 276.66 | 12.82 | 42.4 | 286.08 | 7.78 |
| 15.7 | 279.35 | 2.39* | 42.4 | 286.41 | 9.09 |
| 15.7 | 279.38 | 2.96 | 42.4 | 286.85 | 10.92 |
| 15.7 | 279.50 | 3.61 | 42.4 | 287.00 | 12.16 |
| 15.7 | 279.82 | 4.27 | 42.4 | 287.40 | 13.82 |
| 15.7 | 280.00 | 6.76 | 50.1 | 287.35 | 8.83* |
| 15.7 | 280.55 | 9.09 | 50.1 | 287.45 | 9.85 |
| 15.7 | 281.12 | 12.44 | 50.1 | 287.82 | 10.75 |
| 21.8 | 281.75 | 3.45* | 50.1 | 288.17 | 12.47 |
| 21.8 | 282.08 | 5.23 | 50.1 | 288.52 | 13.82 |
| 21.8 | 282.35 | 6.75 | | | |

* = $Q_2$ quadruple point ($L_W$-H-V-$L_{n-C_4}$)

Hydrate: Methane+n-Butane
Reference: John, V.T., Holder, G.D. (1982)
Phases: $L_W$-H-V, I-H-V-$L_{n-C_4}$, $L_W$-H-V-$L_{n-C_4}$

I-H-V

| $\%C_4H_{10}$ | T(K) | P(kPa) | $\%C_4H_{10}$ | T(K) | P(kPa) |
|---|---|---|---|---|---|
| 0.4 | 251.15 | 1267 | 4.2 | 262.15 | 846 |
| 0.7 | 251.15 | 1011 | 6.3 | 262.15 | 715 |
| 1.1 | 251.15 | 805 | 9.3 | 262.15 | 639 |
| 1.8 | 251.15 | 680 | 12.6 | 262.15 | 570* |
| 3.8 | 251.15 | 522 | 0.55 | 268.15 | 2204 |
| 5.0 | 251.15 | 474 | 0.75 | 268.15 | 1963 |
| 5.9 | 251.15 | 446 | 1.15 | 268.15 | 1728 |
| 8.8 | 251.15 | 391 | 1.50 | 268.15 | 1563 |
| 13.3 | 251.15 | 336* | 2.30 | 268.15 | 1342 |
| 0.5 | 256.15 | 1480 | 3.15 | 268.15 | 1232 |
| 0.6 | 256.15 | 1246 | 3.40 | 268.15 | 1177 |
| 1.0 | 256.15 | 1067 | 5.20 | 268.15 | 1011 |
| 1.5 | 256.15 | 901 | 6.80 | 268.15 | 915 |
| 2.0 | 256.15 | 818 | 7.80 | 268.15 | 880 |
| 3.2 | 256.15 | 680 | 11.00 | 268.15 | 784* |
| 4.5 | 256.15 | 605 | 0.80 | 273.05 | 2611 |
| 6.5 | 256.15 | 536 | 0.95 | 273.05 | 2446 |
| 10.0 | 256.15 | 460 | 1.20 | 273.05 | 2335 |
| 11.4 | 256.15 | 446 | 1.40 | 273.05 | 2142 |
| 12.8 | 256.15 | 432* | 1.95 | 273.05 | 1894 |
| 0.5 | 262.15 | 1784 | 2.75 | 273.05 | 1618 |
| 0.9 | 262.15 | 1529 | 4.10 | 273.05 | 1384 |
| 1.2 | 262.15 | 1356 | 6.60 | 273.05 | 1136 |
| 2.0 | 262.15 | 1108 | 10.10 | 273.05 | 1011* |
| 2.9 | 262.15 | 970 | | | |

* = $Q_1$ Quadruple point (I-H-V-$L_{n-C_4}$)

I-H-V-$L_{n-C_4}$

| T(K) | P(kPa) | T(K) | P(kPa) |
|---|---|---|---|
| 255.3 | 417.1 | 264.5 | 661.9 |
| 256.1 | 430.9 | 255.1 | 682.6 |

| | | | |
|---|---|---|---|
| 257.2 | 458.5 | 265.85 | 710.2 |
| 258.0 | 479.2 | 269.05 | 813.6 |
| 259.0 | 499.9 | 269.25 | 827.4 |
| 259.9 | 524.0 | 270.3 | 875.6 |
| 260.8 | 544.7 | 271.0 | 903.2 |
| 261.9 | 575.7 | 271.7 | 937.7 |
| 263.6 | 620.5 | 272.5 | 979.0 |
| 263.85 | 634.0 | | |

$\underline{L_W\text{-}H\text{-}V\text{-}L}_{n\text{-}C_4}$

| T(K) | P(kPa) | T(K) | P(kPa) |
|---|---|---|---|
| 273.3 | 1046.0 | 275.1 | 1342.0 |
| 273.85 | 1073.0 | 275.9 | 1504.0 |
| 274.3 | 1177.0 | 276.8 | 1722.0 |

<u>Hydrate</u>: Methane+Nitrogen
<u>Reference</u>: Jhaveri, J., Robinson, D.B. (1965)
<u>Phases</u>: $L_W$-H-V

| %$CH_4$ | T(K) | P(MPa) | %$CH_4$ | T(K) | P(MPa) |
|---|---|---|---|---|---|
| 87.3 | 282.78 | 7.40 | 50.25 | 291.83 | 33.19 |
| 87.3 | 284.56 | 9.31 | 27.2 | 273.17 | 7.96 |
| 87.3 | 287.72 | 14.52 | 27.2 | 277.06 | 10.16 |
| 87.3 | 289.50 | 17.11 | 27.2 | 280.00 | 12.64 |
| 87.3 | 290.39 | 17.49 | 27.2 | 282.89 | 17.04 |
| 87.3 | 291.00 | 19.53 | 27.2 | 283.17 | 17.50 |
| 87.3 | 291.50 | 19.99 | 27.2 | 285.11 | 20.72 |
| 87.3 | 292.89 | 22.94 | 27.2 | 286.78 | 25.15 |
| 87.3 | 293.44 | 24.66 | 27.2 | 288.00 | 28.49 |
| 87.3 | 295.22 | 31.31 | 24.0 | 273.17 | 8.62 |
| 73.1 | 273.17 | 3.90 | 24.0 | 274.56 | 9.15 |
| 73.1 | 283.33 | 8.95 | 24.0 | 278.28 | 12.96 |
| 73.1 | 286.83 | 13.22 | 24.0 | 282.06 | 17.44 |
| 73.1 | 289.94 | 19.55 | 24.0 | 285.11 | 24.34 |
| 73.1 | 292.33 | 25.99 | 24.0 | 287.61 | 31.99 |

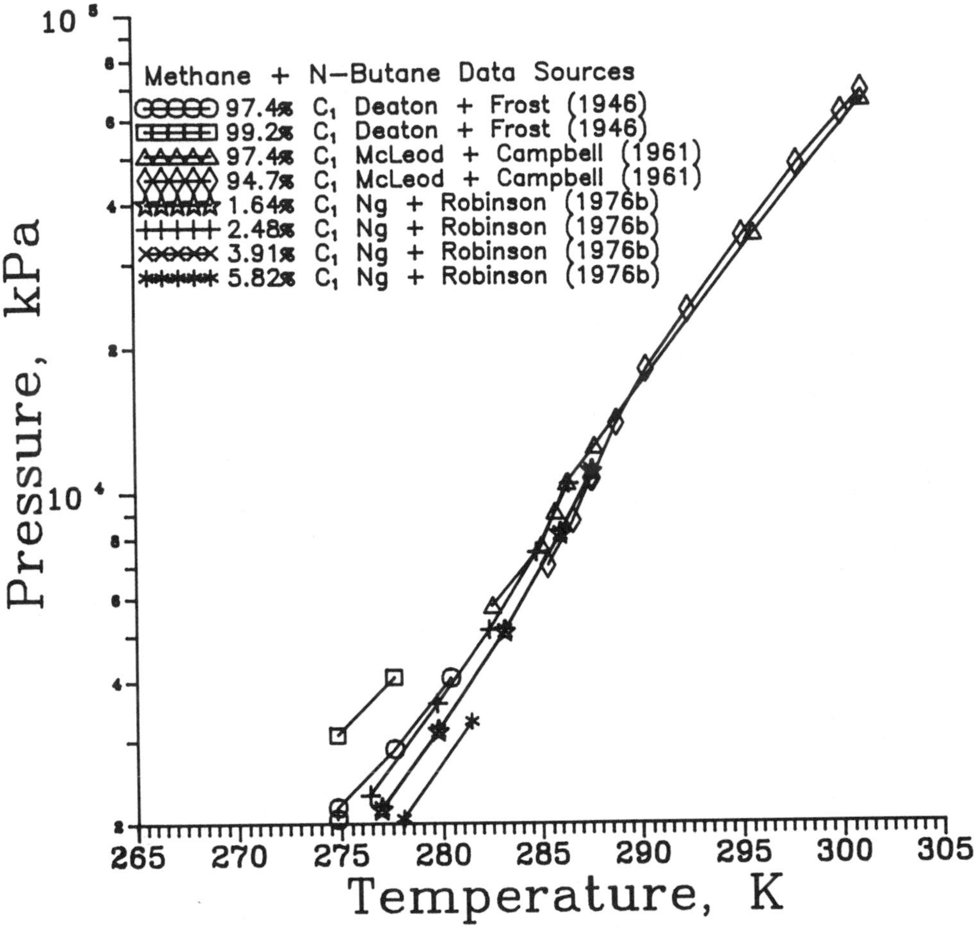


6-17a Methane+N-Butane Mixture ($L_W$-H-V) Data

| | | | | | |
|---|---|---|---|---|---|
| 73.1 | 294.44 | 34.33 | 24.0 | 289.11 | 35.96 |
| 50.25 | 273.17 | 4.96 | 10.8 | 273.17 | 12.55 |
| 50.25 | 277.22 | 6.13 | 10.8 | 277.17 | 15.86 |
| 50.25 | 279.67 | 7.77 | 10.8 | 279.11 | 19.39 |
| 50.25 | 282.33 | 10.49 | 10.8 | 280.94 | 22.52 |
| 50.25 | 287.33 | 17.90 | 10.8 | 282.06 | 25.82 |
| 50.25 | 289.83 | 24.99 | 10.8 | 283.17 | 28.79 |

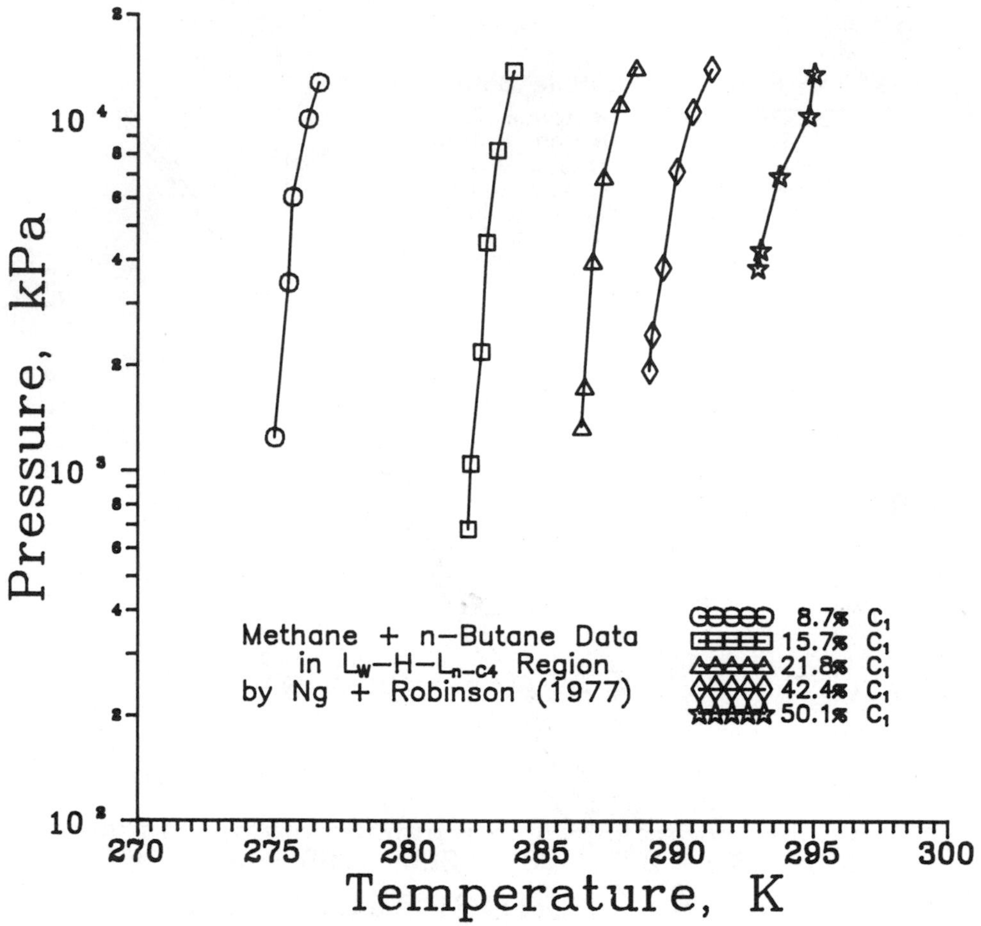


6-17b Methane+N-Butane Mixture ($L_W$-H-$L_{NB}$) Data

Isothermal P-x-y Data

| 273.2K | | | 277.4K | | | 279.8K | | |
|---|---|---|---|---|---|---|---|---|
| P(MPa) | $y_{N_2}$ | $x_{N_2}$ | P(MPa) | $y_{N_2}$ | $x_{N_2}$ | P(MPa) | $y_{N_2}$ | $x_{N_2}$ |
| 2.64 | 0.00 | 0.00 | 3.86 | 0.00 | 0.00 | 5.14 | 0.00 | 0.00 |
| 3.62 | 0.16 | 0.065 | 5.20 | 0.44 | 0.18 | 7.14 | 0.35 | 0.091 |
| 4.31 | 0.31 | 0.098 | 8.11 | 0.63 | 0.31 | 8.37 | 0.46 | 0.224 |
| 5.35 | 0.53 | 0.20 | 10.34 | 0.74 | 0.47 | 15.55 | 0.75 | 0.55 |
| 6.55 | 0.645 | 0.35 | 12.06 | 0.78 | 0.56 | 20.67 | 0.84 | 0.68 |

| | | | | | | | | |
|---|---|---|---|---|---|---|---|---|
| 7.75 | 0.725 | 0.425 | 25.18 | 1.00 | 1.00 | 25.23 | 0.914 | 0.802 |
| 10.64 | 0.815 | 0.62 | | | | 32.42 | 1.00 | 1.00 |
| 11.65 | 0.88 | 0.71 | | | | | | |
| 12.77 | 0.90 | 0.765 | | | | | | |
| 13.32 | 0.925 | 0.81 | | | | | | |
| 14.59 | 0.94 | 0.86 | | | | | | |
| 16.21 | 1.00 | 1.00 | | | | | | |

$x_{N_2}$ = mol fraction nitrogen in hydrate phase
$y_{N_2}$ = mol fraction nitrogen in vapor phase

Hydrate: Methane+Carbon Dioxide
Reference: Unruh, C.H., Katz, D.L. (1949)
Phases: $L_W$-H-V

| $\%CH_4$* | T(K) | P(MPa) | $\%CH_4$* | T(K) | P(MPa) |
|---|---|---|---|---|---|
| 66.0 | 277.0 | 2.84 | 91.5 | 278.4 | 3.95 |
| 70.0 | 278.9 | 3.46 | 93.0 | 281.0 | 5.10 |
| 64.0 | 278.9 | 3.43 | 94.5 | 283.8 | 6.89 |
| 68.0 | 280.9 | 4.24 | 29.0 | 279.6 | 3.00 |
| 72.0 | 282.9 | 5.17 | 39.0 | 282.2 | 4.27 |
| 77.0 | 284.7 | 6.47 | 48.0 | 283.8 | 5.27 |
| 40.0 | 275.5 | 1.99 | 59.0 | 285.5 | 6.89 |
| 56.0 | 279.2 | 3.08 | 59.0 | 285.7 | 7.00 |
| 87.5 | 276.4 | 3.20 | | | |

* vapor composition calculated in reference

Hydrate: Methane+Hydrogen Sulfide
Reference: Noaker, L.J., Katz, D.L. (1954)
Phases: $L_W$-H-V

| $\%CH_4$* | T(K) | P(MPa) | $\%CH_4$* | T(K) | P(MPa) |
|---|---|---|---|---|---|
| 91.77/93.0 | 288.7 | 4.83 | 96.22/96.10 | 276.5 | 2.03 |
| 90.49/91.0 | 284.3 | 2.59 | 99.0 | 278.4 | 3.24 |
| 93.7 | 282.3 | 3.03 | 98.96 | 282.3 | 4.62 |
| 93.5 | 287.1 | 4.79 | 98.89/98.94 | 284.8 | 6.69 |

| | | | | | |
|---|---|---|---|---|---|
| 93.0 | 290.1 | 6.79 | 78.0 | 287.6 | 2.10 |
| 94.27/93.5 | 279.3 | 2.21 | 80.20/79.0 | 295.4 | 5.07 |
| 93.40/93.0 | 290.1 | 6.38 | 78.60/78.0 | 279.8 | 1.03 |
| 97.0 | 278.7 | 2.83 | 90.5 | 281.5 | 2.07 |
| 96.9 | 282.9 | 4.27 | 89.0 | 287.3 | 3.59 |
| 97.08/97.05 | 287.6 | 6.65 | 88.50/88.50 | 292.1 | 6.00 |

* if two values of vapor composition are provided the first was measured and the second calculated; if only value of vapor composition is given, it was calculated in the reference

## Binary Guest Mixtures Containing Ethane and Heavier (or Non-Combustible) Compounds

Hydrate: Ethane+Propane
Reference: Holder, G.D., Hand, J.H. (1982)
Phases: $L_W$-H-V and $L_W$-H-$L_{HC}$

$L_W$-H-V

| $\%C_2H_6$ | T(K) | P(kPa) | $\%C_2H_6$ | T(K) | P(kPa) |
|---|---|---|---|---|---|
| 72.0 | 277.9 | 660 | 32.2 | 281.1 | 1630 |
| 72.0 | 276.9 | 530 | 27.1 | 273.4 | 490 |
| 72.0 | 276.5 | 460 | 27.1 | 273.9 | 540 |
| 55.7 | 275.9 | 500 | 27.1 | 274.3 | 610 |
| 55.7 | 276.4 | 570 | 27.1 | 274.6 | 600 |
| 55.7 | 276.7 | 610 | 27.1 | 275.3 | 770 |
| 55.7 | 277.0 | 650 | 27.1 | 275.6 | 870 |
| 55.7 | 277.4 | 720 | 27.1 | 275.8 | 920 |
| 54.1 | 275.8 | 500 | 26.0 | 274.5 | 630 |
| 54.1 | 276.4 | 590 | 26.0 | 274.7 | 690 |
| 54.1 | 277.0 | 660 | 26.0 | 275.2 | 790 |
| 54.1 | 277.6 | 770 | 26.0 | 276.4 | 940 |
| 54.1 | 278.0 | 850 | 26.0 | 277.1 | 1020 |
| 34.2 | 273.9 | 440 | 26.0 | 277.7 | 1120 |
| 34.2 | 274.2 | 470 | 18.6 | 273.1 | 540 |
| 34.2 | 275.1 | 590 | 18.6 | 273.8 | 640 |
| 34.2 | 275.8 | 690 | 18.6 | 273.8 | 640 |

| | | | | | |
|---|---|---|---|---|---|
| 34.2 | 276.2 | 830 | 18.6 | 274.3 | 660 |
| 34.2 | 276.3 | 850 | 18.6 | 274.7 | 710 |
| 34.2 | 276.5 | 870 | 18.6 | 276.8 | 940 |
| 34.2 | 277.6 | 1060 | 18.6 | 278.9 | 1210 |
| 32.2 | 275.6 | 750 | 18.6 | 279.6 | 1300 |
| 32.2 | 276.1 | 870 | 15.0 | 275.7 | 740 |
| 32.2 | 277.1 | 1140 | 15.0 | 277.2 | 900 |
| 32.2 | 277.2 | 1160 | 15.0 | 280.6 | 1370 |
| 32.2 | 277.9 | 1220 | 14.3 | 279.7 | 1190 |
| 32.2 | 278.6 | 1300 | 14.3 | 280.2 | 1300 |
| 32.2 | 279.4 | 1430 | 14.3 | 282.1 | 1670 |
| 32.2 | 280.4 | 1560 | 14.3 | 283.2 | 2030 |
| 32.2 | 280.6 | 1610 | 14.3 | 283.3 | 2020 |

$\underline{L_W\text{-}H\text{-}L_E}$

| | | | | | |
|---|---|---|---|---|---|
| 16.8 | 278.1 | 910* | 43.5 | 280.2 | 5180 |
| 16.8 | 278.1 | 1440 | 43.5 | 280.6 | 6550 |
| 16.8 | 278.2 | 1550 | 68.9 | 284.3 | 2230* |
| 16.8 | 278.3 | 2560 | 68.9 | 284.3 | 2900 |
| 16.8 | 278.6 | 2790 | 68.9 | 284.4 | 5580 |
| 43.5 | 279.9 | 1470* | 68.9 | 284.5 | 7280 |
| 43.5 | 279.9 | 2300 | | | |

* = $Q_2$ Quadruple Point ($L_W$-H-V-$L_{HC}$)

<u>Hydrates</u>: Ethane+Propane
<u>Reference</u>: Sloan, E.D. et al. (1986)
<u>Phases</u>: $L_{HC}$-H

Isobaric Data at 3.447 MPa

| Liquid Phase Compostion | | T | Water Conc. |
|---|---|---|---|
| $\%C_2H_6$ | $\%C_3H_8$ | (K) | ppm (mol basis) |
| 10.2 | 89.8 | 267.8 | 95 |
| 10.2 | 89.8 | 273.5 | 137 |
| 64.6 | 35.4 | 263.5 | 64 |
| 64.6 | 35.4 | 276.2 | 161 |
| 91.5 | 8.5 | 264.0 | 57 |
| 91.5 | 8.5 | 275.0 | 134 |

## Binary Guest Mixtures Containing Propane and Heavier (or Non-Combustible) Components

Hydrate: Propane+Isobutane
Reference: Kamath, V.A., Holder, G.D. (1982)
Phases: I-H-V

| $\%C_3H_8$ | T(K) | P(kPa) | $\%C_3H_8$ | T(K) | P(kPa) |
|---|---|---|---|---|---|
| 0.0 | 272.12 | 101.33 | 90.7 | 272.23 | 149.43 |
| 12.5 | 272.23 | 108.22 | 90.9 | 272.18 | 149.24 |
| 12.6 | 272.23 | 108.54 | 95.8 | 272.13 | 152.04 |
| 50.7 | 272.23 | 123.96 | 95.8 | 272.08 | 152.13 |
| 50.5 | 272.23 | 124.55 | 95.9 | 272.18 | 153.70 |
| 49.8 | 272.23 | 130.05 | 100.0 | 272.12 | 171.32 |
| 81.0 | 272.13 | 137.09 | | | |

Hydrate: Propane+Isobutane
Reference: Paranjpe, S.G. et al. (1987)
Phases: $L_W$-H-V

$\%C_3H_8$ = vapor composition

| $\%C_3H_8$ | T(K) | P(kPa) | $\%C_3H_8$ | T(K) | P(kPa) |
|---|---|---|---|---|---|
| 11.2 | 275.25 | 213.0 | 48.8 | 276.65 | 365.4 |
| 27.1 | 275.85 | 282.7 | 65.3 | 277.15 | 426.1 |
| 47.5 | 276.55 | 355.1 | 79.4 | 277.85 | 490.0 |

Hydrate: Propane+n-Butane
Reference: Kamath, V.A., Holder, G.D. (1982)
Phases: I-H-V

| $\%C_3H_8$ | T(K) | P(kPa) | $\%C_3H_8$ | T(K) | P(kPa) |
|---|---|---|---|---|---|
| 90.3 | 260.15 | 110.10 | 83.5 | 248.15 | 69.03 |
| 90.3 | 257.85 | 99.26 | 83.5 | 245.85 | 66.78 |
| 90.3 | 254.35 | 83.86 | 83.5 | 242.45 | 59.59 |
| 90.3 | 248.65 | 67.09 | 70.0 | 253.75 | 97.16 |
| 90.3 | 245.05 | 61.23 | 70.0 | 250.95 | 85.86 |

| | | | | | |
|---|---|---|---|---|---|
| 90.3 | 242.05 | 49.20 | 70.0 | 245.65 | 70.07 |
| 83.5 | 250.55 | 76.36 | 70.0 | 241.85 | 55.38 |
| 83.5 | 249.55 | 73.37 | 70.0 | 238.25 | 46.90 |

Hydrate: Propane+n-Butane
Reference: Paranjpe, S.G. et al. (1987)
Phases: I-H-V, $L_W$-H-V, $L_W$-H-V-$L_{n-C_4}$, I-H-V-$L_{n-C_4}$

$\%C_3H_8$ = vapor composition

| $\%C_3H_8$ | T(K) | P(kPa) | $\%C_3H_8$ | T(K) | P(kPa) |
|---|---|---|---|---|---|
| | | I-H-V | | | |
| 86.4 | 271.15 | 153.1 | 80.3 | 271.15 | 191.7 |
| 87.9 | 271.15 | 170.9 | 76.0 | 271.15 | 204.1 |
| 86.1 | 271.15 | 177.9 | 67.6 | 271.15 | 217.9 |
| | | $L_W$-H-V | | | |
| 99.6 | 275.15 | 269.6 | 81.7 | 274.15 | 269.6 |
| 92.4 | 275.15 | 281.3 | 100.0 | 273.15 | 169.6 |
| 89.4 | 275.15 | 302.0 | 96.9 | 273.15 | 171.7 |
| 87.3 | 275.15 | 308.9 | 95.3 | 273.15 | 177.9 |
| 88.0 | 275.15 | 317.2 | 93.8 | 273.15 | 183.4 |
| 86.3 | 275.15 | 339.2 | 89.5 | 273.15 | 193.1 |
| 100.0 | 274.15 | 219.3 | 86.4 | 273.15 | 208.2 |
| 96.1 | 274.15 | 228.2 | 82.8 | 273.15 | 220.6 |
| 93.3 | 274.15 | 240.6 | 78.8 | 273.15 | 227.5 |
| 90.6 | 274.15 | 244.8 | 72.5 | 273.15 | 244.1 |
| | | I-H-V-$L_{n-C_4}$ | | | |
| 64.9 | 273.05 | 215.1 | 64.9 | 267.35 | 174.4 |
| 67.4 | 270.95 | 211.7 | 63.9 | 266.75 | 168.2 |
| 67.3 | 270.05 | 201.3 | 64.8 | 264.05 | 157.9 |
| 62.1 | 269.65 | 184.8 | 66.5 | 261.95 | 146.2 |
| 66.5 | 268.95 | 192.4 | 65.1 | 260.55 | 133.8 |
| | | $L_W$-H-V-$L_{n-C_4}$ | | | |
| 92.6 | 277.85 | 511.6 | 81.6 | 275.15 | 324.1 |
| 84.6 | 276.65 | 399.2 | 71.0 | 273.65 | 242.7 |

Mixture: Propane+Nitrogen
Reference: Ng, H.-J. et al. (1977/78)
Phases: $L_W$-H-V, $L_W$-H-V-$L_{C_3}$, $L_W$-H-$L_{C_3}$

$L_W$-H-V

$\%C_3H_8$ = vapor composition

| $\%C_3H_8$ | T(K) | P(MPa) | $\%C_3H_8$ | T(K) | P(MPa) |
|---|---|---|---|---|---|
| 0.94 | 275.33 | 4.59 | 13.00 | 281.49 | 2.74 |
| 0.94 | 279.55 | 8.16 | 13.00 | 283.15 | 3.54 |
| 0.94 | 283.00 | 13.68 | 13.00 | 286.20 | 5.54 |
| 0.94 | 284.28 | 18.09 | 28.30 | 274.64 | 0.569 |
| 2.51 | 276.29 | 3.03 | 28.30 | 277.02 | 0.889 |
| 2.51 | 279.28 | 4.51 | 28.30 | 279.18 | 1.31 |
| 2.51 | 282.74 | 7.35 | 28.30 | 280.77 | 1.72 |
| 2.51 | 287.10 | 13.64 | 54.20 | 274.16 | 0.332 |
| 6.18 | 274.51 | 1.72 | 54.20 | 276.76 | 0.570 |
| 6.18 | 278.26 | 2.85 | 54.20 | 280.27 | 1.19 |
| 6.18 | 283.05 | 5.50 | 75.00 | 274.54 | 0.256 |
| 6.18 | 286.95 | 9.47 | 75.00 | 275.90 | 0.359 |
| 6.18 | 289.17 | 13.71 | 75.00 | 277.41 | 0.517 |
| 13.00 | 275.10 | 1.10 | 75.00 | 278.67 | 0.676 |
| 13.00 | 278.41 | 1.72 | | | |

$L_W$-H-V-$L_{C_3}$

$\%C_3H_8$ = vapor composition

| $\%C_3H_8$ | T(K) | P(MPa) | $\%C_3H_8$ | T(K) | P(MPa) |
|---|---|---|---|---|---|
| 71.1 | 278.98 | 0.76 | 18.8 | 289.65 | 8.58 |
| 47.7 | 281.23 | 1.54 | 21.3 | 292.27 | 13.37 |
| 26.5 | 283.17 | 2.88 | 21.6 | 292.29 | 13.51 |
| 22.9 | 284.58 | 3.68 | 23.0 | 292.95 | 14.92 |
| 18.8 | 287.95 | 6.37 | 23.7 | 293.80 | 16.99 |

$L_W$-H-$L_{C_3}$

$\%C_3H_8$ = $L_{C_3H_8}$ composition

| $\%C_3H_8$ | T(K) | P(MPa) | $\%C_3H_8$ | T(K) | P(MPa) |
|---|---|---|---|---|---|
| 99.0 | 280.07 | 1.17 | 89.6 | 287.38 | 5.83 |
| 99.0 | 280.13 | 1.98 | 89.6 | 287.86 | 8.14 |
| 99.0 | 280.33 | 5.65 | 89.6 | 288.48 | 11.06 |
| 99.0 | 280.58 | 9.39 | 89.6 | 288.81 | 12.82 |

| | | | | | |
|---|---|---|---|---|---|
| 99.0 | 280.86 | 13.89 | 88.6 | 287.70 | 6.17 |
| 96.8 | 282.74 | 2.34 | 88.6 | 287.95 | 7.34 |
| 96.8 | 282.80 | 2.85 | 88.6 | 288.15 | 8.20 |
| 96.8 | 283.25 | 7.10 | 88.6 | 288.36 | 9.11 |
| 96.8 | 283.68 | 10.44 | 83.1 | 289.80 | 8.86 |
| 93.5 | 285.25 | 4.08 | 83.1 | 290.00 | 9.82 |
| 93.5 | 285.49 | 5.16 | 83.1 | 290.46 | 11.25 |
| 93.5 | 286.06 | 8.41 | 83.1 | 290.87 | 13.29 |
| 93.5 | 286.32 | 11.55 | 83.1 | 291.30 | 15.31 |
| 93.5 | 286.71 | 13.33 | | | |

<u>Hydrate</u>: Propane+Carbon Dioxide
<u>Reference</u>: Robinson, D.B., Mehta, B.R. (1971)
<u>Phases</u>: $L_W$-H-V

$\%C_3H_8$ = vapor composition

| $\%C_3H_8$ | T(K) | P(kPa) | $\%C_3H_8$ | T(K) | P(kPa) |
|---|---|---|---|---|---|
| 5.5 | 284.83 | 4268 | 25.0 | 283.8 | 1917 |
| 6.0 | 276.3 | 1151 | 25.0 | 280.2 | 979 |
| 7.0 | 273.8 | 814 | 26.0 | 281.8 | 1303 |
| 8.0 | 283.7 | 3179 | 42.0 | 283.7 | 1655 |
| 8.0 | 281.7 | 2186 | 42.2 | 275.7 | 414 |
| 8.0 | 273.9 | 676 | 47.5 | 278.6 | 689 |
| 9.0 | 283.5 | 3034 | 48.0 | 281.1 | 1069 |
| 9.0 | 280.4 | 1772 | 60.0 | 279.7 | 793 |
| 9.0 | 278.9 | 1455 | 60.0 | 274.8 | 324 |
| 10.0 | 278.3 | 1255 | 61.0 | 276.4 | 434 |
| 13.0 | 280.9 | 1572 | 63.0 | 279.6 | 752 |
| 13.0 | 273.8 | 517 | 65.0 | 278.3 | 579 |
| 14.0 | 279.4 | 1207 | 72.0 | 275.2 | 303 |
| 15.0 | 275.4 | 827 | 82.0 | 279.1 | 593 |
| 15.0 | 276.5 | 758 | 83.0 | 278.2 | 503 |
| 16.0 | 286.2 | 3378 | 83.0 | 279.1 | 641 |
| 21.0 | 274.0 | 359 | 84.0 | 277.8 | 476 |
| 23.0 | 278.1 | 710 | 86.0 | 276.0 | 338 |
| 24.0 | 277.4 | 655 | | | |

## Binary Guest Mixtures Containing Isobutane and Normal Butane Components

Hydrates: Isobutane+n-Butane
Reference: Godbole, S.J. (1981)
Phases: $L_W$-H-V

| %i-$C_4H_{10}$ | T(K) | P(kPa) | %i-$C_4H_{10}$ | T(K) | P(kPa) |
|---|---|---|---|---|---|
| 0.77 | 267.3 | 68.8 | 0.748 | 265.3 | 80.8 |
| 0.77 | 261.1 | 56.2 | 0.748 | 263.4 | 73.2 |
| 0.835 | 264.3 | 63.2 | | | |

### 6.2.1.c. Equilibria of Ternary Guest Mixtures

Due to the diversity of gas compositions for ternary guest systems, duplicate data sets have not been obtained by various investigators. The variety of gas compositions make plots of the data meaningless, except for self-consistency within an individual investigation. Consequently plots of ternary (and higher) guest data sets are not provided. Frequently such plots are given in the original reference with the data set.

Hydrate: Methane+Ethane+Propane
Reference: Holder, G.D. and Hand, J.H. (1982)
Phases: $L_W$-H-V-$L_{HC}$
(compositions are mol fractions in the $L_{HC}$ phase)

$x_{C_1}/(x_{C_1} + x_{C_3}) = 0.0$

| $x_{C_2}$ | T(K) | P(kPa) | $x_{C_2}$ | T(K) | P(kPa) |
|---|---|---|---|---|---|
| 0.168 | 278.1 | 900 | NM | 278.6 | 1300 |
| NM | 278.1 | 970 | NM | 279.4 | 1430 |
| 0.280 | 277.8 | 1080 | NM | 280.4 | 1560 |
| NM | 278.2 | 700 | NM | 280.6 | 1610 |
| 0.435 | 279.9 | 1480 | NM | 281.1 | 1630 |
| 0.523 | 281.8 | 1740 | 1.00 | 288.3 | 3330 |
| 0.689 | 284.3 | 2230 | | | |

NM ≡ not measured

$x_{C_1}/(x_{C_1} + x_{C_3}) = 0.0255$

| | | | | | |
|---|---|---|---|---|---|
| 0.000 | 282.8 | 1030 | 0.716 | 284.8 | 2320 |
| 0.189 | 281.7 | 1240 | 0.799 | 286.1 | 2730 |
| 0.314 | 280.1 | 1580 | 1.000 | 288.3 | 3330 |
| 0.517 | 282.8 | 2020 | | | |

$x_{C_1}/(x_{C_1} + x_{C_3}) = 0.051$

| | | | | | |
|---|---|---|---|---|---|
| 0.082 | 284.9 | 1500 | 0.549 | 283.7 | 2180 |
| 0.181 | 284.2 | 1610 | 0.726 | 285.7 | 2730 |
| 0.231 | 283.5 | 1680 | 1.000 | 288.3 | 3330 |

$x_{C_1}/(x_{C_1} + x_{C_3}) = 0.092$

| | | | | | |
|---|---|---|---|---|---|
| 0.054 | 287.4 | 2120 | 0.506 | 283.8 | 2380 |
| 0.148 | 286.9 | 2170 | 0.616 | 285.1 | 2570 |
| 0.212 | 285.7 | 2210 | 0.677 | 285.9 | 2820 |
| 0.365 | 284.9 | 2280 | 1.000 | 288.3 | 3330 |

Hydrate: Methane+Propane+Isobutane
Reference: Paranjpe, S.G., et al. (1987)
Phases: $L_W$-H-V-$L_{HC}$

| T (K) | Mol Fraction In Vapor $CH_4$ | $C_3H_8$ | $i\text{-}C_4H_{10}$ | Pressure kPa |
|---|---|---|---|---|
| 276.15 | 0.261 | 0.000 | 0.739 | 220.6 |
| 276.15 | 0.155 | 0.106 | 0.739 | 248.2 |
| 276.15 | 0.123 | 0.194 | 0.683 | 262.0 |
| 276.15 | 0.000 | 0.369 | 0.631 | 303.4 |
| 279.15 | 0.511 | 0.000 | 0.489 | 398.0 |
| 279.15 | 0.201 | 0.085 | 0.714 | 458.5 |
| 279.15 | 0.092 | 0.279 | 0.629 | 495.0 |
| 279.15 | 0.128 | 0.542 | 0.330 | 543.3 |
| 279.15 | 0.128 | 0.737 | 0.135 | 576.4 |
| 279.15 | 0.102 | 0.851 | 0.047 | 606.7 |
| 279.15 | 0.010 | 0.990 | 0.000 | 674.0 |
| 281.15 | 0.620 | 0.000 | 0.380 | 585.1 |
| 281.15 | 0.242 | 0.071 | 0.687 | 621.9 |
| 281.15 | 0.187 | 0.393 | 0.420 | 668.8 |
| 281.15 | 0.150 | 0.569 | 0.281 | 723.9 |
| 281.15 | 0.126 | 0.814 | 0.060 | 773.6 |
| 281.15 | 0.030 | 0.970 | 0.000 | 832.0 |

Hydrate: Methane+Propane+n-Butane
Reference: Paranjpe, S.G., et al. (1987)
Phases: I-H-V-$L_{HC}$ and $L_W$-H-V-$L_{HC}$

| T (K) | Mol Fraction In Vapor $CH_4$ | $C_3H_8$ | $n-C_4H_{10}$ | Pressure kPa |
|---|---|---|---|---|
| 268.15 | 0.000 | 0.660 | 0.340 | 181.4 |
| 268.15 | 0.349 | 0.407 | 0.244 | 216.5 |
| 268.15 | 0.736 | 0.082 | 0.182 | 270.3 |
| 268.15 | 0.840 | 0.027 | 0.127 | 362.7 |
| 268.15 | 0.890 | 0.000 | 0.110 | 784.0 |
| 275.15 | 0.000 | 0.834 | 0.166 | 315.8 |
| 275.15 | 0.327 | 0.258 | 0.415 | 478.5 |
| 275.15 | 0.922 | 0.051 | 0.027 | 523.4 |
| 275.15 | 0.935 | 0.012 | 0.053 | 551.6 |
| 275.15 | 0.921 | 0.000 | 0.079 | 1330.7 |
| 281.15 | 0.111 | 0.889 | 0.000 | 821.9 |
| 281.15 | 0.686 | 0.206 | 0.108 | 1048.0 |
| 281.15 | 0.836 | 0.090 | 0.074 | 1206.6 |
| 281.15 | 0.942 | 0.018 | 0.040 | 1834.0 |
| 281.15 | 0.965 | 0.005 | 0.031 | 2643.8 |
| 281.15 | 0.985 | 0.000 | 0.015 | 3441.0 |

Hydrate: Methane+Propane+n-Decane
Reference: Verma, V.K. (1974)
Phases: $L_W$-H-$L_{HC}$ and $L_W$-H-V-$L_{HC}$

Compositions are given for hydrocarbon liquid phase

| mol% $CH_4$ | mol% $C_3H_8$ | mol% $C_{10}H_{22}$ | T(K) | P(kPa) |
|---|---|---|---|---|
| | | $L_W$-H-$L_{HC}$ | | |
| 14.51 | 27.09 | 58.4 | 287.52 | 2875 |
| 14.51 | 27.09 | 58.4 | 287.80 | 4544 |
| 14.51 | 27.09 | 58.4 | 288.08 | 6888 |
| 14.51 | 27.09 | 58.4 | 288.71 | 10225 |
| 14.51 | 27.09 | 58.4 | 289.22 | 13707 |
| | | $L_W$-H-V-$L_{HC}$ | | |
| 0.00 | 96.55 | 3.45 | 278.31 | 539 |

| | | | | |
|---|---|---|---|---|
| 0.00 | 94.88 | 5.12 | 278.17 | 525 |
| 0.00 | 91.81 | 8.19 | 277.94 | 501 |
| 0.00 | 80.71 | 19.29 | 277.38 | 443 |
| 0.00 | 80.32 | 19.68 | 277.49 | 465 |
| 0.00 | 72.16 | 27.84 | 276.74 | 391 |
| 0.00 | 63.45 | 36.55 | 276.08 | 343 |
| 0.00 | 49.90 | 50.10 | 275.01 | 269 |
| 10.32 | 74.47 | 15.21 | 288.48 | 2241 |
| 8.26 | 59.76 | 31.97 | 286.21 | 1806 |
| 6.41 | 46.40 | 47.19 | 283.87 | 1338 |
| 5.01 | 32.55 | 59.44 | 281.38 | 1014 |
| 3.92 | 27.91 | 68.17 | 278.75 | 758 |
| 33.64 | 57.19 | 9.17 | 295.97 | 7122 |
| 29.38 | 51.10 | 19.52 | 295.12 | 6585 |
| 25.78 | 45.55 | 28.68 | 293.84 | 5702 |
| 22.39 | 39.91 | 37.70 | 292.18 | 4826 |
| 19.32 | 34.87 | 45.81 | 290.62 | 3951 |
| 17.01 | 31.06 | 51.93 | 289.21 | 3427 |
| 14.51 | 27.09 | 58.40 | 287.52 | 2875 |
| 44.66 | 38.46 | 16.88 | 297.81 | 11597 |
| 42.74 | 24.93 | 32.33 | 296.27 | 11370 |
| 34.95 | 20.48 | 44.58 | 294.00 | 8763 |
| 29.09 | 17.08 | 53.83 | 291.73 | 6764 |
| 24.68 | 14.51 | 60.82 | 289.46 | 5426 |
| 20.75 | 12.23 | 67.02 | 287.43 | 4344 |
| 72.08 | 25.72 | 2.20 | 297.89 | 15031* |
| 59.53 | 21.01 | 19.46 | 298.18 | 16886 |
| 53.41 | 19.00 | 25.59 | 297.74 | 16251 |
| 45.37 | 16.20 | 38.42 | 296.34 | 12763 |
| 39.44 | 14.07 | 46.49 | 294.34 | 10336 |
| 32.86 | 11.74 | 55.40 | 291.49 | 7936 |
| 26.83 | 9.68 | 63.53 | 288.88 | 6019 |
| 22.72 | 8.19 | 69.09 | 286.66 | 4847 |
| 86.60 | 8.60 | 4.80 | 297.22 | 21788* |
| 55.42 | 5.62 | 38.96 | 292.90 | 13121 |
| 41.00 | 4.02 | 54.98 | 288.70 | 8019 |
| 32.12 | 3.16 | 64.71 | 285.88 | 5868 |
| 25.36 | 2.50 | 72.14 | 282.22 | 4302 |
| 19.36 | 1.90 | 78.74 | 279.19 | 3151 |
| 91.50 | 0.00 | 8.50 | 295.44 | 31027* |
| 58.48 | 0.00 | 41.52 | 290.44 | 17079 |
| 42.47 | 0.00 | 57.53 | 285.25 | 9350 |

| | | | | |
|---|---|---|---|---|
| 34.52 | 0.00 | 65.48 | 282.86 | 7116 |
| 26.90 | 0.00 | 73.10 | 278.91 | 4702 |

*quadruple-cum-critical point

Hydrate: Methane+Propane+Hydrogen Sulfide
Reference: Schroeter, et al. (1983)
Phases: $L_W$-H-V

| $\%C_3H_8$ | $\%CH_4$ | $\%H_2S$ | T(K) | P(kPa) |
|---|---|---|---|---|
| 7.172 | 88.654 | 4.174 | 275.95 | 561 |
| 7.172 | 88.654 | 4.174 | 277.75 | 706 |
| 7.172 | 88.654 | 4.174 | 284.15 | 1419 |
| 7.172 | 88.654 | 4.174 | 287.35 | 2024 |
| 7.172 | 88.654 | 4.174 | 291.15 | 3367 |
| 7.016 | 81.009 | 11.975 | 275.85 | 339 |
| 7.016 | 81.009 | 11.975 | 283.55 | 817 |
| 7.016 | 81.009 | 11.975 | 292.65 | 2813 |
| 7.402 | 60.888 | 31.71 | 280.35 | 368 |
| 7.402 | 60.888 | 31.71 | 286.25 | 686 |
| 7.402 | 60.888 | 31.71 | 292.25 | 1444 |
| 7.402 | 60.888 | 31.71 | 297.45 | 2555 |
| 7.402 | 60.888 | 31.71 | 300.95 | 4275 |

Hydrate: Methane+Carbon Dioxide+Hydrogen Sulfide
Reference: Robinson, D.B., Hutton, J.M. (1967)
Phases: $L_W$-H-V

| $\%CH_4$ | $\%CO_2$ | $\%H_2S$ | T(K) | P(MPa) | $\%CH_4$ | $\%CO_2$ | $\%H_2S$ | T(K) | P(MPa) |
|---|---|---|---|---|---|---|---|---|---|
| 78.5 | 13.9 | 7.6 | 281.1 | 1.675 | 68.6 | 24.9 | 6.5 | 279.2 | 1.475 |
| 80.3 | 13.0 | 6.7 | 282.5 | 2.275 | 69.9 | 24.1 | 6.0 | 282.1 | 2.034 |
| 81.0 | 13.0 | 6.0 | 286.6 | 3.868 | 70.5 | 23.5 | 6.0 | 284.0 | 2.771 |
| 82.0 | 12.6 | 5.4 | 287.3 | 4.558 | 71.5 | 22.8 | 5.7 | 286.4 | 3.744 |
| 82.0 | 12.6 | 5.4 | 289.4 | 5.888 | 72.5 | 22.0 | 5.5 | 288.4 | 4.930 |
| 82.0 | 12.6 | 5.4 | 290.8 | 6.881 | 72.5 | 22.0 | 5.5 | 290.0 | 6.185 |
| 82.0 | 12.5 | 5.5 | 292.2 | 8.653 | 72.5 | 22.0 | 5.5 | 290.9 | 7.550 |

| | | | | | | | | | |
|---|---|---|---|---|---|---|---|---|---|
| 82.0 | 12.6 | 5.4 | 292.9 | 9.632 | 72.0 | 22.3 | 5.7 | 293.0 | 11.225 |
| 82.0 | 12.6 | 5.4 | 293.6 | 10.790 | 72.3 | 22.2 | 5.5 | 293.7 | 12.011 |
| 82.5 | 12.1 | 5.4 | 294.7 | 12.341 | 69.9 | 12.7 | 17.4 | 287.4 | 2.020 |
| 82.0 | 12.6 | 5.4 | 295.6 | 14.079 | 70.0 | 12.3 | 16.7 | 289.9 | 2.648 |
| 82.0 | 12.6 | 5.4 | 296.4 | 15.707 | 72.0 | 12.0 | 16.0 | 291.5 | 3.330 |
| 81.0 | 11.8 | 7.2 | 284.2 | 1.903 | 72.0 | 12.0 | 16.0 | 293.3 | 4.392 |
| 81.0 | 11.8 | 7.2 | 285.4 | 2.765 | 72.0 | 12.0 | 16.0 | 294.7 | 5.123 |
| 80.0 | 12.0 | 8.0 | 289.1 | 4.254 | 71.1 | 11.9 | 17.0 | 295.4 | 6.495 |
| 80.0 | 12.0 | 8.0 | 290.4 | 4.978 | 70.8 | 12.1 | 17.1 | 296.5 | 7.384 |
| 80.0 | 12.0 | 8.0 | 292.1 | 5.943 | 72.5 | 11.9 | 15.6 | 297.6 | 8.405 |
| 81.6 | 11.1 | 7.3 | 293.1 | 6.984 | 68.8 | 13.6 | 17.6 | 297.6 | 8.005 |
| 83.9 | 9.4 | 6.7 | 294.0 | 7.529 | | | | | |

### 6.2.1.d Equilibria of Multicomponent Guest Mixtures

Hydrate: Natural Gases
Reference: Wilcox, W.I. et al. (1941)
Phases: $L_W$-H-V

Compositions of Gases

| Gas | %$N_2$ | %$CO_2$ | %$CH_4$ | %$C_2H_6$ | %$C_3H_8$ | %i-$C_4H_{10}$ | %n-$C_4H_{10}$ | %$C_5H_{12}^+$ |
|---|---|---|---|---|---|---|---|---|
| B | 0.64 | | 86.41 | 6.47 | 3.57 | 0.99 | 1.14 | 0.78 |
| C | 0.43 | 0.51 | 93.20* | 4.25 | 1.61 | | | |
| D | | | 88.36 | 6.82 | 2.54 | 0.38 | 0.89 | 1.01 |

* includes nitrogen

Phase Equilibrium Data

| T(K) | P(MPa) | T(K) | P(MPa) | T(K) | P(MPa) |
|---|---|---|---|---|---|
| | | Gas B | | | |
| 278.8 | 1.255 | 292.2 | 6.964 | 296.1 | 16.84 |
| 282.9 | 1.924 | 294.1 | 9.680 | 297.1 | 20.68 |
| 288.7 | 4.123 | 295.1 | 12.270 | 298.3 | 27.32 |

Gas C

| | | | | | |
|---|---|---|---|---|---|
| 277.7 | 1.600 | 289.2 | 6.964 | 293.3 | 14.13 |
| 283.9 | 3.392 | 292.1 | 10.501 | | |

Gas C (Second Series)

| | | | | | |
|---|---|---|---|---|---|
| 279.1 | 1.924 | 291.2 | 9.481 | 294.3 | 17.16 |
| 281.8 | 2.648 | 293.3 | 13.780 | 296.7 | 27.50 |
| 286.7 | 4.820 | 295.0 | 20.264 | 295.6 | 22.99 |
| 290.3 | 8.136 | | | | |

Gas D

| | | | | | |
|---|---|---|---|---|---|
| 276.8 | 1.207 | 284.9 | 3.516 | 291.2 | 8.205 |
| 281.6 | 2.186 | 288.7 | 5.447 | 293.2 | 12.000 |

Gas D (Second Series)

| | | | | | |
|---|---|---|---|---|---|
| 289.2 | 6.171 | 293.9 | 14.07 | 296.9 | 26.55 |
| 292.1 | 9.308 | 295.0 | 18.20 | 295.9 | 22.55 |

Hydrates: Natural Gases
Reference: Deaton, W.M., Frost, E.M., Jr. (1946)
Phases: $L_W$-H-V

Compositions of Gases (mol per cent)

| Gas↓ Cmpnt→ | $CO_2$ | $H_2S$ | $N_2$ | $CH_4$ | $C_2H_6$ | $C_3H_8$ | $C_4H_{10}$ |
|---|---|---|---|---|---|---|---|
| A | 0.20 | | 7.7 | 65.4 | 12.7 | 10.3 | 3.7 |
| B | 0.20 | | 1.1 | 87.9 | 4.4 | 4.9 | 1.5 |
| C | 0.20 | | 9.4 | 78.4 | 6.0 | 3.6 | 2.4 |
| D | 0.30 | | 9.5 | 79.4 | 5.8 | 3.6 | 1.4 |
| E | 3.25 | 0.25 | 1.1 | 87.8 | 4.0 | 2.1 | 1.5 |
| F | 0.40 | | 0.3 | 91.0 | 3.2 | 2.0 | 3.1 |
| G | | | 1.0 | 90.8 | 3.0 | 2.1 | 3.2 |
| H | 0.20 | | 14.3 | 75.2 | 5.9 | 3.3 | 1.1 |
| I | | | 3.4 | 88.5 | 4.3 | 2.0 | 1.7 |
| J | 0.9 | | 1.2 | 90.6 | 3.8 | 1.5 | 2.0 |
| K | 0.8 | | 25.0 | 67.4 | 3.7 | 1.9 | 1.2 |
| L | 0.6 | | 0.2 | 96.5 | 0.9 | 1.8 | |

Phase Equilibrium Data

| T(K) | P(kPa) | T(K) | P(kPa) | T(K) | P(kPa) | T(K) | P(KPa) |
|---|---|---|---|---|---|---|---|
| | | | Gas A | | | | |
| 274.8 | 627 | 285.9 | 2571 | 290.1 | 4296 | 292.6 | 6302 |
| 280.3 | 1262 | 288.7 | 3592 | 291.5 | 5364 | 294.0 | 8536 |
| 283.2 | 1806 | | | | | | |
| | | | Gas B | | | | |
| 273.7 | 600 | 280.4 | 1338 | 283.7 | 2089 | 285.9 | 2668 |
| 275.4 | 738 | 282.6 | 1779 | 284.3 | 2248 | 286.5 | 2861 |
| 277.6 | 993 | | | | | | |
| | | | Gas C | | | | |
| 273.7 | 724 | 275.9 | 903 | 277.6 | 1172 | 279.8 | 1462 |
| 274.8 | 807 | 276.5 | 972 | 278.2 | 1241 | 283.2 | 2213 |
| | | | Gas D | | | | |
| 273.7 | 752 | 276.5 | 1089 | 280.4 | 1675 | | |
| 274.8 | 883 | 278.7 | 1400 | 282.1 | 2096 | | |
| | | | Gas E | | | | |
| 275.4 | 945 | 280.3 | 1717 | 285.9 | 3454 | 289.3 | 5254 |
| | | | Gas F | | | | |
| 273.7 | 765 | 280.4 | 1731 | 288.7 | 4909 | 292.1 | 8653 |
| 277.6 | 1241 | 285.9 | 3461 | 288.7 | 4895 | 292.7 | 9391 |
| 280.4 | 1731 | | | | | | |
| | | | Gas G | | | | |
| 273.7 | 758 | 280.4 | 1723 | 288.7 | 5033 | 291.5 | 7729 |
| 277.6 | 1234 | 285.9 | 3468 | 288.7 | 5033 | | |
| | | | Gas H | | | | |
| 274.2 | 758 | 277.6 | 1255 | 280.4 | 1758 | 282.1 | 2131 |
| 275.4 | 945 | | | | | | |
| | | | Gas I | | | | |
| 273.7 | 793 | 275.3 | 972 | 277.6 | 1310 | 280.4 | 1813 |
| | | | Gas J | | | | |
| 273.7 | 883 | 277.6 | 1455 | 285.9 | 4034 | 287.6 | 5068 |

| | | | | | | | |
|---|---|---|---|---|---|---|---|
| 274.3 | 952 | 280.4 | 2027 | 285.9 | 4027 | 288.2 | 5433 |
| 274.8 | 1020 | 280.4 | 1993 | 285.9 | 4054 | 288.7 | 5812 |
| 274.8 | 1027 | 280.4 | 2006 | 286.5 | 4364 | 289.8 | 6985 |
| 275.9 | 1172 | 283.2 | 2841 | 287.1 | 4675 | 290.9 | 8384 |
| 277.6 | 1441 | 285.4 | 3765 | 287.1 | 4682 | | |

Gas K

| | | | | | | | |
|---|---|---|---|---|---|---|---|
| 274.3 | 1069 | 277.6 | 1607 | 283.2 | 3165 | 286.0 | 4592 |
| 275.4 | 1220 | 280.4 | 2248 | | | | |

Gas L

| | | | | | | | |
|---|---|---|---|---|---|---|---|
| 273.7 | 1262 | 277.6 | 2027 | 283.2 | 4047 | 289.8 | 10329 |
| 274.8 | 1427 | 280.4 | 2855 | 287.7 | 7425 | 289.8 | 10439 |
| 274.8 | 1420 | | | | | | |

Hydrate: Natural Gases
Reference: Kobayashi, R., et al. (1951)
Phases: $L_W$-H-V

Compositions of Gases

| Gas | %$N_2$ | %$CH_4$ | %$C_2H_6$ | %$C_3H_8$ | %i-$C_4$ | %n-$C_4$ | %$C_5H_{12}$ | %$C_6H_{14}^+$ |
|---|---|---|---|---|---|---|---|---|
| Hugoton | 15.0 | 73.29 | 6.70 | 3.90 | 0.36 | 0.55 | 0.20 | 0.00 |
| Michigan | 6.8 | 79.64 | 9.38 | 3.22 | 0.18 | 0.58 | 0.15 | 0.05 |

Phase Equilibrium Data

| T(K) | P(MPa) | T(K) | P(MPa) | T(K) | P(MPa) |
|---|---|---|---|---|---|
| | | Hugoton Gas | | | |
| 281.6 | 1.765 | 287.7 | 3.847 | 290.9 | 5.833 |
| 283.9 | 2.517 | 288.9 | 4.461 | | |
| | | Michigan Gas | | | |
| 283.3 | 2.186 | 287.2 | 3.578 | 291.0 | 5.661 |
| 285.7 | 2.930 | 289.4 | 4.613 | | |

Hydrate: 90.6%$CH_4$,6.6%$C_2H_6$,1.8%$C_3H_8$,0.5%i$C_4H_{10}$,0.5%n$C_4H_{10}$
Reference: McLeod, H.O., Campbell, J.M. (1961)
Phases: $L_W$-H-V

| T(K) | P(MPa) | T(K) | P(MPa) | T(K) | P(MPa) |
|---|---|---|---|---|---|
| 293.6 | 13.55 | 301.7 | 52.16 | 295.8 | 20.24 |
| 297.5 | 27.68 | 303.1 | 62.85 | 298.6 | 33.75 |
| 300.0 | 41.34 | | | | |

Hydrates: 3 Natural Gases
Reference: Lapin, A., Cinnamon, S.J. (1969)
Phases: $L_W$-H-V

Compositions of Gases

| Gas | %$N_2$ | %$CH_4$ | %$C_2H_6$ | %$C_3H_8$ | %n-$C_4H_{10}$ | %n-$C_5H_{12}$ | %n-$C_6H_{14}$ |
|---|---|---|---|---|---|---|---|
| 1 | 1.70 | 72.78 | 14.50 | 7.63 | 2.65 | 0.63 | 0.11 |
| 2 | 1.58 | 67.69 | 13.50 | 14.10 | 2.45 | 0.58 | 0.10 |
| 3 | 1.47 | 62.49 | 12.47 | 20.62 | 2.26 | 0.58 | 0.11 |

Formation Temperature at 4.207 MPa

| Gas 1 | Gas 2 | Gas 3 |
|---|---|---|
| 288.75 | 289.85 | 289.85 |

Hydrates: Hydrate Denuding of $CH_4$ and $C_3H_8$ from Condensates or Crudes
Reference: Verma, V.K. et al. (1975)
Phases: $L_W$-H-V-$L_{HC}$

Initial Properties of Condensates and Crude

| Property | Prudhoe Bay Condensate | Ray Field Condensate | North Slope Crude |
|---|---|---|---|
| Molecular Wt ($C_6H_6$ Freezing Point Method) | 136 | 110 | 260 |
| Specific Gravity | 0.75 | 0.73 | 0.932 |

Properties Before and After Hydrate Formation

| | Liquid Molar Composition | | | Hydrate Composition | | Quadruple Point | |
|---|---|---|---|---|---|---|---|
| | %$CH_4$ | %$C_3H_8$ | %Cond | %$CH_4$ | %$C_3H_8$ | T K | P MPa |
| **Prudhoe Bay Condensate** | | | | | | | |
| Initial | 32.2 | 18.2 | 49.6 | Not Applicable | | 293.9 | 6.825 |
| Final | 21.2 | 19.6 | 59.2 | 92.5 | 7.5 | 291.3 | 4.875 |
| **Ray Field Condensate** | | | | | | | |
| Initial | 49.2 | 16.2 | 34.6 | Not Applicable | | 297.2 | 15.27 |
| Final | 46.1 | 13.1 | 40.8 | 64.8* | 35.2* | 295.4 | 13.66 |
| **North Slope Crude Oil** | | | | | | | |
| Initial | 32.8 | 14.5 | 52.7 | Not Applicable | | 296.7 | 10.17 |
| Final | 17.0 | 14.2 | 68.8 | 85.9 | 14.1 | 290.5 | 6.41 |

* calculated by material balance

Hydrates: Natural Gases, Hexane, Decane, and Crude Oils
Reference: Holder, G.D. (1976)
Phases: $L_W$-H-V-$L_{HC}$

Overall Mol Per Cent of Hydrocarbon Components

| Expt | $CH_4$ | $C_2H_6$ | $C_3H_8$ | $CO_2$ | n-$C_6H_{14}$ | c-$C_6H_{12}$ | n-$C_{10}H_{22}$ | Crude (320MW) |
|---|---|---|---|---|---|---|---|---|
| I | 45.51 | 1.82 | 1.44 | | | 31.87 | | 19.36 |
| II | 64.16 | 3.65 | 2.48 | 0.62 | 11.10 | | 9.57 | 8.42 |

Experiment I

| T(K) | P(MPa) |
|---|---|
| 285.9 | 8.412 |
| 286.3 | 8.826 |
| 288.1 | 9.998 |
| 288.4 | 10.204 |
| 288.7 | 10.342 |
| 288.8 | 10.928 |

Experiment II

| T(K) | P(MPa) |
|---|---|
| 289.0 | 8.688 |
| 289.2 | 8.908 |
| 289.4 | 9.412 |
| 289.9 | 10.025 |
| 290.6 | 10.770 |
| 290.8 | 10.722 |

| | | | |
|---|---|---|---|
| 289.2 | 11.308 | 291.5 | 10.963 |
| 289.9 | 11.997 | | |
| 290.3 | 12.893 | | |

<u>Hydrates</u>: Natural Gas Liquids
<u>Reference</u>: Ng and Robinson, 1976a
<u>Phases</u>: $L_W$-H-$L_{HC}$ and $L_W$-H-V-$L_{HC}$

Composition of Liquids
Concentration, mol %

| Gas → Component | I | II | III | IV | V | VI |
|---|---|---|---|---|---|---|
| Nitrogen | 0.3 | 0.2 | 0.2 | | | |
| Methane | | 2.2 | 21.9 | | | |
| Ethane | 31.3 | 30.6 | 24.7 | 23.4 | 21.5 | 17.0 |
| Propane | 51.5 | 50.8 | 40.8 | 30.4 | 48.9 | 38.6 |
| Isobutane | 16.9 | 16.2 | 12.4 | 19.6 | 23.8 | 18.9 |
| n-Pentane | | | | 26.6 | | |
| Carbon Dioxide | | | | | 5.8 | 25.5 |

| T(K) | P(kPa) | T(K) | P(kPa) | T(K) | P(kPa) |
|---|---|---|---|---|---|
| Gas I | | II | | III | |
| 277.7 | 1158* | 281.2 | 1565* | 291.7 | 4799* |
| 277.7 | 1186 | 281.2 | 1620 | 291.8 | 5240 |
| 277.7 | 1462 | 281.3 | 2461 | 292.1 | 5902 |
| 277.7 | 2234 | 281.4 | 3813 | 292.5 | 7212 |
| 277.7 | 4523 | 281.7 | 6102 | 293.2 | 9797 |
| 277.8 | 7074 | 281.9 | 8232 | 293.9 | 13623 |
| 277.8 | 9714 | 282.2 | 11224 | | |
| Gas IV | | V | | VI | |
| 274.8 | 689* | 280.1 | 1207* | 283.9 | 2344* |
| 274.8 | 1730 | 280.1 | 1351 | 284.0 | 2496 |
| 275.1 | 4054 | 280.1 | 1427 | 284.2 | 3316 |

| | | | | | |
|---|---|---|---|---|---|
| 275.3 | 6964 | 280.1 | 2399 | 284.6 | 5130 |
| 275.6 | 11893 | 280.2 | 3689 | 284.9 | 8163 |
| | | 280.4 | 6129 | 285.7 | 14700 |
| | | 280.7 | 9377 | | |
| | | 280.9 | 14403 | | |

* = $L_W$-H-V-$L_{HC}$

Hydrates: Natural Gases
References: Aoyagi, K., Kobayashi, R. (1978)
Phases: V-H

Compositions

| Gas | %$CH_4$ | %$C_2H_6$ | %$C_3H_8$ | %$CO_2$ |
|---|---|---|---|---|
| I | 75.02 | 7.95 | 3.99 | 13.04 |
| II | 87.06 | 7.96 | 3.88 | 1.10 |

Gas I

| T K | P MPa | $H_2O$ ppm(mol) | T K | P MPa | $H_2O$ ppm(mol) |
|---|---|---|---|---|---|
| 267.1 | 4.499 | 98.9 | 249.0 | 4.499 | 20.6 |
| 267.1 | 5.857 | 87.1 | 249.8 | 12.078 | 18.4 |
| 267.1 | 12.068 | 63.0 | 243.2 | 4.479 | 10.5 |
| 260.9 | 4.458 | 58.8 | 243.7 | 5.847 | 10.3 |
| 261.2 | 5.836 | 56.7 | 243.2 | 12.048 | 10.5 |
| 260.9 | 12.048 | 41.6 | 237.2 | 12.088 | 4.52 |
| 251.8 | 5.857 | 25.2 | 233.9 | 12.068 | 2.50 |

Gas II

| T K | P MPa | $H_2O$ ppm(mol) | T K | P MPa | $H_2O$ ppm(mol) |
|---|---|---|---|---|---|
| 277.6 | 10.345 | 118.0 | 260.9 | 3.445 | 63.0 |
| 277.6 | 3.445 | 252.0 | 249.8 | 10.345 | 10.0 |
| 260.9 | 10.345 | 28.4 | 249.8 | 3.445 | 19.7 |

Hydrates: Methane+Ethane+Propane+Pentane
Reference: Sloan, E.D., et al. (1986)
Phases: $L_{HC}$-H

Isobaric Data at 3.447 MPa

| Liquid Phase Composition | | | | T | Water Conc. |
|---|---|---|---|---|---|
| $\%CH_4$ | $\%C_2H_6$ | $\%C_3H_8$ | $\%n\text{-}C_5H_{12}$ | K | ppm (mol basis) |
| 10.1 | 4.4 | 26.1 | 59.4 | 264.1 | 56 |
| 10.1 | 4.4 | 26.1 | 59.4 | 270.7 | 124 |

Hydrate: Gas Liquids and Condensate
Reference: Ng, H.-J., et al. (1987a)

Composition of Hydrocarbon Liquids
Concentration, mol %

| Component | A | B | C | D |
|---|---|---|---|---|
| Nitrogen | 0.00 | 0.04 | 0.16 | 0.64 |
| Methane | 2.49 | 12.48 | 26.19 | 73.03 |
| Carbon Dioxide | 0.48 | 12.01 | 2.10 | 3.11 |
| Ethane | 4.22 | 8.88 | 8.27 | 8.04 |
| Propane | 8.63 | 10.57 | 7.50 | 4.28 |
| Isobutane | 2.85 | 2.14 | 1.83 | 0.73 |
| n-Butane | 7.02 | 5.63 | 4.05 | 1.50 |
| Isopentane | 3.39 | 1.74 | 1.85 | 0.54 |
| n-Pentane | 4.59 | 2.85 | 2.45 | 0.60 |
| Hexanes plus | 66.33 | 53.66 | 45.60 | 7.53 |
| M.Wt. | 123.0 | 113.0 | 90.2 | 32.4 |
| Saturation Pressure (MPa) at Ts | 0.71* | 4.34* | 8.95* | 43.94** |
| Ts(K) | 281.15 | 310.95 | 310.95 | 416.15 |

* = Bubble Point Temperature
** = Retrograde Dew Point Pressure

| T K | P MPa | Phases | T K | P MPa | Phases |
|---|---|---|---|---|---|

Liquid A

| | | | | | |
|---|---|---|---|---|---|
| 273.85 | 0.64 | $L_W$-H-V-$L_{HC}$(V=0) | 274.75 | 15.95 | $L_W$-H-$L_{HC}$ |
| 273.95 | 5.27 | $L_W$-H-$L_{HC}$ | 275.75 | 20.78 | $L_W$-H-$L_{HC}$ |
| 274.45 | 10.34 | $L_W$-H-$L_{HC}$ | | | |

Liquid B

| | | | | | |
|---|---|---|---|---|---|
| 279.65 | 1.52 | $L_W$-H-V-$L_{HC}$ | 286.75 | 12.00 | $L_W$-H-$L_{HC}$ |
| 286.15 | 3.54 | $L_W$-H-V-$L_{HC}$(V=0) | 288.35 | 20.00 | $L_W$-H-$L_{HC}$ |

Liquid C

| | | | | | |
|---|---|---|---|---|---|
| 287.05 | 4.00 | $L_W$-H-V-$L_{HC}$ | 291.75 | 12.00 | $L_W$-H-$L_{HC}$ |
| 291.35 | 7.74 | $L_W$-H-V-LHC(V=0) | 293.05 | 20.00 | $L_W$-H-$L_{HC}$ |

Liquid D

| | | | | | |
|---|---|---|---|---|---|
| 290.75 | 6.01 | $L_W$-H-V-$L_{HC}$ | 294.95 | 15.01 | $L_W$-H-V-$L_{HC}$ |
| 293.75 | 11.07 | $L_W$-H-V-$L_{HC}$ | 296.25 | 9.99 | $L_W$-H-V-$L_{HC}$ |

Hydrates: Alaskan West Sak Crude
Reference: Paranjpe, S.G., et al (1988)
Phases: $L_W$-H-V-$L_{HC}$

Live West Sak Crude Compositon

| Comp | mol % | Comp | mol % | Comp | mol % |
|---|---|---|---|---|---|
| $CO_2$ | 0.016 | $C_7$ | 0.016 | $C_{15}$ | 1.944 |
| $N_2$ | 0.032 | $C_8$ | 0.008 | $C_{16}$ | 1.795 |
| $C_1$ | 38.333 | $C_9$ | 0.823 | $C_{17}$ | 1.570 |
| $C_2$ | 0.857 | $C_{10}$ | 1.496 | $C_{18}$ | 1.795 |
| $C_3$ | 0.359 | $C_{11}$ | 1.720 | $C_{19}$ | 2.468 |
| $C_4$ | 0.179 | $C_{12}$ | 1.346 | $C_{20}$ | 2.841 |
| $C_5$ | 0.064 | $C_{13}$ | 1.496 | $C_{21}^+$ | 39.037 |
| $C_6$ | 0.200 | $C_{14}$ | 1.795 | Avg MolWt $C_{21}^+$ | = 455 |

West Sak Crude PVT Properties

| Press MPa | GOR SCF/STB | Density g/cc | Viscosity cp | Press MPa | GOR SCF/STB | Density g/cc | Viscosity cp |
|---|---|---|---|---|---|---|---|
| 11.65 | 210 | 0.902 | 35.4 | 4.93 | 93 | 0.917 | 68.4 |
| 10.34 | 188 | 0.905 | 40.0 | 3.45 | 68 | 0.920 | 82.5 |
| 8.96 | 165 | 0.908 | 45.2 | 2.07 | 42 | 0.923 | 102.1 |
| 7.58 | 141 | 0.911 | 51.2 | 0.69 | 15 | 0.927 | 127.6 |
| 6.21 | 117 | 0.914 | 58.5 | | | | |

Analysis of West Sak Separator Gas at 2.165MPa and 291.5K

| Comp | mol % | Comp | mol % | Comp | mol % |
|---|---|---|---|---|---|
| $H_2S$ | 0.00 | $C_2$ | 0.24 | $i\text{-}C_5$ | 0.03 |
| $CO_2$ | 0.02 | $C_3$ | 0.31 | $n\text{-}C_5$ | 0.03 |
| $N_2$ | 0.16 | $i\text{-}C_4$ | 0.07 | $C_6$ | 0.02 |
| $C_1$ | 98.33 | $n\text{-}C_4$ | 0.12 | $C_7^+$ | 0.02 |

Constant Composition Expansion for West Sak Crude at 299.8 K

| Press MPa | RelLiqVol | Press MPa | RelLiqVol | Press MPa | RelLiqVol |
|---|---|---|---|---|---|
| 11.75 | 1.0000 | 8.68 | 0.9339 | 4.87 | 0.7615 |
| 9.98 | 0.9742 | 7.04 | 0.8777 | 4.10 | 0.6849 |
| 9.46 | 0.9544 | 6.26 | 0.8441 | 3.17 | 0.6054 |

Rel Liq Vol ≡ (Vol of Liq)/(Vol of Liq at Bub Pt Press)

Saturation Pressures of West Sak Crude

| T(K) | P(MPa) | T(K) | P(MPa) | T(K) | P(MPa) | T(K) | P(MPa) |
|---|---|---|---|---|---|---|---|
| 299.8 | 11.75 | 310.9 | 14.12 | 366.5 | 16.24 | 394.3 | 17.20 |

$L_W$-H-V-$L_{HC}$ Data for West Sak Crude

| T(K) | P(MPa) | T(K) | P(MPa) | T(K) | P(MPa) |
|---|---|---|---|---|---|
| 279.75 | 5.363 | 281.55 | 6.385 | 282.45 | 7.033 |
| 280.85 | 6.026 | 282.05 | 6.791 | 284.05 | 8.301 |

Hydrate: Crude Oil
Reference: Avlonitis, D.A. (1988)
Phases: $L_W$-H-V-$L_{HC}$

Characterization of Crude Oil

| Comp | mol % | Comp | mol % | Comp | mol % |
|---|---|---|---|---|---|
| $N_2$ | 0.61 | $C_3H_8$ | 8.06 | n-$C_5H_{12}$ | 2.57 |
| $CO_2$ | 2.01 | i-$C_4H_{10}$ | 1.34 | n-$C_6H_{14}$ | 3.15 |
| $CH_4$ | 35.56 | n-$C_4H_{10}$ | 4.26 | $C_7^+$ | 31.26 |
| $C_2H_6$ | 9.90 | i-$C_5H_{12}$ | 1.28 | | |

Note: the i= $C_7^+$ fraction is characterized by $T_c$ = 770K, $P_c$ = 103.42 kPa, $\omega$ = 0.690 and Equation-of-State $\delta_{ij}$ interaction constants (see Appendix A) for the Soave Equation of $\delta_{C_1}$= 0.055; $\delta_{C_2}$=0.030; $\delta_{N_2}$=0.150; $\delta_{CO_2}$=0.120

$L_W$-H-V-$L_{HC}$ Data

| T(K) | P(MPa) | T(K) | P(MPa) | T(K) | P(MPa) | T(K) | P(MPa) |
|---|---|---|---|---|---|---|---|
| 280.15 | 1.262 | 283.65 | 2.179 | 288.25 | 4.592 | 290.55 | 5.730 |
| 280.25 | 1.427 | 284.65 | 2.627 | 289.65 | 5.144 | 292.35 | 8.032 |
| 283.05 | 2.075 | 285.75 | 3.179 | | | | |

### 6.2.1.e. Equilibria with Inhibitors

The phase equilibria data for hydrates with inhibitors are presented below. As in the previous results, data plots are provided for those systems which have been considered by more than one investigation, as a first order means of data evaluation. Individual investigators usually include plots of their data in the original reference; where this is not the case, a semilogarithmic plot has been provided. Unless otherwise indicated the concentration of the inhibitor in the aqueous phase is included in the column marked "wt %."

Simple Methane Hydrates with Inhibitors

Hydrate: Methane with Methanol
Reference: Ng, H.-J., Robinson, D.B. (1985)
Phases: $L_W$-H-V

| wt % | T(K) | P(MPa) | wt % | T(K) | P(MPa) |
|---|---|---|---|---|---|
| 10.0 | 266.23 | 2.14 | 20.0 | 263.34 | 2.83 |
| 10.0 | 271.24 | 3.41 | 20.0 | 267.51 | 4.20 |
| 10.0 | 275.87 | 5.63 | 20.0 | 270.08 | 5.61 |
| 10.0 | 280.31 | 9.07 | 20.0 | 273.55 | 8.41 |
| 10.0 | 283.67 | 13.32 | 20.0 | 277.56 | 13.30 |
| 10.0 | 286.40 | 18.82 | 20.0 | 280.17 | 18.75 |

Hydrate: Methane with Methanol
Reference: Robinson, D.B., Ng, H.-J.(1986)
Phases: $L_W$-H-V

| wt % | T(K) | P(MPa) | wt % | T(K) | P(MPa) |
|---|---|---|---|---|---|
| 35 | 250.9 | 2.38 | 35 | 270.1 | 20.51 |
| 35 | 256.3 | 3.69 | 50 | 233.1 | 1.47 |
| 35 | 260.3 | 6.81 | 50 | 240.1 | 2.95 |
| 35 | 264.6 | 10.16 | 50 | 247.4 | 7.24 |
| 35 | 267.8 | 13.68 | 50 | 250.4 | 10.54 |
| 35 | 268.5 | 17.22 | 50 | 255.3 | 16.98 |

Hydrate: Methane with Methanol
Reference: Ng, H.-J., et al. (1987b)
Phases: $L_W$-H-V

| wt % | T(K) | P(MPa) | wt % | T(K) | P(MPa) |
|---|---|---|---|---|---|
| 50.0 | 232.8 | 1.17 | 65.0 | 238.4 | 9.03 |
| 50.0 | 244.0 | 3.54 | 65.0 | 244.3 | 20.49 |
| 50.0 | 251.4 | 6.98 | 73.7 | 223.2 | 6.73 |
| 50.0 | 255.5 | 12.26 | 73.7 | 227.4 | 12.37 |
| 50.0 | 259.5 | 19.93 | 73.7 | 229.9 | 20.30 |
| 65.0 | 214.1 | 0.76 | 85.0 | 194.6 | 11.49 |
| 65.0 | 224.6 | 2.10 | 85.0 | 195.4 | 11.51 |
| 65.0 | 227.6 | 2.89 | 85.0 | 197.6 | 20.42 |
| 65.0 | 231.6 | 4.14 | | | |

Hydrate: Methane with Ethylene Glycol
Reference: Robinson, D.B., Ng, H.-J.(1986)
Phases: $L_W$-H-V

| wt % | T(K) | P(MPa) | wt % | T(K) | P(MPa) |
|---|---|---|---|---|---|
| 10 | 270.24 | 2.42 | 30 | 274.36 | 7.86 |
| 10 | 273.49 | 3.40 | 30 | 280.14 | 16.14 |
| 10 | 280.19 | 6.53 | 30 | 279.89 | 16.38 |
| 10 | 287.10 | 15.61 | 50 | 263.43 | 9.89 |
| 30 | 267.59 | 3.77 | 50 | 266.32 | 14.08 |
| 30 | 269.73 | 4.93 | 50 | 266.48 | 15.24 |

Hydrate: Methane with 15 wt % Ethanol Solution
Reference: Kobayashi, R., et al. (1951)
Phases: $L_W$-H-V

| wt % | T(K) | P(MPa) | wt % | T(K) | P(MPa) |
|---|---|---|---|---|---|
| 15 | 273.3 | 3.38 | 15 | 281.1 | 8.36 |
| 15 | 277.2 | 5.47 | 15 | 284.7 | 13.67 |
| 15 | 279.4 | 7.06 | | | |

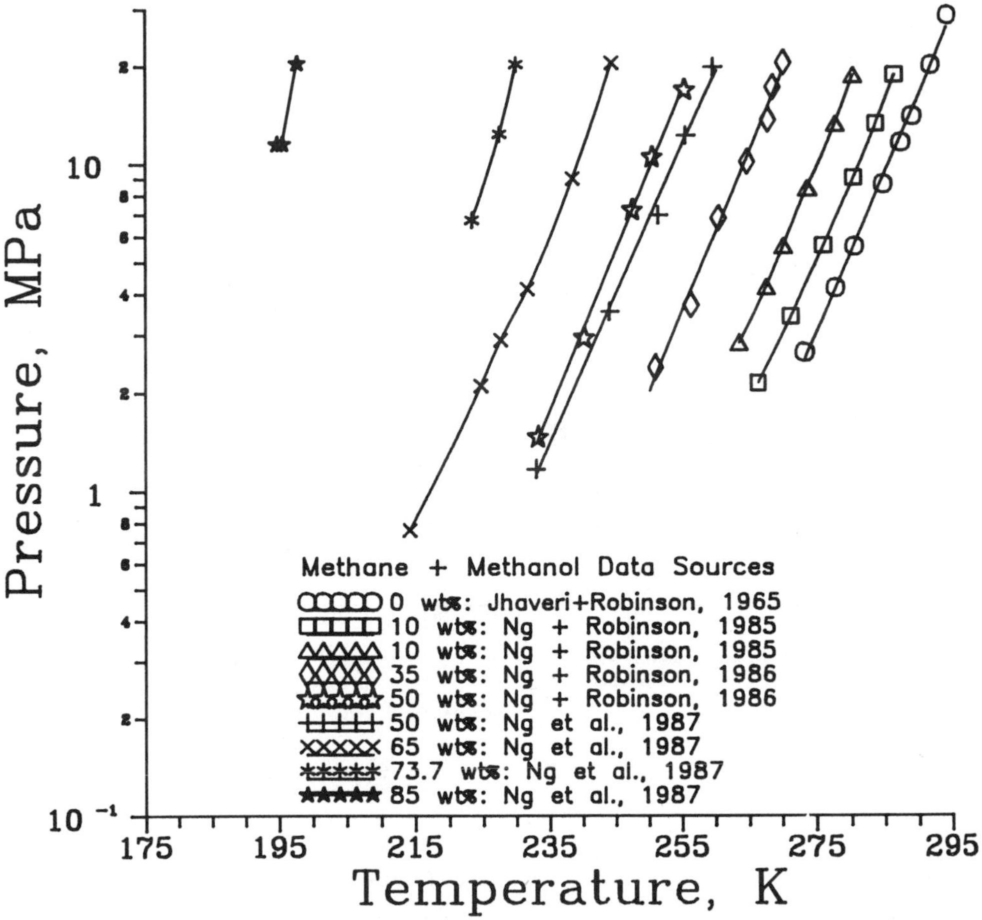


6-18 Methanol Inhibition of Simple Methane Hydrates

Hydrate: Methane with 10 wt % Sodium Chloride Solution
Reference: Kobayashi, R., et al. (1951)
Phases: $L_W$-H-V

| wt % | T(K) | P(MPa) | wt % | T(K) | P(MPa) |
|---|---|---|---|---|---|
| 10 | 270.0 | 2.59 | 10 | 276.9 | 4.89 |
| 10 | 271.0 | 2.80 | 10 | 278.6 | 6.38 |
| 10 | 272.7 | 3.58 | 10 | 282.1 | 10.03 |
| 10 | 274.5 | 3.65 | 10 | 284.3 | 13.42 |

Hydrate: Methane with 20 wt % Sodium Chloride Solution
Reference: Kobayashi, R., et al. (1951)
Phases: $L_W$-H-V

| wt % | T(K) | P(MPa) | wt % | T(K) | P(MPa) |
|---|---|---|---|---|---|
| 20 | 265.9 | 3.78 | 20 | 275.7 | 11.09 |
| 20 | 267.8 | 4.63 | 20 | 276.4 | 10.82 |
| 20 | 272.3 | 6.18 | 20 | 276.3 | 13.66 |
| 20 | 272.3 | 7.19 | | | |

Hydrate: Methane with Sodium Chloride Solutions
Reference: de Roo, J.L. et al. (1983)
Phases: $L_W$-H-V

| mol fract NaCl | T(K) | P(MPa) | mol fract NaCl | T(K) | P(MPa) |
|---|---|---|---|---|---|
| 0.03936 | 268.30 | 2.69 | 0.07785 | 264.45 | 4.06 |
| 0.03936 | 271.05 | 3.53 | 0.07785 | 266.45 | 5.05 |
| 0.03936 | 273.25 | 4.50 | 0.07785 | 268.15 | 6.12 |
| 0.03936 | 274.75 | 5.29 | 0.07785 | 269.45 | 7.05 |
| 0.03936 | 275.90 | 5.98 | 0.07785 | 271.15 | 8.83 |
| 0.03936 | 278.05 | 7.55 | 0.07785 | 272.85 | 11.00 |
| 0.05976 | 263.35 | 2.39 | 0.08909 | 263.05 | 4.78 |
| 0.05976 | 265.75 | 3.13 | 0.08909 | 264.55 | 5.78 |
| 0.05976 | 268.85 | 4.26 | 0.08909 | 266.25 | 7.11 |
| 0.05976 | 271.95 | 6.12 | 0.08909 | 267.55 | 8.36 |
| 0.05976 | 274.95 | 8.57 | 0.08909 | 268.65 | 9.55 |
| 0.07785 | 261.85 | 2.94 | | | |

## Simple Ethane Hydrates with Inhibitors

Hydrate: Ethane with Methanol
Reference: Ng, H.-J., Robinson, D.B. (1985)
Phases: $L_W$-H-V, $L_W$-H-V-$L_{C_2H_6}$, $L_W$-H-$L_{C_2H_6}$

$L_W$-H-V

| wt % | T(K) | P(MPa) | wt % | T(K) | P(MPa) |
|---|---|---|---|---|---|
| 10.0 | 268.28 | 0.417 | 20.0 | 263.53 | 0.550 |
| 10.0 | 272.10 | 0.731 | 20.0 | 264.86 | 0.614 |

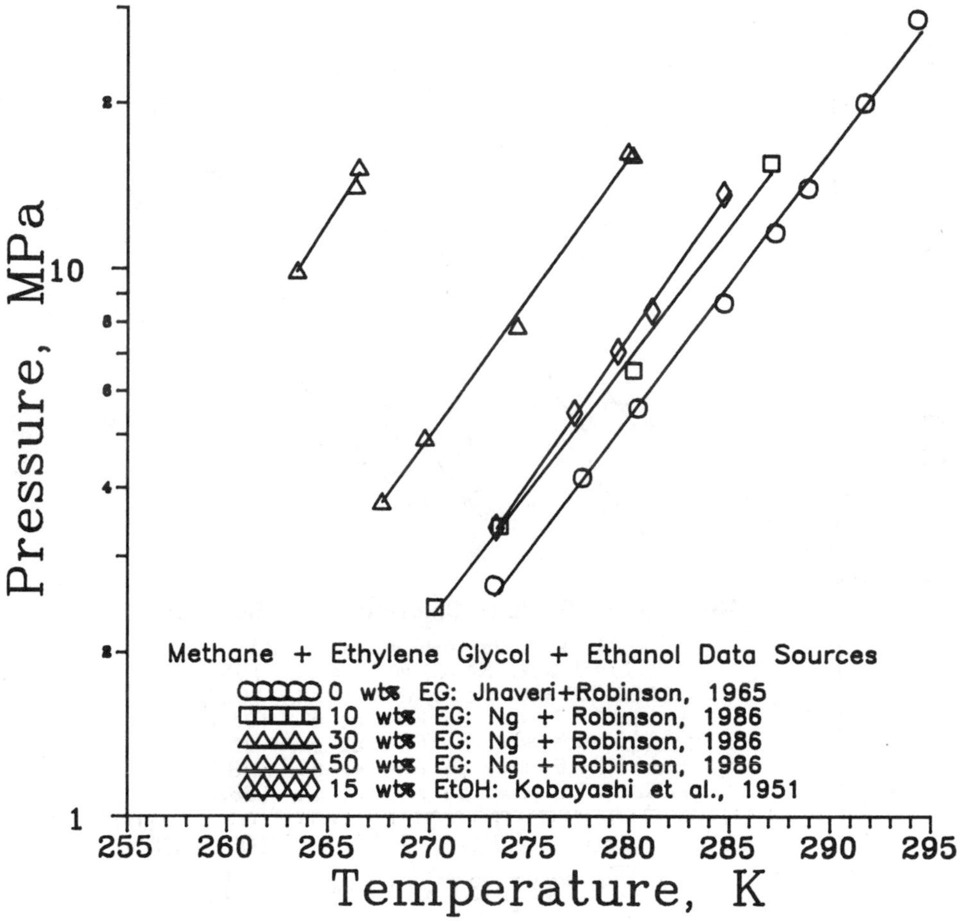


6-19 Ethanol and Ethylene Glycol Inhibition of Simple Methane Hydrates

| | | | | | |
|---|---|---|---|---|---|
| 10.0 | 276.01 | 1.160 | 20.0 | 267.72 | 0.869 |
| 10.0 | 278.43 | 1.630 | 20.0 | 268.81 | 1.030 |
| 10.0 | 280.40 | 2.160 | 20.0 | 271.75 | 1.520 |
| 10.0 | 281.41 | 2.800 | 20.0 | 274.07 | 2.060 |
| 10.0 | 281.89 | 2.820 | | | |

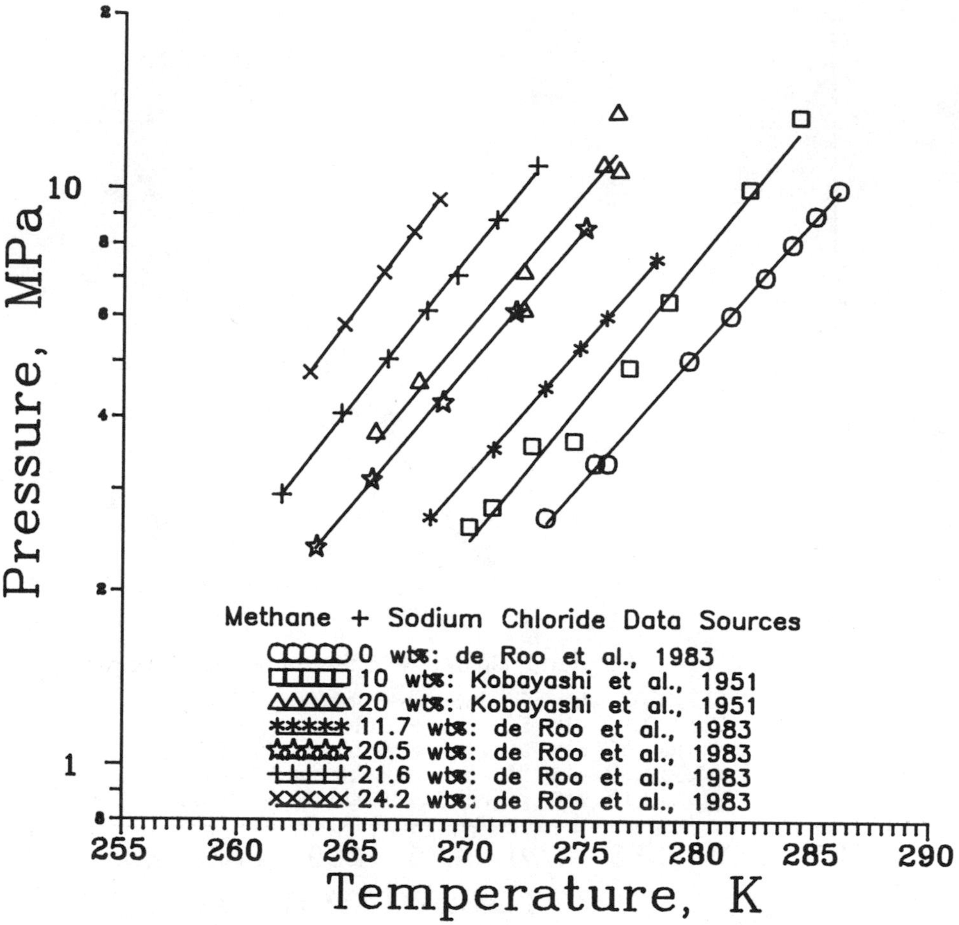


6-20 Sodium Chloride Inhibition cf Simple Methane Hydrates

| | | | | | |
|---|---|---|---|---|---|
| | | $\underline{L_W\text{-H-V-L}}_{C_2H_6}$ | | | |
| 10.0 | 282.20 | 2.910 | 20.0 | 275.69 | 2.650 |
| | | $\underline{L_W\text{-H-L}}_{C_2H_6}$ | | | |
| 10.0 | 282.05 | 3.990 | 10.0 | 283.65 | 13.760 |
| 10.0 | 282.36 | 4.220 | 10.0 | 284.45 | 20.200 |
| 10.0 | 282.00 | 5.650 | 20.0 | 276.30 | 5.890 |
| 10.0 | 281.97 | 6.590 | 20.0 | 277.02 | 10.030 |
| 10.0 | 282.74 | 7.300 | 20.0 | 277.84 | 15.120 |
| 10.0 | 282.92 | 10.360 | 20.0 | 278.61 | 20.400 |

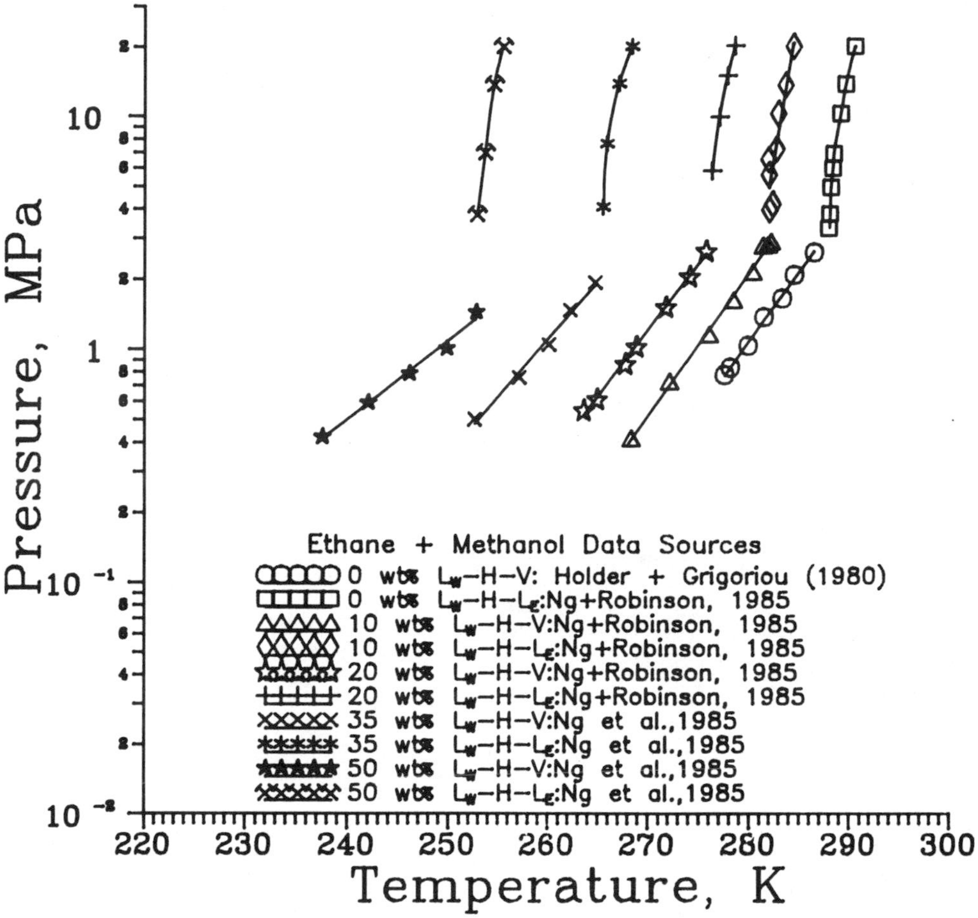


6-21 Methanol Inhibition of Simple Ethane Hydrates

<u>Hydrate</u>: Ethane with Methanol
<u>Reference</u>: Ng, H.-J. et al., (1985)
<u>Phases</u>: $L_W$-H-V, $L_W$-H-V-$L_{C_2H_6}$, $L_W$-H-$L_{C_2H_6}$

| wt % | T(K) | P(MPa) | wt % | T(K) | P(MPa) |
|---|---|---|---|---|---|
| | | $L_W$-H-V | | | |
| 35 | 252.6 | 0.502 | 50 | 237.5 | 0.423 |

| | | | | | |
|---|---|---|---|---|---|
| 35 | 257.1 | 0.758 | 50 | 242.0 | 0.592 |
| 35 | 260.1 | 1.048 | 50 | 246.1 | 0.786 |
| 35 | 262.2 | 1.475 | 50 | 249.8 | 1.007 |

$\underline{L_W\text{-H-V-L}_{C_2H_6}}$

| | | | | | |
|---|---|---|---|---|---|
| 35 | 264.7 | 1.937 | 50 | 252.8 | 1.441 |

$\underline{L_W\text{-H-L}_{C_2H_6}}$

| | | | | | |
|---|---|---|---|---|---|
| 35 | 265.5 | 4.095 | 50 | 252.9 | 3.820 |
| 35 | 265.9 | 7.695 | 50 | 253.7 | 7.074 |
| 35 | 267.1 | 13.930 | 50 | 254.6 | 13.890 |
| 35 | 268.4 | 20.180 | 50 | 255.5 | 20.350 |

<u>Simple Propane Hydrates with Inhibitors</u>

<u>Hydrate</u>: Propane with Methanol
<u>Reference</u>: Ng, H.-J., Robinson, D.B. (1985)
<u>Phases</u>: $L_W$-H-V, $L_W$-H-$L_{C_3H_8}$

| wt % | T(K) | P(MPa) | wt % | T(K) | P(MPa) |
|---|---|---|---|---|---|
| | | $\underline{L_W\text{-H-V}}$ | | | |
| 5.00 | 272.12 | 0.234 | 10.39 | 269.23 | 0.228 |
| 5.00 | 272.58 | 0.259 | 10.39 | 270.93 | 0.360 |
| 5.00 | 273.28 | 0.316 | 10.39 | 271.03 | 0.352 |
| 5.00 | 274.18 | 0.405 | 10.39 | 271.59 | 0.415 |
| 5.00 | 274.79 | 0.468 | 10.39 | 271.82 | 0.434 |
| 10.39 | 268.30 | 0.185 | | | |
| | | $\underline{L_W\text{-H-L}_{C_3H_8}}$ | | | |
| 5.00 | 275.02 | 0.794 | 10.39 | 272.10 | 0.984 |
| 5.00 | 275.09 | 1.720 | 10.39 | 272.07 | 2.737 |
| 5.00 | 274.97 | 6.340 | 10.39 | 272.08 | 6.510 |

Hydrate: Propane with Methanol
Reference: Ng, H.-J., Robinson, D.B. (1984)
Phases: $L_W$-H-V, $L_W$-H-$L_{C_3H_8}$

| wt % | T(K) | P(MPa) | wt % | T(K) | P(MPa) |
|---|---|---|---|---|---|
| | | | $L_W$-H-V | | |
| 35 | 248.0 | 0.137 | 50 | 229.7 | 0.090 |
| 35 | 250.2 | 0.207 | | | |
| | | | $L_W$-H-$L_{C_3H_8}$ | | |
| 35 | 250.6 | 0.876 | 50 | 229.9 | 1.970 |
| 35 | 251.1 | 6.090 | 50 | 229.3 | 7.830 |
| 35 | 251.0 | 9.770 | 50 | 229.3 | 19.710 |
| 35 | 251.3 | 20.38 | | | |

Hydrate: Propane with Sodium Chloride
Reference: Kobayashi, R., et al. (1951)
Phases: $L_W$-H-V

| wt % | T(K) | P(MPa) | wt % | T(K) | P(MPa) |
|---|---|---|---|---|---|
| 10 | 268.3 | 0.122 | 10 | 272.4 | 0.370 |
| 10 | 269.7 | 0.170 | 10 | 272.8 | 0.479 |
| 10 | 271.8 | 0.278 | 10 | 273.0 | 1.118 |
| 10 | 272.0 | 0.309 | 10 | 273.1 | 1.911 |

Hydrate: Propane with Sodium Chloride
Reference: Patil, S.L. (1987)
Phases: $L_W$-H-V

| wt % | T(K) | P(kPa) | wt % | T(K) | P(kPa) |
|---|---|---|---|---|---|
| 3.0 | 272.15 | 179 | 5.0 | 274.45 | 324 |
| 3.0 | 274.25 | 290 | 5.0 | 275.55 | 448 |
| 3.0 | 275.45 | 366 | 10.0 | 270.75 | 191 |
| 3.0 | 276.25 | 455 | 10.0 | 272.15 | 259 |
| 5.0 | 271.15 | 185 | 10.0 | 273.65 | 355 |
| 5.0 | 272.65 | 241 | 10.0 | 274.65 | 450 |

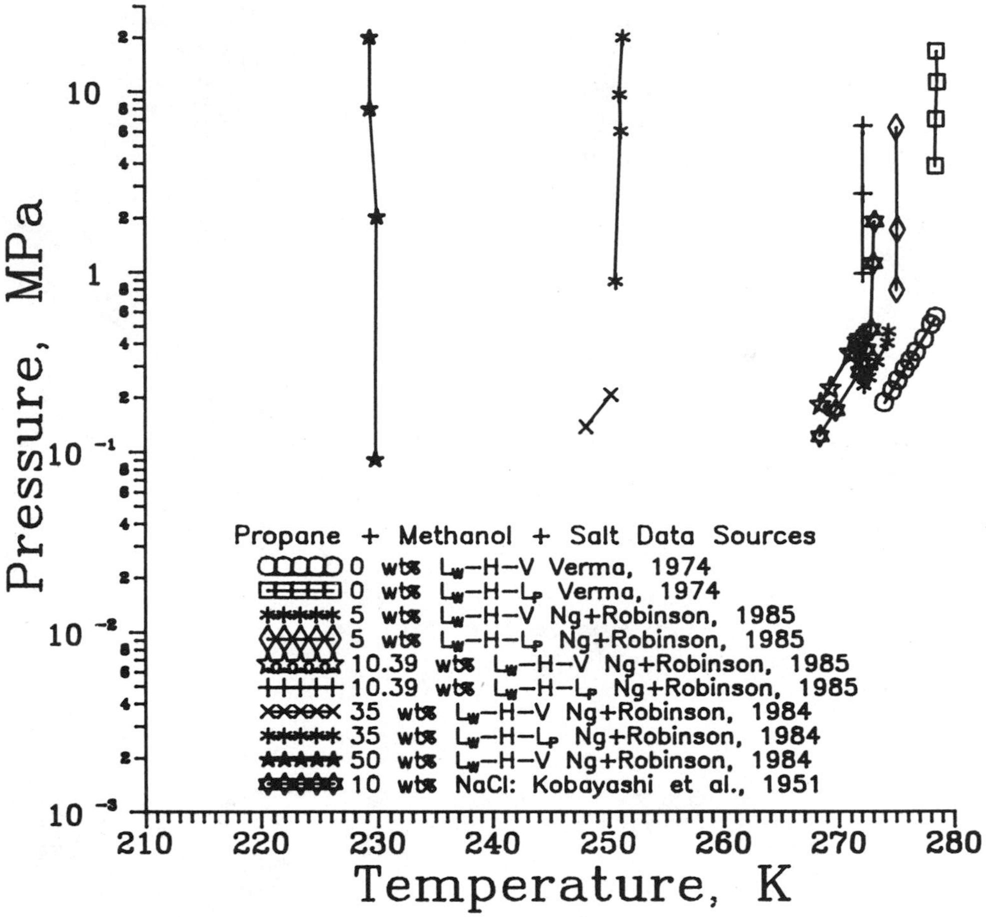


6-22 Methanol and Salt Inhibition of Simple Propane Hydrates

<u>Simple Isobutane Hydrates with Inhibitors</u>

<u>Hydrate</u>: Isobutane with Sodium Chloride
<u>Reference</u>: Schneider, G.R., and Farrar, J. (1968)
<u>Phases</u>: $L_W$-H-V

| NaCl wt % | T K | P kPa | NaCl wt % | T K | P kPa |
|---|---|---|---|---|---|
| 1.10 | 273.2 | 127.2 | 1.08 | 274.2 | 159.5 |
| 1.10 | 273.4 | 134.8 | 9.93 | 268.2 | 116.6 |

| | | | | | |
|---|---|---|---|---|---|
| 1.10 | 273.7 | 140.6 | 9.93 | 268.0 | 111.0 |
| 1.09 | 273.9 | 149.1 | 9.93 | 267.5 | 100.3 |
| 1.10 | 274.1 | 156.4 | | | |

Hydrate: Isobutane with Sodium Chloride
Reference: Rouher, O.S. (1968)
Phases: $L_W$-H-V

| NaCl wt % | T K | P kPa | NaCl wt % | T K | P kPa |
|---|---|---|---|---|---|
| 5.00 | 270.5 | 110.3 | 3.05* | 271.7 | 119.2 |
| 5.00 | 270.8 | 120.3 | 3.05* | 271.8 | 124.6 |
| 5.00 | 271.1 | 123.5 | 3.05* | 271.9 | 124.9 |
| 5.00 | 271.3 | 135.0 | 3.05* | 271.9 | 127.0 |
| 5.00 | 270.0 | 104.6 | 3.32* | 271.0 | 101.4 |
| 5.00 | 270.4 | 108.2 | 3.32* | 271.2 | 104.4 |
| 5.00 | 270.5 | 112.1 | 3.32* | 272.0 | 126.0 |
| 5.00 | 270.9 | 120.2 | 3.27* | 271.1 | 101.5 |
| 5.00 | 271.1 | 123.4 | 3.21* | 272.0 | 123.2 |
| 5.00 | 271.2 | 132.9 | 3.16* | 271.3 | 105.4 |
| 5.00 | 271.3 | 132.1 | 3.16* | 272.2 | 126.5 |
| 5.00 | 271.4 | 138.5 | 10.0 | 266.7 | 102.0 |
| 5.00 | 271.6 | 142.3 | 10.0 | 266.9 | 105.3 |
| 3.05* | 270.9 | 104.8 | 10.0 | 267.2 | 110.5 |
| 3.05* | 271.0 | 107.0 | 10.0 | 267.3 | 113.3 |
| 3.05* | 271.1 | 109.9 | 10.0 | 267.6 | 118.8 |
| 3.05* | 271.2 | 112.9 | 10.0 | 266.8 | 94.9 |
| 3.05* | 271.3 | 114.3 | 10.0 | 267.2 | 101.8 |
| 3.05* | 271.4 | 117.3 | 10.0 | 267.4 | 105.4 |
| 3.05* | 271.5 | 120.0 | 10.0 | 267.7 | 113.8 |
| 3.05* | 271.6 | 124.3 | 10.0 | 267.7 | 111.4 |
| 3.05* | 271.8 | 128.1 | 10.0 | 267.9 | 114.9 |
| 3.05* | 271.9 | 131.6 | 10.0 | 268.2 | 122.0 |
| 3.05* | 272.1 | 137.0 | 10.55 | 267.0 | 101.2 |

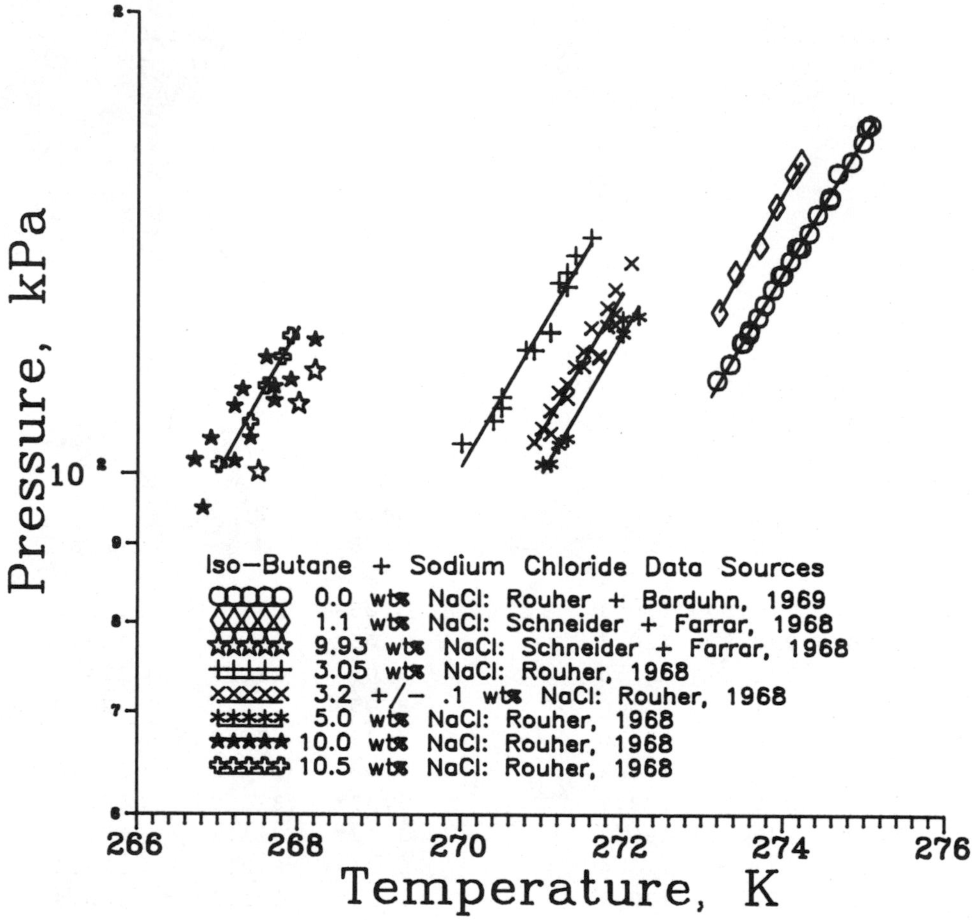


6-23 Sodium Chloride Inhibition of Simple Iso-Butane Hydrates

| | | | | | |
|---|---|---|---|---|---|
| 3.05* | 271.1 | 106.2 | 10.55 | 267.4 | 107.9 |
| 3.05* | 271.3 | 112.0 | 10.5 | 267.6 | 113.8 |
| 3.05* | 271.5 | 117.4 | 10.5 | 267.8 | 118.9 |
| 3.05* | 271.7 | 118.9 | 10.5 | 267.9 | 122.6 |

* = sea water with 3.05 wt% NaCl equivalent activity

## Simple Carbon Dioxide Hydrate with Inhibitors

Hydrate: Carbon Dioxide with Hydrogen Chloride
Reference: Larson, S.D. (1955)
Phases: $L_W$-H-V,

| HCl | T(K) | P(kPa) | HCl | T(K) | P(kPa) |
|---|---|---|---|---|---|
| 0.1N | 274.2 | 1379 | 0.5N | 281.0 | 3716 |
| 0.1N | 278.5 | 2351 | 0.5N | 282.0 | 4371 $Q_2$ |
| 0.1N | 282.15 | 3916 | 1.0N | 273.5 | 1565 |
| 0.1N | 283.1 | 4489 $Q_2$ | 1.0N | 276.6 | 2262 |
| 0.5N | 272.6 | 1310 | 1.0N | 278.5 | 2903 |
| 0.5N | 277.0 | 2172 | 1.0N | 281.2 | 4289 $Q_2$ |

$Q_2$ = inhibited upper quadruple point ($L_W$-H-V-$L_{CO_2}$)

Hydrate: Carbon Dioxide with Sodium Hydroxide
Reference: Larson, S.D. (1955)
Phases: $L_W$-H-V

| NaOH | T(K) | P(kPa) | NaOH | T(K) | P(kPa) |
|---|---|---|---|---|---|
| 0.1N | 273.5 | 1324 | 0.5N | 279.0 | 2868 |
| 0.1N | 276.3 | 1834 | 0.5N | 281.0 | 3827 |
| 0.1N | 279.1 | 2592 | 0.5N | 282.0 | 4378 $Q_2$ |
| 0.1N | 282.2 | 3923 | 1.0N | 273.9 | 1600 |
| 0.1N | 283.0 | 4482 $Q_2$ | 1.0N | 276.6 | 2220 |
| 0.5N | 273.3 | 1393 | 1.0N | 279.6 | 3358 |
| 0.5N | 276.3 | 2000 | 1.0N | 281.3 | 4296 $Q_2$ |

$Q_2$ = inhibited upper quadruple point ($L_W$-H-V-$L_{CO_2}$)

Hydrate: Carbon Dioxide with Sodium Chloride
Reference: Larson, S.D. (1955)
Phases: $L_W$-H-V

| NaCl | T(K) | P(kPa) | NaCl | T(K) | P(kPa) |
|---|---|---|---|---|---|
| 1.0M | 273.6 | 1655 | 1.0M | 277.5 | 2586 |
| 1.0M | 275.2 | 1931 | 1.0M | 279.7 | 3619 |

Hydrate: Carbon Dioxide with Methanol
Reference: Ng, H.-J., Robinson, D.B. (1985)
Phases: $L_W$-H-V, $L_W$-H-$L_{CO_2}$

| wt % | T(K) | P(MPa) | wt % | T(K) | P(MPa) |
|---|---|---|---|---|---|
| | | $L_W$-H-V | | | |
| 10.00 | 269.49 | 1.59 | 20.02 | 264.54 | 1.83 |
| 10.00 | 269.57 | 1.58 | 20.02 | 265.15 | 1.98 |
| 10.00 | 271.33 | 2.06 | 20.02 | 266.36 | 2.21 |
| 10.00 | 273.80 | 2.89 | 20.02 | 267.18 | 2.53 |
| 10.00 | 273.78 | 2.85 | 20.02 | 268.07 | 2.74 |
| 10.00 | 274.92 | 3.48 | 20.02 | 268.86 | 2.94 |
| 20.02 | 263.96 | 1.59 | | | |
| | | $L_W$-H-$L_{CO_2}$ | | | |
| 10.00 | 276.01 | 4.60 | 20.02 | 270.07 | 7.68 |
| 10.00 | 276.75 | 7.23 | 20.02 | 270.50 | 11.27 |
| 10.00 | 277.43 | 10.09 | 20.02 | 270.68 | 11.40 |
| 10.00 | 278.12 | 13.98 | 20.02 | 271.59 | 15.89 |
| 20.02 | 269.15 | 3.34 | 20.02 | 271.84 | 16.09 |
| 20.02 | 269.65 | 5.50 | | | |

Hydrate: Carbon Dioxide with Methanol
Reference: Robinson, D.B., Ng, H.-J.(1986)
Phases: $L_W$-H-V, $L_W$-H-$L_{CO_2}$

| wt % | T(K) | P(MPa) | wt % | T(K) | P(MPa) |
|---|---|---|---|---|---|
| | | $L_W$-V-H | | | |
| 35 | 242.0 | 0.379 | 50 | 232.6 | 0.496 |
| 35 | 247.6 | 0.724 | 50 | 235.5 | 0.676 |
| 35 | 250.1 | 1.030 | 50 | 241.3 | 1.310 |
| 35 | 252.4 | 1.390 | | | |
| 35 | 255.1 | 1.770 | | | |

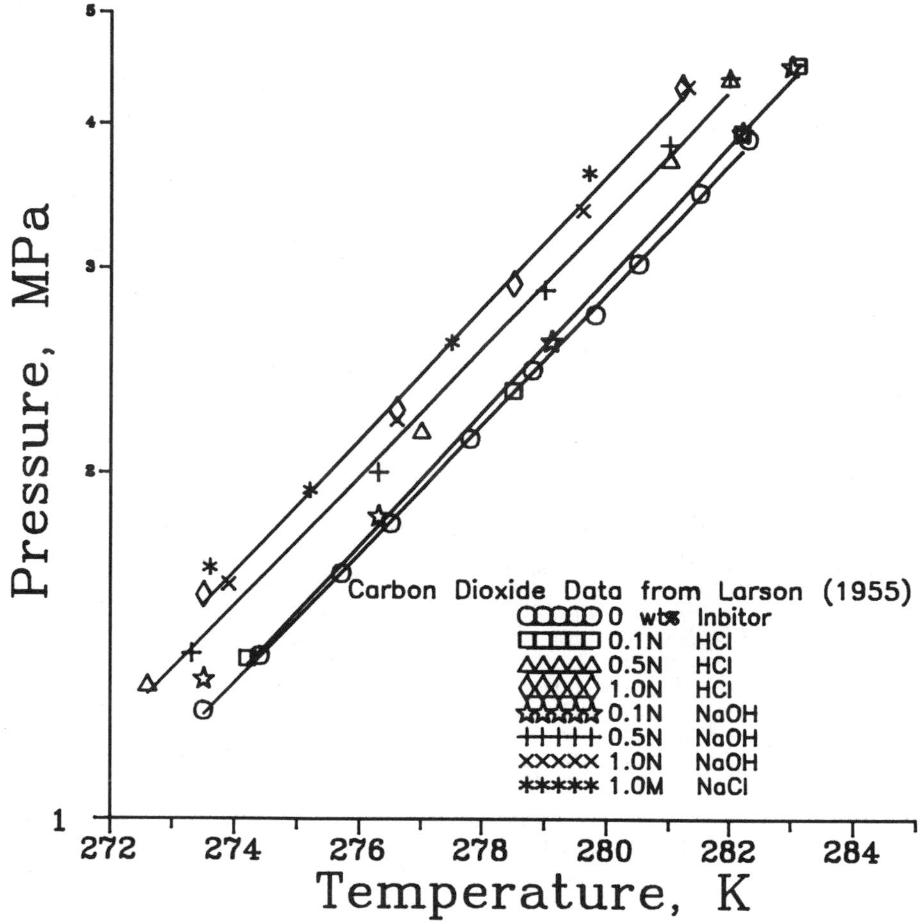


6-24 Acid, Base, and Salt Inhibition of Simple Carbon Dioxide Hydrates

$L_W$-H-$L_{CO_2}$

| | | | | | |
|---|---|---|---|---|---|
| 35 | 256.9 | 2.180 | 50 | 241.1 | 8.83 |
| 35 | 257.5 | 2.870 | 50 | 241.8 | 12.36 |
| 35 | 257.8 | 5.910 | 50 | 241.3 | 14.62 |
| 35 | 257.9 | 6.870 | 50 | 241.1 | 19.53 |
| 35 | 258.0 | 13.340 | | | |
| 35 | 258.5 | 20.700 | | | |

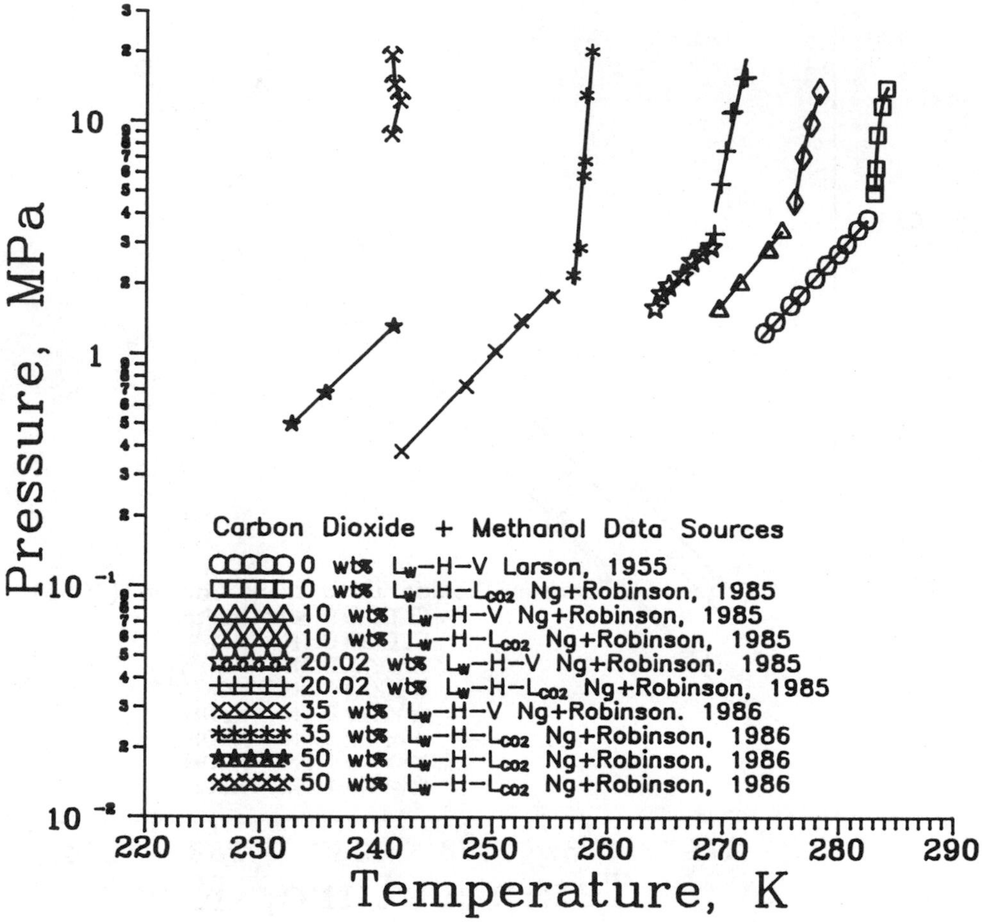


6-25 Methanol Inhibition of Simple Carbon Dioxide Hydrates

## Simple Hydrogen Sulfide Hydrates with Inhibitors

Hydrate: Hydrogen Sulfide with Methanol
Reference: Bond, D.C., Russell, N.B. (1949)
Phases: $L_W$-H-V

| wt % | T(K) | P(kPa) | wt % | T(K) | P(kPa) |
|---|---|---|---|---|---|
| 16.5 | 273.2 | 275.80 | 16.5 | 290.1 | 1496.20 |
| 16.5 | 283.2 | 730.86 | | | |

<u>Hydrate</u>: Hydrogen Sulfide with Methanol
<u>Reference</u>: Ng, H.-J., Robinson, D.B. (1985)
<u>Phases</u>: $L_W$-H-V, $L_W$-H-V-$L_{H_2S}$, $L_W$-H-$L_{H_2S}$

| wt % | T(K) | P(MPa) | wt % | T(K) | P(MPa) |
|---|---|---|---|---|---|
| | | | $L_W$-H-V | | |
| 10.0 | 265.69 | 0.068 | 10.0 | 285.37 | 0.541 |
| 10.0 | 267.54 | 0.084 | 10.0 | 291.77 | 1.080 |
| 10.0 | 272.98 | 0.148 | 20.0 | 271.79 | 0.221 |
| 10.0 | 278.90 | 0.270 | 20.0 | 281.15 | 0.593 |
| | | | $L_W$-H-V-$L_{H_2S}$ | | |
| 10.0 | 297.47 | 1.900 | 20.0 | 291.13 | 1.630 |
| | | | $L_W$-H-$L_{H_2S}$ | | |
| 10.0 | 297.71 | 3.850 | 20.0 | 291.70 | 3.090 |
| 10.0 | 298.80 | 7.230 | 20.0 | 292.57 | 7.290 |
| 10.0 | 299.07 | 10.350 | 20.0 | 293.55 | 10.460 |
| 10.0 | 299.47 | 14.710 | 20.0 | 293.50 | 14.570 |
| 10.0 | 299.52 | 18.160 | 20.0 | 294.27 | 18.260 |

<u>Hydrate</u>: Hydrogen Sulfide with Methanol
<u>Reference</u>: Ng, H.-J., et al. (1985)
<u>Phases</u>: $L_W$-H-V, $L_W$-H-V-$L_{H_2S}$, $L_W$-H-$L_{H_2S}$

| wt % | T(K) | P(MPa) | wt % | T(K) | P(MPa) |
|---|---|---|---|---|---|
| | | | $L_W$-V-H | | |
| 35 | 263.2 | 0.217 | 50 | 255.9 | 0.283 |
| 35 | 268.9 | 0.361 | 50 | 262.1 | 0.426 |
| 35 | 274.2 | 0.579 | 50 | 264.8 | 0.517 |
| 35 | 284.5 | 1.351 | 50 | 267.6 | 0.642 |
| 50 | 251.6 | 0.177 | 50 | 272.1 | 0.920 |
| | | | $L_W$-H-V-$L_{H_2S}$ | | |
| 50 | 277.8 | 1.220 | | | |
| | | | $L_W$-H-$L_{H_2S}$ | | |
| 35 | 284.5 | 1.834 | 50 | 278.0 | 1.951 |

| | | | | | |
|---|---|---|---|---|---|
| 35 | 285.1 | 6.812 | 50 | 278.7 | 5.516 |
| 35 | 285.8 | 13.840 | 50 | 279.4 | 12.360 |
| 35 | 286.3 | 19.650 | 50 | 279.9 | 18.480 |

<u>Hydrate</u>: Hydrogen Sulfide with Sodium Chloride
<u>Reference</u>: Bond, D.C., Russell, N.B. (1949)
<u>Phases</u>: $L_W$-H-V

| wt % | T(K) | P(kPa) | wt % | T(K) | P(kPa) |
|---|---|---|---|---|---|
| 10.0 | 274.8 | 206.85 | 26.4 | 276.17 | 668.81 |
| 10.0 | 287.1 | 648.12 | 26.4 | 278.17 | 1020.45 |
| 10.0 | 294.8 | 1875.42 | 26.4 | 280.17 | 1303.14 |
| 26.4 | 269.2 | 420.59 | 26.4 | 280.17 | 1447.94 |

<u>Hydrate</u>: Hydrogen Sulfide with Calcium Chloride
<u>Reference</u>: Bond, D.C., Russell, N.B. (1949)
<u>Phases</u>: $L_W$-H-V

| wt % | T(K) | P(kPa) | wt % | T(K) | P(kPa) |
|---|---|---|---|---|---|
| 10.0 | 274.8 | 172.37 | 21.1 | 277.2 | 655.02 |
| 10.0 | 288.4 | 723.97 | 21.1 | 281.2 | 917.03 |
| 10.0 | 295.4 | 1896.11 | 21.1 | 284.2 | 1489.31 |
| 21.1 | 271.7 | 365.43 | 21.1 | 284.3 | 1565.15 |
| | | | 36.0 | 265.4 | 882.55 |

<u>Hydrate</u>: Hydrogen Sulfide with Ethanol
<u>Reference</u>: Bond, D.C., Russell, N.B. (1949)
<u>Phases</u>: $L_W$-H-V

| wt % | T(K) | P(kPa) | wt % | T(K) | P(kPa) |
|---|---|---|---|---|---|
| 16.5 | 280.7 | 386.12 | 16.5 | 291.8 | 1482.41 |
| 16.5 | 287.9 | 882.55 | | | |

Hydrate: Hydrogen Sulfide with Dextrose
Reference: Bond, D.C., Russell, N.B. (1949)
Phases: $L_W$-H-V

| wt % | T(K) | P(kPa) | wt % | T(K) | P(kPa) |
|---|---|---|---|---|---|
| 50.0 | 284.8 | 627.44 | 50.0 | 292.6 | 1758.21 |
| 50.0 | 289.3 | 999.77 | | | |

Hydrate: Hydrogen Sulfide with Sucrose
Reference: Bond, D.C., Russell, N.B. (1949)
Phases: $L_W$-H-V

| wt % | T(K) | P(kPa) | wt % | T(K) | P(kPa) |
|---|---|---|---|---|---|
| 50.0 | 292.1 | 827.39 | 50.0 | 295.9 | 1930.58 |
| 50.0 | 293.7 | 1372.09 | 50.0 | 295.9 | 1909.90 |

## Binary Mixtures of Methane+Ethane with Inhibitors

Hydrate: 89.51 Mol % Methane and 10.49 % Ethane with Methanol
Reference: Ng, H.-J., Robinson, D.B. (1983)
Phases: $L_W$-H-V

| wt % | T(K) | P(MPa) | wt % | T(K) | P(MPa) |
|---|---|---|---|---|---|
| 10.02 | 268.66 | 1.40 | 20.01 | 263.93 | 1.49 |
| 10.02 | 270.93 | 1.78 | 20.01 | 266.99 | 2.11 |
| 10.02 | 273.63 | 2.32 | 20.01 | 272.27 | 3.76 |
| 10.02 | 277.14 | 3.28 | 20.01 | 275.19 | 5.49 |
| 10.02 | 280.45 | 4.78 | 20.01 | 278.07 | 8.34 |
| 10.02 | 284.87 | 8.42 | 20.01 | 281.29 | 13.22 |
| 10.02 | 287.51 | 13.21 | 20.01 | 283.61 | 19.02 |
| 10.02 | 289.44 | 18.89 | | | |

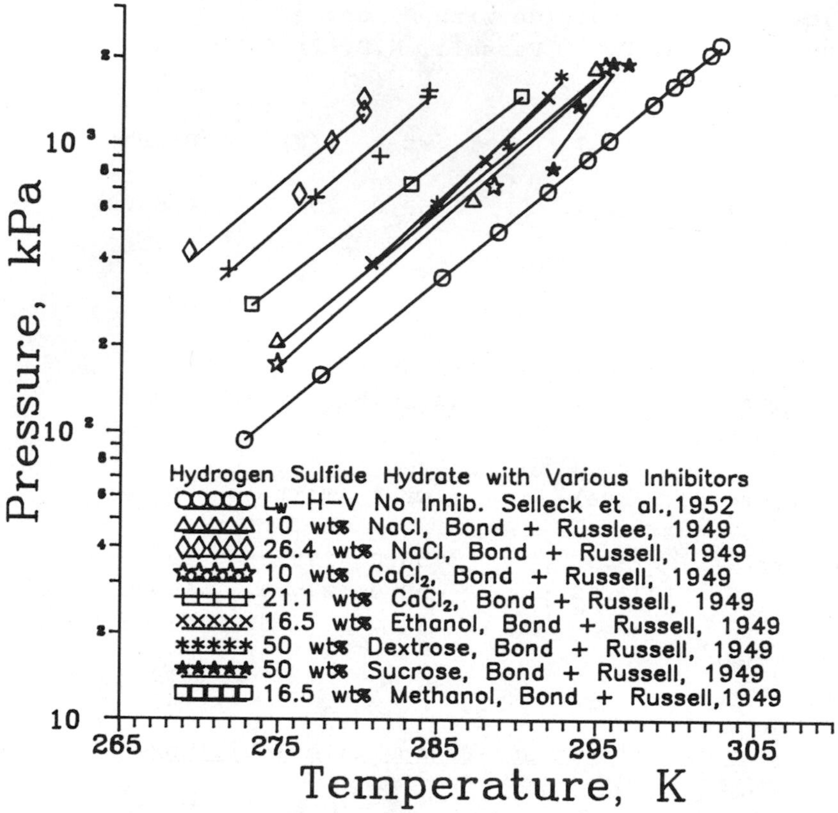


6-26 Simple Hydrogen Sulfide Hydrates with Various Inhibitors

## Binary Mixtures of Methane+Propane with Inhibitors

Hydrate: 95.01 Mol % Methane and 4.99 % Propane with Methanol
Reference: Ng, H.-J., Robinson, D.B. (1983)
Phases: $L_W$-H-V

| wt % | T(K) | P(MPa) | wt % | T(K) | P(MPa) |
|---|---|---|---|---|---|
| 10.0 | 265.51 | 0.532 | 20.0 | 265.17 | 0.938 |
| 10.0 | 270.06 | 0.903 | 20.0 | 270.47 | 1.772 |
| 10.0 | 274.54 | 1.544 | 20.0 | 275.69 | 3.144 |

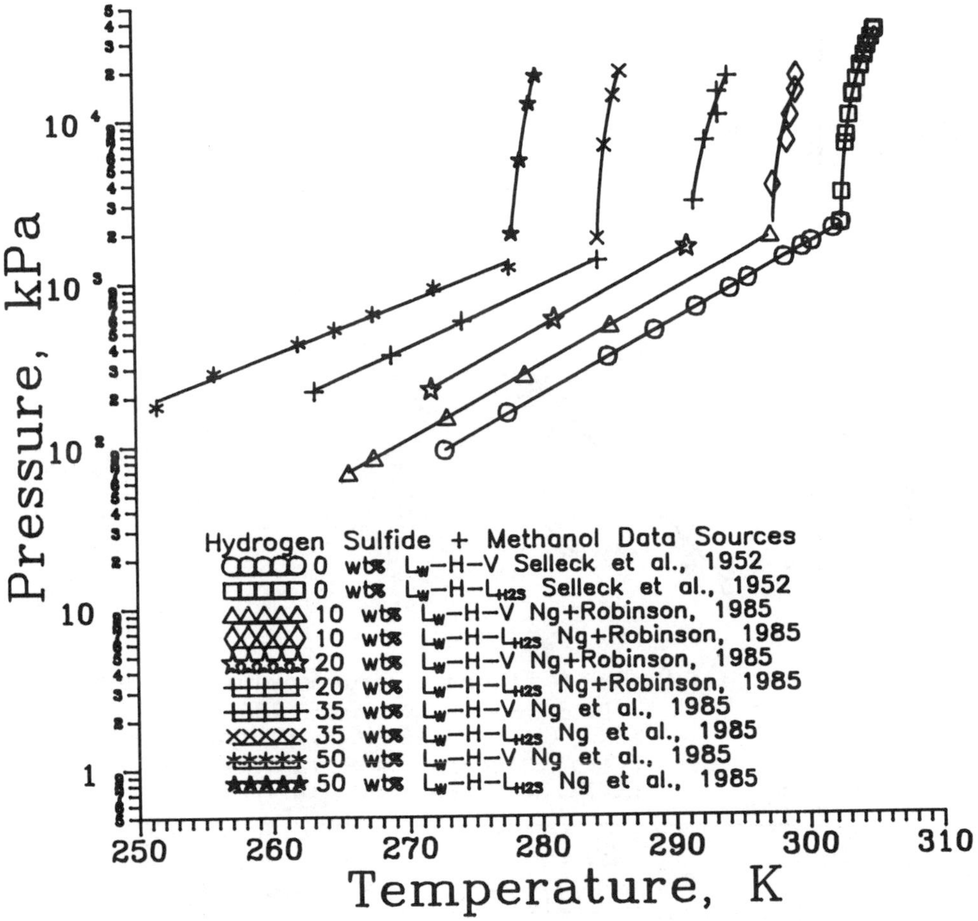


6-27 Methanol Inhibition of Simple Hydrogen Sulfide Hydrates

| | | | | | |
|---|---|---|---|---|---|
| 10.0 | 280.35 | 3.006 | 20.0 | 281.93 | 6.846 |
| 10.0 | 286.87 | 6.950 | 20.0 | 286.47 | 14.100 |
| 10.0 | 291.23 | 13.831 | | | |

<u>Hydrate</u>: 91.12 Mol % Methane and 8.88% Propane with Methanol
<u>Reference</u>: Ng, H.-J., Robinson, D.B. (1984)
<u>Phases</u>: $L_W$-H-V

| wt % | T(K) | P(MPa) | wt % | T(K) | P(MPa) |
|---|---|---|---|---|---|
| 35 | 253.1 | 0.621 | 50 | 249.5 | 1.690 |
| 35 | 261.3 | 1.570 | 50 | 255.2 | 3.760 |

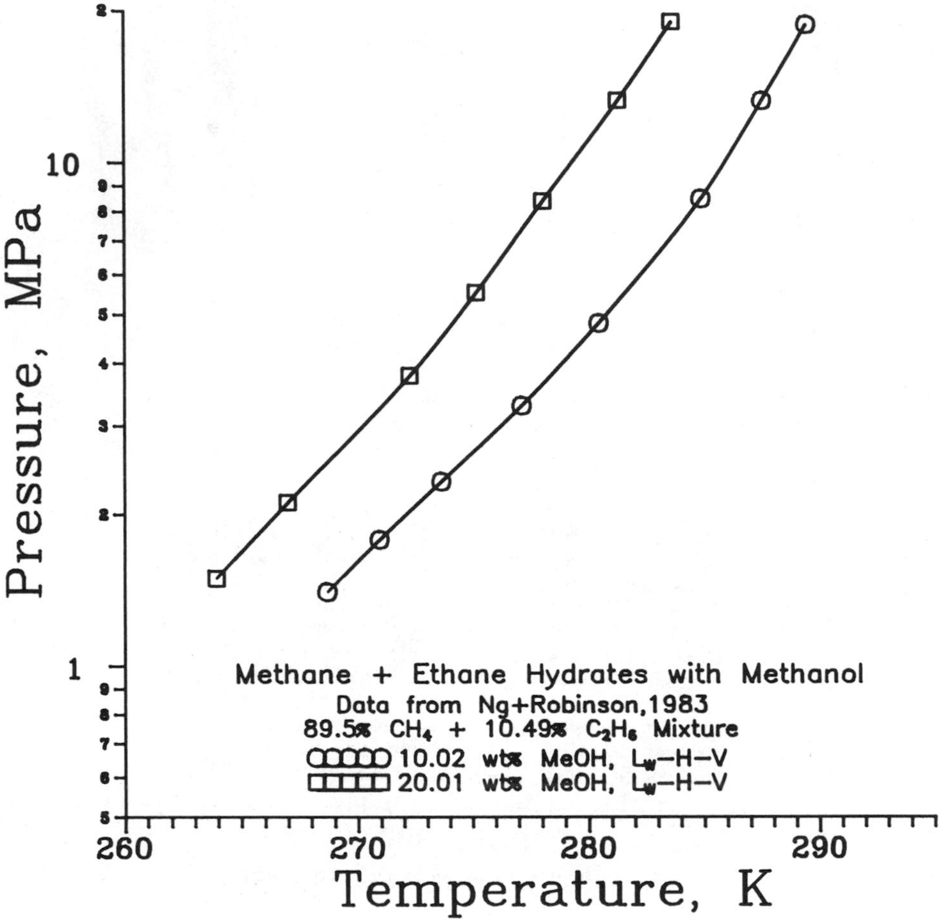


6-28 Methanol Inhibition of Methane+Ethane Hydrates

| | | | | | |
|---|---|---|---|---|---|
| 35 | 268.8 | 3.890 | 50 | 259.4 | 8.240 |
| 35 | 272.9 | 7.150 | 50 | 260.8 | 13.580 |
| 35 | 276.3 | 14.070 | 50 | 260.5 | 13.860 |
| 35 | 276.6 | 20.110 | 50 | 262.6 | 20.420 |
| 50 | 241.2 | 0.689 | | | |

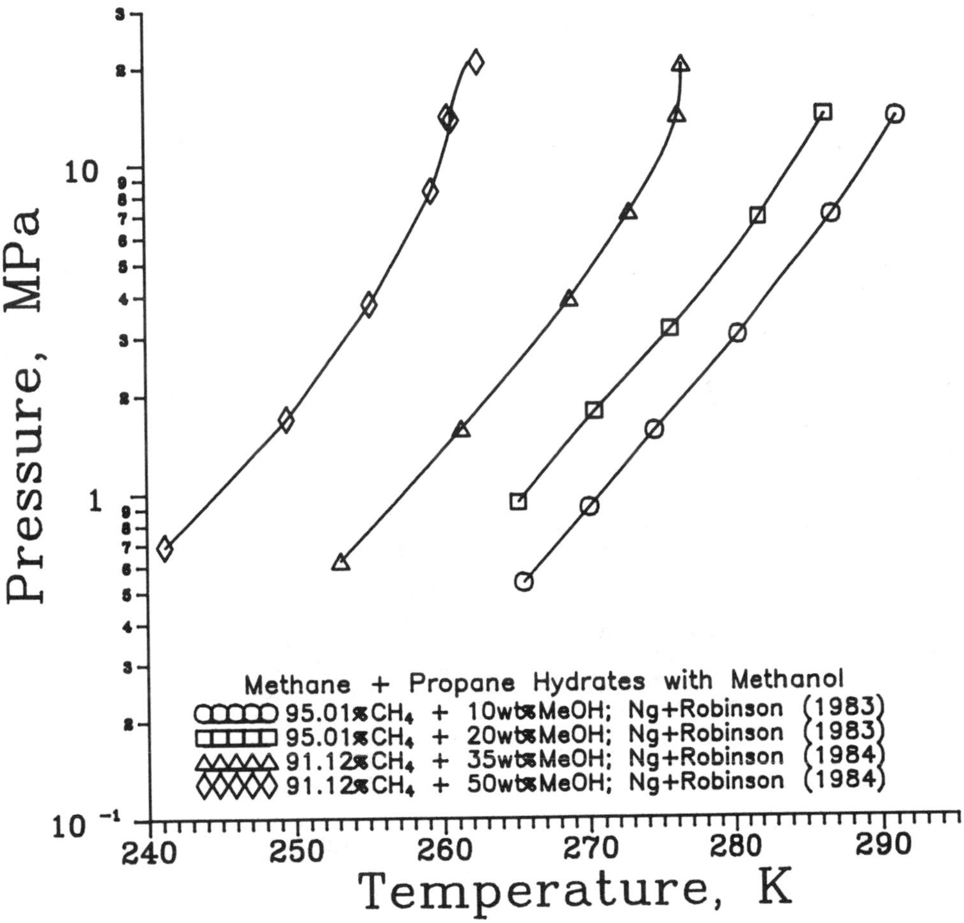


6-29 Methanol Inhibition of Methane+Propane Hydrates

<u>Binary Mixtures of Methane+Carbon Dioxide with Inhibitors</u>

Hydrate: 90.09 Mol % Methane and 9.91 % Carbon Dioxide with Methanol
Reference: Ng, H.-J., Robinson, D.B. (1983)
Phases: $L_W$-H-V

| wt % | T(K) | P(MPa) | wt % | T(K) | P(MPa) |
|---|---|---|---|---|---|
| 10.0 | 265.39 | 1.49 | 10.0 | 286.95 | 18.95 |

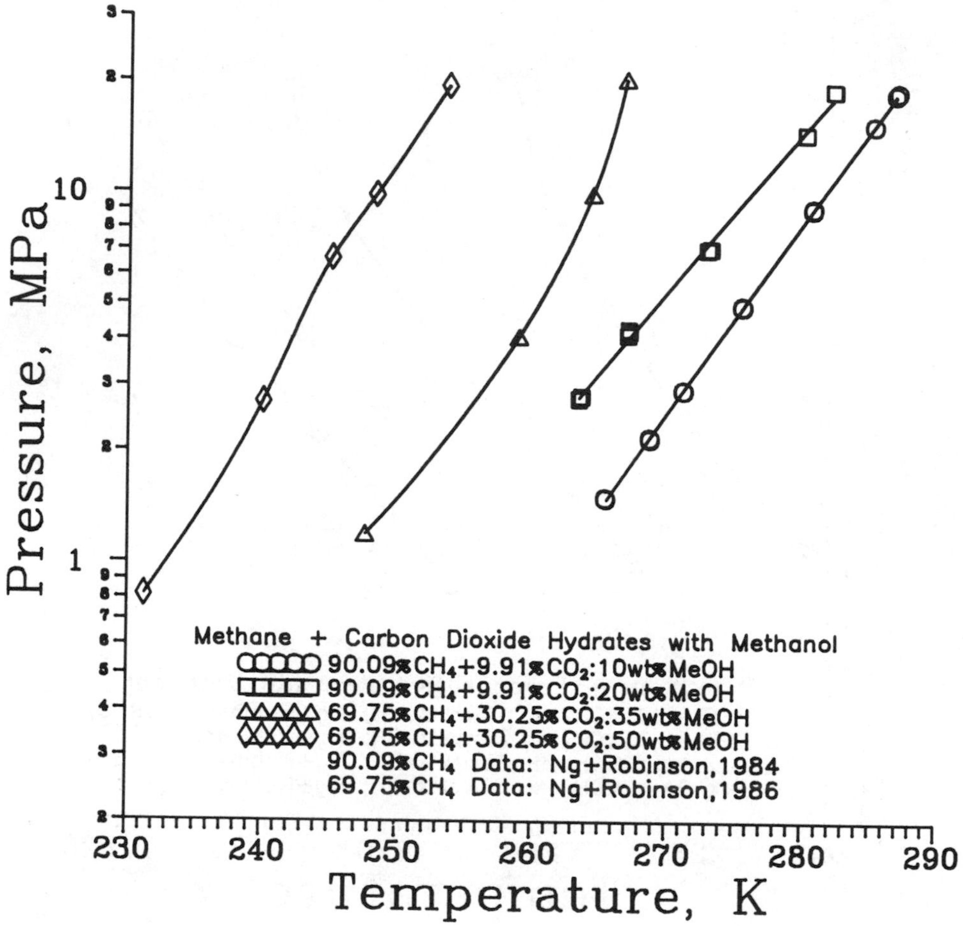


6-30 Methanol Inhibition of Methane+Carbon Dioxide Hydrates

| | | | | | |
|---|---|---|---|---|---|
| 10.0 | 265.35 | 1.50 | 20.0 | 263.35 | 2.76 |
| 10.0 | 268.70 | 2.16 | 20.0 | 263.55 | 2.81 |
| 10.0 | 268.67 | 2.18 | 20.0 | 267.03 | 4.12 |
| 10.0 | 271.20 | 2.92 | 20.0 | 267.11 | 4.21 |
| 10.0 | 271.15 | 2.92 | 20.0 | 267.07 | 4.27 |
| 10.0 | 275.52 | 4.91 | 20.0 | 272.94 | 6.98 |

| | | | | | |
|---|---|---|---|---|---|
| 10.0 | 275.49 | 4.93 | 20.0 | 273.16 | 7.03 |
| 10.0 | 280.63 | 8.98 | 20.0 | 280.11 | 14.36 |
| 10.0 | 280.67 | 9.05 | 20.0 | 280.11 | 14.40 |
| 10.0 | 285.15 | 15.28 | 20.0 | 282.21 | 19.00 |
| 10.0 | 285.15 | 15.29 | 20.0 | 282.13 | 19.01 |
| 10.0 | 286.75 | 18.66 | | | |

Hydrate: 69.75 Mol % Methane and 30.25 % Carbon Dioxide with Methanol
Reference: Robinson, D.B., Ng, H.-J. (1986)
Phases: $L_W$-H-V

| wt % | T(K) | P(MPa) | wt % | T(K) | P(MPa) |
|---|---|---|---|---|---|
| 35 | 247.6 | 1.190 | 50 | 240.1 | 2.710 |
| 35 | 258.9 | 4.070 | 50 | 245.1 | 6.640 |
| 35 | 264.3 | 9.890 | 50 | 248.3 | 9.830 |
| 35 | 266.8 | 20.270 | 50 | 253.6 | 19.430 |
| 50 | 231.3 | 0.814 | | | |

## Binary Mixtures of Methane+Hydrogen Sulfide with Inhibitors

Hydrate: Methane and Hydrogen Sulfide with 20 wt% Methanol
Reference: Ng, H.-J., et al. (1985)
Phases: $L_W$-H-V

| Vapor Phase Compositon Mol Fract. (Dry Basis) | | Temperature | Pressure |
|---|---|---|---|
| $CH_4$ | $H_2S$ | T(K) | P(MPa) |
| 0.896 | 0.104 | 264.5 | 0.945 |
| 0.897 | 0.103 | 271.8 | 2.130 |
| 0.837 | 0.163 | 281.4 | 5.750 |
| 0.780 | 0.220 | 287.0 | 11.670 |
| 0.741 | 0.259 | 290.3 | 18.710 |

## Binary Mixtures of Ethane+Carbon Dioxide with Inhibitors

Hydrate: 75% Ethane and 25 Mol % Carbon Dioxide with Methanol
Reference: Ng, H.-J., et al. (1985)
Phases: $L_W$-H-V, $L_W$-H-V-$L_{MIX}$, $L_W$-H-$L_{MIX}$

| wt % | T(K) | P(MPa) | wt % | T(K) | P(MPa) |
|---|---|---|---|---|---|
| | | $L_W$-H-V | | | |
| 20 | 266.4 | 0.738 | 35 | 266.8 | 2.110 |
| 20 | 271.1 | 1.400 | 50 | 237.0 | 0.319 |
| 20 | 275.3 | 2.689 | 50 | 242.1 | 0.494 |
| 35 | 251.4 | 0.422 | 50 | 248.4 | 0.862 |
| 35 | 256.9 | 0.724 | 50 | 251.5 | 1.172 |
| 35 | 260.2 | 0.993 | 50 | 254.3 | 1.593 |
| 35 | 263.9 | 1.586 | 50 | 255.3 | 1.800 |
| | | $L_W$-H-V-$L_{MIX}$ | | | |
| 20 | 276.9 | 3.523 | 50 | 255.5 | 1.855 |
| 35 | 267.8 | 2.627 | 50 | 255.8 | 2.096 |
| 35 | 268.1 | 2.806 | | | |
| | | $L_W$-H-$L_{MIX}$ | | | |
| 20 | 277.1 | 4.254 | 35 | 270.4 | 13.960 |
| 20 | 278.3 | 9.080 | 35 | 271.6 | 20.170 |
| 20 | 279.3 | 14.470 | 50 | 256.5 | 3.999 |
| 20 | 280.5 | 20.770 | 50 | 257.9 | 8.267 |
| 35 | 268.6 | 4.192 | 50 | 259.4 | 15.010 |
| 35 | 269.2 | 6.998 | 50 | 260.6 | 20.420 |

## Binary Mixtures of Propane+n-Butane with Inhibitors

Hydrate: Propane+n-Butane with Sodium Chloride
Reference: Paranjpe, S.G., et al. (1987)
Phases: $L_W$-H-V and $L_W$-H-V-$L_{HC}$

| wt % NaCl | mol % $C_3H_8$ | T(K) | P(kPa) | wt % NaCl | mol % $C_3H_8$ | T(K) | P(kPa) |
|---|---|---|---|---|---|---|---|
| | | | $L_W$-H-V | | | | |
| 3.0 | 0.999 | 275.2 | 350.3 | 5.0 | 0.984 | 273.2 | 299.2 |
| 3.0 | 0.955 | 275.2 | 373.7 | 5.0 | 0.952 | 273.2 | 307.3 |

| | | | | | | | |
|---|---|---|---|---|---|---|---|
| 3.0 | 0.906 | 275.2 | 391.6 | 5.0 | 0.927 | 273.2 | 317.2 |
| 3.0 | 0.997 | 273.2 | 224.8 | 5.0 | 0.894 | 273.2 | 330.9 |
| 3.0 | 0.921 | 273.2 | 246.8 | 5.0 | 0.906 | 273.2 | 336.5 |
| 3.0 | 0.854 | 273.2 | 273.7 | 5.0 | 0.981 | 272.2 | 242.7 |
| 3.0 | 1.000 | 272.2 | 197.2 | 5.0 | 0.930 | 272.2 | 255.1 |
| 3.0 | 0.964 | 272.2 | 209.6 | 5.0 | 0.905 | 272.2 | 262.7 |
| 3.0 | 0.833 | 272.2 | 221.3 | 5.0 | 0.878 | 272.2 | 272.3 |
| 3.0 | 0.826 | 272.2 | 238.6 | | | | |

$L_W$-H-V-$L_{HC}$

| | | | | | | | |
|---|---|---|---|---|---|---|---|
| 3.0 | NA | 275.2 | 373.7 | 5.0 | NA | 274.2 | 412.3 |
| 3.0 | NA | 274.2 | 332.3 | 5.0 | NA | 273.2 | 350.3 |
| 3.0 | NA | 273.2 | 281.3 | 5.0 | NA | 272.2 | 284.1 |
| 3.0 | NA | 272.2 | 241.3 | 5.0 | NA | 271.2 | 241.3 |
| 3.0 | NA | 271.3 | 214.4 | 5.0 | NA | 270.2 | 209.6 |

## Ternary and Multicomponent Mixtures with Inhibitors

Hydrate: Methane+Carbon Dioxide+Hydrogen Sulfide with Methanol
Reference: Ng, H.-J. et al. (1985)
Phases: $L_W$-V-H

| Methanol Conc. | Vapor Phase Composition Mol Fraction | | | Temp. | Pressure |
|---|---|---|---|---|---|
| wt % | $CH_4$ | $CO_2$ | $H_2S$ | T(K) | P(MPa) |
| 10 | 0.7647 | 0.1639 | 0.0714 | 264.9 | 0.556 |
| 10 | 0.7648 | 0.1654 | 0.0698 | 267.7 | 0.738 |
| 10 | 0.6980 | 0.2059 | 0.0961 | 275.7 | 1.410 |
| 10 | 0.6941 | 0.1998 | 0.1061 | 280.9 | 2.230 |
| 10 | 0.6635 | 0.2167 | 0.1198 | 285.6 | 3.900 |

| | | | | | |
|---|---|---|---|---|---|
| 10 | 0.6106 | 0.2350 | 0.1544 | 291.0 | 6.540 |
| 20 | 0.6938 | 0.2161 | 0.0901 | 267.5 | 1.080 |
| 20 | 0.6463 | 0.2334 | 0.1203 | 273.5 | 1.630 |
| 20 | 0.6263 | 0.2338 | 0.1399 | 277.8 | 2.450 |
| 20 | 0.6119 | 0.2408 | 0.1473 | 280.7 | 3.340 |
| 20 | 0.6060 | 0.2408 | 0.1532 | 283.6 | 4.190 |
| 20 | 0.5873 | 0.2434 | 0.1693 | 285.7 | 6.710 |
| 20 | 0.5924 | 0.2470 | 0.1606 | 284.5 | 5.630 |
| 20 | 0.5370 | 0.1988 | 0.2642 | 270.0 | 0.575 |

Hydrate: Synthetic Natural Gas Mixture with Methanol
Reference: Ng, H.-J., Robinson, D.B. (1983)
Phases: $L_W$-H-V

Gas Composition

| Component | Mol % | Component | Mol % |
|---|---|---|---|
| $N_2$ | 7.00 | $C_3H_8$ | 2.34 |
| $CH_4$ | 84.13 | $nC_4H_{10}$ | 0.93 |
| $C_2H_6$ | 4.67 | $nC_5H_{12}$ | 0.93 |

| wt % | T(K) | P(MPa) | wt % | T(K) | P(MPa) |
|---|---|---|---|---|---|
| 10.0 | 267.67 | 0.90 | 20.0 | 264.83 | 1.26 |
| 10.0 | 273.47 | 1.80 | 20.0 | 269.99 | 2.38 |
| 10.0 | 279.15 | 3.57 | 20.0 | 275.04 | 4.66 |
| 10.0 | 283.54 | 6.78 | 20.0 | 279.22 | 8.92 |
| 10.0 | 286.31 | 10.86 | 20.0 | 281.39 | 13.73 |
| 10.0 | 288.55 | 17.19 | 20.0 | 283.29 | 18.82 |
| 10.0 | 288.93 | 18.82 | | | |

Hydrate: Synthetic Natural Gas Mixture containing Carbon Dioxide with Methanol
Reference: Ng, H.-J., Robinson, D.B. (1983)
Phases: $L_W$-H-V

Gas Composition

| Component | Mol % | Component | Mol% |
|---|---|---|---|
| $N_2$ | 5.96 | $nC_4H_{10}$ | 0.79 |
| $CH_4$ | 71.60 | $nC_5H_{12}$ | 0.79 |
| $C_2H_6$ | 4.73 | $CO_2$ | 14.19 |
| $C_3H_8$ | 1.94 | | |

| wt % | T(K) | P(MPa) | wt % | T(K) | P(MPa) |
|---|---|---|---|---|---|
| 10.0 | 268.27 | 1.04 | 20.0 | 264.41 | 1.41 |
| 10.0 | 270.96 | 1.46 | 20.0 | 270.40 | 2.83 |
| 10.0 | 276.56 | 2.76 | 20.0 | 274.13 | 4.77 |
| 10.0 | 281.46 | 5.52 | 20.0 | 276.26 | 6.94 |
| 10.0 | 283.88 | 8.42 | 20.0 | 277.95 | 9.53 |
| 10.0 | 285.03 | 10.73 | 20.0 | 278.85 | 12.14 |
| 10.0 | 286.47 | 13.91 | 20.0 | 279.54 | 15.04 |
| 10.0 | 287.67 | 17.44 | 20.0 | 280.31 | 16.75 |
| 10.0 | 288.34 | 19.03 | 20.0 | 280.97 | 19.15 |

Hydrate: Seven-Component Mixture with Methanol
Reference: Robinson, D.B., Ng, H.-J.(1986)
Phases: $L_W$-H-V, $L_W$-H-V-$L_{HC}$, $L_W$-H-$L_{HC}$

| Component | Mol % | Component | Mol % |
|---|---|---|---|
| $N_2$ | 5.26 | $C_3H_8$ | 2.02 |
| $CO_2$ | 13.37 | $nC_4H_{10}$ | 0.80 |
| $CH_4$ | 73.90 | $nC_5H_{10}$ | 0.80 |
| $C_2H_6$ | 3.85 | | |

| wt % | T(K) | P(MPa) | wt % | T(K) | P(MPa) |
|---|---|---|---|---|---|
| | | | Lw-H-V | | |
| 35 | 244.9 | 0.362 | 35 | 248.2 | 0.600 |
| | | | $L_W$-H-V-$L_{HC}$ | | |
| 35 | 256.1 | 1.390 | 50 | 241.5 | 1.410 |
| 35 | 262.6 | 3.610 | 50 | 250.1 | 3.450 |
| 35 | 266.8 | 7.290 | 50 | 254.5 | 7.250 |
| 50 | 234.4 | 0.518 | | | |
| | | | $L_W$-H-$L_{HC}$ | | |
| 35 | 269.9 | 13.820 | 50 | 256.8 | 14.220 |
| 35 | 273.1 | 20.350 | 50 | 258.5 | 20.280 |
| 50 | 256.3 | 13.570 | | | |

Hydrate: Gas Liquid and Condensate with Methanol and Mono-Ethylene Glycol
Reference: Ng, H.-J., et al. (1987a)

Composition of Hydrocarbon Liquids
Concentration, mol %

| Component | C | D |
|---|---|---|
| Nitrogen | 0.16 | 0.64 |
| Methane | 26.19 | 73.03 |
| Carbon Dioxide | 2.10 | 3.11 |
| Ethane | 8.27 | 8.04 |
| Propane | 7.50 | 4.28 |
| Isobutane | 1.83 | 0.73 |
| n-Butane | 4.05 | 1.50 |
| Isopentane | 1.85 | 0.54 |
| n-Pentane | 2.45 | 0.60 |
| Hexanes plus | 45.60 | 7.53 |
| M.Wt. | 90.2 | 32.4 |
| Saturation Pressure (MPa) at Ts | 8.95* | 43.94** |
| Ts(K) | 310.95 | 416.15 |

* = Bubble Point Temperature
** = Retrograde Dew Point Pressure

| Aqueous Solution wt% | T K | P MPa | Phases |
|---|---|---|---|
| | Liquid C | | |
| 13% MeOH | 281.05 | 3.91 | $L_W$-H-V-$L_{HC}$ |
| | 285.85 | 7.40 | $L_W$-H-V-$L_{HC}$ (V=0) |
| | 286.15 | 12.00 | $L_W$-H-$L_{HC}$ |
| | 287.55 | 20.00 | $L_W$-H-$L_{HC}$ |
| 24% MeOH | 275.25 | 4.02 | $L_W$-H-V-$L_{HC}$ |
| | 278.85 | 6.97 | $L_W$-H-V-$L_{HC}$ (V=0) |
| | 279.05 | 11.99 | LW-H-$L_{HC}$ |
| | 280.35 | 19.99 | $L_W$-H-$L_{HC}$ |
| 20% MEG | 281.25 | 3.94 | $L_W$-H-V-$L_{HC}$ |
| | 285.65 | 7.37 | $L_W$-H-V-$L_{HC}$ (V=0) |
| | 286.35 | 12.00 | $L_W$-H-$L_{HC}$ |
| | 287.75 | 20.00 | $L_W$-H-$L_{HC}$ |

| | | | |
|---|---|---|---|
| 39% MEG | 274.35 | 4.08 | $L_W$-H-V-$L_{HC}$ |
| | 278.35 | 6.92 | $L_W$-H-V-$L_{HC}$(V=0) |
| | 278.75 | 12.06 | $L_W$-H-$L_{HC}$ |
| | 280.05 | 19.99 | $L_W$-H-$L_{HC}$ |
| | Liquid D | | |
| 16% MeOH | 283.25 | 6.00 | $L_W$-H-V-$L_{HC}$ |
| | 286.35 | 11.03 | $L_W$-H-V-$L_{HC}$ |
| | 288.05 | 15.01 | $L_W$-H-V-$L_{HC}$ |
| | 288.95 | 19.99 | $L_W$-H-V-$L_{HC}$ |
| 29% MeOH | 275.95 | 6.03 | $L_W$-H-V-$L_{HC}$ |
| | 278.75 | 10.99 | $L_W$-H-V-$L_{HC}$ |
| | 280.05 | 14.97 | $L_W$-H-V-$L_{HC}$ |
| | 281.55 | 20.00 | $L_W$-H-V-$L_{HC}$ |

Hydrate: Lean Gas and Rich Gas with Methanol
Reference: Ng, H.-J., et al. (1987b)
Phases: $L_W$-H-V-$L_{LG}$, $L_W$-H-$L_{LG}$, $L_W$-H-V-$L_{RG}$, $L_W$-H-$L_{RG}$

Mol Fraction

| Component | Lean Gas | Rich Gas |
|---|---|---|
| $CH_4$ | 0.9351 | 0.8999 |
| $C_2H_6$ | 0.0458 | 0.0631 |
| $C_3H_8$ | 0.0131 | 0.0240 |
| $iC_4H_{10}$ | 0.0010 | 0.0030 |
| $nC_4H_{10}$ | 0.0020 | 0.0050 |
| $iC_5H_{12}$ | 0.0010 | 0.0010 |
| $nC_5H_{12}$ | 0.0010 | 0.0010 |
| $nC_6H_{14}$ | 0.0010 | 0.0030 |

Lean Gas

| wt % | T(K) | P(MPa) | wt % | T(K) | P(MPa) |
|---|---|---|---|---|---|
| | | $L_W$-H-V-$L_{LG}$ | | | |
| 65.0 | 225.8 | 0.66 | 73.7 | 229.2 | 2.90 |
| 65.0 | 235.0 | 1.61 | 73.7 | 235.1 | 5.79 |
| 65.0 | 244.4 | 4.25 | 85.0 | 199.8 | 2.08 |
| 73.7 | 221.9 | 1.37 | 85.0 | 204.4 | 4.63 |

$L_W$-$H$-$L_{LG}$

| | | | | | |
|---|---|---|---|---|---|
| 65.0 | 248.9 | 10.29 | 73.7 | 237.2 | 20.07 |
| 65.0 | 250.4 | 19.75 | 85.0 | 205.8 | 11.34 |
| 73.7 | 236.4 | 9.51 | 85.0 | 206.8 | 20.37 |
| 73.7 | 235.9 | 10.17 | | | |

Rich Gas

$L_W$-$H$-$V$-$L_{RG}$

| | | | | | |
|---|---|---|---|---|---|
| 65.0 | 229.4 | 0.77 | 73.7 | 228.9 | 2.41 |
| 65.0 | 238.2 | 1.70 | 73.7 | 236.4 | 5.72 |
| 65.0 | 246.0 | 4.28 | 85.0 | 198.1 | 1.37 |
| 73.7 | 218.2 | 0.76 | 85.0 | 207.4 | 3.56 |

$L_W$-$H$-$L_{RG}$

| | | | | | |
|---|---|---|---|---|---|
| 65.0 | 250.0 | 10.54 | 73.7 | 238.6 | 19.70 |
| 65.0 | 251.2 | 20.17 | 85.0 | 211.1 | 7.92 |
| 73.7 | 237.7 | 10.23 | 85.0 | 212.4 | 19.70 |

### 6.2.2 Thermal Property Data

In this section the hydrate measurements for heat capacity, enthalpy of dissociation and thermal conductivity are presented. The other thermal properties such as thermal expansion are presented in Chapter 2 with their analogs for ice.

#### 6.2.2.a. Heat Capacity and Heat of Dissociation

Since there is a paucity of data, the data for the heat capacity and enthalpy of dissociation are presented by investigation without discrimination, other than the accuracy statements presented in Section 6.1.2.

Compounds: Methane, Ethane, and Propane
Reference: Handa, Y.P. (1986b)
Properties: Molar Heat Capacity and Enthalpy of Dissociation

| T | Constant Pressure Molar Heat Capacity, $C_p$ ($J \cdot K^{-1} \cdot mol^{-1}$) | | |
|---|---|---|---|
| (K) | $CH_4 \cdot 6.0H_2O$ | $C_2H_6 \cdot 7.67H_2O$ | $C_3H_8 \cdot 17.0H_2O$ |
| 85 | 107.7 | 149.6 | 281.7 |
| 90 | 112.1 | 156.0 | 294.0 |
| 100 | 121.4 | 167.2 | 318.8 |
| 110 | 131.5 | 177.2 | 342.0 |
| 120 | 140.3 | 188.6 | 366.5 |
| 130 | 149.0 | 199.4 | 392.4 |
| 140 | 156.8 | 210.0 | 415.9 |
| 150 | 164.2 | 219.6 | 437.6 |
| 160 | 171.1 | 229.0 | 459.3 |
| 170 | 178.6 | 237.9 | 481.0 |
| 180 | 186.3 | 248.2 | 502.4 |
| 190 | 194.1 | 259.1 | 524.8 |
| 200 | 201.4 | 269.2 | 548.3 |
| 210 | 209.8 | 277.4 | 573.5 |
| 220 | 219.3 | 292.8 | 599.5 |
| 230 | 225.9 | 301.7 | 617.7 |
| 240 | 233.7 | 310.9 | 644.0 |
| 250 | 240.4 | 323.0 | 674.4 |
| 260 | 248.4 | 337.8 | 710.2 |
| 270 | 257.6 | | |

| Compound | T Range K | Standard (273.15K, 101.32 kPa) Molar Enthalpy of Dissociation $\Delta H^0$ ($kJ \cdot mol^{-1}$) | |
|---|---|---|---|
| | | H → I+G | H → L+G |
| $CH_4$ | 160-210 | 18.13±0.27 | 54.19±0.28 |
| $C_2H_6$ | 190-250 | 25.70±0.37 | 71.80±0.38 |
| $C_3H_8$ | 210-260 | 27.00±0.33 | 129.2±0.4 |

Compounds: Isobutane and Two Naturally Occurring Hydrates
Reference: Handa, Y.P. (1986c, 1988)
Properties: Enthalpy of Dissociation

Composition of Naturally Occurring Hydrates
(mol per cent)

| Cmpnt↓ Sample Origin→ | Gulf of Mexico | Mid-America Trench |
|---|---|---|
| $CH_4$ | 66.0 | 99.93 |
| $C_2H_6$ | 2.9 | 0.01 |
| $C_3H_8$ | 14.7 | 0.01 |
| i-$C_4H_{10}$ | 3.7 | 0.05 |
| n-$C_4H_{10}$ | 0.08 | |
| neo-$C_5H_{12}$ | 0.13 | |
| i-$C_5H_{12}$ | 0.01 | |
| cy-$C_5H_{12}$ | 0.01 | |
| $CO_2$ | 7.8 | |
| $H_2S$ | 0.49 | |
| $N_2$ | 4.0 | |
| wt% Sediment | 7.4 | 0.764 |

Standard Molar Enthalpy of Dissociation $\Delta H^0$

| Hydrate Sample | H→I+G | H→L+G |
|---|---|---|
| Isobutane | 18.13±0.27(KJ•mol$^{-1}$) | 54.19±0.28(KJ•mol$^{-1}$) |
| Gulf of Mexico | 27.8(J•g$^{-1}$)(assuming all water as hydrate) | |
| | 33.1(J•g$^{-1}$) (assuming 20% water as ice) | |
| Mid-America Trench | 17.5 (kJ•mol$^{-1}$) (19.2 mass% water was ice) | |

Compound: Methane
Reference: Lievois, J.S. (1987)
Property: Enthalpy of Dissociation

| T(K) | $\Delta H_{dis}$ (J/gmol $CH_4$) |
|---|---|
| 278.15 | 57,739 |
| 278.15 | 57,358 |
| 278.15 | 57,697 |
| 283.15 | 52,798 |
| 283.15 | 53,610 |

Compound: Methane
Reference: Rueff, R.M. et al. (1988)
Property: Constant Volume Heat Capacity and Enthalpy of Dissociation

| T(K) | $C_v$(J/g-K) |
|---|---|
| 245 | 1.62 |
| 255 | 1.56 |
| 255 | 1.61 |
| 256 | 1.70 |
| 257 | 1.66 |
| 259 | 1.61 |

| T(K) | $\Delta H_{dis}$(J/g) |
|---|---|
| 285 | 429.66 |

### 6.2.2.b. Thermal Conductivity

Hydrate: Methane
Reference: Cook, J.G., Leaist, D.G. (1983)
Property: Thermal Conductivity

T = 216.15 K Thermal Conductivity = 0.45 $W \cdot m^{-1}K^{-1}$

Hydrate: Propane
Reference: Stoll,R.D., Bryan, G.M. (1979)
Property: Thermal Conductivity

T = 275.15K Thermal Conductivity = 0.393 $W \cdot m^{-1}K^{-1}$

## 6.3 Summary and Relationship to Chapters Which Follow

The object of this chapter is to provide an overview of the experimental methods, the phase equilibria data, and the thermal property data available on hydrates since the publication of Hammerschmidt (1934) on hydrates in natural gas systems. The tabulations and associated plots illustrate that more data are needed, particularly for phase equilibria mixtures of non-combustible components and for thermal property data.

Chapters 7 and 8 consider two applications of the data in this chapter and the correlation methods of Chapters 4 and 5. Chapter 7 regards hydrates as an energy resource for exploration and recovery in the earth, as examples of phase equilibria and thermal property data application. In Chapter 8 problems caused by hydrates in gas processing and production are considered. In both chapters perturbations on phase equilibria are considered, such as the effects of third surfaces. The differences between those equilibria and the equilibria presented in this chapter find application in such things as hydrate formation in porous media and in drilling muds.

# 7

# Hydrates in the Earth

Only in the last quarter-century have we recognized that the formation of **in situ** hydrates in the geosphere predated their artificial formation (1810) by several million years. In addition to their age, it appears that hydrates in nature are ubiquitous, with some probability of occurrence wherever methane and water are in close proximity at low temperature and elevated pressures. Because hydrates concentrate methane by a factor of 170, and because as little as 10% of the recovered energy is required for dissociation, hydrate reservoirs have been considered as a substantial future energy resource. There is agreement between researchers in both the eastern (Makogon, 1988) and western (Kvenvolden, 1988a) hemispheres that the total amount of gas in this solid form may surpass the energy content of the total fossil fuel reserves by as much as a factor of two.

The initial section of this chapter deals with the estimates of the extent of natural gas in **in situ** hydrates, as well as the location and requirements for their formation. The second section concerns the mechanism for the generation of hydrates, and the corresponding geochemical and geophysical markers. The final section of the chapter deals with the mathematical models for the recovery of gas from these solid compounds, a case study of hydrate production, and the economics of recovery of gas from hydrates.

## 7.1 Extent and Location of Natural Gas Hydrate Deposits

In 1965, Makogon (1965) announced the observation of gas hydrates in permafrost regions of the Soviet Union. Since that time there have been two extreme views of **in situ** hydrate reserves. In one view, they have been ignored, presumably because they were considered to be too dispersed and difficult to recover. In the other view, they were thought to be pervasive in all regions of the earth with permafrost (23 per cent of the land mass) and in the thermodynamically stable regions of the oceans (90 per cent of their areal extent). With further exploration and the production of gas from a hydrate reservoir, a third, more realistic view has gradually evolved. Today it is recognized not only that hydrates are widespread in outer continental margins of the oceans and a portion of the region of permafrost, but also that economics dictate that recovery of gas from hydrates must be postponed for a number of years.

### 7.1.1 Extent of the Occurrence of In Situ Gas Hydrates

Knowledge of the occurrence of **in situ** gas hydrates is very incomplete, and is obtained from both indirect and direct evidence. In permafrost regions, hydrate evi-

**Table 7-1**
**Estimates of Methane In In Situ Hydrates**

| Permafrost Hydrates $m^3$ | Oceanic Hydrates $m^3$ | Reference |
|---|---|---|
| $5.7\times10^{13}$ | $5\text{-}25\times10^{15}$ | Trofimuk et al.(1977) |
| $3.1\times10^{13}$ | $3.1\times10^{15}$ | McIver (1981) |
| $3.4\times10^{16}$ | $7.6\times10^{18}$ | Dobrynin et al.(1981) |
| $1.4\times10^{13}$ | | Meyer (1981) |
| $1.0\times10^{14}$ | $1.0\times10^{16}$ | Makogon (1988) |
| | $1.8\times10^{16}$ | Kvenvolden, (1988a) |

dence is limited to drillings and associated well logs and cores, whereas the sonic velocity technology obtained after World War II has developed seismic methods as a less direct but relatively inexpensive means of sensing on continental margins. Economics, and to some extent time, have been the determining factors for the exploration of these two regions of the earth. These factors have promoted increasing interest in the larger, oceanic hydrate reservoirs.

Table 7-1 provides a decade of estimates of natural gas in hydrates in the geosphere. These estimates range from the maximum value of Dobrynin et al. (1981), who apparently assumed that hydrates would occur wherever satisfactory thermodynamic conditions exist, to the minimum values of McIver (1981) and Meyer (1981) who considered more limiting factors such as availability of methane, limited porosity and percentages of organic matter, thermal history of various regions, etc. All of the estimates of natural gas hydrates are not well defined, and therefore somewhat speculative. It appears however that the most recent estimates, made by independent investigators through different methods are converging on very large values of gas reserves in hydrated form.

The most recent appraisal of the amount of **in situ** hydrates (Kvenvolden, 1988a) was obtained by scaling a gross estimate of the amount of hydrates in the continental margin of northern Alaska to the total length of continental margins worldwide. In Table 7-1, note that each investigator determined the hydrate reserve in the ocean to be at least two orders of magnitude greater than that in the permafrost. The estimates of the oceanic hydrate reserves are so large that any error in those approximations would encompass the entire permafrost hydrate reserves.

Even the most conservative estimates of gas in hydrates in Table 7-1 indicate their enormous energy potential. Kvenvolden (1988a) indicated that the 10,000 Gigatons (1 Gt = $10^{15}$ g) or $1.8\times10^{16}$ $m^3$ of carbon in hydrates may surpass the available carbon in the global cycle by a factor of two. He noted, based on Bolin (1983) and Moore and Bolin (1987), the global carbon cycle amounts to be as follows: fossil fuel deposits (5,000 Gt), terrestrial soil, detritus, and peat (1,960 Gt), marine dissolved materials (980 Gt), terrestrial biota (830 Gt), the atmosphere (3.6 Gt), and marine biota (3Gt). Relative to hydrates, the only larger pool of inorganic and organic carbon is that disseminated in and throughout sediments and rocks (20,000,000 Gt), which is unrecoverable as an energy resource.

The major difficulty in considering hydrates as actual reserves stems from the solid, dispersed character of hydrates themselves. The recovery of gas from hydrates is more complex than that from a normal gas reservoir.In addition, while an energy balance of the dissociation of pure hydrates is very favorable, the hydrates may be dispersed in sediment, which will mitigate their economic recovery. However, before turning to the recovery of gas from hydrates, consider first the locations of hydrate reserves, and requirements for their formation.

### 7.1.2 Location and Requirements for In Situ Hydrate Formation

Table 7-2, modified after Kvenvolden (1988a), lists forty sites around the world for which there is some evi-

**Table 7-2 Gas Hydrate Locations**
(After Kvenvolden, 1988a)

| Number in Figure 7-1 | Location | Evidence | Reference |
|---|---|---|---|
| | | Oceanic Hydrates | |
| 1 | Pacific Ocean off Panama | BSR | Shipley et al.(1979) |
| | | LitRvw | Krason&Ciesnik(1986a) |
| 2 | Middle America (MAT) Trench off Costa Rica | BSR | Shipley et al.(1979) |
| | | Lit Rvw | Finley&Krason(1986a) |
| 3 | MAT off Nicaragua | Samples | Kvenvolden & McDonald(1985) |
| | | Lit Rvw | Finley & Krason (1986a) |
| 4 | MAT off Guatemala | BSR | Shipley et al.(1979) |
| | | Samples | Harrison & Curiale(1982) |
| | | Samples | Kvenvolden and McDonald(1985) |
| | | Lit Rvw | Finley & Krason (1986a) |
| 5 | MAT off Mexico | BSR | Shipley et al.(1979) |
| | | Lit Rvw | Finley & Krason (1986a) |
| 6 | Eel River Basin off California | BSR | Field&Kvenvolden(1985) |
| | | Lit Rvw | Krason & Ciesnik (1986b) |
| | | Samples | Kennicut et al.(1989) |
| 7 | E. Aleutian Trench off Alaska | BSR | Kvenvolden & vonHeune(1985) |
| 8 | Beringian Margin off Alaska | BSR | Marlow et al.(1981) |
| | | Lit Rvw | Krason & Ciesnik (1987) |
| 9 | Middle Aleutian Trench off Alaska | BSR | McCarthy & others (1984) |
| | | Lit Rvw | Krason & Ciesnik (1987) |
| 10 | Nankai Trough off Japan | BSR | Aoki et al.(1983) |
| 11 | Timor Trough off Australia | Gas | McKirdy and Cook(1980) |
| 12 | Hikurangi Trough off New Zealand | BSR | Katz(1981) |
| 13 | Wilkes Land Margin off Antartica | BSR | Kvenvolden et al.(1987) |

**Table 7-2. (continued) Gas Hydrate Locations**

| | | | |
|---|---|---|---|
| 14 | W. Ross Sea off Antarctica | Gas | McIver (1975) |
| 15 | Peru-Chile Trench off Peru | BSR | Shepard(1979) |
| | | Samples | Kvenvolden&Kastner(1988) |
| 16 | Barbados Ridge Complex off Barbados | BSR | Ladd et al.(1982) |
| 17 | Colombia Basin off Panama and Colombia | BSR | Shipley et al.(1979) |
| | | Lit Rvw | Finley & Krason (1986b) |
| 18 | W. Gulf of Mexico off Mexico | BSR | Shipley et al.(1979) |
| | | | Hedberg(1980) |
| | | Lit Rvw | Krason et al.(1985) |
| 19 | Gulf of Mexico off S. USA | Samples | Brooks et al.(1984) |
| | | Lit Rvw | Brooks & Bryant (1985) |
| 20 | Blake Outer Ridge off SE USA | BSR | Markl et al.(1970) |
| | | | Shipley et al.(1979) |
| | | | Dillon et al. (1980) |
| | | Samples | Kvenvolden & Barnard(1983) |
| | | Lit Rvw | Krason&Ridley(1985a) |
| 21 | Continental Rise off E. USA | BSR | Tucholke et al.(1977) |
| | | Lit Rvw | Krason & Ridley(1985b) |
| 22 | Labrador Shelf off Newfoundland | BSR | Taylor et al.(1979) |
| | | Lit Rvw | Krason&Rudloff(1985) |
| 23 | Beaufort Sea off Alaska | BSR | Grantz et al.(1976) |
| 24 | Beaufort Sea off Canada | Log | Weaver & Stewart(1982) |
| 25 | Svedrup Basin off Canada | Log | Judge (1982) |

**Table 7-2. (continued) Gas Hydrate Locations**

| | | | |
|---|---|---|---|
| 26 | Continent Slope off W. Norway | BSR | Bugge et al.(1987) |
| 27 | Cont Slope off SW Africa | Slides Slumps | Summerhayes et al.(1979) |
| 28 | Makran Margin, Gulf of Oman | BSR | White(1979) |
| 29 | Black Sea, USSR | Samples | Yefremova and Zhizhchenko(1972) |
| | | Lit Rvw | Ciesnik & Krason(1987) |
| 30 | Caspian Sea, USSR | Sample | Yefremova and Gritshina(1981) |
| 31 | Okhotsk Sea, USSR | Sample | Makogon(1988b) |
| 32 | Baikal Lake, USSR | Sample | Makogon(1988b) |
| | <u>Continental Hydrates</u> | | |
| A | North Slope, Alaska | Logs | Collett(1983) |
| | | Sample | Collett & Kvenvolden(1987) |
| | | Lit Rvw | Collett et al.(1988) |
| B | Mackenzie Delta, Canada | Logs | Bily&Dick(1974) |
| C | Arctic Isands, Canada | Logs | Davidson et al.(1978) Judge(1982) |
| D | Timan-Pechora Province,USSR | Gas | Cherskiy et al.(1985) |
| E | Messokayha Field, USSR | Sample | Makogon et al.(1972) |

**Table 7-2. (continued) Gas Hydrate Locations**

| | | | |
|---|---|---|---|
| F | E. Siberian Craton, USSR | Gas | Cherskiy et al.(1985) |
| G | NE Siberia, USSR | Gas | Cherskiy et al.(1985) |
| H | Kamchatka, USSR | Gas | Cherskiy et al.(1985) |

dence of **in situ** hydrates. Along with each listing is the type of evidence and the original citation for the hydrate occurrence. The numbers of the sites in Table 7-2 correspond to the locations shown in the global map of Figure 7-1. While the number of the listings is small, the sitings to date are considered to be only a small portion of the actual occurrence of gas hydrates, with substantial exploration yet to be done.

As one considers the table and the map, two aspects are striking, as follows:
(a) that the nature of hydrate indications is usually either from a small body of direct evidence obtained through drilling, or from a much larger body of indirect geophysical evidence based on the occurrence of Bottom Simulating Reflectors (BSRs) on marine seismic records, and
(b) that while a number of occurrences are in the permafrost regions of the globe, thrice that number have been sited in outer continental margins. A surprising number of occurrences are situated in oceans around the equator, where one might not anticipate ice-like compounds. This is due to the phase equilbrium property of hydrates which enable their stability at temperatures both below and above the ice point, when the pressure is elevated. The deep bottom waters are at temperatures ranging typically between 272 and 277 K.

As an initial heuristic for hydrate formation, only four conditions are necessary: (1) adequate gas molecules to stabilize most of the hydrate cavities, (2) sufficient water molecules to form the cavity, (3) a temperature

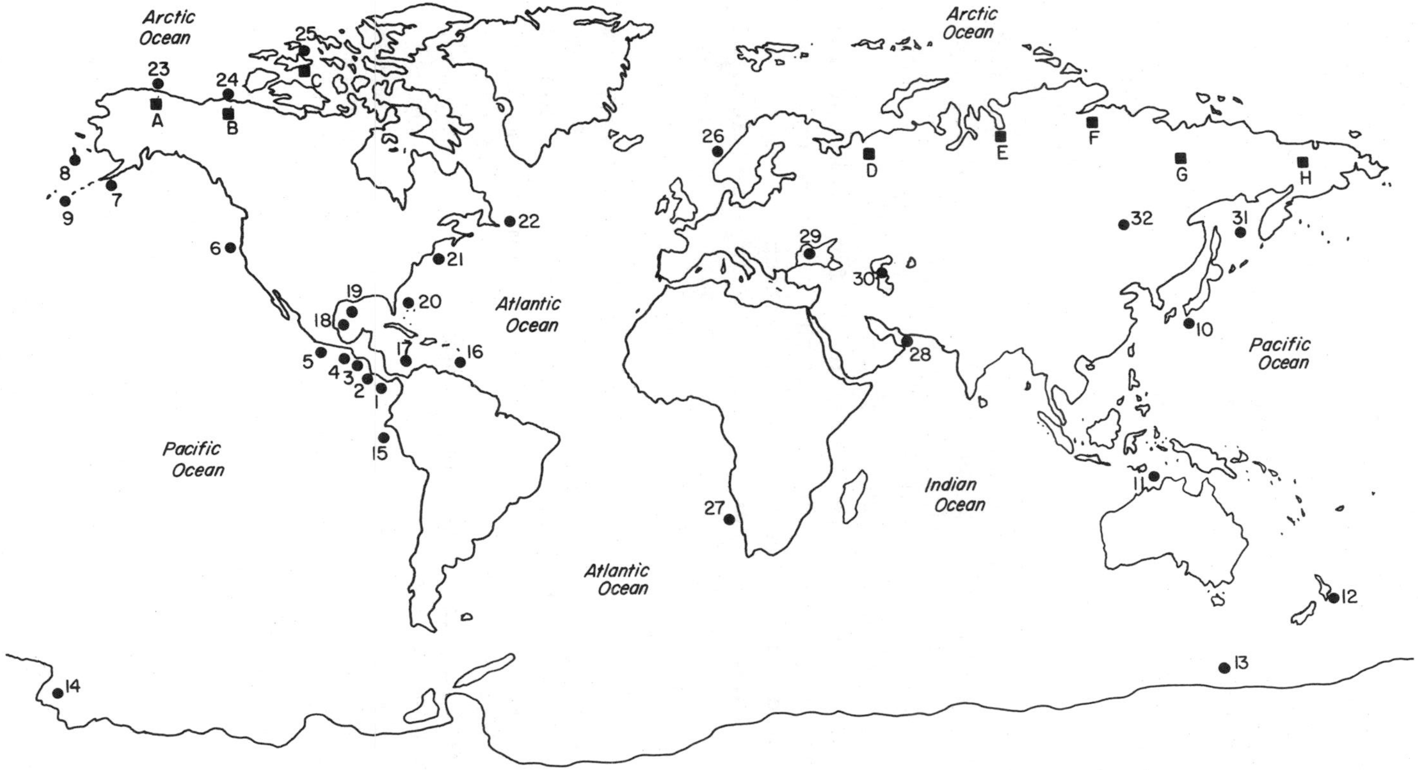


7-1 The Earth Showing Locations of Known and Inferred Hydrate Deposits in Oceans (•) and Permafrost (■) see text for nomenclature (Reproduced, with permission, from <u>Chem Geology</u> (Kvenvolden, 1988a), by Elsevier Science Publishers)

within the hydrate phase equilibrium region, and (4) a pressure within the hydrate phase equilibrium region. The first two criteria are discussed in the following section on biogenic and thermogenic methane generation. To a good approximation, the final two criteria for equilibrium properties may be determined as they were in Chapters 4 and 5; refinement is discussed in Section 7.2.3 on the physical geochemistry of hydrate formation.

Figures 7-2a,b are arbitrary examples of the depths of hydrate phase stability in permafrost and in oceans, respectively. In each figure the dashed lines represent the thermal gradients as a function of depth. The slopes of the dashed lines are discontinuous both at the base of the permafrost and the water/sediment interface, where new sediment thermal conductivities cause new thermal gradients. The solid lines were drawn from the methane hydrate phase equilibria data, with the pressure converted to depth assuming hydrostatic conditions in both the water and sediment. In each diagram the intersections of the solid and dashed lines bound the depths of the hydrate stability fields.

From Figure 7-2b one would not expect hydrates to be stable in the region above the sediment due to an absence of a means of concentrating gas in water and a means of retaining the hydrates, because their density is much less than that of water. In both Figure 7-2a and Figure 7-2b, a small addition of heavier natural gas components, such as propane or isobutane, will cause the hydrate stability region to increase due to a displacement of the phase boundary line away from the geothermal gradient line. In neither case, however have hydrates been found at sediment depths greater than about 2000 m, due to the high temperatures resulting from the geothermal gradient.

### 7.1.3 Hydrate Formation from Crude Oil Reservoirs

While the formation of hydrates from hydrocarbon vapor appears to be the most common case in nature, there are instances of hydrate formation from hydrocarbon li-

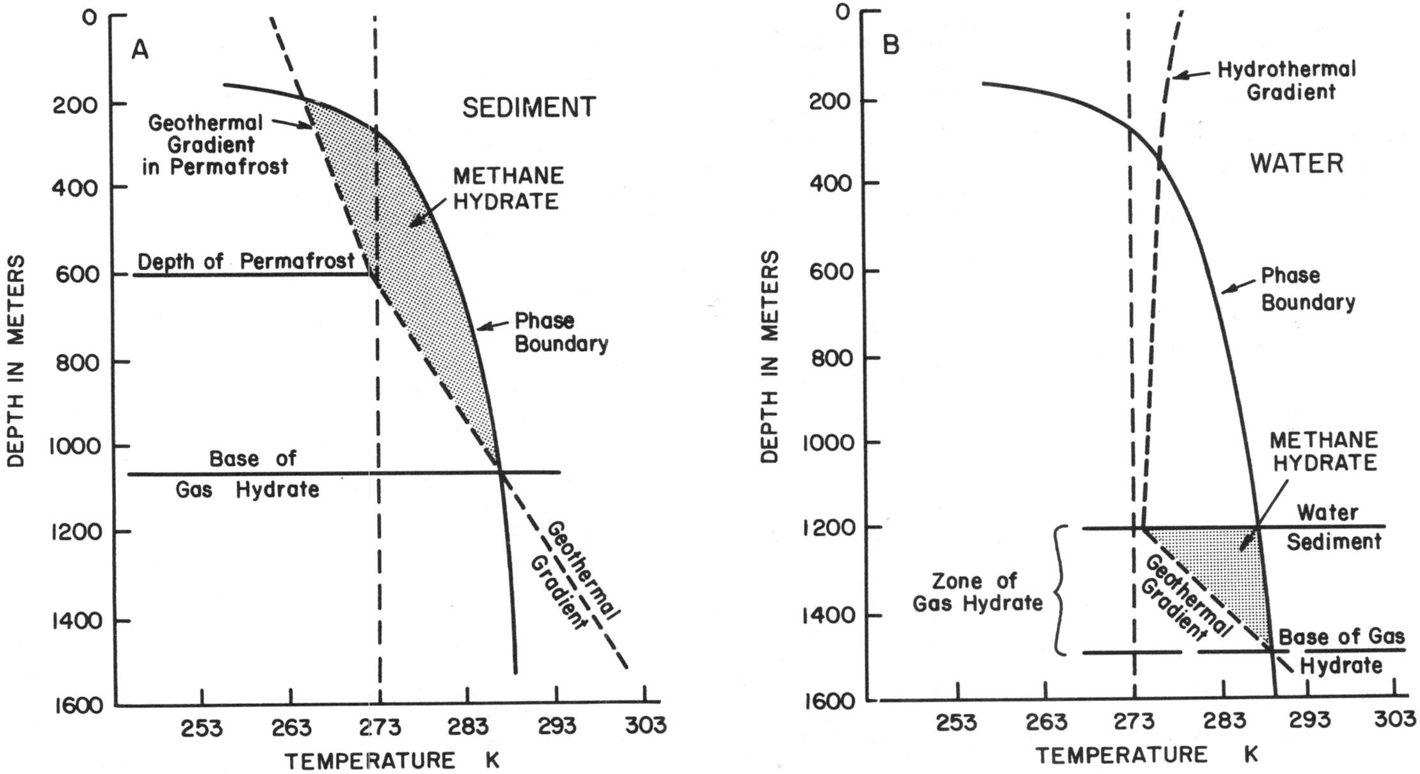


7-2 Envelopes of Methane Hydrate Stability (A) in Permafrost and (B) in Ocean Sediment
(Reproduced, with permission, from <u>Chem Geology</u> (Kvenvolden, 1988a), by Elsevier Science Publishers)

quids, with no vapor phase present. Shortly after Makogon's announcement of **in situ** hydrates, Katz (1972) suggested that hydrates could form from crude oils which have only small quantities of hydrate forming components. Katz's analysis of these systems at their bubble points is discussed in Section 4.4.2. In doctoral studies at the University of Michigan Verma (1974) and Holder (1976) verified Katz's suggestion with data presented in Section 6.2.1. These studies showed conclusively that water reservoirs could denude crude oil of its lighter components through hydrate formation, thereby leaving the "dead" or highly viscous oil at a substantially lower vapor pressure. Further studies by Verma et al.(1975), Holder et al.(1976), and Collett (1983) indicate that hydrates are present in the immediate vicinity of such "heavy" oils in the North Slope of Alaska at depths less than 1200m.

The mechanism for hydrate formation from these oils is similar to the mechanism of hydrate formation from gases which have high contents of large molecules. In both cases, molecules larger than normal butane do not enter the hydrate lattice, but instead act as diluents in the fluid phase. The effect of these diluents is to require lower temperatures or higher pressures for hydrate formation than if the fluid were composed only of smaller molecules. In the case of oils the formation of hydrates will remove those smaller components which cause low oil viscosity and higher bubble point pressures. The result can be very high shear requirements for flow, or pore pluggage if the oil is in sediment.

## 7.2 Generation Mechanism and Geophysical/Geochemical Markers

With the four criteria of Section 7.1.2 as initial determinants of hydrate formation, consider the slightly more sophisticated criteria for determining hydrates in the geosphere. In the thermodynamic hydrate regime of temperature and pressure, water is seldom the limiting chemical; therefore hydrate generation (and several of the markers of hydrate formation) are related to the generation of methane. The origins of the gases in hydrate are both thermogenic and biogenic. After considering the source of the gas, the mechanism is investigated before turning to the physical geochemical effects of hydrate formation in sediment.

### 7.2.1 Generation of Gases for Hydrate Formation

The thermogenic generation and character of gases is more common to natural gas reservoirs and will not be considered in detail here. It has been reviewed thoroughly in a number of references (see for example Hunt, 1979, p.163ff). The thermogenic gases are produced by a catagenesis process characterized by high temperatures (>373 K), producing relatively large amounts of ethane and higher hydrocarbons. Organic diagenesis however, will be briefly reviewed because it is slightly different from the "usual" gas generation scheme, due to its low temperature mechanism.

The words "organic diagenesis" are used to denote the low temperature chemical or biogenic conversion of organic matter to methane. Organic diagenesis is usually enhanced by high values of the clastic/organic flux. Kvenvolden (1985a) suggests sedimentation rates between $30 m/10^6 yr$ and $300 m/10^6 yr$ are necessary for hydrate formation; however Finley and Krason (1986a) put these limits between 5 and 1,000 $m/10^6 yr$, based upon evidence from Deep Sea Drilling Project (DSDP) Legs 67 and 84.

The following discussion is derived almost entirely from Hesse (1986) who presented a thorough review of the process of organic diagenesis. He modified the work of Claypool and Kaplan (1974), to suggest the six stages diagrammed in Figure 7-3 which organic matter passes through in anaerobic sediments. At each stage the change in the standard Gibbs Free Energy of reaction is more negative (favorable) than the reaction mechanism(s) of the stages above and below it. In the first stage the organic matter (containing carbon, nitrogen and phosphorus in the ratio 106:16:1) is oxidized by dissolved oxy-

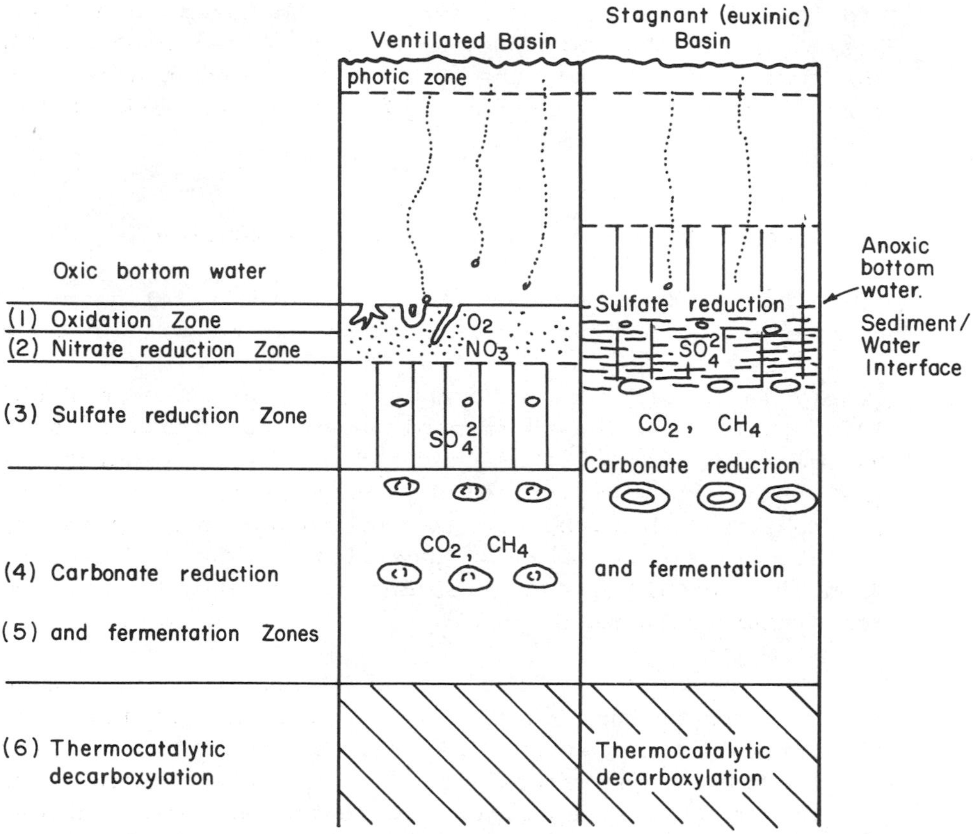


7-3 Stages of Organic Matter Oxidation in Anoxic Sediments (Reproduced, with permission, from Geoscience Canada, (Hesse, 1986), by the Geological Association of Canada)

gen. In somewhat deeper sediments, the organic matter is reduced in the second stage by nitrates, primarily in the form of nitric acid. The third stage is characterized by sulfate ion reduction. Immediately below is the fourth stage of carbonate reduction which generates the methane for hydrate formation. Without giving the complete carbonate mechanism, the overall reaction for the production of methane from an organic is:

$$(CH_2O)_{106}(NH_3)_{16}(H_3PO_4) \rightarrow 53CO_2 + 53CH_4 + 16NH_3 + H_3PO_4 \qquad (7.1)$$

Below the stage of methane production is the fermentation (fifth) stage, followed by the sixth, thermocatalytic stage at still greater depths. The initial five stages require the presence of bacteria; below stage (5) bacterial activity ceases and the more usual thermocatalytic reactions associated with hydrocarbon production begin. The main chemical species released to the pore water from the microbial breakdown of organic matter are the carbonate species listed as $\sum CO_2$, which include $CO_2$, $H_2CO_3$, $HCO_3^-$, and $CO_3^=$.

Carbonates are generated in all of the zones, but they are consumed only below the third stage. After the depletion of more than 80 per cent of the sulfate, the components of $\sum CO_2$ become oxidants, leading to the production of methane. It is possible to distinguish between the $CO_2$ produced and the $CO_2$ consumed, determining the extent of carbon isotopic fractionation, particularly associated with the methane production in stage (4).

The isotope $^{13}C$ is distributed through sediments of all geological ages, and its difference in mass relative to $^{12}C$ is brought about through fractionation by both biological and physical processes. The fractionation is measured relative to a standard sample of Peedee Belemnite (PDB), with the units of parts per thousand (‰). The ratio difference ($\delta$) of $^{13}C$ to $^{12}C$ may be measured spectrometrically, and is defined by:

$$\delta^{13}C \equiv \left[\frac{(^{13}C/^{12}C)_{sample}}{(^{13}C/^{12}C)_{PDB}} - 1\right] \times 10^3 \qquad (7.2)$$

Methane formed by biogenic processes ranges in $\delta^{13}C$ from about -55‰ to -85‰, while methane formed from thermogenic processes ranges from -25‰ to -60‰ (Hunt, 1979, pg. 25.)

During the first three stages of organic matter decomposition, negligible carbon isotopic fractionation occurs. However, in the fourth stage, the methane generated from carbonates has a isotopic composition which is

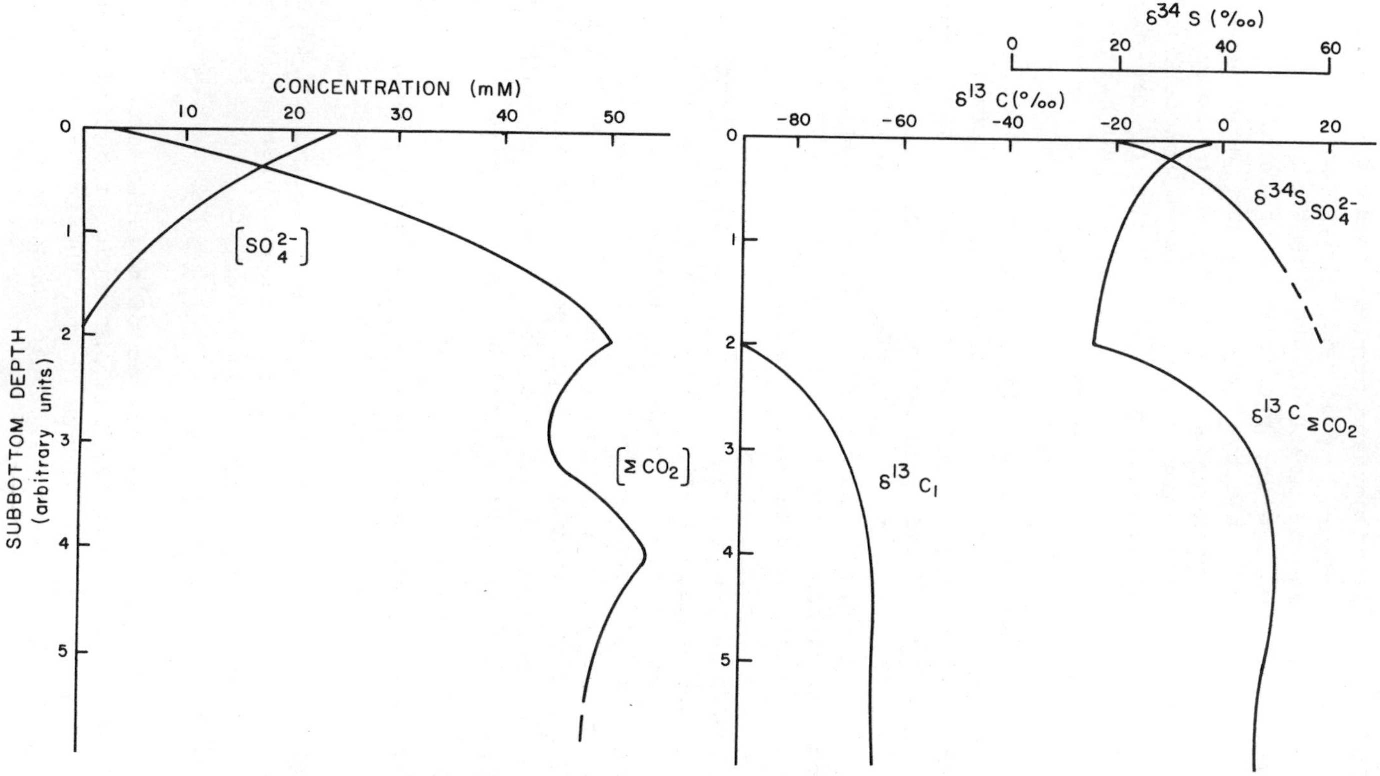


7-4 Generalized Profiles of Concentration and Isotope Ratio Changes for Dissolved Sulfur and Carbon Species in Anoxic Marine Sediments. Depth Scale is Arbitrary with Depth Units Ranging from $10^{-1}$ to $10^{2}$m.
(Reproduced, with permission, from Ann. Rev. Earth Planet. Sci., (Claypool and Kvenvolden, 1983), by Annual Reviews, Inc.)

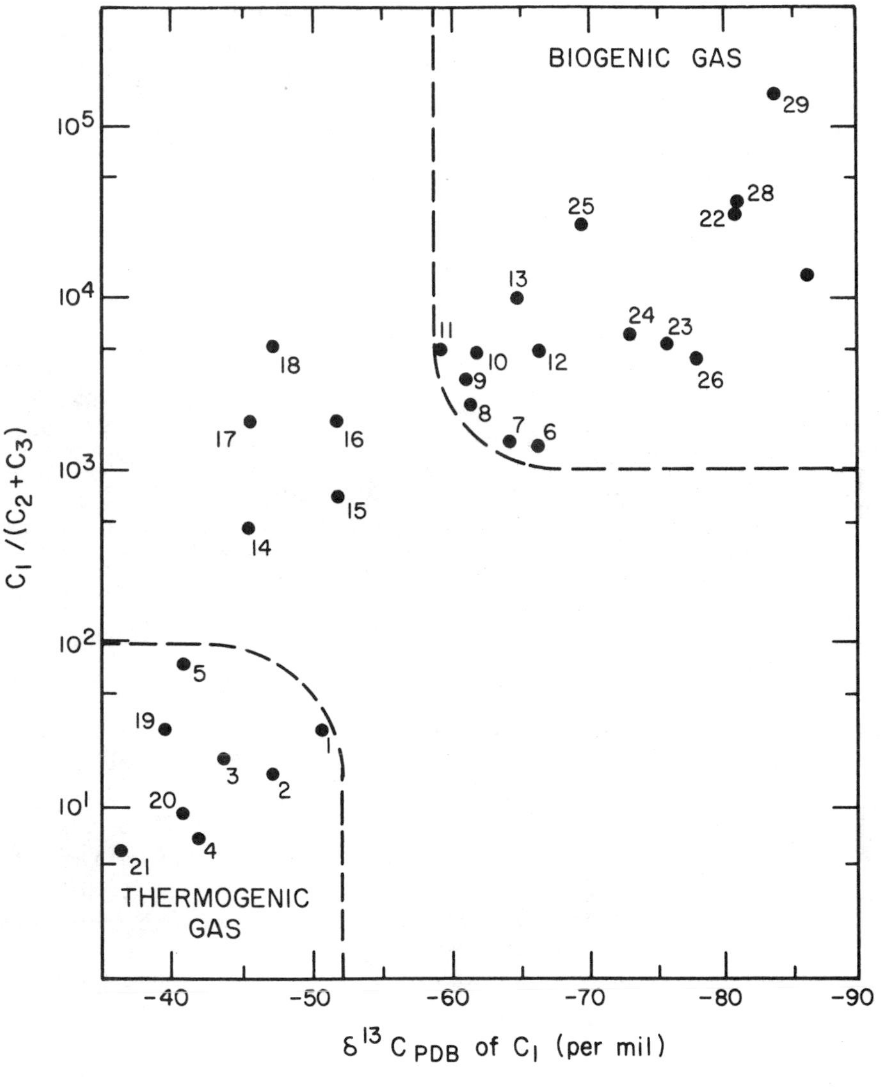


7-5 Crossplot of $C_1/(C_2+C_3)$ Against Isotopic Composition of $C_1$ to Distinguish Biogenic and Thermogenic Gas (Reproduced, with permission, from Ann. Rev. Earth Planet. Sci., (Claypool and Kvenvolden, 1983), by Annual Reviews, Inc.)

about 70 per cent lighter than the carbon of the parent material (Hesse, 1986). The residual $CO_2$ consequently becomes enriched in $\delta^{13}C$ reaching positive values as high as +15‰ to + 25‰. Figure 7-4 shows the reduction in

**Table 7-3 Gas Characteristics of Gulf of Mexico Hydrate Samples (after Brooks et al. (1986))**

| Site | $\delta^{13}C$ $CH_4$ | Composition of Gas, mol% $C_1$ | $C_2$ | $C_3$ | $i\text{-}C_4$ | $n\text{-}C_4$ | $CO_2$ | $\frac{C_1}{C_2+C_3}$ |
|---|---|---|---|---|---|---|---|---|
| | | | **Biogenic Hydrate Samples** | | | | | |
| Orca Basin | -71.3 | 99.1 | 0.34 | 0.28 | | | 0.24 | 159 |
| Garden Banks-388 | -70.4 | 99.5 | 0.12 | | | | 0.26 | 829 |
| Green Canyon-257 | -69.2 | 99.7 | | | | | 0.26 | >1000 |
| Green Canyon-320 | -66.5 | 99.7 | 0.08 | | | | 0.12 | 1246 |
| | | | **Thermogenic Gas Hydrates** | | | | | |
| Green Canyon-184 | -44.6 | 70.9 | 4.7 | 15.6 | 4.4 | 0.3 | 4.1 | 3.2 |
| Green Canyon-204 | -56.5 | 61.9 | 9.2 | 22.8 | 4.5 | 1.3 | 0.2 | 1.9 |
| Green Canyon-234 | -43.2 | 74.3 | 4.0 | 13.0 | 3.2 | 0.86 | 4.6 | 4.4 |
| Mississippi Cnyn | -48.2 | 93.4 | 1.2 | 1.3 | | | 4.0 | 37.4 |

sulfates and the corresponding growth of both the parent carbonates and the offspring methane with subbottom depth. The methane production is seen to be parallel and lower in isotope content than the carbonates. The sulfur isotope ($\delta^{34}S$) content is defined in a manner similar to Equation (7.2) with the replacement of the fraction $^{13}C/^{12}C$ by $^{34}S/^{32}S$ in both the numerator and the denominator, using Cañon Diablo meteoritic troilite as a standard. The $\delta^{34}S$ value increases from 20 to 60‰ before substantial biogenic methane is produced.

A second characteristic enabling the differentiation of biogenic gas and thermogenic gas is the ratio of methane to heavier hydrocarbons. Biogenic gas typically has values greater than $10^3$ for the ratio of methane to the sum of ethane and propane [$C_1/(C_2+C_3)$], while for thermogenic gas this ratio is usually less than 100 (Bernard, et al., 1976). Figure 7-5 presents a typical cross plot of the ratio of $C_1/(C_2+C_3)$ against isotopic composition for several gas seeps as a valuable means for distinguishing the gas source. Table 7-3, taken from Brooks et al. (1986), quantifies this difference for hydrate samples. Note that there is usually between 10 and 20 mol per cent propane in the thermogenic hydrates, compared against its relative absence in biogenic hydrates. These amounts are consistent with the phase equilibria heuristics of Chapters 4 and 5 which suggest that propane should be concentrated in the hydrate phase. One would not expect a small (e.g. ppm) hydrate propane concentration from a thermogenic gas.

### 7.2.2 Mechanisms for Generation of Hydrates

Methane, either from biogenic or thermogenic sources, comes into contact with water in sediments to form hydrates. In the northern Gulf of Mexico for example Brooks (1988) estimated that approximately equal numbers of hydrate samples have been recovered from each type gas source. Due to the high temperature requirements for its production, the thermogenic gas must migrate from its source at depth to the hydrate stability region. This gas migration occurs via channels such as faults associated with salt diapirs, which are common to regions such as the Gulf of Mexico. For Prudhoe Bay permafrost hydrates, Collett et al. (1988) suggested that thermogenic gas migrated along faults from deeper regions, where it was mixed with biogenic gas and either directly converted to gas hydrate or first concentrated in existing structural/stratigraphic traps, and later converted to hydrate. It appears that most of the gas in the hydrate core recovered from the Northwest Eileen State Well Number Two (see Section 7.2.6) was of biogenic origin; the biogenic source is likely to predominate for hydrates in permafrost (Kvenvolden, 1988b).

Biogenic gas however, may form in the hydrate stability region because the temperature required for gas production is within the pressure-temperature envelope just below the water-sediment interface of many continental margins, as shown in Figure 7-2b. Secondly, if channels are available, biogenic gas may migrate to regions within the hydrate stability envelope. It is possible to have both means of supplying biogenic gas, indicated by Kvenvolden et al.(1984) and Kvenvolden and Claypool (1985) at DSDP Site 570 in the Middle America Trench.

Because the gas is less dense than either water or sediments, it will percolate upward in the region of hydrate stability. Kvenvolden suggested that a minimum residual methane concentration of 10 ml/l of wet sediment was necessary for hydrate formation. The upward gas motion may be sealed by a relatively impermeable layer of sediment, such as an upper dolomite layer (Finley and Krason, 1986a) or the upper siltstone sequence in the North Slope of Alaska (Collett et al.,1988). Additionally permafrost itself may act as an upper gas seal. The formation of a layer of hydrates themselves can act as another mechanism for a gas seal. These seals can act to provide traps for free gas, and they can subsequently act to provide sites for hydrate formation from the free gas.

Hydrates will likely form at the two-phase (gas-liquid) interface, as observed in laboratory experiments. As indicated in Chapter 3 initial hydrate formation serves as a nucleation site for additional formation from the gas and liquid phases. However, many geochemists (Claypool and Kaplan, 1974, Finley and Krason, 1987, etc.) suggest that hydrates will form from gas (either at equilibrium or supersaturated) dissolved in the liquid phase. While hydrate formation from a single phase may be possible, it seems unlikely based on the necessity of filling about 90 per cent of the cages with guest molecules for hydrate stability. Hydrate stability requirements result in a need to provide a gas concentration in the hydrate two orders of magnitude greater than the gas solubility in the liquid phase. The ability of gas hydrates to form from the liquid phase is addressed further in Chapter 4, with regard to the question of solubility data and the availability of sufficient numbers of gas molecules.

Based upon an overview of several years of research sponsored by the U.S. Department of Energy, R.D. Malone

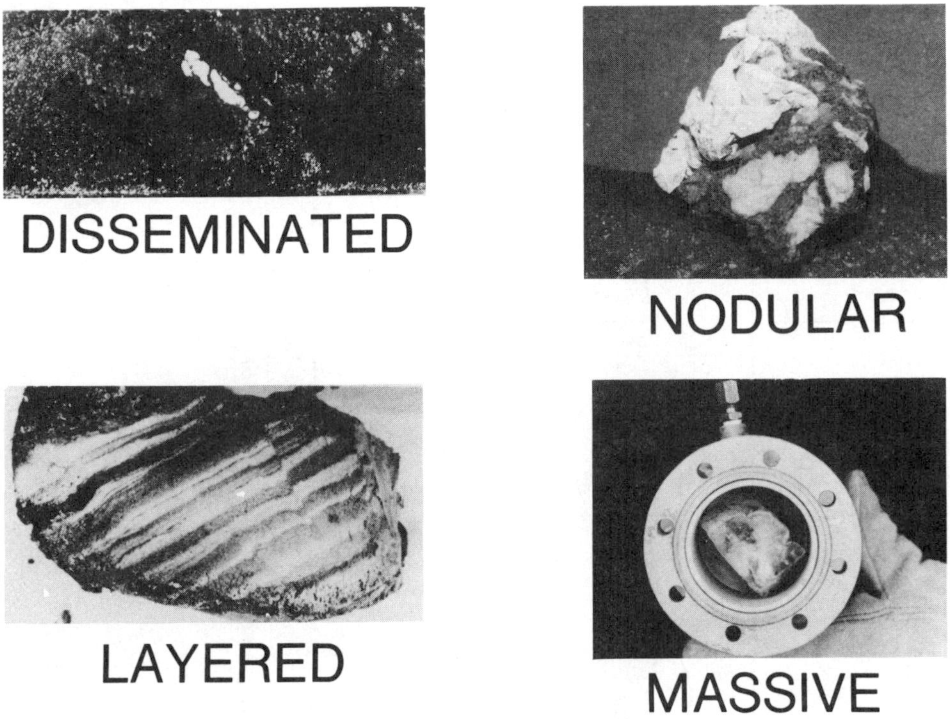


7-6 Photographs of Four Types of Hydrates
(Courtesy of W.F. Lawson, U.S. Department of Energy, 1988)

(1985) suggested that hydrates occur in four types, each of which is depicted in Figure 7-6:

(1) The first type of hydrates are finely disseminated, perhaps formed in place, such as those samples found in the Orca and Mississippi Canyon areas of the Gulf of Mexico.

(2) Nodular hydrates, up to 5 cm in diameter, may occur, such as found in the Green Canyon, Gulf of Mexico; the gas in these hydrates may be of thermogenic origin which migrated from some depth.

(3) Layered hydrates are separated by thin layers of sediments, such as the cores recovered from the Blake-Bahama Ridge. Examples of these hydrates probably occur both offshore and in permafrost regions.

(4) Massive hydrates, such as the one recovered from site 570 of DSDP Leg 84 off the Middle America Trench, may be

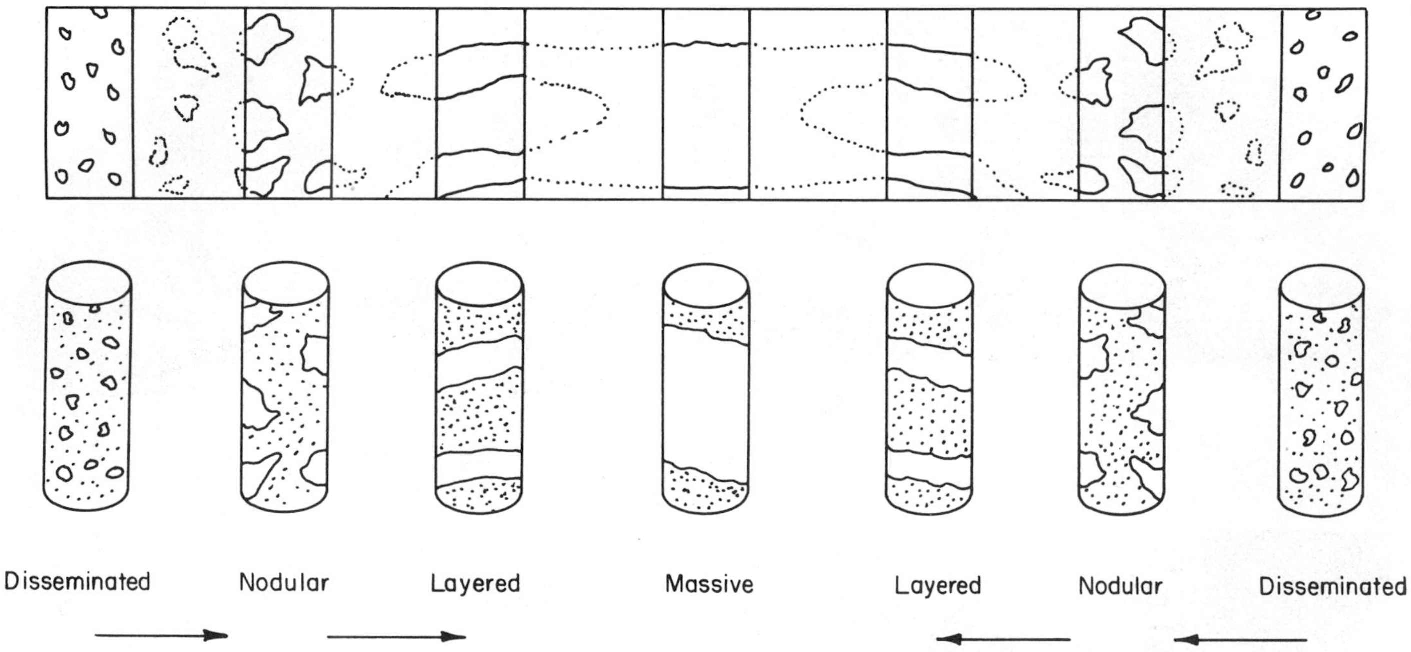


7-7 A Mechanism for Hydrate Consolidation
(Courtesy of U.S. Dept. of Energy, (R.D. Malone, 1985))

as thick as 3-4 m and contain 95% hydrate and less than five percent sediments. While there is some dispute as to whether this sample is of biogenic origin (Kvenvolden and Claypool, 1985) or thermogenic origin (Finley and Krason, 1986a), it appears that much of the gas migrated to the hydrate site, and perhaps either formed along a fault or pushed aside sediments as the massive hydrate grew.

The observations of the above four types of hydrates led Malone (1985) to postulate a simple model for hydrate formation in deep-sea sediments, as depicted in Figure 7-7, with arrows to indicate the direction of hydrate evolution. The initial formation of disseminated hydrate crystals grow to nodules, then to layers, and finally to a massive hydrate. Brooks et al. (1985) suggested that, since most Gulf of Mexico hydrates were found to be biogenic if disseminated and thermogenic if more massive, then the thermogenic hydrates may be more mature, in which a greater supply of gas was available through faults and diapirs. They note however, that hydrate formation is influenced by a number of other factors, among which is sediment texture, formation of authigenic carbonate rubble, and shallow faulting and fracturing of the sediments.

As a second mechanism for formation of marine hydrates, Wendlandt and Harrison (1988) hypothesized that massive hydrates (nodular, layered, and massive) might result from episodic melting-recrystallization induced by pressure (and related temperature) fluctuations arising from sea level changes. By analogy, pyrite occurrences also show textures ranging from disseminated through nodular to layered and massive, that have been attributed to recrystallization with and/or without additional sulfur being introduced into the system. Eight such periodic sea level changes of as much as 100m, illustrated in Figure 7-8, have resulted from continental glaciation in the last 700,000 years. The resulting long-range periodic changes of pressure and temperature could cause the melting and recrystallization of hydrates in marine sediments. Wendlandt and Harrison further suggested that

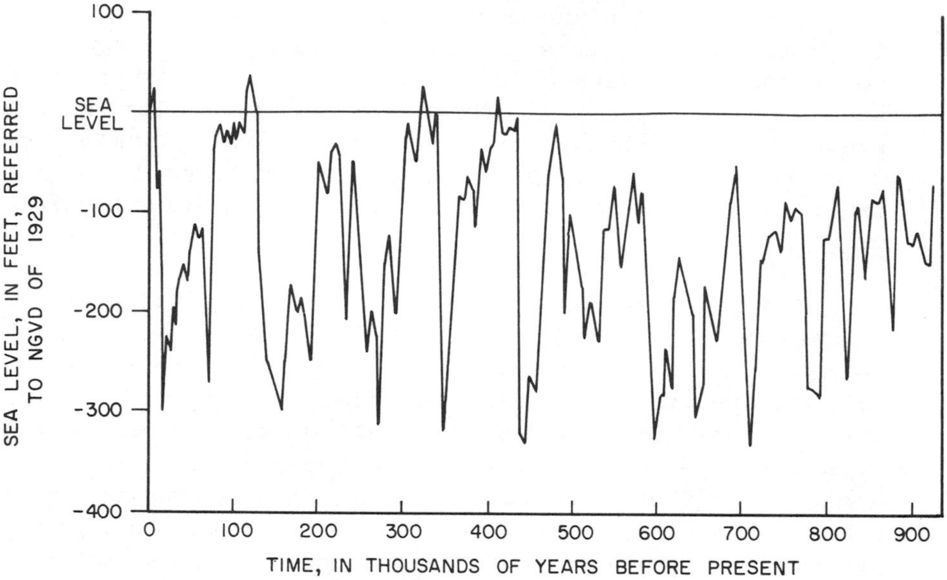


7-8 Changes in Sea Level Over the Last Million Years

glaciation could also cause the pressure and temperature to fluctuate for onshore hydrates as well. Experimental support for the recrystallization mechanism has been obtained in this laboratory by Yousif et al. (1989) who have successfully used such an annealing/recrystallization process as a means for growing hydrate masses in sediments.

### 7.2.3 Geochemical Indicators of Hydrates in Sediments

Regardless of the mechanism of hydrate formation, we recall from the Chapters 2 and 4 that, when the hydrates form they exclude ions of all sorts; the ions have effective radii of aqueous water molecule hydration (in the normal sense) which cause them both to be too large for the hydrate cages and to compete with clathrate hydrates for the available water molecules. So it is logical that hydrates formed in salt water would be depleted in chloride ions, while the water around them would be enriched. The chloride ion enrichment of the hydrate formation environment will act to depress the hydrate formation temperature, as discussed in Chapter 4. Evenros (1971) demonstrated the affect of salt on hydrate depression in sediments.

Many of the hydrate geologists and geochemists have chosen to ignore the hydrate temperature depression by salts, indicating that it was approximately offset by the temperature increase caused by hydrocarbons heavier than methane. As a consequence, the hydrate equilibrium condition is often taken as that of simple methane hydrate formation from pure water. There may be four separate errors introduced by such an approximation:

(1) While the salt water concentration is nominally only 3.5 wt%, the salt enrichment may be as high as the solubility limit (about 26 wt %) as the hydrates are formed and the salt excluded. The aqueous salt concentration depends upon several factors such as the amount of hydrate formed, salt diffusion and convection, etc. The experimental data of de Roo et al. (1983) for simple methane hydrates shows that the temperature depression limit can be as large as 13 K for a 24 wt% aqueous concentration of salt.

(2) A small amount of propane in the gas will cause a crystal structure change. As indicated in Chapter 5 (see Figure 5-9 and accompanying discussion) less than 1 mol% propane in the gas can change the structure from I to II and reduce the required pressure to stabilize hydrates to less than one-half that for pure methane. Put another way, thermogenic hydrates have so much propane that their temperature stability range at depth is greater than simple methane hydrates by several hundred meters. For biogenic hydrates the methane stability curve is not affected by the small amounts of other heavy hydrocarbons.

(3) There may be additional pressure required for hydrate equilibration in porous media due to capillary forces. Yousif et al. (1989) determined that hydrates could be stable at pressures as great as 70 kPa beyond the equilibrium value, when they were formed in porous media with a porosity of 0.2 (400 mD).

(4) Certain solids such as clays appear to stabilize hy-

drates to higher temperatures than in free liquids. In this laboratory Rueff and Sloan (1985) determined that silt from Blake Bahama Ridge cores had natural contaminants which affect the heat of dissociation. Cha et al. (1988) measured an elevation of hydrate stability temperature by as much as 3K using a montmorillonite commonly found in sediments.

While some of the above four effects on the phase equilibria conditions may offset each other, none should be discounted **a priori** to obtain the most accurate estimate of the stability region of a hydrate reservoir.

As a fifth physical geochemical effect, water fractionates and distributes $H_2{}^{18}O$ and $H_2{}^{16}O$ differently in hydrates. The $^{18}O$ isotope concentrates slightly in the hydrate cages, relative to that in water. Davidson et al. (1983) measured the hydrate $^{18}O/^{16}O$ enrichment factor, $\alpha$, defined as

$$\alpha \equiv \frac{[^{18}O]/[^{16}O]_{solid}}{[^{18}O]/[^{16}O]_{liquid}} \quad (7.3)$$

They found the value of $\alpha = 1.0026$ for hydrates to be similar to that for ice. As a consequence of hydrate formation, $^{18}O$ will be concentrated in the hydrate phase and depleted in the surrounding water. This isotopic fractionation can be measured, but the change is too small to affect the phase equilibrium, either due to concentrations in the hydrate or the surrounding liquid.

### 7.2.4 Summary of Geological and Geochemical Indicators

The above discussion of the hydrate mechanism enables a summary of the means to distinguish the presence of hydrates. These means may be both geological and geochemical, listed as follows:

1. For organic diagenesis to occur, the sedimentation rate should be rapid (on the order of 30 m/$10^6$yr) for the burial of organics. The total organic content of the

sediments should be greater than 0.5%. The residual methane concentration should be greater than 10 ml/l of wet sediment in order for hydrates to form.

2. The methane in hydrates which is formed from biogenic carbon will have $\delta^{13}C$ values from -55‰ to - 85‰, while those from thermogenic sources will have $\delta^{13}C$ values from -25‰ to -60‰.

3. Methane in hydrates which is formed from biogenic sources will have $[C_1/(C_2+C_3)]$ values greater than $10^3$, while the corresponding ratio for thermogenic methane will be less than $10^2$.

4. Two mechanisms have been hypothesized for hydrate formation in sediments: (a) growth from disseminated to massive deposits, and (b) recrystallization or annealing with continental glaciation resulting from pressure and temperature changes. There are also two theories for hydrate formation, either at the gas-liquid interface or from the liquid phase itself; laboratory evidence appears to favor the former.

5. The pressure-temperature phase equilibria envelope of hydrates may be significantly affected by the enrichment of salt concentration, heavier hydrocarbons, pore diameters, and makeup of the clays/sediments themselves.

6. Hydrates will be depleted in chloride ions and enriched in $^{18}O$, while the water environment of hydrate formation should be enriched in chloride ions and depleted in $^{18}O$.

### 7.2.5 Geophysical Indicators and Means of Detection

Where no hydrate cores are available as direct evidence, the indirect geophysical evidence for hydrate determination is bifurcated. In permafrost regions, below the permafrost layer well logs determine the presence of hydrates, whereas in the deep oceans acoustic reflections within the bottom sediments, called Bottom Simulating Reflectors (BSRs), are used. Although well logs have been obtained at some of the ocean hydrate sites, BSR measurements have not been obtained in permafrost regions; sonic measurements of permafrost do not differ substantially from those of hydrate.

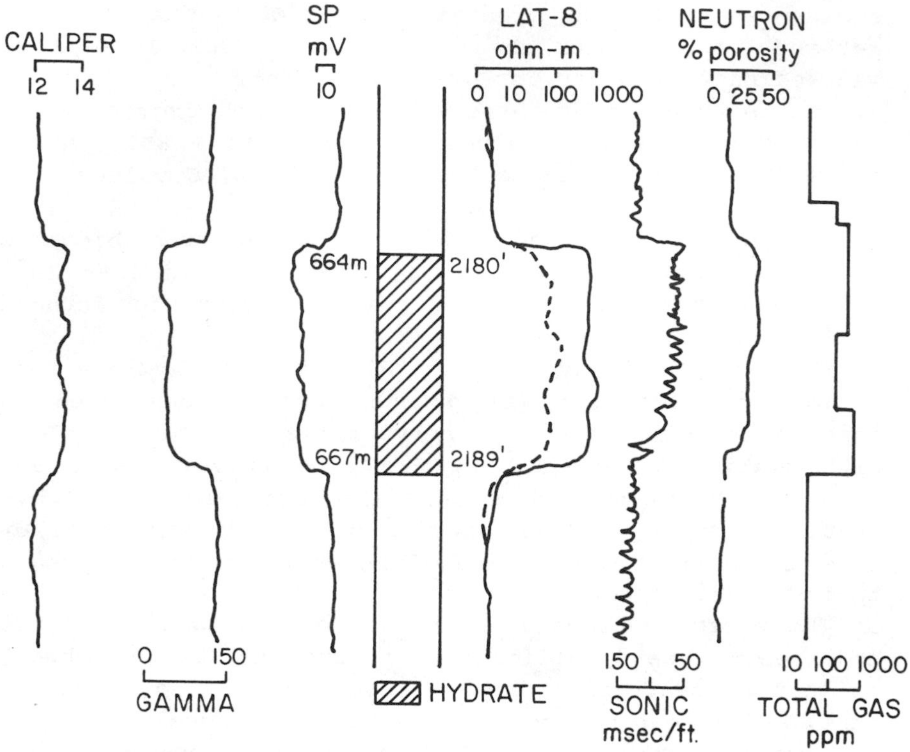


7-9 Typical Well Logs Responses to Hydrates in Permafrost (Reproduced, with permission, from (Collett, 1983))

Collett (1983) and Collett and Ehlig-Economides (1983) studied the logs of 125 wells in the Prudhoe Bay region, determining 102 hydrate occurrences in 32 wells. The analyses of other permafrost wells were based upon the logs of a known hydrate well, the Northwest Eileen State Well Number Two, in which Arco/Exxon recovered hydrate cores in 1972. Other studies on the effects of hydrates in wells and logs in the permafrost have been done by Weaver and Stewart (1982), Collett et al. (1984), Kamath et al. (1987), and Mathews (1986). In general, all studies indicated difficulty in distinguishing hydrates based on single logs, particularly in the permafrost region. To have some confidence in hydrate determination through logs, it is necessary to consider a

**Table 7-4. Summary of Well Log Responses to Hydrates**
From Collett (1983)

| Type of Log | Response to Hydrates |
|---|---|
| **Mud Log** | The dissociation of hydrates causes a significant increase of gas in the drilling mud, which is encountered at the top of the well. Cold or dense drilling fluids may suppress dissociation resulting in lower gas |
| **Dual Induction** | In the shallow penetration log, a smaller response is obtained relative to a free gas zone due to hydrate dissociation. The deep induction log shows high resistivity and mimics an ice-bearing reservoir |
| **Spontaneous Potential** | Compared to a free gas bearing zone, the spontaneous potential log is less negative, but similar to that of ice |
| **Caliper Log** | An oversized drill hole is indicated by hydrate dissociation. This may also occur with ice in the permafrost |
| **Acoustic Transit Time** | The acoustic wave time decreases relative to either water or free gas; however the acoustic transit time is like ice bearing sediments |
| **Neutron Porosity** | An increase is observed relative to ice and relative to the decrease in a free gas zone; see the cross plot versus acoustic transit time in Figure 7-11 |
| **Density** | The density decrease apparent in hydrates can be distinguished from water but not from ice |
| **Drilling Rate** | The drilling rate also decreases in the hydrate region relative to that in a fluid-saturated sediment, but not significantly different from that of ice |

suite of logs. Figure 7-9 shows that the simultaneous responses of logs corroborate the interpretations of hydrates. Table 7-4 provides a summary of the individual responses of well logs to the presence of hydrates. The drilling mud log is the most responsive to hydrates, but its response may not be very different from a log for free gas.

Figure 7-10 shows a particularly helpful crossplot of the deep induction resistivity log and the sonic transit time log for seventeen intervals below the base of the permafrost at the Northwest Eileen State Well Number Two. Hydrate saturation units fall into a region of relatively higher resistivity and faster transit times whereas free gas saturated units fall into an area of lower resistivity and slower transit times. This plot is useful only below the base of the permafrost, because of the similarity in resistivity and transit time velocities in hydrates and in permafrost. Later Collett et al. (1988) extended the work to determine that a cross plot of neutron porosity and acoustic transit time can be used to discern hydrates above the permafrost as shown in Figure 7-11 for the Kuparuk River Unit 1B-1 production well. In principle, it should also be possible to use a thermal conductivity measurement, such as shown for laboratory sediments by Asher (1987), as a second means of determining hydrates above the base of the permafrost.

In addition to identifying the presence of hydrates, well logs can be used to determine such difficult properties as reservoir porosity and degree of hydrate saturation. To determine the porosities of reservoir rocks with gas hydrates, Collett et al.(1988) performed neutron porosity calculations within the identified gas hydrate region of seven wells. The porosity values, ranging from 22 to 48 per cent, are plotted in Figure 7-12. A series of Pickett crossplots such as Figure 7-13 were made for twenty seven hydrate-bearing wells. The degree of hydrate saturation ranged from 70 to 95 per cent, with an average value of 85% for the North Slope wells.

For hydrates in ocean sediments, the technology for detecting the Bottom Simulating Reflector (BSR), was ba-

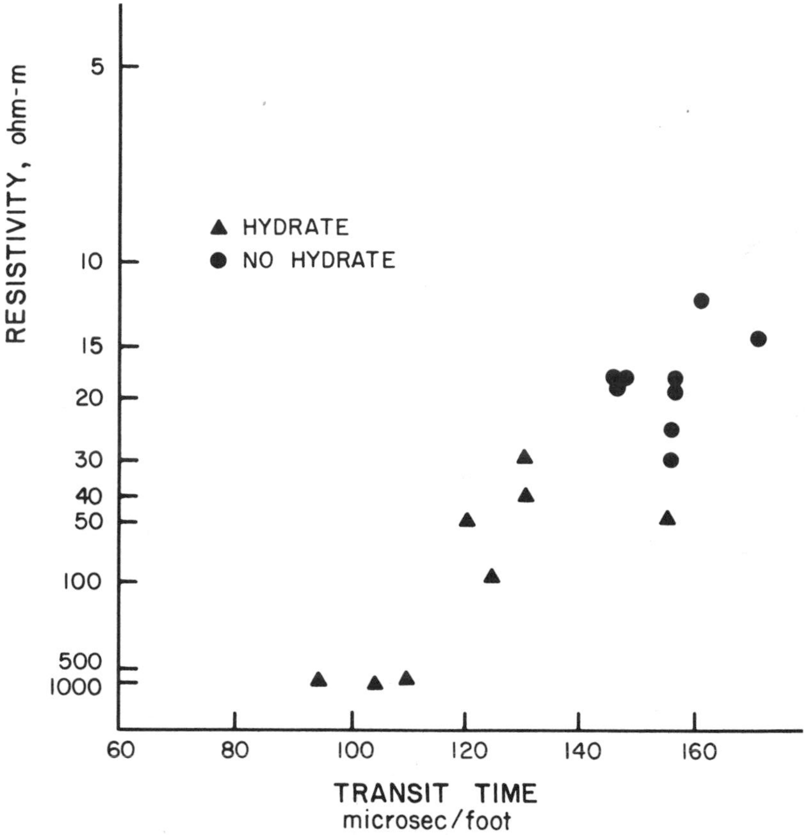


7-10 Resistivity-Transit Time Crossplot from Northwest Eileen State Well Number Two (Reproduced, with permission, from (Collett, 1983))

sically determined in 1953 with the development of a precision ocean depth recorder (Hamblin, 1985 p.11). In this technique a sonic wave is bounced off the ocean floor and the time recorded for the return of the reflected wave to the source. Velocity contrasts beneath the ocean floor mark a change in material density, such as would be obtained by hydrate-filled sediments overlying a gas. BSRs related to hydrates are normally taken as indications of the base of the hydrates, marked by a sharp decrease in sonic velocity. Figure 7-14a shows the most famous BSR related to hydrates, in the Blake-Bahama

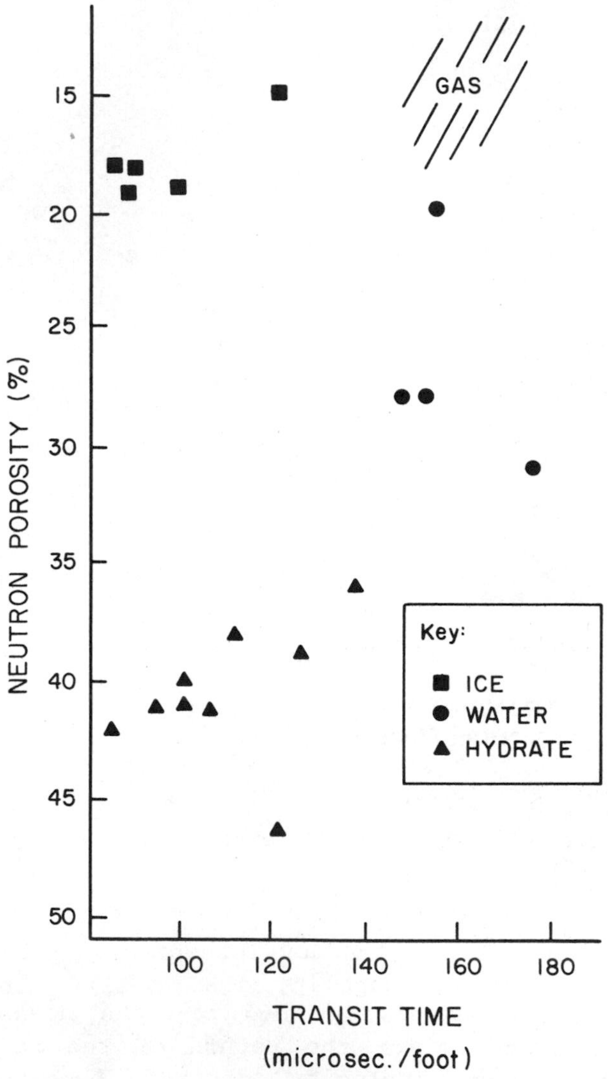


7-11 Neutron Porosity-Transit Time Crossplot to Distinguish Ice, Water, and Hydrate in Permafrost (Reproduced, courtesy of U.S. Dept. of Energy, (Collett et al., 1988))

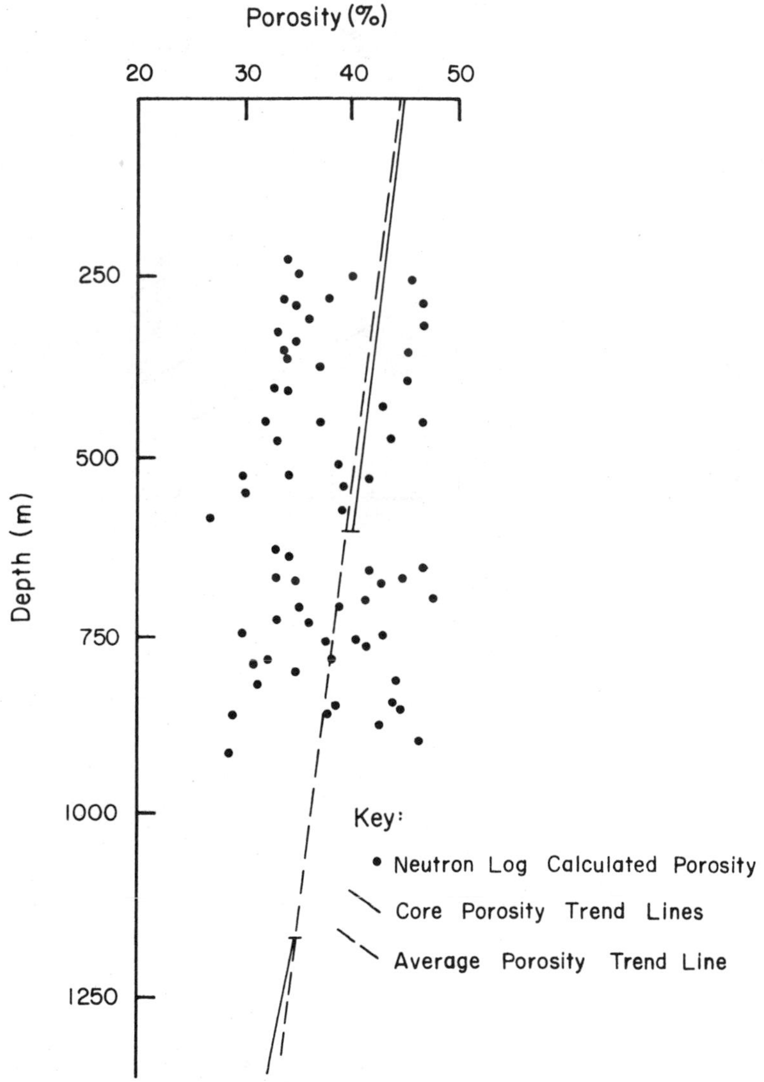


7-12 Neutron Porosity in Hydrate Intervals of Seven Wells (Reproduced, courtesy of U.S. Dept. of Energy, (Collett et al., 1988))

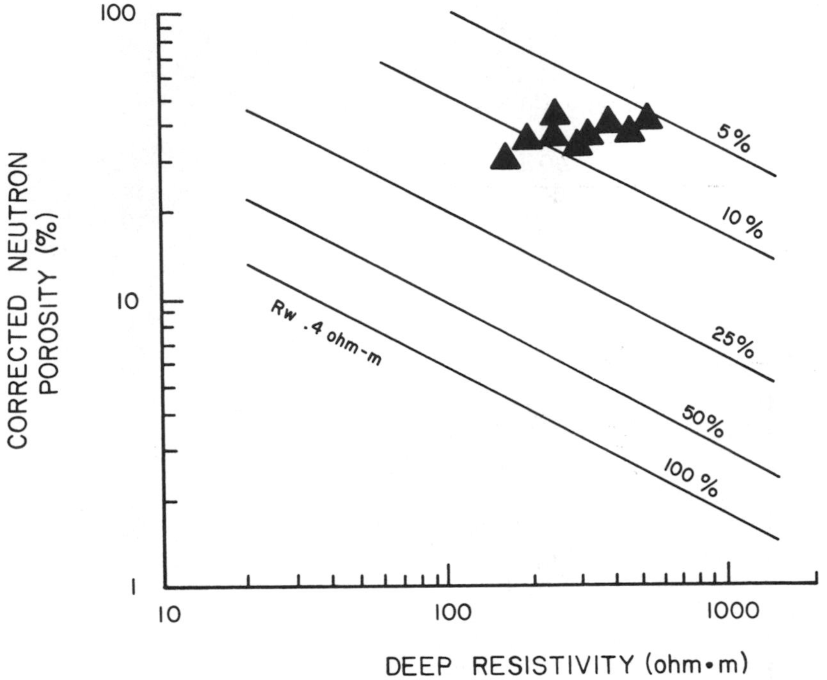


7-13 Pickett Crossplot for Hydrates in Permafrost (Reproduced, courtesy of U.S. Dept. of Energy, (Collett et al., 1988))

Ridge, Figure 7-14b diagrams velocity of the wave in the sediment (at the arrow in 7-14a) as a function of depth. Table 7-5 shows the areal extent of many of the BSRs cited in the hydrate locations of Table 7-2.

While Table 7-5 suggests that hydrates are commonly found in the oceans, some care must be exercised, because BSRs are not reliable as sole indicators of hydrates. For example, Finley and Krason (1986a) indicate Sites 490, 498, 565, and 570 on DSDP Leg 84 in the Middle America Trench where hydrates were recovered with no BSR's present; conversely BSRs existed beneath Sites 496 and 569, yet no hydrates were recovered by coring to within 200 m (vertical) of the BSR.

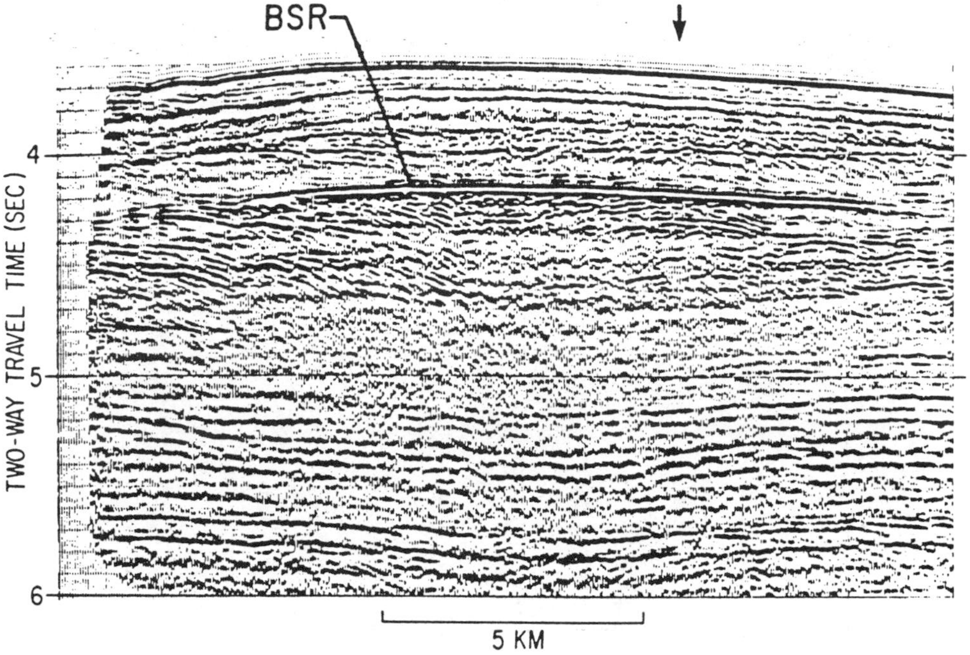


7-14a Bottom Simulating Reflector (BSR) for Hydrate Deposit In Blake-Bahama Ridge
(Reproduced, courtesy of U.S. Geological Survey, (Dillon and Paull, 1983))

### 7.2.6 Two Case Studies of Geochemistry and Geophysics of Hydrates

In order to bring together the concepts of the chapter thus far, it is useful to consider two case studies representing the most prominent Western Hemisphere corings of hydrates, from DSDP Site 570, Leg 84 for marine hydrates, and from Northwest Eileen State Well Number Two for permafrost hydrates. Because each site is the best exemplar of the direct evidence for hydrates, the indirect evidence gathered at each site is validated to a greater extent.

#### 7.2.6.a. Hydrate Evidence at Site 570 DSDP Leg 84

The largest hydrates samples recovered to date from an ocean environment have been those from the Middle

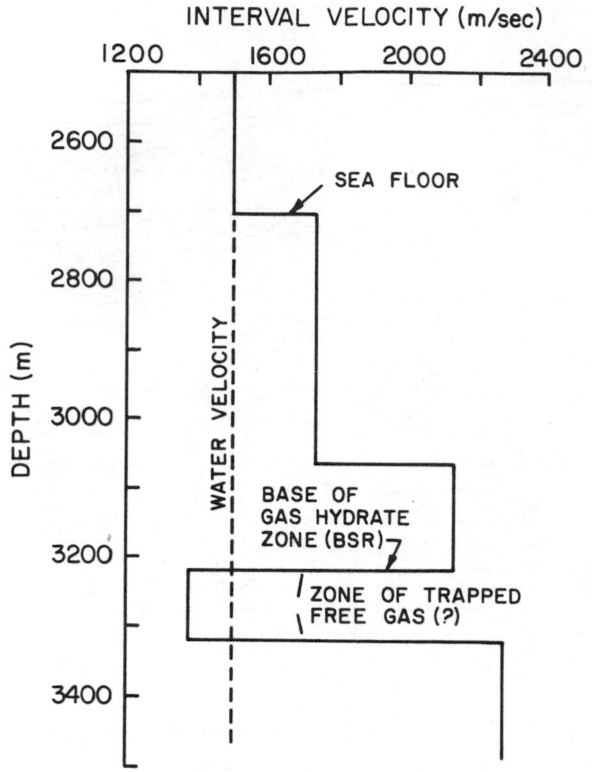


7-14b Velocity Analysis at Location of Arrow in BSR of Figure 7-14a
(Reproduced, courtesy of U.S. Geological Survey, (Dillon and Paull, 1983))

America Trench, DSDP Leg 84, Site 570 ($13^{0}17.1'$ latitude N, $91^{0}23.6'$longitude W), in eight core sections from the site. In section 570-27-1 a 1.05m long white massive gas hydrate was recovered in a core between 249.1m and 258.8m sub-bottom depth. Logging tools indicated that the massive hydrate was actually 3 to 4m thick and that there were hydrates in a 15m interval from 1958.5m to 1975.5m (water plus sediment depth). A portion of the massive hydrate was returned to land and subdivided in this laboratory for onshore experiments (Sloan, 1985); a photograph, courtesy of the U.S. Department of Energy, of a section of that hydrate is shown at lower right in Figure 7-6.

**Table 7-5 Gas Hydrate BSR Extent**

| Number in Figure 7-1 | Location | Extent & Reference Area/Depth |
|---|---|---|
| | | Oceanic Hydrates |
| 1 | Pacific Ocean off Panama | 4,500 $km^2$ poor BSR data<br>250-550 subbottom depth<br>Krason & Ciesnik (1986a) |
| 2 &3 | MAT off Costa Rica and Nicaragua | 14,000 $km^2$ variable quality BSR data<br>600 m subbottom depth<br>Finley & Krason (1986a) |
| 4 | MAT off Guatemala | 23,000 $km^2$ variable quality BSR data<br>0-450 $km^2$ depth<br>Finley & Krason (1986a) |
| 5 | MAT off Mexico | 19,000 $km^2$ variable quality BSR data<br>0-640 m subbottom depth<br>Finley & Krason (1986a) |
| 6 | Eel River Basin off California | 3,000 $km^2$<br>200 m subbottom depth<br>Krason & Ciesnik (1986b) |
| 7 | E. Aleutian Trench off Alaska | Not Available |
| 8 | Beringian Margin off Alaska | Not Available |
| 9 | Middle Aleutian Trench off Alaska | Not Available |
| 10 | Nankai Trough off Japan | Not Available |
| 12 | Hikurangi Trough off New Zealand | Not available |
| 13 | Wilkes Land Margin off Antartica | Not Available |
| 15 | Peru-Chile Trench off Peru | Areal BSR Extent not Available<br>350-570m subbottom depth |
| 16 | Barbados Ridge Complex off Barbados | Not available |

**Table 7-5 (continued) Gas Hydrate BSR Extent**

| | | |
|---|---|---|
| 17 | Colombia Basin off Panama and Colombia | 30,000 $km^2$<br>60-200 m sobbottom depth<br>Finley & Krason (1986b) |
| 18 | W. Gulf of Mexico off Mexico | 8,000 $km^2$ good BSR data<br>100-600m subbottom depth<br>Krason et al.(1985) |
| 20 | Blake Outer Ridge off SE USA | 31,000 $km^2$ good BSR data<br>22,000 $km^2$ acceptable BSR<br>245m subbottom depth at Site(533); Krason & Ridley(1985a) |
| 21 | Baltimore Canyon off E. USA | 12,600 $km^2$ good BSR data<br>19,150 $km^2$ acceptable BSR<br>450-600 m subbottom depth<br>Krason&Ridley(1985b) |
| 22 | Labrador Shelf off Newfoundland | $<100km^2$ poor BSR data<br>20-600 m subbottom depth<br>Krason&Rudloff(1985) |
| 23 | Beaufort Sea off Alask | areal extent not available<br>100-800m subbottom depth<br>Grantz et al.(1976) |
| 26 | Continent Slope off W. Norway | Not Available |
| 28 | Makran Margin, Gulf of Oman | undetermined area<br>350-700 m subbottom depth<br>White(1979) |

Shipboard scientists Kvenvolden and McDonald (1985) presented data such as found in Table 7-6 to describe the hydrates, depth, and gas compositions for the hydrate cores found in Site 570. The site was drilled in 1718m of water; the entire sub-bottom depth (402m) of the hole was within 300m above the lower end of the zone of gas hydrate stability such as indicated in Figure 7-2b. Other cores retrieved around the massive hydrate showed

**Table 7-6**
**Hydrates and Gases from DSDP Leg 84 Site 570**
from Kvenvolden and McDonald, 1985

| Site-Section | Subbottom depth,m | Comment | $C_1$ % | $C_2$ ppm | $C_3$ ppm | $i\text{-}C_4$ ppm | $C_5$(tot) ppm |
|---|---|---|---|---|---|---|---|
| 570-21-1 | 192 | Ash laminated with hydrate at top of core | 44 | 390 | 0.9 | 0.7 | |
| 570-26-5 | 246 | White hydrate in fractures of mudstone | No Gas Analysis | | | | |
| 570-27-1 | 249 | Massive | 74 | 1200 | 31 | 10 | 9.6 |
| | | Hydrate, 1.05m | 58 | 2600 | 79 | 37 | 49.5 |
| | | | 69 | 1600 | 48 | 23 | 26 |
| | | | 74 | 1200 | 31 | 10 | 9.6 |
| 570-28 | 259-268 | White hydrate | 75 | 890 | 11 | 2 | 6.06 |
| | | in fractures of mudstone | 75 | 1200 | 14 | 3 | 9.16 |
| 570-29-3 | 273 | White hydrate in fractures of mudstone | | No Gas Analysis | | | |
| 570-32-4 | 303 | Hydrate with ash | | No Gas Analysis | | | |
| 570-36-1 | 338 | Hydrate with sand | | No Gas Analysis | | | |

other hydrates occurred above and below, in fractures of mudstone, and in laminated ash, somewhat as Figure 7-7 would suggest, were the diagram viewed in the vertical direction. Thus there were three types of hydrates at the site i.e. massive, in fractures, and laminated.

Site 570 had the largest amount of total organic carbon available for methanogenesis of all the sites on Leg 84, ranging between 1.1 and 3.2%. The organic carbon averaged over 1.5% and the organic carbon in the hydrated intervals had a mean TOC value of 2.6%. Unfortunately, the resulting gas composition data in Table 7-6 are somewhat sparse for cores other than the preserved massive core. Analyses were limited because the shipboard party did not expect hydrates at the site due to the absence of a BSR. The methane to ethane ratios were in the 400-450 range from 246m to 274m in the gas hydrate zone, whereupon it increased to 1,900 at 317 m. Recall that these ratios in the hydrate zone are indeterminate relative to the heuristics for thermogenic (<100) and biogenic (>1000) gas. In cores immediately above and below the massive hydrate, the propane content also increased over 40 times the average background value in the hydrate region, although the highest propane concentration found was small (400 ppm) in any core.

Table 7-6 indicates that the hydrates at the massive and fractured intervals contained more heavy ($C_3^+$) hydrocarbons than the surrounding hydrated cores; however Kvenvolden (1988b) noted that the comparison with only one sample is insufficient evidence for generalization. It is interesting to note that such smaller concentrations of larger hydrocarbons are probably insufficient to cause the conversion of the hydrate structure from I to II, according to the methods of Chapter 5. One would expect, for example, much higher propane concentrations in structure II hydrate, on the order of 10-50%, if the hydrates were derived from a thermogenic gas which had as little as 1% propane.

Curiously, Finley and Krason (1986a) noted that in the hydrated cores themselves the propane values were only a fraction of the concentrations in the surrounding sediment cores; in the massive and fracture hydrates, the propane contents are only 20-40% and 5-8%, respectively, of the surrounding cores. The content of butanes and pentanes also showed this same trend of decreasing concentration in the hydrates relative to the surrounding non-hydrated sediments.

The carbon isotope data from Site 570 are presented in Figure 7-15. The $\delta^{13}C$ increases from -80‰ to -55‰, normal biogenic values, until core 27 at which depth it increases abruptly to the -40 to -44 ‰ range, which are typical thermogenic values. The gas present is so slightly diluted with heavier components, that biogenic gas is suggested to have been produced from anomalously heavy carbonates (Claypool et al., 1985, Jeffrey et al., 1985, Kvenvolden and McDonald, 1985). Kvenvolden and McDonald also note that the heavier components in gas from the serpentite layer below the hydrate indicated a thermogenic gas source and that some structure II gas hydrates may have been present. Von Huene et al. (1985) suggested that the isotopic carbon break in Figure 7-15 corresponds to a geologic unconformity of $0.5 \times 10^6$ years, as well as the zone of massive hydrate formation.

In Figure 7-15, the parallel lines for carbon dioxide, total carbonate, and methane, indicate the biogenic gas mechanism previously suggested, wherein the carbonates serve as the source for methane. Finley and Krason (1986a) do suggest an anomaly, in that the difference between the $\delta^{13}C$ values for $CO_2$ and $CH_4$ does deviate substantially from the normal value of 70‰ in the massive hydrate region; those authors take this decrease as further evidence that the massive hydrate may be of different origin than the other biogenic gas generated in place at Site 570.

Fewer pore water samples are available from Site 570, because hydrates were not expected at the site. The available samples tend to confirm the decrease in chlorinity in a hydrate core. There were no $\delta^{18}O$ water samples obtained at depths greater than 183m, to encompass the region of hydrate occurrence. Chlorinity samples were obtained however, and the results are shown in Figure 7-16. A slight decrease is shown to a depth of 234m. The three deeper samples 18.75‰ (234m), 14.3‰ (280 m), and 8.95 ‰ (337m), corresponding to the decrease in chlorinity associated with a dissociated hydrate core. For the dissociated massive gas hydrate itself, Brooks et al. (1985) reported a mean $\delta^{18}O$ value of 2.3‰; Claypool et al. (1985) reported 3.72‰, and Barraclough (in Sloan, 1985) reported 2.0±0.2‰.

A suite of ten well logs were obtained at Site 570 by Mathews and von Huene (1985) which define both the massive hydrate and the thin dolomite layer beneath the massive hydrate. Taken together, the hydrate zone was

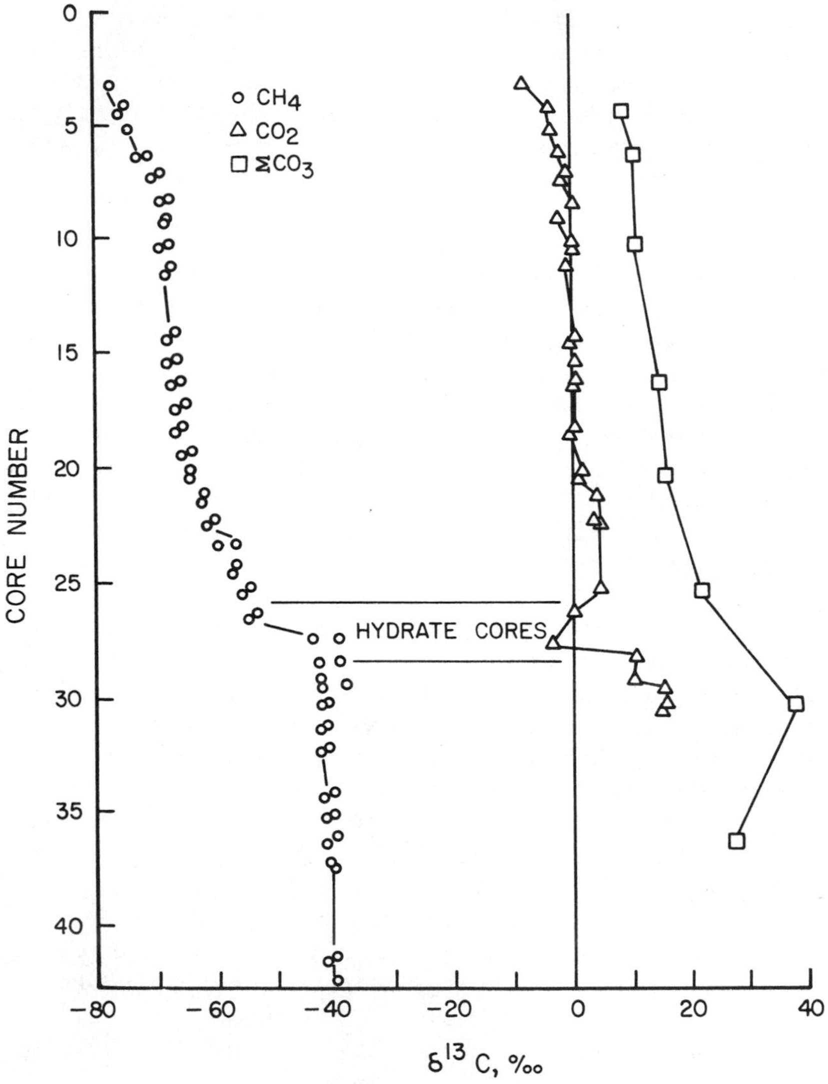


7-15 Carbon Isotope Content of Methane, Carbon Dioxide, and Total Carbonate From DSDP Site 570 Cores (Reproduced, courtesy of U.S. Dept. of Energy, (Finley and Krason, 1986a))

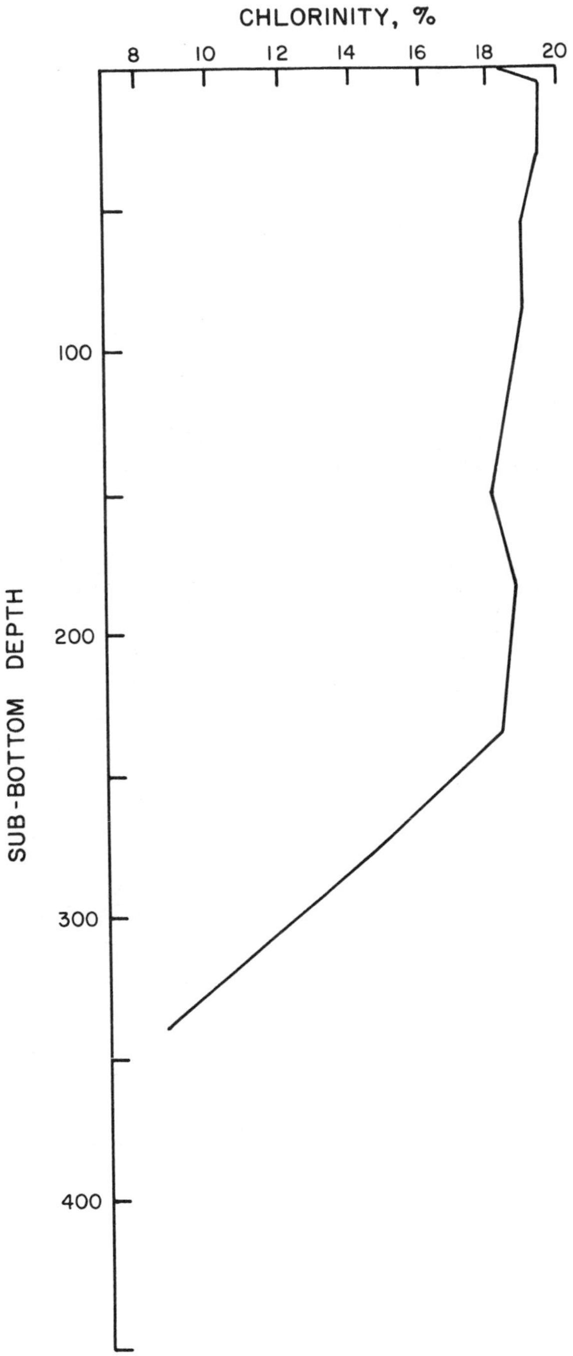


7-16 Chlorinity Values at DSDP Site 570
(Reproduced, courtesy of U.S. Govt., (Kvenvolden and McDonald, 1985))

clearly defined by the low apparent density (ca. 1.05 $g/cm^3$), the high resistivities (155 ohm-m), the high velocity (ca. 3.6 km/s), the high neutron porosity (ca. 67%) and the low gamma-ray response (15-25 API units). Among the suite of logs, the responses of the resistivity, velocity, and density logs are most significant; those responses as well as the calculated acoustic impedance (density times velocity) are shown in Figure 7-17. Particularly noteworthy are the responses of the logs to the massive hydrate zone and to the dolomite layer below that zone at 1975m. Table 7-7 from Mathews and von Huene describes the log responses to both the methane hydrate zone in Site 570 and the massive methane hydrate within that zone.

Figure 7-18 summarizes the lithostratigraphy of Site 570. Volpe et al. (1985) presented seismic evidence such as Figure 7-19, to show unconformities and suggest faulting in the immediate area of Site 570. Given the evidence, Mathews and von Huene (1985) first suggested a mechanism for the formation of the massive hydrate formation, namely that the massive hydrate was formed in a fault and that free gas migration hydrated as it filled the void left during fault movement.

Finley and Krason (1986a) indicated that the discontinuity in the $\delta^{13}C$ plot in Figure 7-15 could be explained by the removal of 140m of section, such as might be accomplished by faulting. If a subbottom section between 200m and 340m were removed and the remaining sections rejoined, one would obtain a close representation of the isotopic $\delta^{13}C$ profile. They also suggest that a fault would enable mobility for hydrocarbons, such as the high values of propane, butanes, and pentanes found in the hydrated core regions. Finley and Krason note that this one Site may be unique in the formation of a hydrate in a fault zone.

DSDP Leg 84 represents the best documented Western Hemisphere hydrate sites. Kvenvolden and McDonald (1985) pointed out that the results of Legs 66, 67, and 84 show that gas hydrates are common in landward slope sediments of the Middle America Trench from Mexico to Costa Rica.

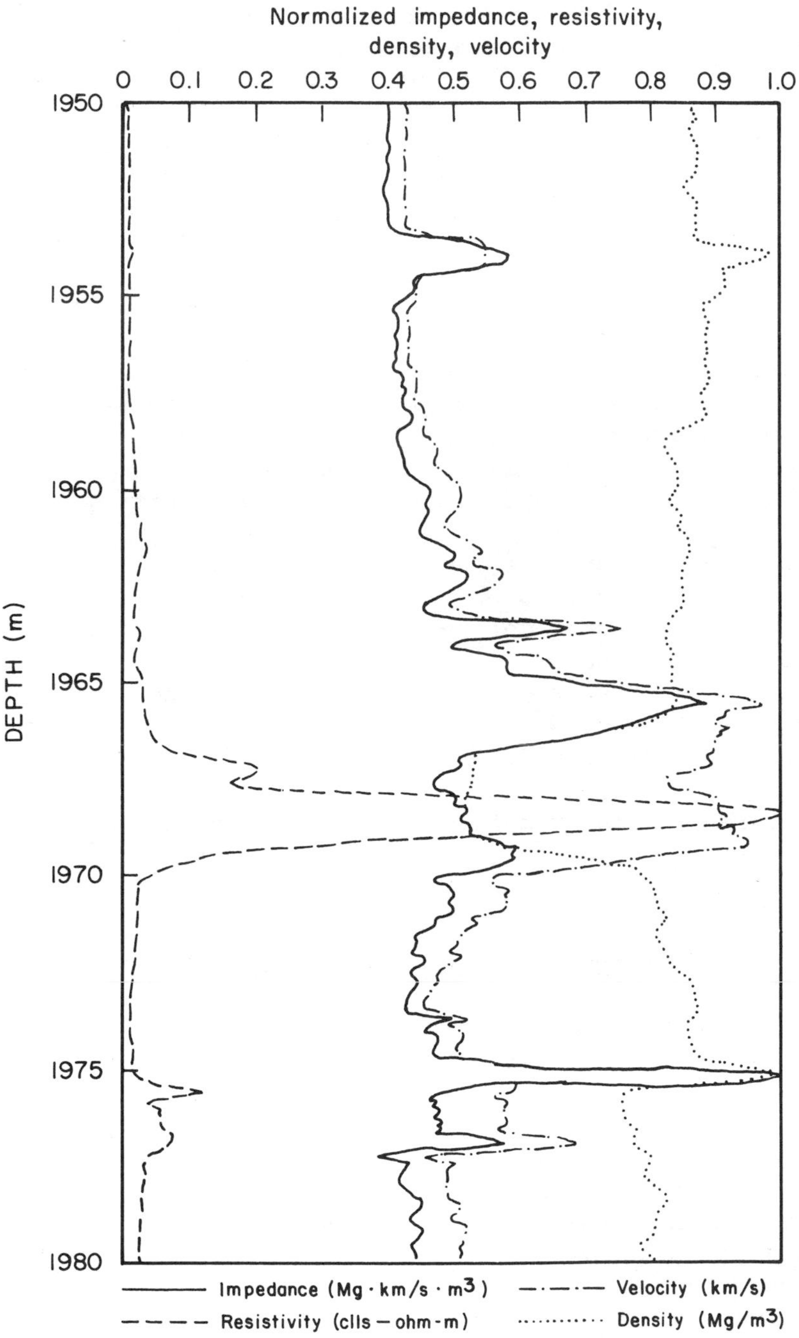


7-17 Some Well Logs at DSDP Site 570
(Reproduced, courtesy of U.S. Govt., (Mathews and von Huene, 1985))

**Table 7-7 Log Responses to Hydrate at DSDP Leg 84 Site 570**
(from Mathews and von Huene)

| Log Data | Hydrate Layer (1958.5-1973.5m) | Massive Hydrate Zone (1965.4-1969.4m) |
|---|---|---|
| p-wave sonic | defines layer | defines zone 4 m thick |
| Density | defines layer | defines zone 2.7m thick |
| Resistivity | both LLs and LLd define layer | both LLs and LLd define zone - 0.6m deeper |
| Full-WaveSonic | wiggle trace defines layer variable density defines | wiggle trace defines layer variable density defines |
| Gamma-Ray | doesn't define layer | defines zone 0.5m deeper |
| Neutron Porosity | doesn't define layer | defines zone 0.5m deeper |
| Spontan. Potent. | doesn't define layer | doesn't define zone |
| Caliper | doesn't define layer | doesn't define zone |
| Temperature | doesn't define layer | doesn't define zone |

### 7.2.6.b. Northwest Eileen State Well Number Two

Katz (1971) was the first to suggest that hydrates could be stable up to depths of 1,200 meters on the North Slope of Alaska. Of the wells drilled in the permafrost regions of the western hemisphere, the only hydrate core reported was at the Northwest Eileen State Number Two Well at the Prudhoe Bay field on March 15, 1972. The well was drilled as a joint venture by Arco/Exxon, and only a limited amount of non-proprietary data is available (Kvenvolden and McMenamin (1980) and Collett (1983)). The analysis of this well has seminal importance however. The logs from this well were used as the basis for analyses of other wells in the region, which enabled a hydrated reserve estimate equivalent to one-third of the conventional gas reserve of Prudhoe Bay.

For permafrost hydrate formation Collett et al.(1988) indicated that the most important reservoir and fluid variables are geothermal gradients, gas chemistries, pore-fluid salinities, pore pressures, and reservoir rock grain sizes. Of these variables the geothermal gradient and gas chemistry appear to be eminent.

The Northwest Eileen State Well Number Two was drilled with cooled drilling fluid to prevent thawing of the permafrost and/or hydrate, and the hydrate sample was recovered in a pressure core barrel. The well was drilled in permafrost which had a base at 532m, in which the thermal gradient was 1.9 K/100m. Collett et al.(1988) indicated that the mean annual temperature at the well is 262K and the base of the permafrost is 272K. The geothermal gradient below the base of the permafrost was calculated to be 3.2K/100m, so that a plot of the hydrate phase stability region, similar to Figure 7-2a, could be determined as Figure 7-20. Based on the analysis of many such curves, Collett et al. (1988) concluded that simple methane hydrates cannot be present in North Slope sediments when the permafrost thickness is less than 274m.

The hydrate cores were recovered from 666m, that is, 134m below the base of the permafrost. The hydrate cores were not removed from the pressure core barrel for visual inspection, but the pressure from the core was allowed to replenish itself several times at a constant temperature of 274K, indicating the presence of hydrates rather than either a soluble or free gas phase. Barker (cited in Collett, et al. 1988) reported the gas compositions given in Table 7-8, in the hydrate intervals. Neither the gas nor the water from the dissociated cores were subjected to isotopic analyses. The gas is believed to be a mixture of both biogenic and thermogenic sources, with biogenic sources dominating (Kvenvolden, 1988b).

Salinity measurements were obtained in the Northwest Eileen State Number Two well in the hydrate region, rang-

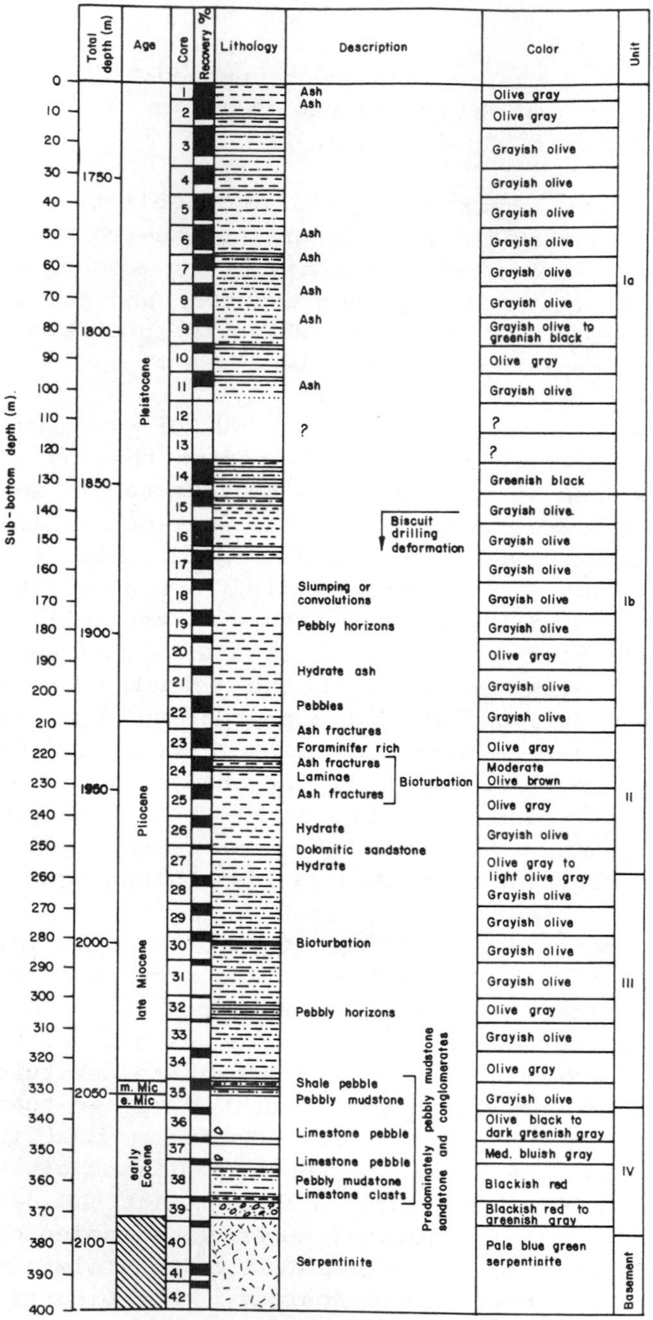


7-18 Lithostratigraphic Summary of DSDP Site 570 (Reproduced, courtesy of U.S. Govt., (von Huene et al., 1985))

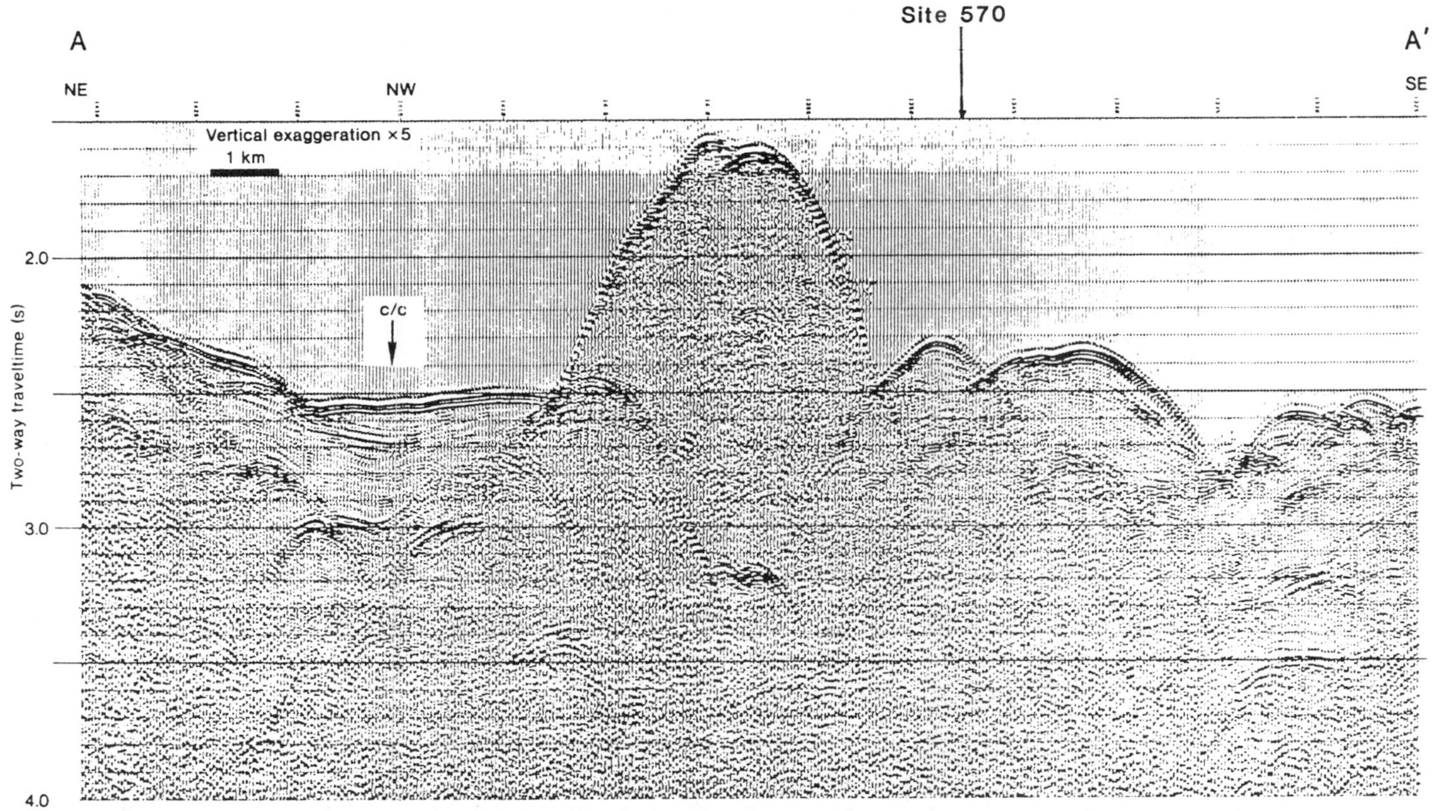


Note large topographic high composed of serpentinite adjacent to Site 570

7-19 Seismic Record of Continental Slope Near DSDP Site 570
(Reproduced, courtesy of U.S. Govt., (Volpe et al., 1985))

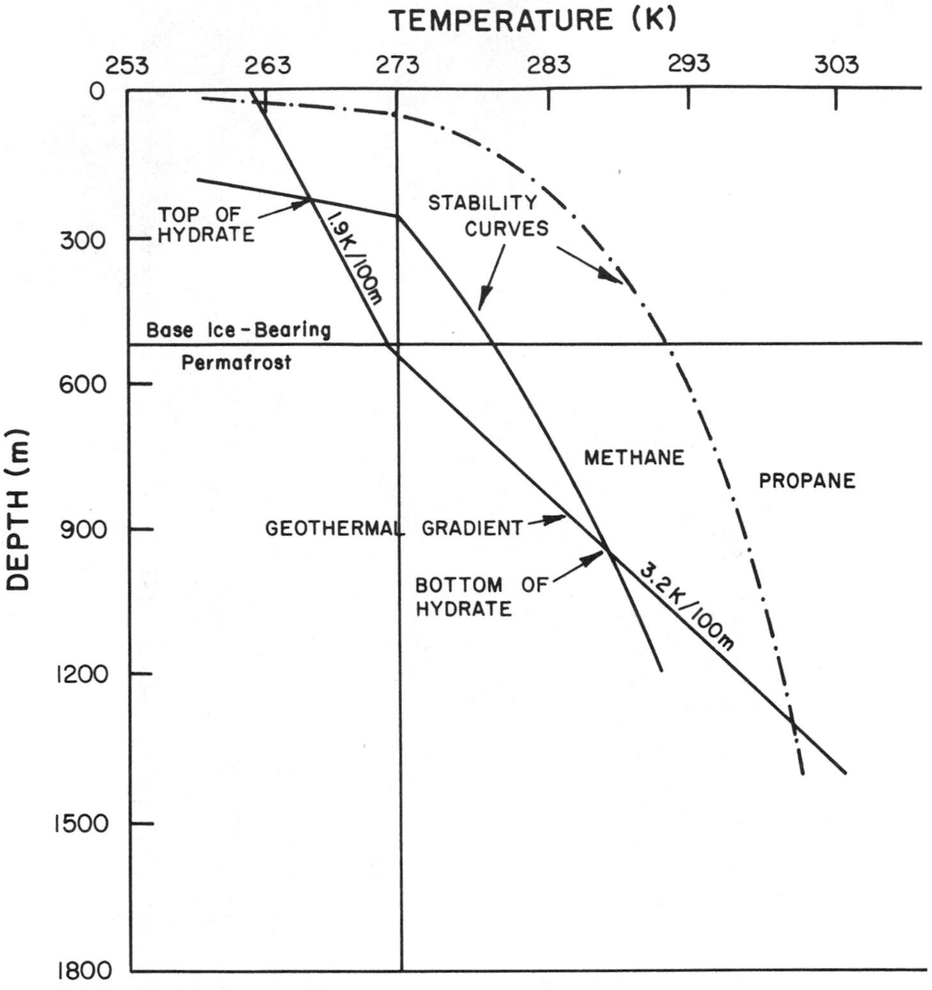


7-20 Hydrate Stability Envelope for Northwest Eileen State Well Number Two
(Reproduced, courtesy of U.S. Dept. of Energy, (Collett et al., 1988))

ing from 0.9 to 3.5 ‰ salinity, with an average of 1.82 ‰ for an average temperature depression of 0.11K. The hydrostatic pore pressure was assumed to be 9.795 kPa/m. The effect of the particle grain size on hydrate stability was assumed to be negligible.

**Table 7-8 Gas Analyses from Cores and Flow Tests at Northwest Eileen State Well Number Two**
(from Collett et al. 1988)

| Sample Type | Depth m | $CH_4$ % | $C_2H_6$ % | $C_3H_8$ % | $CO_2$ % | $O_2$ % | $N_2$ % |
|---|---|---|---|---|---|---|---|
| Core | 577-580 | 84.04 | Trace | -- | -- | 1.26 | 14.70 |
| Core | 577-580 | 97.60 | Trace | -- | Trace | 0.14 | 2.26 |
| Core | 577-580 | 98.76 | Trace | -- | Trace | 0.15 | 1.09 |
| Core | 577-580 | 97.59 | Trace | -- | -- | 0.40 | 2.01 |
| Core | 664-667 | 86.95 | Trace | -- | -- | 0.52 | 12.53 |
| Core | 664-667 | 99.14 | Trace | -- | -- | 0.02 | 0.84 |
| Core | 664-667 | 99.17 | Trace | -- | -- | 0.03 | 0.80 |
| Core | 664-667 | 98.49 | Trace | -- | -- | 0.05 | 1.46 |
| Flow Test | 663-671 | 92.79 | 0.02 | Trace | Trace | -- | 7.19 |
| Flow Test | 663-671 | 92.76 | 0.01 | Trace | Trace | -- | 7.23 |

Collett et al. (1988) characterized the geology in which the hydrates were found in the Prudhoe Bay area as follows:

> "The rock package overlying the Ugnu sands is characterized by numerous coarsening-upward and fining-upward sandstone and siltstaone sequences. ...The coarsening-upward units may be distributary mouth bar deposits and it is likely the fining-upward sequences are channel and overbank deposits. Overlying this nonmarine deltaic package is a marine siltstone and mudstone (which forms an upper impermeable boundary of the hydrate region)...In most of the Prudhoe Bay-Kuparuk River area, the upper boundary of the siltstone an mudstone unit is an erosional unconformity ...remnant of a northeasterly-migrating deltaic channel sequence which has cut into the underlying marine sequence."

All of the Prudhoe Bay hydrates appear to occur within relatively permeable sandstone rocks, with indi-

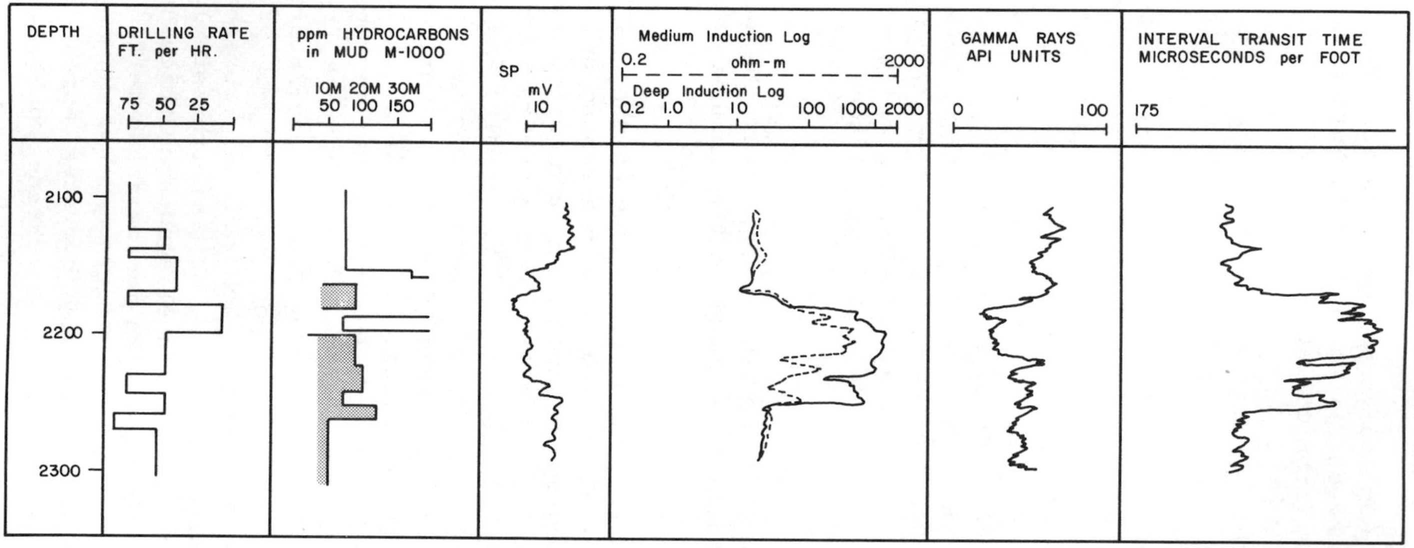


7-21 Logs in Hydrate Region from Northwest Eileen State Well Number Two
(Reproduced, with permission, from (Collett, 1983))

vidual occurrence ranging in thickness from 2m to 28m. The hydrate occurrences are capped by a relatively impermeable siltstone, deposited during a basin-wide marine transgression in the Eocene. As indicated by Collett et al. (1988) the most likely mechanism for hydrate formation in the Northwest Eileen State Well Number Two is migration of thermogenic gas from reservoirs of Prudhoe Bay upward along the Eileen fault into the overlying sediments which contained **in situ** biogenic gas. The gas was either converted to hydrate upon entering the stability zone, or a free gas collected and concentrated in structural traps before converting to hydrates.

The logs obtained from the well are indicated in Figure 7-21. These logs, together with the cross plot of Figure 7-10 form the bases for the interpretation of all of the well logs in the North Slope. The responses of the logs are specific exemplars of the more general responses discussed in Section 7.2.5. However, because hydrates cannot be conclusively identified with well logs, Collett et al. (1988) determined four hydrate criteria both above and below the permafrost based upon their study of the Northwest Eileen State Well Number Two and others:

1. The resistivity should be high (by a factor of approximate 50) relative to water,
2. The acoustic transit times should be short (an increase of approximately 40 $\mu$s per foot) relative to water,
3. Significant release of free-gas during drilling (as indicated on the gas chromatograph of the mud log)
4. A unit suspected to be gas hydrate bearing must be laterally continuous in two or more wells, in areas with high density of wells.

## 7.3 The Recovery of Gas from Hydrated Reservoirs

With the recognized reservoirs of hydrates in continental margins and in permafrost regions, the question arises about the feasibility and economics of recovery of gas from these substances. Here the question is divided into three components: a) mathematical models and laboratory data for the recovery of gas from hydrates, b) a field case study of the recovery of gas from hydrates, and c) the economics of recovery.

### 7.3.1 Methods and Models

There are many proposed methods for gas recovery from **in situ** hydrates. The largest amount of field experience with recovery of gas from hydrates has been obtained in the U.S.S.R.. Makogon (1974) and Krason and Ciesnik (1985) provide several examples of recovery schemes including injection of hot water, reduction of the reservoir pressure, inhibitor injection, dissolution of gas released from hydrates, and the injection of non-hydrating gases such as air at the reservoir conditions of temperature and pressure. In addition to the Russian work, there are novel ideas such as fire flooding (Halleck et al., 1982), the use of microwaves, and burial of nuclear wastes (Malone, 1985, p. 27) to obtain the energy required to dissociate hydrates.

However, all proposed schemes to date may be classified as one (or more) of three basic types. The first type, thermal stimulation, proposes the use of a source of energy to raise the reservoir temperature, breaking the hydrate hydrogen bonds and releasing the gas. The second type of scheme, depressurization, involves reducing the pressure at the face of the hydrate below the equilibrium value so that the hydrate decomposes. Inhibition, the third scheme, proposes to displace the hydrate equilibrium conditions beyond the reservoir temperature and pressure through injection of some inhibitor in the liquid phase adjacent to the hydrate. A variation on the third scheme involves the displacement of the equilibrium conditions via reduction of the concentration of the hydrate forming gases adjacent to the hydrate.

All schemes are common however, in that they must obey the First Law of Thermodynamics. This principle may be realized best on a microscopic scale by noting that the dissociated gas and water molecules have more energy (more translational freedom) than they have in the hydrate crystal structure. This means that in all cases, energy must flow from the surroundings to the system of hydrates which is being dissociated. Fortunately, as shown in the example calculation below, the energy balance is favorable, i.e. the energy obtained from the gas in the hydrate is fifteen times greater than the energy required to dissociate the hydrate, assuming no energy losses. The frontispiece of this book is a photograph of a burning hydrate mass, supporting its own combustion.

Example 7-1: A 30% porosity reservoir with a rock volumetric heat capacity of 2.24 MJ/($m^3$-K) contains structure I hydrates at 281K and 10MPa. Calculate the energy needed to dissociate the hydrate and the energy content of the gas which the hydrate contains.

Solution:

Basis: As a basis use 1$m^3$ of reservoir.

Assume:

1. that the entire porosity of the rock is filled with hydrate, with neither free water or gas.
2. that the gas which evolves has a nominal heating value of 37,250 kJ/$m^3$ (1000 BTU/$ft^3$)

Step 1. Calculate the Energy Output (Gained) from the dissociation.

$$\frac{37{,}250\ \text{kJ}}{\text{m}^3\ \text{gas}} \times \frac{170\ \text{m}^3\ \text{gas}}{\text{m}^3\ \text{hydrate}} \times \frac{0.3\ \text{m}^3\ \text{hydrate}}{\text{m}^3\ \text{reservoir}} = \frac{1.9 \times 10^6\ \text{kJ}}{\text{m}^3\ \text{reservoir}}$$

Step 2. Calculate the Energy Input to the dissociation.

2a. Calculate the sensible heat to the hydrate and reservoir

In order to bring the reservoir from 281K to the hydrate dissociation temperature (286K) at the prevailing pressure of 10MPa, it must be heated. Using a volumetric heat capacity of 2,500 kJ/($m^3$•K) for the hydrate we obtain:

$$\left[\frac{0.7m^3\ \text{rock}}{m^3\ \text{reservoir}}\times\frac{2{,}240\ \text{kJ}}{m^3\bullet K}+\frac{0.3m^3\ \text{hydrate}}{m^3\ \text{reservoir}}\times\frac{2{,}500\ \text{kJ}}{m^3\bullet K}\right](286-281K)=$$
$$11{,}600\ \text{kJ}/(m^3\ \text{reservoir})$$

2b. Calculate the energy of dissociation

Once at the dissociation temperature the enthalpy of dissociation (368,803 kJ/$m^3$) must be input to the hydrate. No energy input is associated with the reservoir here because the temperature is constant at 286K.

$$\frac{368{,}803\ \text{kJ}}{m^3\ \text{hydrate}}\times\frac{0.3\ m^3\ \text{hydrate}}{m^3\ \text{reservoir}}=\frac{110{,}640\ \text{kJ}}{m^3\ \text{reservoir}}$$

2c. Calculate the total energy input as the sensible heat and the heat of dissociation. Note that the sensible heat is only about 10% of the heat of dissociation.

Total Energy Input = (11,600 + 110,640) = 122,240 kJ/$m^3$ reservoir.

The energy obtained from the process is 15.5 times the energy required for the process.

With the above example as an initial indication that further estimates should be made, consider some of the models which better approximate the different methods of hydrate dissociation. In the following subsections the thermal stimulation and the depressurization models are discussed. Surprisingly, no mathematical models have been generated for the use of an inhibitor such as methanol, the third type of hydrate dissociation.

Holder et al. (1984) note that depressurization and hot water injection seem to be the most promising techniques for further evaluation because they have lower heat losses than steam injection, they do not involve the gas dilution, and they do not require large inhibitor quantities or expenses. A comparison was made between thermal stimulation and depressurization by Lewin, Associates, and Consultants (1983, p. 155). For similar cases, the heat required for dissociation was calculated to be 0.122 $MJ/m^3$ and 0.097 $MJ/m^3$ of reservoir for thermal stimulation and depressurization, respectively. The energy gained from thermal stimulation was 1.72 $MJ/m^3$, also less favorable than the energy gained from depressurization (1.85 $MJ/m^3$) of a reservoir. The primary disadvantage of thermal stimulation is the requirement that sensible heat be injected before dissociation occurs, which in turn requires that a substantial amount of heat be input to the non-hydrate bearing rock. This secondary heating may cause 10% to 75% of the injected heat to be lost in thermal stimulation. Neither heat loss nor sensible heat increase is a factor in depressurization. Section 7.3.3 provides an economic confirmation of these factors; depressurization appears to be a more viable recovery method.

All of the models presented here are for schemes of gas recovery in permafrost regions. While the physical principles are nominally the same for recovery in deep oceans, changes must be made to account for the overlying hydrostatic water in that case. Due to the challenges in recovering gas from deep water, only conceptual models have been presented to date. Such schemes have been proposed by Rose and Pfannkuch (1982) for thermal stimulation by heated overlying sea water or underlying brine, and by Soviet workers (Trofimuk et al., 1980) who have suggested dredging of hydrates from the ocean floor.

### 7.3.1.a. Thermal Stimulation Models

The initial thermal stimulation models of P.L. McGuire (1981) indicated two heat transfer controlled techniques to produce the hydrates: a) by a frontal sweep of hot water, similar to a conventional steam flooding, diagrammed in Figure 7-22a and b) a fracture flow of hot water in a vertical fracture linking the injection well to a single production well as shown in Figure 7-22b. In McGuire's first technique the Marx-Langenheim (1959)

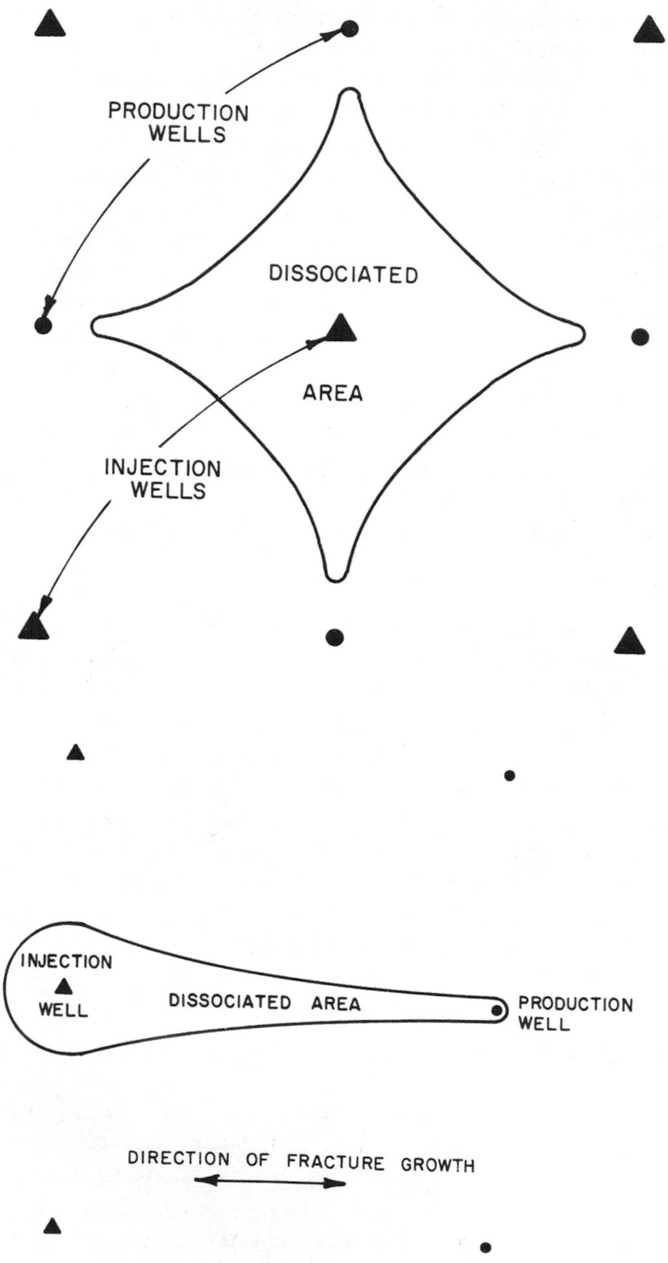


7-22 McGuire's Schematic for Gas Recovery from Hydrate
(Top) Conventional Steam Flooding Model
(Bottom) Fracture Flow Model
(Reproduced, with permission, from Proc. 4th Can Permafrost Conf, (McGuire, 1982) by National Res. Council of Canada

equation was used to calculate the area covered by the heat front, from which all hydrate must be produced. The model assumes that the gas evolved at the dissociation front will flow to the production well, and that the permeability is sufficient to allow continuous water or steam injection at a constant temperature. Parametric studies on the frontal sweep model indicates that the hydrate reservoir should be of 15% porosity (minimum) and 7.5m thick, using injection fluids between 340K and 395K for economic feasibility.

The second model of McGuire, the fracture flow model, assumed laminar slug flow and was considered to be a lower bound to the production problem. In the most favorable case of the parametric study, only 16 percent of the energy input to the reservoir was recovered. In both of the McGuire models, no account was taken of flow in porous media. Recent experimental studies by de Boer et al. (1985), Yousif et al. (1989), and Nadem et al.(1988) indicate that the porosity undergoes a tremendous reduction upon the formation of hydrates, which can sustain a pressure gradient of over 12MPa/m. Based upon such experiments, it seems difficult to conceive of a hydrate-containing reservoir with a porosity to support either the frontal sweep or the fracture flow assumptions. Nevertheless these early models led the way to further work and indicated that hydrates could be produced at sufficient rates to extend the lives of fields with both hydrates and free gas.

Kamath (1984) considered the steady-state, heat transfer limited dissociation of hydrates by hot water to be similar to nucleate boiling phenomena. The heat transfer coefficients for hydrate dissociation were determined under a layer of circulating hot water. He also showed that the rate of heat transfer and the rate of dissociation are governed by a power law function of the temperature drop across the hydrate interface, identical to the case of nucleate boiling. In addition, Kamath performed hydrate dissociation experiments in unconsolidated sand with 0.1, 0.2, and 0.5 volume fractions of sand. He arranged his apparatus so that heat transfer was through vertical sand to the undissociated hydrate. His model assumes that heat is transferred only by conduction from the hot water through the sand to the hydrate, and that the temperature profile obtained in the sand zone is linear.

In subsequent work, Kamath and Godbole (1987) suggested the use of hot brine as a more efficient alternative than steam in the cyclic injection technique of Bayles et al. (1986). Since brine acts as a hydrate inhibitor, it lowers the dissociation temperature, in addition to providing heat of dissociation. In this scheme, a brine volume equivalent to the pore space is injected. Heat losses to the underburden and overburden, and well bore heat losses were all calculated using standard petroleum engineering techniques. A hydrate dissociation stage follows the injection stage during which the dissociation front and the temperature profiles in the dissociated and undissociated zones were calculated. The radial heat conduction equations in both zones were solved numerically as a moving boundary problem. At the end of the dissociation stage, gas production is commenced and the gas volume produced is determined from a material balance. The parametric study shows that an acceptable gas production rate is obtained if the reservoir has minimum properties as follows: a porosity of 15%, a thickness of 25 feet, brine salinity as high as possible at a temperature between 395K and 475K, and a brine injection rate of at least 33 $m^3$/hr. This model suffers from a similar problem to some of the others in that the effect of flow in porous media is not considered.

Bayles et al. (1986) generated a model for cyclical steam injection in the permafrost region, to generate upper and lower bounds on the energy efficiency ratio [≡(energy out)/(energy in)]. The proposed process had three steps: a) steam injection, b)a soak period, c)a production period of dissociated gas and water. A macroscopic model was determined for the prediction of the amount of steam injected, the average reservoir pressure and water saturation during the steam injection and soaking steps. Heat losses were considered in the injection phase, with the remainder of the heat consumed in dissociating the hydrates and heating the overburden and underburden strata.

The sensitivity test of Bayles et al. indicated a higher energy efficiency ratio when the reservoir porosity, thickness and depth were increased. In all of the parametric case studies for reservoirs below the permafrost, the energy efficiency ratios varied between 4 and 9.6. This model was an improvement on the previous one, in that a more realistic recovery technique was used. However the heat transfer controlled model was batchwise, time independent, and did not determine thermal, velocity, or pressure gradients in the porous media.

The Selim and Sloan (1985, 1987, 1989) model evolved from a hydrate thermal dissociation model without porous media to one with porous media. The 1985 model was confirmed by Ullerich et al. (1987) to predict a priori the dissociation of hydrates with no adjustable parameters. A schematic of the 1987 model in porous media, shown in Figure 7-23, indicates that hydrate dissociation occurs at a moving boundary X(t) which separates the undissociated hydrate region from the dissociated (gas and water) region in the porous media. Analytical, time dependent solutions were obtained for temperature and pressure profiles as well as for global quantities such as gas produced. A parametric study showed that, while the rate of dissociation is a strong function of the thermal diffusivities of each zone and the porosity of the porous medium, an energy efficiency of 9 could be expected. In order to obtain an analytical solution for the model without adjustable parameters, the flow of water in the porous media had to be neglected. Field work (Makogon, 1981 p.169ff) and later laboratory research (Yousif et al., 1988) indicated that neglecting the water flow was questionable.

### 7.3.1.b. Depressurization Models

As the reservoir pressure is reduced below the three phase ($L_W$-H-V) equilibrium value, hydrates dissociate,

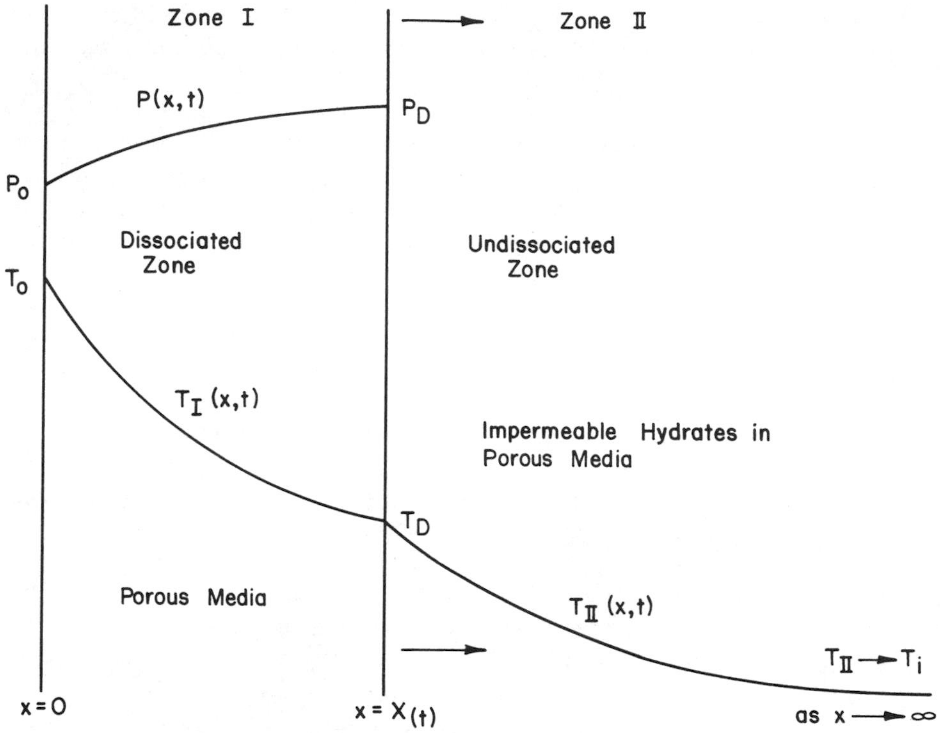


7-23 Typical Pressure and Temperature Profiles for Hydrate Thermal Dissociation in Moving Boundary [X(t)] Model (Selim and Sloan, 1987)

absorbing energy from the surroundings, so that a decrease in reservoir temperature is observed. A thermal gradient is obtained so that heat flows to the dissociating hydrate interface by conduction. Hydrates will continue to dissociate until sufficient gas evolves to achieve the equilibrium pressure at the lower temperature. The depressurization technique requires a controlled thermal gradient for continued dissociation of hydrates. The depressurization model has the advantage of not explicitly requiring heat input from the surroundings. Also, as shown in the Messoyakha Field case study in the following subsection, depressurization is the only technique which has been used to produce a hydrate field for an extended period of time.

The work of Verigin et al. (1980) modelled depressurization as a unidimensional process in a semi-infinite block of porous media, initially at thermal and pressure equilibrium. The pressure is reduced below the equilibrium value at one face of the block, to begin the dissociation process. These workers used a moving boundary to separate the dissociated and undissociated zone, with gas as the only mobile phase in either zone. The momentum equation is used for the gas in each zone, coupled with the mass balance at the moving boundary. This model assumes that sufficient heat instantaneously flows from the surroundings to the hydrate dissociation face, so that the temperature is constant everywhere. The model solution gives the pressure profile in each region, together with the location of the moving boundary front and the gas production. A second assumption in the model is that the water phase is immobile after hydrate dissociation.

McGuire (1981) suggested that depressurization with either of two reservoir fracturing methods be used to cause increased flow capacity of a hydrate zone in permafrost regions. He showed that, when hydrates are depressurized, the resulting ice occupies 22% less volume than the parent hydrates, increasing the flow capacity of the reservoir. As in his thermal stimulation model, McGuire used standard petroleum engineering flow techniques.

In the Holder and Angert (1982) numerical model, a gas reservoir is overlain by a massive impermeable methane hydrate layer. As diagrammed in Figure 7-24, gas is produced from the free gas zone through a single well to lower the reservoir pressure. Below the equilibrium pressure hydrate begins to replenish the free gas zone through its dissociation. It is assumed that the sensible heat of the reservoir provides the necessary energy for hydrate dissociation. Gas and water flows are calculated in the free gas zone, but they are assumed not to flow in the hydrate zone. The flow equations and the heat conduction equation are solved in the vertical direction. By a parametric study, the hydrate zone contributes 20-30% of the total produced gas and the reservoir temperature does not change appreciably during dissociation.

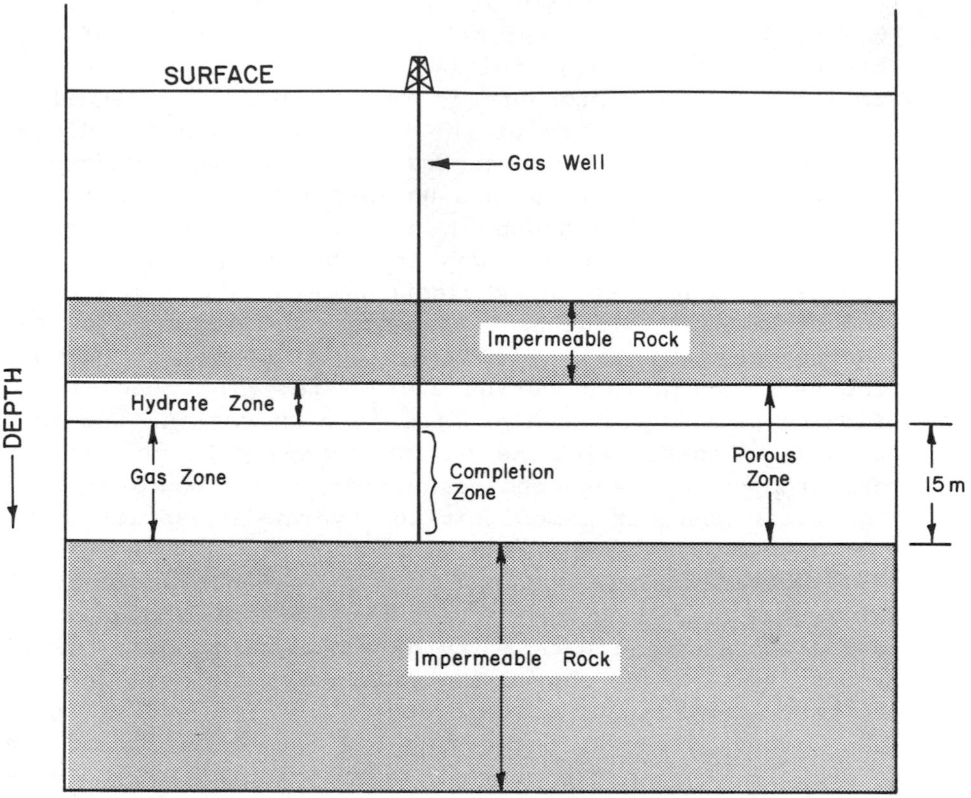


7-24 Producing a Gas Zone Which is Replenished by Hydrate Dissociation
(Reproduced, with permission, from J. Petrol Tech., (Holder and Angert, 1982) by Society of Petroleum Engineers)

The Burshears et al.(1986) model is an extension of the above model by Holder and Angert. The model differs in that the previous simple hydrate was replaced by a multicomponent hydrate. The parametric study shows that lower gas specific gravity, higher porosity, permeability and reservoir thickness increase the gas productivity of the hydrate zone. The dissociated water was assumed not to affect the gas production rate.

In this laboratory Yousif et al. (1989) formulated a depressurization model similar to the Soviet work of

Verigin et al. (1980) in using a moving boundary to model the semi-infinite hydrate layer. An isothermal solution was assumed in order to simulate a very slow depressurization process such as in the Messoyakha field case study of Section 7.3.2. The pressure of the system is assumed to be only slightly below that of hydrate dissociation. The mass and momentum transport equations were solved analytically, in conjunction with the moving boundary condition. The solution yields the pressure profiles within each zone, the gas production rate, and the location of the dissociation front with time. As in the previous isothermal model from this laboratory, the water phase was assumed immobile in the dissociated zone so that the flow equation can be neglected. If the hydrate layer is of finite dimensions, the model results can be taken as applicable at small times. The model predicted the gas production and the location of the dissociation front to be proportional to the square root of time; the work of Verigin et al. (1980) predicted that these conditions vary with the inverse of the square root of time. The model of Yousif et al. (1989) was confirmed by experiments on hydrate dissociation in Berea sandstone cores.

### 7.3.1.c. Details of a Depressurization Model

Because the work of Yousif et al.(1988a) represents the state-of-the-art in **a priori** depressurization prediction and laboratory measurement of unidimensional dissociation in a consolidated core, a detailed exposition of the mathematics and results are presented.

Consider a hydrate reserve in consolidated sediment at a uniform temperature slightly above the ice point, and at a uniform pressure $P_i$ above that for three-phase ($V$-$L_W$-$H$) equilibria, $P_D$. The hydrate consists of pure methane for these initial estimations, and the entire system is maintained at isothermal conditions by an adequate flow of heat from the surrounding media, assuming slow dissociation. The medium has a uniform porosity $\varepsilon$ and occupies the semi-infinite region $0<x<\infty$. Let $S_H$ and $S_g$ denote the uniform hydrate and gas saturations, re-

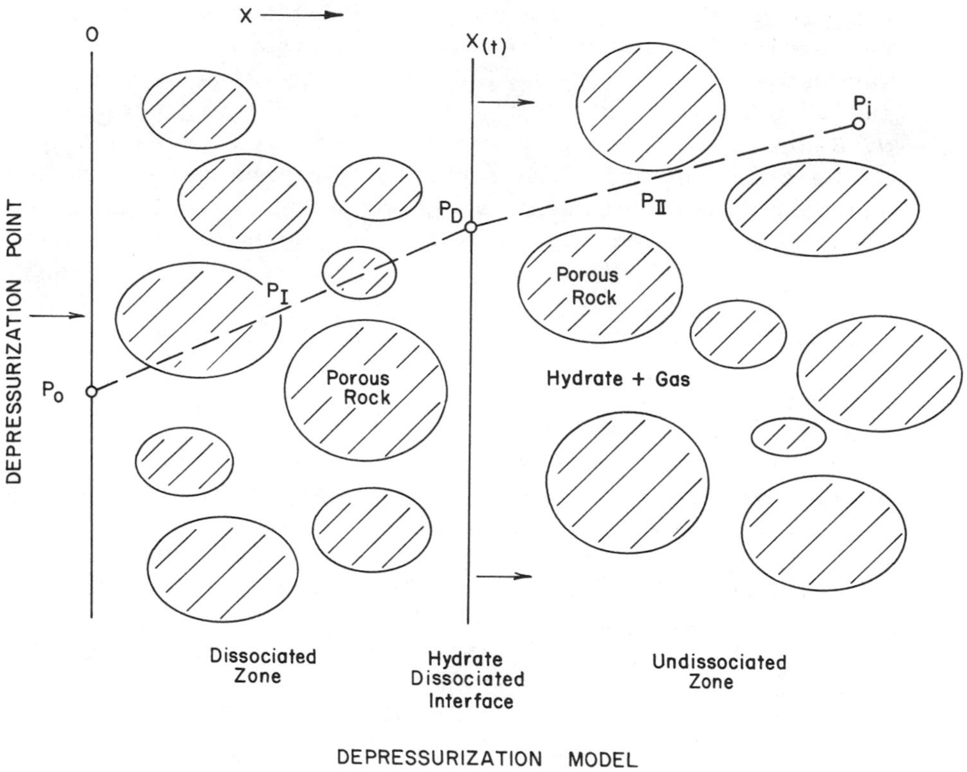


7-25 Depressurization Model for Hydrate Dissociation in Sediment with Moving Boundary [X(t)] Model
(Reproduced, with permission, from J. Inclus. Phenom., (Yousif et al., 1989), by Kluwer Academic Publishers

spectively, so that $\varepsilon S_H$ is the volume fraction of the hydrate and $\varepsilon S_g$ is the volume fraction of the gas in the medium.

At time zero, the pressure at the boundary x=0 is suddenly lowered to a constant value $P_o$ just less than the three-phase equilibrium value, $P_D$. The hydrate at the gas-hydrate interface will dissociate and there will be a moving dissociation interface at some distance x=X(t) which separates the water-gas dissociated zone I from the hydrate-gas undissociated zone II. As shown in Figure 7-25 at any time t>0, dissociated zone I occupies the region 0<x<X(t) while the undissociated zone II occu-

pies the region $X(t)<x<\infty$. Clearly, the following pressure distributions develop in the system: a pressure distribution $P_I(x,t)$ in region I with $P_o < P_I(x,t) < P_D$ and a pressure distribution $P_{II}(x,t)$ in region II with $P_D < P_{II}(x,t) < P_i$.

In this model, dissociated water is assumed not to flow, but instead contributes to the effective permeability of the system. The unidimensional continuity equation may be used to model the gas flow in the dissociated zone I as:

$$S_g \varepsilon \frac{\partial \rho}{\partial t} + \frac{\partial (\rho v)}{\partial x} = 0 \qquad (7.4)$$

where the density and time are denoted by $\rho$ and t respectively. Darcy's Law may be used to relate the velocity v of the gas to the pressure gradient $(\partial P/\partial x)$ to obtain for one dimension:

$$v = -\frac{\kappa}{\mu}\frac{\partial P}{\partial x} \qquad (7.5)$$

where the permeability and viscosity are denoted by $\kappa$ and $\mu$. When Equation (7.5) is substituted into (7.4), along with the relationship ($\rho \equiv PM/zRT$), between density, pressure, molecular weight, M, and temperature, the resulting equation is obtained for constant viscosity and permeability:

$$\frac{1}{P}\frac{\partial P^2}{\partial t} = \frac{\kappa}{S_g \varepsilon \mu}\frac{\partial^2 P^2}{\partial x^2} \qquad (7.6)$$

In Equation (7.6) the reciprocal pressure factor in the initial term causes the equation to be non-linear. If the pressure change in the system is restricted to a low value, the reciprocal of the average pressure $P_{avg}$ ($\equiv [P_D+P_o]/2$) may be substituted as a constant in the first term of Equation (7.6) to remove the non-linearity.

For purposes of conforming to petroleum engineering nomenclature, we introduce the term for the correction of the gas volume in the pores, $C_g$ which is related to the pressure P and the compressibility z by:

$$C_g = \frac{1}{P} - \frac{1}{z}\left(\frac{\partial z}{\partial P}\right)_T \tag{7.7}$$

The second term on the right is relatively minor at these conditions and may be assumed to be negligible, so that $C_{g_{avg}}$ may replace the first factor in Equation (7.6). Finally, the petroleum nomenclature change $C_t$ ($\equiv S_g C_g$) is used to obtain the equation for flow in the dissociated zone as:

$$\frac{\partial P_I^2}{\partial t} = \alpha_I \left(\frac{\partial^2 P_I^2}{\partial x^2}\right) \qquad \text{for } 0 < x < X(t), \quad t>0 \tag{7.8}$$

where:

$$\alpha_I \equiv \frac{\kappa_I}{\varepsilon C_{t_I} \mu_{g_I}} \tag{7.9}$$

and X(t) is the position of the moving hydrate dissociation boundary. In the undissociated hydrate zone we can keep the same porosity $\varepsilon$ by defining $S_g$ to be the fraction of the pore volume filled with gas. An equation similar to (7.8) is obtained for the undissociated phase:

$$\frac{\partial P_{II}^2}{\partial t} = \alpha_{II} \frac{\partial^2 P_{II}^2}{\partial x^2}, \qquad X(t) < x < \infty, \ t>0 \tag{7.10}$$

where:

$$\alpha_{II} \equiv \frac{\kappa_{II}}{\varepsilon C_{t_{II}} \mu_{g_{II}}} \tag{7.11}$$

Equations (7.8) and (7.10) represent the primary equations for the pressure change with time in the dissociated and undissociated zones, respectively. Accompanying these two equations are four boundary conditions and two initial conditions. The initial conditions are:

$$P_{II} = P_i \qquad \text{for } 0 < x < \infty, \ t=0 \tag{7.12}$$

and $$X(0) = 0 \qquad \text{at } t=0 \quad (7.13)$$

The four boundary conditions are given in Equations (7.14) - (7.17):

$$P_I = P_o \qquad \text{at } x = 0, \quad t>0 \quad (7.14)$$

$$P_I = P_{II} = P_D \qquad \text{at } x = X(t), \quad t>0 \quad (7.15)$$

$$P_{II} = P_i \qquad \text{at } x = \infty, \quad t>0 \quad (7.16)$$

and

$$\chi_I \frac{\partial P_I^2}{\partial x} - \chi_{II} \frac{\partial P_{II}^2}{\partial x} = \beta \frac{\partial x}{\partial t} \qquad \text{at } x = X(t), \ t>0 \quad (7.17a)$$

where: $$\beta \equiv 2\varepsilon \mu_{g_D} P_D \left[ S_H \omega_{CH_4} \frac{\rho_H}{\rho_D} - (S_H - S_W) \right] \quad (7.17b)$$

All of the initial conditions and the boundary conditions are self-evident except (7.17), which requires a brief explanation. Equation (7.17) represents a jump gas mass balance at the dissociation interface (Slattery, 1978); it states that the change in mass flux across the moving hydrate interface is brought about by hydrate dissociation. The subscript D in Equation (7.17b) indicates that the properties should be evaluated in the gas phase at the dissociation interface. The quantity $\omega_g$ in Equation (7.17b) represents the mass of the hydrocarbon gas contained in a unit mass of hydrate, which is given by:

$$\omega_g = \frac{n_H}{n_H M_W + M_g} \quad (7.17c)$$

where $n_H$ is the hydrate number. Using the statistical thermodynamics model (CSMHYD on the floppy disk in the frontpapers) we estimated the occupancy of the cages at 93.5% for $n_H$ of 6.15 at our experimental conditions. The quantity $S_w$ in Equation (7.17b) represents the water saturation in the dissociated zone I, which is easily found as:

$$S_W = \frac{V_W}{V_W + V_{g_F} + V_{g_D}} \tag{7.17d}$$

where: $V_W = \varepsilon S_H(1-\omega_g)\rho_H/\rho_W,$

$$V_{g_F} = \varepsilon(1-S_H)$$

and $V_{g_D} = \varepsilon S_H \omega_g \rho_H/\rho_{g_D}$

In the above equations $V_{g_F}$ represents the volume fraction of the free gas prior to dissociation; $V_{g_D}$ and $V_W$ represent the volume fraction of the gas and water released during dissociation, respectively.

The analytical solution to the boundary-value problem given by Equations (7.8) through (7.17) is similar to the Neumann problem of heat conduction during melting or solidification of a semi-infinite region (Carslaw and Jaeger, 1959). The solution yields the pressure profiles within each zone, as:

$$\frac{P_I^2 - P_o^2}{P_D^2 - P_o^2} = \frac{\text{erf }(a\eta)}{\text{erf }(a\xi)} \tag{7.18}$$

$$\frac{P_{II}^2 - P_i^2}{P_D^2 - P_i^2} = \frac{\text{erfc }\eta}{\text{erfc }\xi} \tag{7.19}$$

In Equations (7.18,19) the following dimensionless groups were used:

$$a \equiv \left(\frac{\alpha_{II}}{\alpha_I}\right)^{1/2} \quad (7.20); \quad \eta \equiv \frac{x}{\sqrt{4\alpha_{II}t}} \quad (7.21); \quad \xi \equiv \frac{X(t)}{\sqrt{4\alpha_{II}t}} \quad (7.22)$$

It may be shown that the value of $\xi$ in Equation (7.22) is a constant which relates the position of the moving boundary X(t) to the square root of time t. The value of $\xi$ is obtained through the differentiation of Equations (7.18) and (7.19) for use in the final boundary Equation

(7.17). As a result, a transcendental equation is obtained for $\xi$, as:

$$a\left[\frac{x_I\,(P_D^2 - P_o^2)}{x_{II}(P_i^2 - P_D^2)}\right]\frac{\exp(-a^2\xi^2)}{\text{erf } a\xi} - \frac{\exp(-\xi^2)}{\text{erfc } \xi} = \frac{\sqrt{\pi}\;\alpha_{II}\;\beta}{x_{II}(P_i^2-P_D{}^2)}\;\xi \tag{7.23}$$

With the iterative solution of Equation (7.23) for $\xi$, that value may be substituted back into Equations (7.18), (7.19), and (7.22) to obtain the pressure profiles and the position of the moving hydrate boundary X(t) with time. The position of the moving boundary with time may be compared against measurements for the time of increase of resistances placed along the core length, as indications of the hydrate front position. The volume of gas produced from the core may also be obtained for comparison with experiment from the value of $\xi$ via the equation:

$$\frac{Q}{A_C} = \frac{a x_I}{\mu_I P_o\sqrt{\pi\alpha_{II}}}\;\frac{P_D^2 - P_o^2}{\text{erf } a\xi}\;\sqrt{t} \tag{7.24}$$

The above solution is applicable to semi-infinite media; in a typical experiment however, a core of finite length L is used and the solution obtained is not strictly applicable. Modification of the solution to allow for a finite medium of length L required replacement of the outer boundary condition (Equation 7.16) by the condition of zero flux ($\partial P_{II}/\partial x = 0$ at $x = L$). Under this circumstance the boundary-value problem does not admit an analytical solution and one must resort to numerical techniques. However, for short times the semi-infinite solution presented here is still applicable.

Figures 7-26 and 7-27 contain predictions (lines) from Equations (7.23,24) for the gas produced and the position of the moving hydrate boundary from experimental data (circles). Table 7-9 provides the a priori parameters used for the model. The reader should refer to the original work for details of the experiment and discussion. The limitations of the model are threefold: 1) the

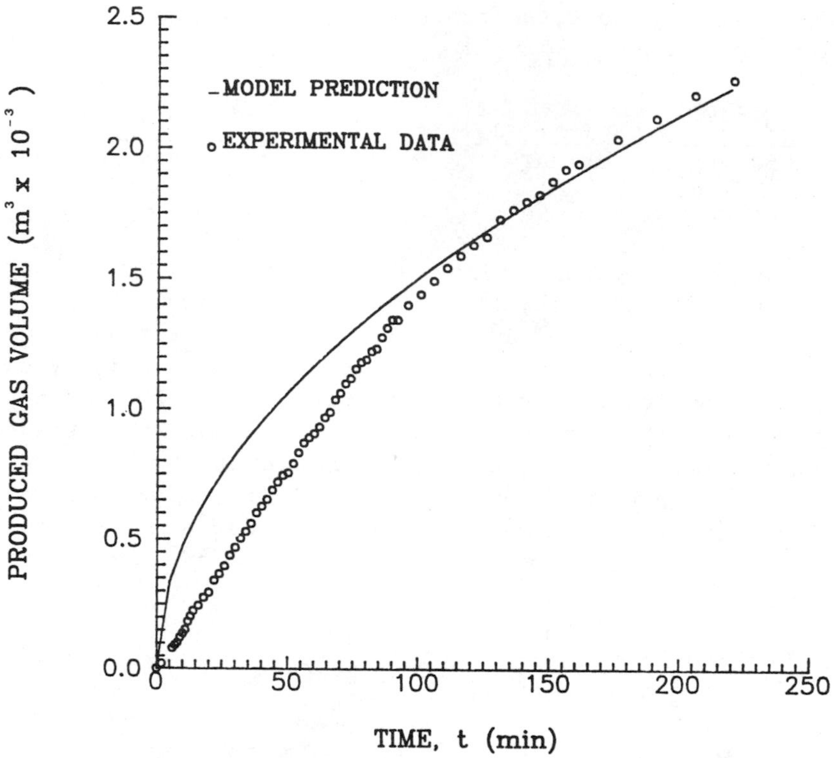


7-26 Produced Gas Volume as a Function of Time for Hydrate Depressurization Compared to Moving Boundary Model (Reproduced, with permission, from J. Inclus. Phenom., (Yousif et al., 1989), by Kluwer Academic Publishers

pressure is assumed to be only slightly below the dissociation value, 2) the hydrated medium is assumed to be infinite (or short time), and 3) the flow of water is neglected. A numerical solution to a similar model has been formulated by Yousif et al.(1988) to overcome these restrictions at the expense of fitted parameters.

#### 7.3.1.d. Conclusions from Recovery Scheme Modelling

The gas recovery models reviewed in this section represent either heat transfer controlled (thermal stimulation) or momentum transfer controlled (depressuriza-

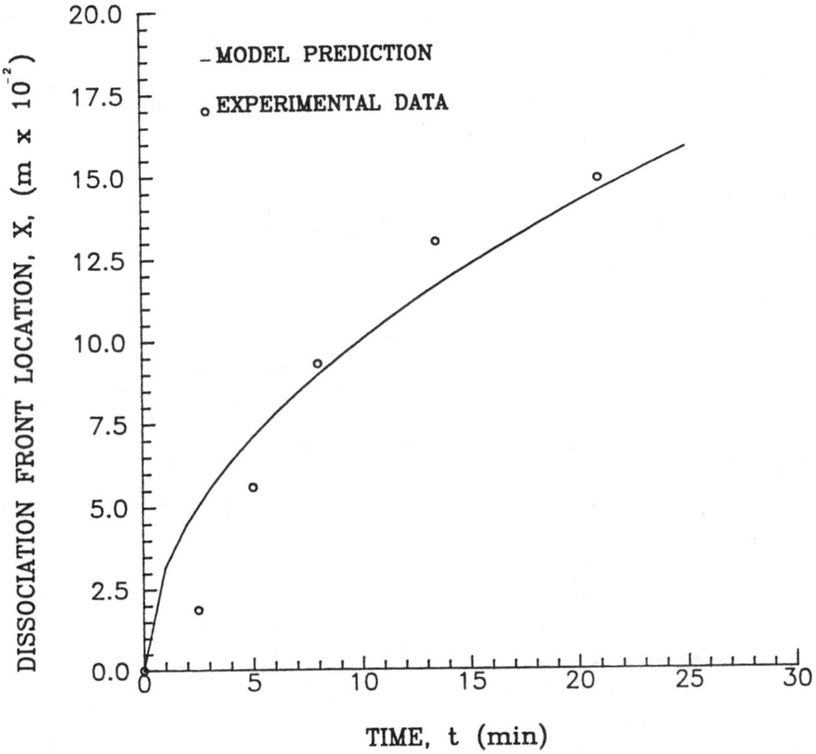


7-27 Location of Hydrate Dissociation Boundary as a Function of Time, Compared to Moving Boundary Model (Reproduced, with permission, from J. Inclus. Phenom., (Yousif et al., 1989), by Kluwer Academic Publishers

**Table 7-9 Parameters Used by Yousif et al. (1988a) in Depressurization Model (Figures 7-26,27)**

| | |
|---|---|
| $P_o$, Withdrawal Pressure, MPa | 2.5 |
| $P_i$, Initial Pressure, MPa | 3.17 |
| $\kappa_I$, Dissociated Zone Permeability, $m^2$ | $5 \times 10^{-17}$ |
| $\kappa_{II}$, Undissociated Zone Permeability, $m^2$ | $1.31 \times 10^{-17}$ |
| $\varepsilon$, Porosity | 0.188 |
| $S_w$, Water Saturation | 0.55 |
| $\kappa_{abs}$, absolute permeability, $m^2$ | $9.869 \times 10^{-14}$ |
| $\xi$, constant in Eqn. (7.22) | 0.0928 |
| $\alpha_{II}$, constant in Eqn. (7.11), $m^2/s$ | $5.055 \times 10^{-5}$ |
| Produced Water | 40% |

tion) models. A truly comprehensive model must solve the mass, heat, and momentum equations simultaneously for hydrate, gas and water phases in porous media. The determination of such a model is tied to the reliable property measurement of hydrated porous media, such as permeability, porosity, and saturation of each phase. Measurement of such properties and the formulation of a comprehensive numerical model is beyond the present state-of-the-art.

The models which are presently available give good agreement for the properties of reservoirs which are helpful for estimating the recovery of hydrates. For good economic efficiency, the reservoir should have high porosity and a substantial thickness of hydrate. The thermal conductivity of the reservoir rock should be high and the heat capacity of the reservoir rock should be low for optimal hydrate production.

There have been only conceptual models for gas recovery from hydrates in marine sediments. The study of hydrate depressurization in the permafrost has been done in detail both in models and in the field, as the following case study illustrates. There is a general consensus that depressurization is the generally the most technically feasible means of recovery of hydrates in the permafrost.

### 7.3.2 Gas Recovery from Hydrates at Messoyakha: A Case Study

While the Messoayaka gas hydrate field is the only available exemplar for gas production from hydrate, the hydrates have been produced from this field semi-continuously for over 17 years. The field is located in the north-east of western Siberia, close to the junction of the Messoyakha River and the Yenisei River, 250 km west of the town of Norilsk, as shown in Figure 7-28.

Figure 7-29 provides a cross-section of the field, showing the hydrate deposit to be overlying the free gas zone. The depth-temperature plot of Figure 7-30 from

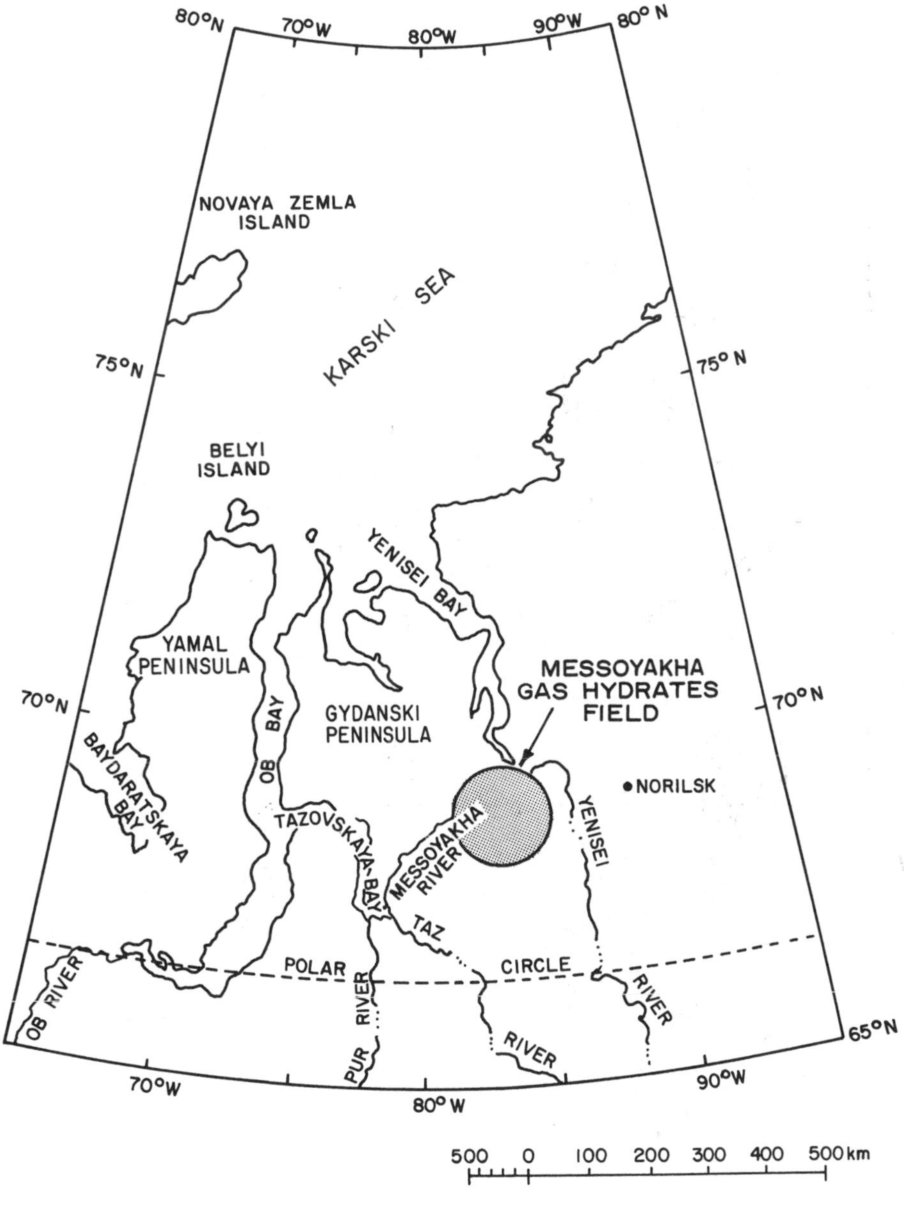


7-28 Map Showing Location of Messoyakha Gas Hydrates Field (Reproduced courtesy of U.S. Dept. of Energy, (Krason and Ciesnik, 1985))

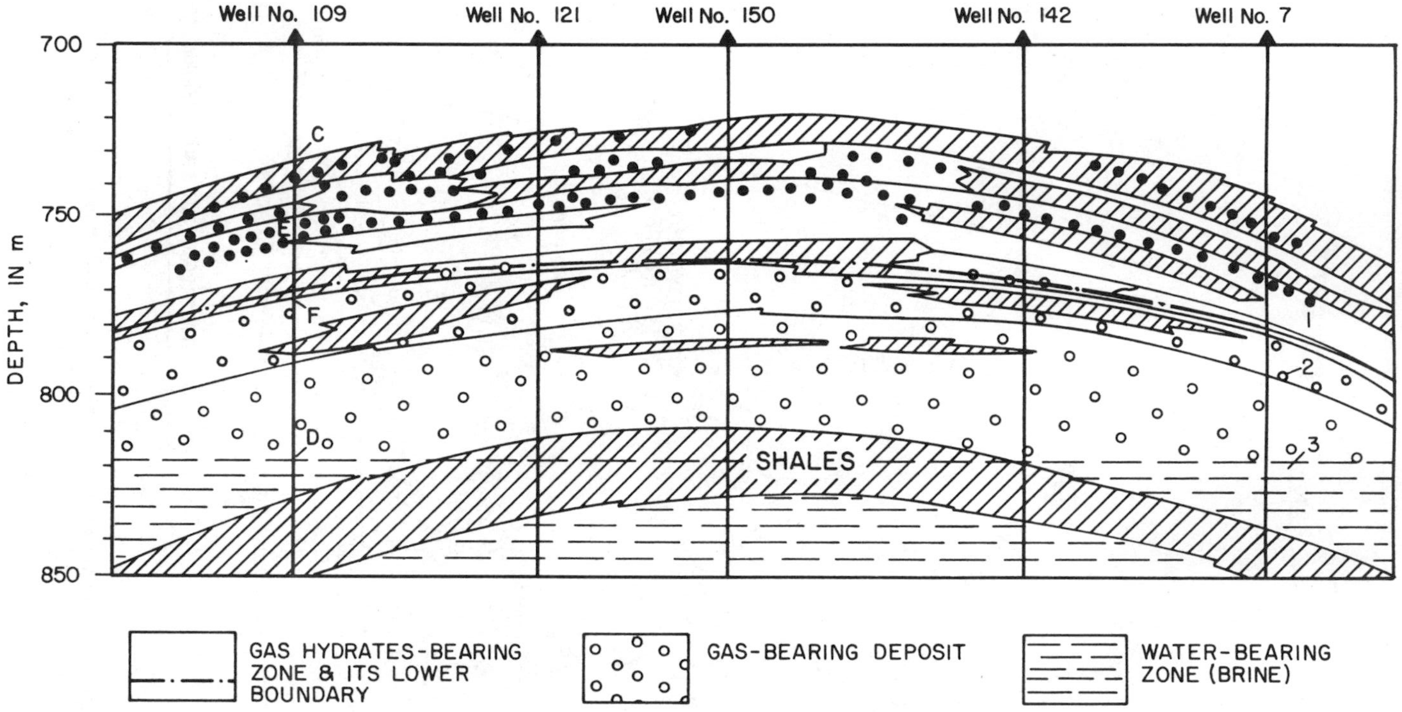


7-29 Cross Section of Messoyakha Gas Hydrate Field (Reproduced, with permission, from (Makogon, 1988b))

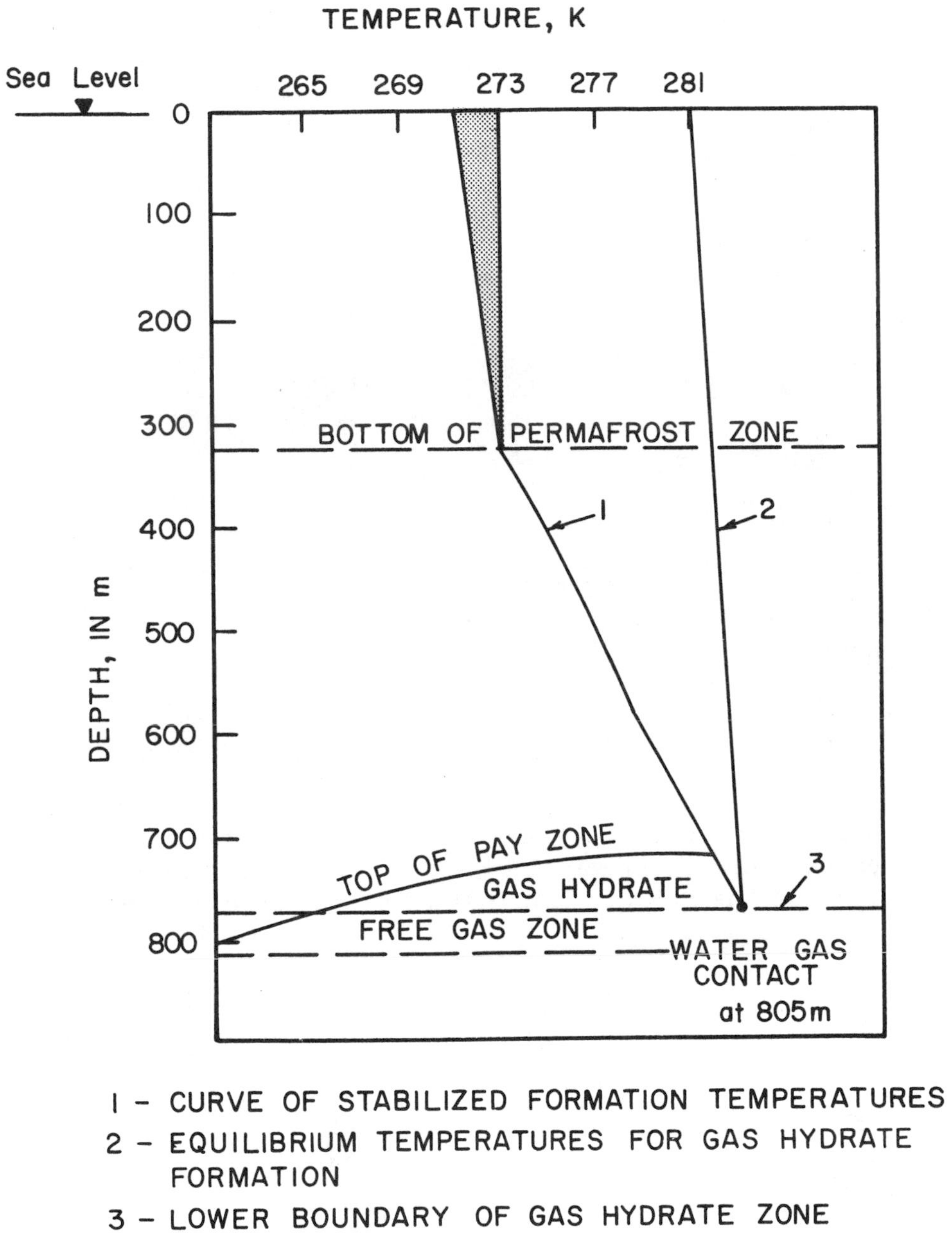


7-30 Gas Hydrate Stability Envelope at Messoyakha (Reproduced courtesy of U.S. Dept. of Energy, (Sheshukov, 1972))

**Table 7-10**
**Properties of Messoyakha Gas/Hydrate Field**
(after Makogon, 1988)

| | |
|---|---|
| Area of the pay zone, | 12.5km × 19km |
| Thickness of the pay zone | 84 m |
| Open porosity | 16-38% (average 25%) |
| Residual water saturation | 29-50% (average 40%) |
| Initial reservoir pressure | 7.8 MPa |
| Reservoir temperature range | 281 - 285K |
| Reservoir water salinity | < 1.5 wt% |
| Water-free gas composition | 98.6% $CH_4$ |
| | 0.1% $C_2H_6$ |
| | 0.1% $C_3H_8^+$ |
| | 0.5% $CO_2$ |
| | 0.7% $N_2$ |

Sheshukov et al. (1972) shows the hydrate layer to extend to the intersection of the 281K geotherm, later determined to be closer to 283K by Makogon (1988). The gas in the hydrate zone is both in the free and in the hydrated state. Makogon (1988) provided the most recent information regarding the properties of the field, as tabulated in Table 7-10.

A suite of well logs from Well Number 136 of the Messoyakha field is presented in Figure 7-31 by Sapir et al. (1973). The Soviet work indicated the need to use a suite of well logs rather than a single log to indicate hydrates, similar to the later findings in the Western hemisphere as previously discussed in Section 7.2.5.

The Messoyakha field has been produced through both inhibitor injection and depressurization, as well as combinations of the two. The inhibitor injection tests, presented in Table 7-11 from the combined results by Sumetz (1974) and Makogon (1974), frequently gave dramatic short-term increases in production rates, due to hydrate dissociation in the vicinity of each injected well bore. In the table methanol and mixtures of methanol and calcium chloride were injected under pressure, using a

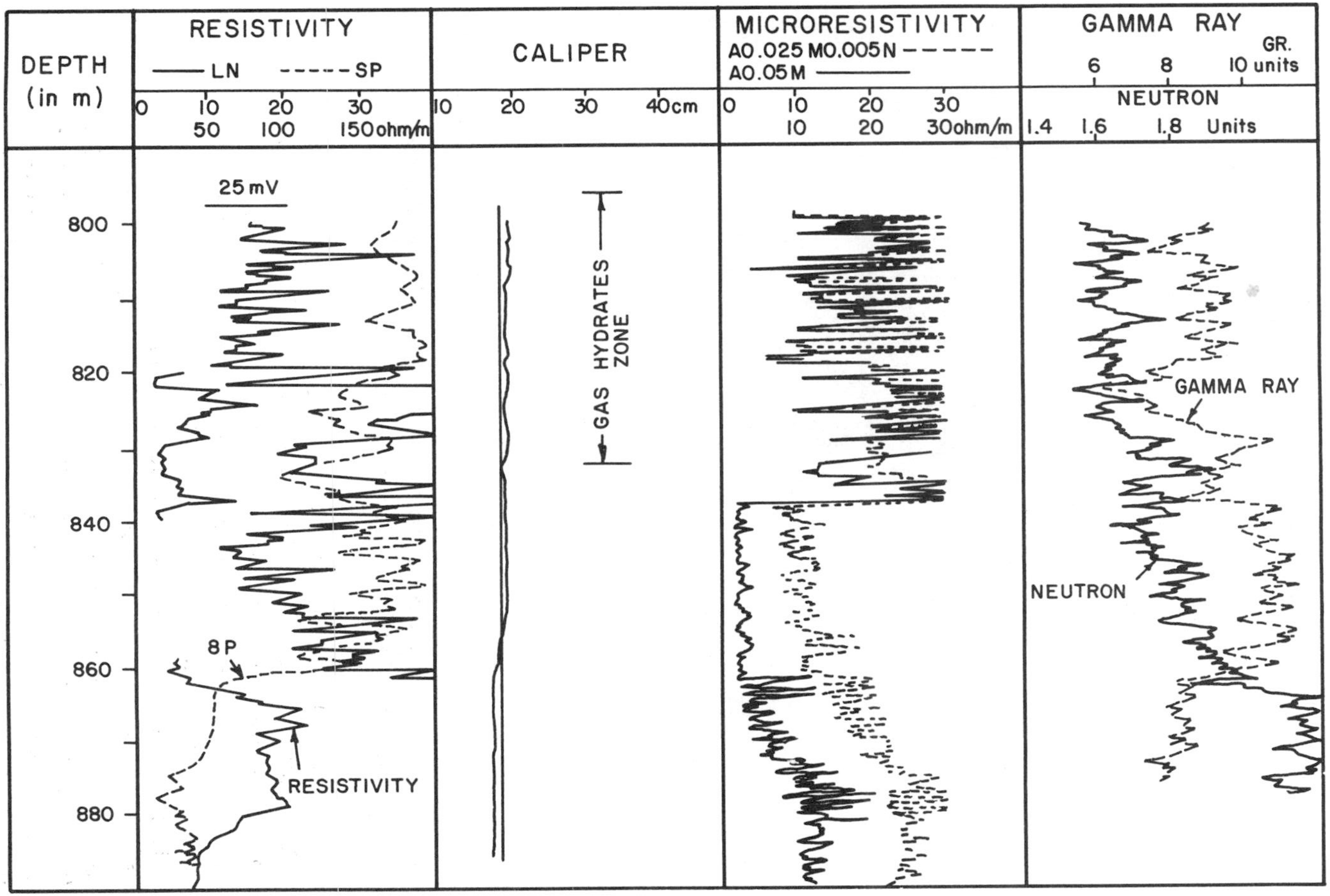


7-31 Well Logs from Well Number 136 at Messyahkha (Reproduced courtesy of U.S. Dept. of Energy, (Sapir et al., 1973))

**Table 7-11 Test Results from Inhibitor Injection in the Messoyakha Field**
from Sumetz (1974) and Makogon (1981,p.174)

| Well No. | Type of Inhibitor | Volume of Inhibitor $m^3$ | Gas Flow Before Treatment $1000m^3$/day | Gas Flow After Treatment $1000m^3$/day |
|---|---|---|---|---|
| 2 | 96wt% methanol | 3.5 | Expected results Not Achieved | |
| 129 | 96wt% methanol | 3.5 | 30 | 150 |
| 131 | 96wt% methanol | 3.0 | 175 | 275 |
| 133 | methanol | unknown | 25 | 50 |
| | | | 50 | 100 |
| | | | 100 | 150 |
| | | | 150 | 200 |
| | | | 200 | 250 |
| 138 | Mixture 10%MeOH 90 vol% of 30wt% $CaCl_2$ | 4.8 | 200 | 300 |
| 139 | Mix of Well 138 | 2.8 | 120 | 180 |
| 141 | Mix of Well 138 | 4.8 | 150 | 200 |
| 142 | methanol | unknown | 5 | 50 |
| | | | 10 | 100 |
| | | | 25 | 150 |
| | | | 50 | 200 |

"cement aggregate". In order to promote long-term dissociation of hydrates, depressurization was used.

Makogon (1988) indicates that of all the complex studies obtained during the 19 years of the production life of the Messoyakha field, the most informative results came from an analysis of the reservoir pressure change as a function of the gas withdrawal rate. A diagram showing pressure and gas production as a function of time is shown in Figure 7-32, with the accompanying pressure-temperature relationship in Figures 7-33a,b. While Figure 7-33a may represent the measured pressure-temperature values (Makogon, 1988) far away from the hydrate, Poettmann (1988) suggested that the values in the neighborhood of the hydrate interface are better represented

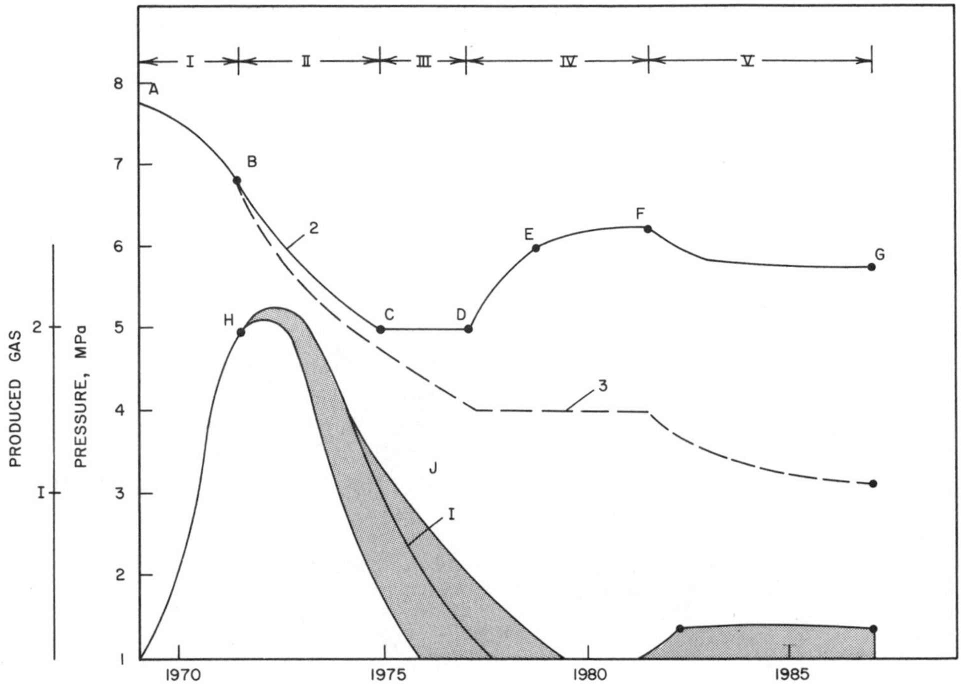


7-32 Pressure (Upper Curves) without Hydrates (Dashed) and with Hydrates (Solid) as a Function of Time; Produced Gas (Lower Curves) both With and Without Hydrates See Text for labels (Reproduced, with permission, from (Makogon, 1988b)

by Figure 7-33b, for reasons given below. The combination of these three figures represents a classic study in slow depressurization, done via the production of the free gas reservoir over a period of years.

In Figures 7-33, the following initial points are used (with C,D,E,F corresponding to letters on Well No. 109 in the reservoir diagram of Figure 7-29): AB = hydrate equilibrium line; C = temperature at the top of the pay zone; D = temperature at a level of gas and water contact; E = average gas-hydrate temperature; F = temperature at boundary surface between gas and gas-hydrate reserves; G = minimum pressure at which gas hydrates are stable (at the intersection of the geothermal gradient

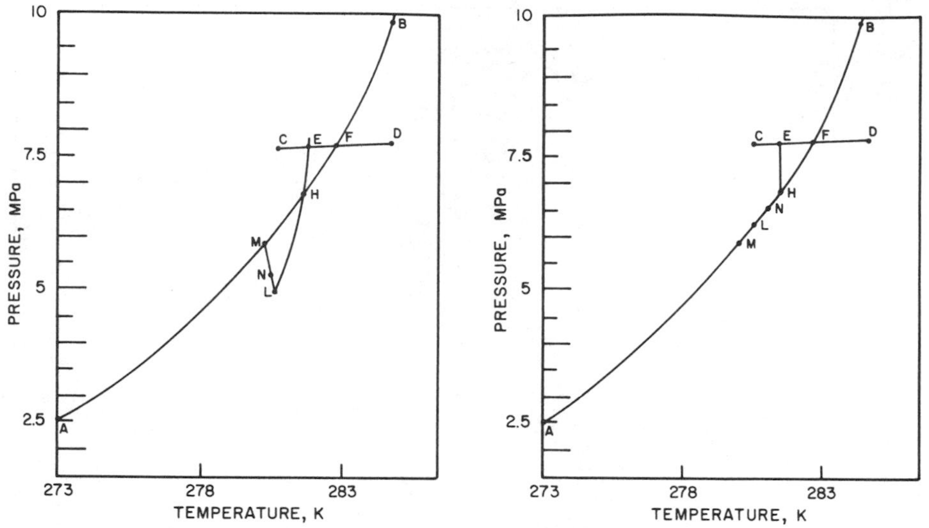


7-33 Messoyakha Pressure and Temperature with Hydrate Production
(Left) Away From Hydrate Face (Makogon, 1988b)
(Right) Hypothesized at Hydrate Face (Poettmann, 1988)

and the phase boundary in Figure 7-30); H = beginning dissociation pressure for gas hydrates.

As the production of the field was begun from 1969 to mid-1971, the pressure decreased from E in Figure 7-33 to just above the hydrate dissociation point at H in Figures 7-32, and 7-33a,b. During this period only free gas was produced because the pressure was above the hydrate equilibrium value.

When the pressure reached the hydrate equilibrium value at point H, the hydrates began to dissociate, adding the shaded portion to the gas production curve at the lower half of Figure 7-32. The top half of Figure 7-32 shows the reservoir pressure was maintained at a higher value (solid line 2) than the expected pressure (dashed line 3) due to the addition of gas to the reservoir by the hydrates. After Point H in Figure 7-33a, Makogon (1988) indicated that the pressure away from the hydrate decreased below the equilibrium value, with a slight decrease in temperature. The pressure at the hydrate interface in Figure 7-33b however, should not deviate from the equilibrium line, unless all of the hydrates were dissociated. Consequently the pressure and temperature at the hydrate interface decreased along the equilibrium curve as the reservoir was depressurized. Between the third and sixth year of the field development there was probably a pressure drop between the hydrate interface, from H to L in Figure 7-33b, and the measured pressure from H to L in Figure 7-33a, due to the flow of gas in porous media.

As the Messoyakha reservoir attained the pressure at point L (7-33), the average pressure stabilized for two years to point M, indicating that the volume of gas recovered was replenished by gas liberated from hydrate. The difference between curves I and J in Figure 7-32 indicates that the total gas produced (curve I) was slightly less than the gas liberated (curve J) by the hydrates. During this period, from the seventh to the eight year of the life of the reservoir, the pressure away from the hydrate face in Figure 7-33a rose slightly, compensated by a pressure decrease at the hydrate face in Figure 7-33b.

At point M, the reservoir was shut in, while other higher pressure gas reservoirs were produced (Makogon, 1988a). During that period, the average pressure of the reservoir began to rise from point M to point N in Figure 7-33b. The difference between curves I and J in Figure 7-32 indicates that gas continued to be liberated by the hydrates until the reservoir pressure was uniform at point F of the top line in mid-1981. As the average gas pressure approached the equilibrium value the amount of gas produced decreased exponentially. Makogon indicated that the temperature of the reservoir tended to be restored to its original value, after a period of time. The equilibrium pressure itself rose slightly as the high reservoir heat capacity increased the temperature of the hydrate mass.

Since 1982 there has only been a modest production of the Messoyakha reservoir. The amount of gas withdrawn has been equivalent to the amount of gas liberated from the hydrate. The total amount of gas liberated from hydrates thus far has amounted to 36% of the total gas withdrawn from the field. It is noted further that the position of the gas-water interface did not change over the seventeen year period of the production of the field.

Makogon (1988a) stated that the current production of gas from hydrates amounts to 0.2% of the total Soviet production of natural gas, and that the future production of gas from hydrate was forecast to be 1% of the total by 1997, increasing fivefold every five years thereafter.

### 7.3.3 Economic Evaluation of the Production of Gas from Hydrates

Of the several works associated with hydrate recovery economics, only a few have been investigated in sufficient detail to provide some degree of validity. Of these the work of Holder et al. (1984) seems to be the most reliable for schemes of thermal stimulation and depressurization; the summary in this section is abstracted almost entirely from that effort.

Holder et al. (1984) considered the economics of a model for thermal stimulation in the permafrost region of Alaska, which is at an average gas recovery rate between the upper (frontal sweep) and lower (fracture flow) bounds determined by the McGuire (1982) model discussed in Section 7.3.1. For depressurization in both permafrost and offshore reservoirs, the model of Holder and Angert (1982) was considered for hydrate dissociation via depressurization of an underlying free gas. The original authors note that, while such a reservoir has yet to be found in the offshore regions, the calculation should provide a reasonable estimate. The extension of these estimates to more modern models seems straightforward.

In the thermal stimulation model results, an injection rate of 30,000 barrels/day of 433K water was as-

**Table 7.12 Economic Evaluation of Hydrate Recovery from Offshore and Permafrost Reservoirs**
(from Holder et al. 1984)

| | Alaska Free Gas Prodtn. | Permafrost Triple Thermal Rate | Permafrost Depress. | California Free Gas Prodtn. | Offshore Depress. |
|---|---|---|---|---|---|
| Total Investment,K$ | 2,853 | 7,705 | 3,209 | 516 | 516 |
| Total Oprtng Cst,K$ | 1,108 | 3,586 | 2,974 | 26 | 26 |
| Total Ann. Equiv.Cost,K$ | 1,789 | 5,424 | 3,740 | 150 | 150 |
| Total Production MM SCM/yr | 31.2 | 21.24 | 93.45 | 1.29 | 1.61 |
| Net Production MM SCM/yr | 22.6 | 16.99 | 74.76 | 1.08 | 1.29 |
| Annual Cost of Prdtn/M SCM,$ | 71.7 | 319.2 | 50.14 | 141.24 | 115.82 |
| Transportation Cost/ M SCM,$ | 135.6 | 135.6 | 135.6 | 7.06 | 7.06 |
| Break Even Price/M SCM,$ | 207.3 | 454.8 | 185.7 | 148.3 | 122.88 |

sumed, along with a 40% reservoir porosity, and a permeability of 600 millidarcies. In the thermal stimulation study, cases were obtained for two hydrate reservoir recovery rates, the normal rate from a 7.6 m and a triple recovery rate three times that of the normal. In the depressurization model, the production pressure was 690 kPa. The porosity was assumed to be high enough to allow

$31 \times 10^6$ SCM/year to be produced in this way. The results are summarized in Table 7-12, relative to those for free gas production from both Northern Alaska and from offshore California.

Relative to the cost of free gas offshore California, Table 7-12 shows the thermal stimulation technique of permafrost hydrate production to be economically unfeasible, even with the triple rate of production studied. However gas depressurization on the North Slope, and particularly offshore California seems to be economically attractive.

## 7.4 Conclusion

Reservoirs of hydrate gas must be considered as a very significant energy resource in the earth. The recovery of gas from these reservoirs is still highly uncertain however, as only one example of recovery exists. Because these hydrated gas reservoirs are so dispersed and their solid form causes them to be difficult to recover, wide scale recovery will probably not be done until the twenty-first century. The depressurization of free gas reservoirs underlying hydrate reserves appears to be the best method for recovery.

Much research still remains to be done to enable the economic recovery of gas from hydrates. Research in the area of hydrates as a resource is at the exigency of the funding supplied by agencies with long-range vision. Fortunately many of the physical phenomena associated with the recovery of gas from hydrates, are also common to research on hydrates as a problem to the processing and production industry. The processing and production problems of hydrates are much more immediate, so that such funds can serve as a vehicle for more study of hydrates as a resource. The final chapter of this book considers these hydrate problems and their alleviation.

# 8

# Hydrates in Natural Gas Production and Processing

As noted in the previous chapter, hydrates as an energy source will probably not be realized until the twenty-first century. For the last half century however, the hydrocarbon industry has been dealing with hydrates due to their potential for problems in production or processing. Whether considering hydrates as an opportunity or as a problem, each arena has a substantial contribution to make to the other due to the identical nature of the physics of both natural and artificial processes.

For example, at the pressures and temperatures of an **in situ** hydrate formation in deep oceans (Section 7.1.2), one would anticipate severe problems with hydrate formations in water-based drilling fluids in wells. The same thermodynamic phase conditions which promote hydrate formation in offshore pipelines also enable formation as a result of methane production from decaying organic matter in ocean sediments. Evidence of **in situ** hydrate formation from crude oils with no vapor phase present (Section 7.1.3) indicates hydrate formation is possible in crude oil pipelines with no vapor phase present. As another illustration, in permafrost regions, Bily and Dick (1974) first presented drilling problems due to **in situ** hydrate dissociation from heat generated in the drilling process. Finally, the economics of gas recovery from **in situ** hydrates suggest that depressurization of a hydrate process or pipeline plug may be more viable than thermal dissociation, depending upon the value of such things as waste heat and flow disruption.

The purpose of this chapter is to consider the application of the physical and chemical phenomena presented in Chapters 2 through 5 to hydrate formation in hydrocarbon production and processes. In such a diverse industry, an exhaustive study of these applications would be unwieldy. Therefore only a few illustrations of the principles are given, leaving further applications to the reader. Principles and illustrations of formation, prevention and dissociation processes are given in Section 8.1. The special case of hydrate formation and dissociation in well drilling is given in Section 8.2.

## 8.1 Methods of Hydrate Prevention and Dissociation

The current state-of-the-art requires that the methods of hydrate prevention and dissociation be estimated via thermodynamics rather than through kinetic means. The fundamental reason for this is the relative difficulty of measuring time-dependent (kinetic) properties in the field or laboratory, compared to time-independent (thermodynamic) measurements. As noted in Chapter 3, substantial progress is currently being made in the area of hydrate kinetics, but it may be several years before confidence in a standard kinetics calculation method is realized via industrial application. In contrast, most of the phase equilibria methods in Chapters 4 and 5 have attained widespread industrial use. Consequently thermodynamic equilibrium methods are used exclusively in the processes which follow. In these processes the "hydrate formation point" should be taken to mean the point of incipient hydrate formation, or the highest possible temperature at which hydrates exist for a given pressure. As shown in Chapter 3, an as-of-yet undefined degree of metastability may prevent immediate hydrate formation at conditions within the thermodynamically stable region; therefore only the thermodynamically stable point is considered.

The introductory portion of this section discusses a typical gas gathering/processing system as a means for determining where hydrate formation and dissociation conditions are likely to occur. The succeeding portions of the section illustrate the physical and chemical principles associated with individual portions of the processing example provided in Section 8.1.1.

The means of hydrate prevention and dissociation are determined from the phase diagrams. Even the most sophisticated dissociation techniques (e.g. microwave, geothermal, or nuclear warming, etc.) have their foundation in the pressure-temperature and isobaric T-x diagrams such as shown in Section 4.1. Fundamentally, there are four ways to dissociate gas hydrates or to prevent them from forming:

1. remove the water to lower the dew point,
2. control the temperature of the system,
3. control the pressure of the system, and
4. use inhibitors to shift the equilibrium.

The above methods are often used individually, or jointly for prevention and/or more rapid hydrate dissociation. The first of the above methods, discussed in Section 8.1.2, is related to each of the isobaric temperature-composition diagrams of Figure 4-2 by attaining those temperatures, pressures, and water concentrations in the single gas or liquid hydrocarbon phase regions, well away from any phase boundary. The last three techniques, treated in Sections 8.1.3, 4, and 5 may be related to a portion of a pressure-temperature phase diagram such as Figure 4-1 between quadruple points $Q_1$(I-$L_W$-H-V) and $Q_2$($L_W$-H-V-$L_{HC}$) for a constant composition gas mixture in the region well away from the three-phase ($L_W$-V-$L_{HC}$) region. A second type of pressure-temperature diagram, with the intrusion of the three-phase ($L_W$-V-$L_{HC}$) envelope, is given as a special case of the discussion here, in Section 8.1.6. All of the phase diagrams may be obtained with the methods of Chapters 4 and 5.

### 8.1.1. Typical Gathering System with Hydrate Occurrence Points

A simple gas gathering/processing system is shown in Figure 8-1. The gas comes from the well at (point A), and then may be heated before point B, where it passes through a pressure regulation valve to point C. At point C, depending upon the climate and gas conditions the operator may elect to dehydrate or inject inhibitor to prevent hydrate formation at point D, before the gas enters the gathering system. Gases from several wells may be gathered in this manner before they are of sufficient quantity to process economically.

Gas from the gathering line proceeds some distance to the processing plant, where the entire stream may be dehydrated or inhibitor may be injected (E) before compression (F), cooling in a chiller (G), and expansion, via either a valve or a turboexpander, to provide liquid recovery (H) from the gas stream. The gas then goes to the user where it is burned for heating value. The recovered hydrocarbon liquids may be used as chemical feedstocks.

In such a process hydrate formation and blockage jeopardizes operation at each point which has conditions of high pressure, low temperature, or cooling by expansion, such as across the various valves or across a turboexpander. Conversely, if hydrate formation has occurred, it is necessary to decrease the pressure, in-

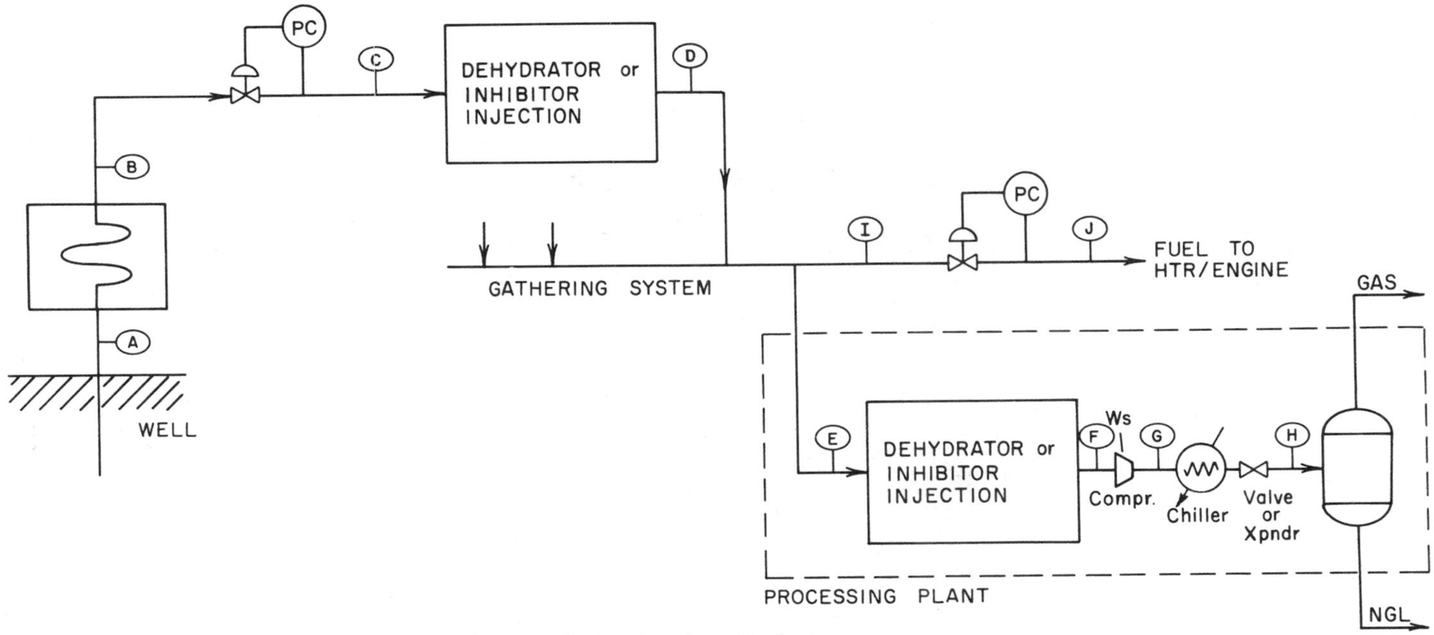


8-1 Typical Gas Production, Gathering, and Processing System crease the temperature, or inject an inhibitor at the point of blockage. Typically hydrate formation and accumulation will occur in a process at the below points:

• low lying points of the equipment, such as pipelines passing underneath a roadway,
• points of change in direction of flow, such as pipe elbows, and
• points of flow obstruction, such as screens preceding heat exchangers.

As an example, to prevent hydrate formation in a pipeline during cold weather, it is common to maintain the temperature and pressure conditions above the hydrate formation conditions and safely within the vapor-liquid two-phase (or all vapor) region. In addition to pipeline burial, a common technique is to install a heater system at the wellhead, between points A and B in Figure 8-1, to raise the gas temperature (Dendy, 1969). The heat exchanger can be as simple as a barrel of hot oil through which a coiled gas pipe passes (Handwerk, 1988). Such a heating system may be a cost effective alternative to inhibitor injection, particularly in cases of short duration of cold weather. However, hydrates may re-form downstream if the pipeline is long enough for heat losses to return the stream to below the hydrate temperature.

Handwerk (1988) provided the following three examples for which hydrate formation was typical in a system such as that shown in Figure 8-1.

<u>At Point J in Figure 8-1</u>. Gas which has not been dehydrated (i.e. wet gas) is taken from a high pressure well or gathering line and used for heater or compressor engine fuel at lower pressures. Typically the gas may pass through two pressure reduction regulators (control valves) in series. The Joule-Thomson effect reduces the temperature enough to allow hydrates to form and they plug the control valve or the piping downstream of the valve at some restriction such as an orifice.

<u>At Point F in Figure 8-1</u>. Cryogenic gas processing units are usually preceded by molecular sieve dehy-

dration systems. Over time, enough water can pass through these to allow hydrates to slowly collect in downstream equipment in the cold sections and cause plugging and resulting loss of flow. The usual places where this becomes a problem are inlet dust-dirt screens at plate-fin heat exchangers and cryogenic pumps or expanders.

The time frequency for such problems is often weeks. However, if regeneration of the dessicant is not adequate or if the dehydration unit is not operating properly, hydrates can plug equipment in a matter of hours. The usual "on-line" short term remedy is to inject a small amount of methanol either periodically or continuously.

At Point G in Figure 8-1. Wet gas being processed for hydrocarbon liquid recovery by refrigeration often forms hydrates in the chiller or related cold gas/warm gas, shell and tube heat exchangers. The wet inlet gas flows through the tubes of the exchangers or chillers and the metal tube wall temperature is cold enough to cause hydrates to "freeze" on the inside surface of the tubes which in turn results in a flow restriction and corresponding high pressure drop.

The above phenomenon is always anticipated and ethylene glycol is continuously sprayed into the feed gas to prevent - or more correctly to dissociate - the hydrate formation and/or accumulation. The hydrate problems that occur in these systems normally result from the following factors:

1. unequal distribution of glycol in the heat exchanger tubes,
2. excessive glycol viscosity so that flow is restricted through the tubes, and
3. an inadequate amount of glycol for inhibition.

With the above overview of a typical process and examples of places where hydrate formation might occur, the remaining portions of Section 8.1 consider the individual means of hydrate control (e.g water removal, temperature control, pressure control, and inhibition).

### 8.1.2. Hydrate Control Through Water Removal (Dew Point Lowering)

The initial technique to control hydrates is to remove either the host molecule or sufficient guest molecules necessary for stability of the cages. The removal of the guest molecules may be thought of as depressurization, an option treated in Section 8.1.4, which is not desirable due to the interruption of the flow process. With respect to removal of host molecules, however, the following passage is still true today, as found by Deaton and Frost (1946, p 41.) almost a half-century ago:

> "The only method found to be completely satisfactory in preventing the formation of hydrates in gas-transmission lines is to dehydrate the gas entering the line to a dew point low enough to preclude formation of hydrates at any point in the system."

Two common misconceptions exist concerning the presence of water to form hydrates in pipelines, both of which are discussed in detail in Section 4.6. The first and most common misconception is that a free water phase is absolutely necessary for the formation of hydrates. The second misconception is that ice is the solid phase in equilibrium with hydrocarbon fluid at temperatures below the hydrate point. Figure 8-2 (originally presented as Figure 4-2g) serves as visual confirmation that it is thermodynamically possible to achieve H-V and H-$L_{HC}$ equilibrium, and that ice is not the correct condensed water phase over the majority of the temperature range.

Regarding the first misconception above, from a practical standpoint, the water concentration of the hydrocarbon phases at the two phase boundary is so low (typically < 1000 ppm[mol]) that substantial time may be required for the water molecules to accumulate into a large enough hydrate mass to be of concern. As indicated in Chapter 3, the time required for hydrate nucleation from a single phase hydrocarbon is in excess of that available in most laboratory studies to date. However

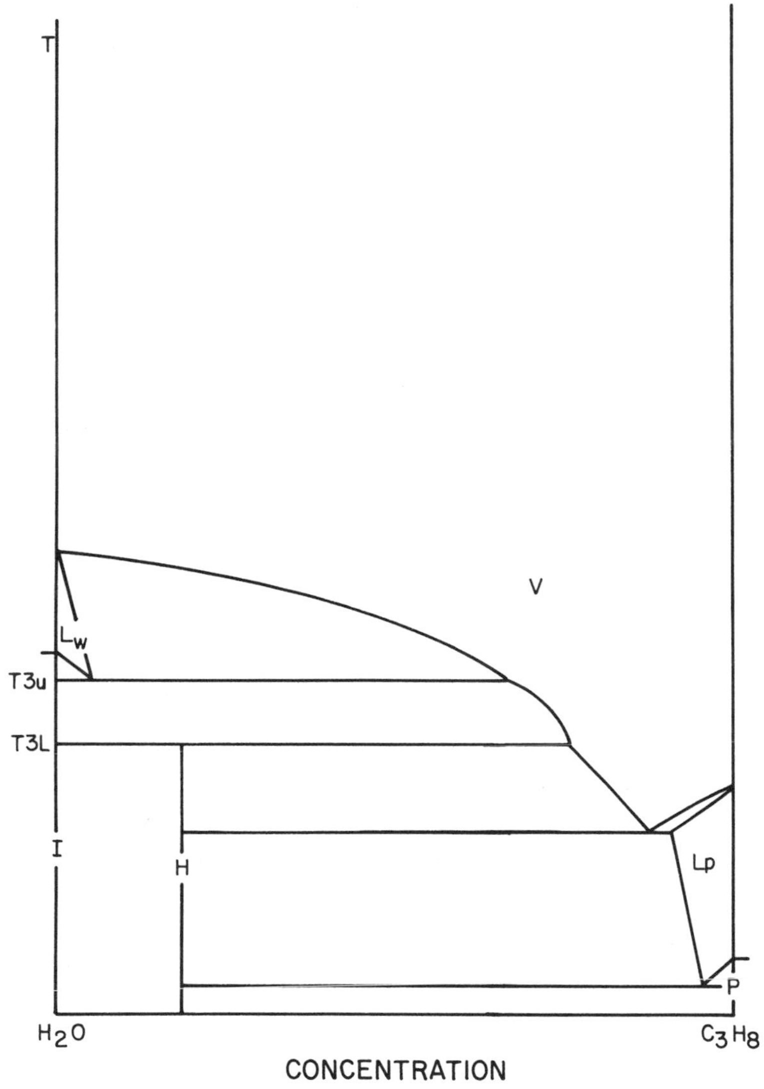


8-2 Isobaric Temperature-Composition Diagram for Propane+Water (Reproduced, with permission, from Can. J. Chem. Eng., (Harmens and Sloan, 1989), by Can. Soc. for Chem. Eng.)

for a process such as a pipeline, which is in continuous use over a number of years, such nucleation may occur. In addition there have been multiple studies (Sloan et al. (1976); Cady (1983a,b); Woolridge et al. (1987);

Kobayashi et al. (1987)) which demonstrate that hydrate growth can easily occur from a hydrocarbon phase if a hydrate nucleus is either already present, at adsorbed sites on a wall, or on a third surface. This implies that if any incident caused free (or adsorbed) water to enter a process, the water could entirely convert to hydrate and continue to grow (in the absence of other free water) to the point of flow obstruction.

The second common misconception is that, if two phases (hydrocarbon and one other) do exist, then the solid phase will be ice (Bucklin et al. 1985). Figure 8-2 (as well as all of the T-x diagrams of Figure 4-2) shows that ice is the correct second condensed phase in equilibrium with hydrocarbon only over a very small temperature range between $T_{3U}$ and $T_{3L}$. Below $T_{3L}$ the second condensed water phase is hydrate, not ice, in equilibrium with vapor hydrocarbons just below $T_{3L}$ or with liquid hydrocarbons (or solid hydrocarbons) at lower temperatures. For the more common solid hydrate dew point, methods are presented in detail in Section 4.6.1, Section 5.4, and in the CSMHYD program on the floppy disk in the frontpapers. Details of dew point calculations for hydrocarbon with the less common solid phase (ice) are given in Section 4.6.2.

Except in arctic conditions, the maximum water content of gas is specified as 4 $lb_m$/MMSCF; typically the water content is 2.25 $lb_m$/MMSCF. After the free water is separated from the gas stream, two processes are commonly used to lower the dew point by removing dissolved water. These two processes have evolved from the earlier three types suggested in 1946 (p.42ff) by Deaton and Frost (hygroscopic solutions, chemical absorption by solids, and physical adsorption by solids) because economics have eliminated the non-regenerative use of solid chemical absorbants. In the first process, the water concentration is lowered by contacting the gas with a compound, typically tri-ethylene glycol (TEG), which removes water through hydrogen bonding. In the second process, the gas is contacted with a solid, such as a molecular sieve, alumina or silica gel, which selectively adsorbs water.

In the first process, the hygroscopic liquid TEG absorbs water from the gas in an isothermal equilibrium stage absorber, and is subsequently heated to remove the water before recycling the TEG fluid to the absorber. The hydrogen bonding of water to TEG is so effective that usually only two or three ideal stages are required, with circulation rates of 2 to 5 $lb_m$ glycol/($lb_m$ water removed). Overall tray efficiency is typically 25% to 35% so that the absorber usually contains 8 to 12 trays. The detailed design and operation of such a process is outside the scope of this book (see Perry, 1960; Hall and Polderman, 1960; Loomer and Welch, 1961; Redus, 1965; Sivalls, 1974; Gas Processors Association Handbook, 1987). Bucklin et al. (1985) have used TEG for control of hydrates in arctic conditions, for water contents as low as 0.05 $lb_m$/MMSCF or a dew point of $-60^0$F.

Within ten years of Hammerschmidt's discovery of hydrates in pipelines, he and his colleagues were publishing details of gas dehydration with solid dessicants. The commercial use of the process dates to just after the Second World War. In general, solid dessicants are placed in parallel packed bed adsorbers through which gas is passed. One adsorber is used to remove water from the gas while the parallel unit is being regenerated at high temperature or cooling. There are a number of solid dessicants, but the more common ones, in order of increasing drying efficiency, are silica gel, alumina, and molecular sieves. Table 8-1 presents a comparison of the obtainable dew points and properties of those three solid dessicants.

In recent years molecular sieves have gained popularity due to their advantages of providing extremely low dew points, high adsorption of water (rather than large amounts of hydrocarbon) and, unlike silica gel, they are not damaged by liquid water. Molecular sieves are crystalline solids which provide the lowest dew points obtainable and they are commonly used in processes such as turboexpander plants, where the low temperature requirements are the most stringent, perhaps as low as 0.1 ppm. Molecular sieves are designed to adsorb particular mole-

**Table 8-1**
**Properties of Solid Dehydration Agent**
Modified from Hales (1982)

| Property | Molecular Sieve | Silic Gel | Activated Alumina |
|---|---|---|---|
| Range of Dew Points, $^0F$ | -150→150 | -80→20 | -100→130 |
| Typical Water Capacity ($lb_mH_2O$/100 $lb_m$ adsorbent) at 298K | | | |
| 10% Relative Humidity | 20 | 6 | 5 |
| 20% Relative Humidity | 21 | 10 | 7 |
| 30% Relative Humidity | 22 | 15 | 9 |
| 40% Relative Humidity | 22.5 | 20 | 11 |
| Major Adsorbent Constituents (% dry basis) | | | |
| $SiO_2$ | 40-42 | 99.82 | |
| $Al_2O_3$ | 33-36 | | $95^+$ |
| Surface Area ($m^2/g$) | 600-800 | 750-800 | 200-350 |
| Pore Volume ($cm^3/g$) | 0.28 | 0.43 | 0.21-0.34 |
| Average Pore Diameter (Å) | 3,4,5,6,8,&10 | 22 | 26-41 |
| Cost (\$/$lb_m$,1977) | 1.3-1.8 | 0.48 | 0.21-0.605 |

cules: based upon polarity they adsorb with decreasing selectivity water, alcohols, glycols, hydrogen sulfide, carbon dioxide, while they adsorb mercaptans and heavy organics somewhat less, based upon mass.

The design and operations of solid dessicant processes are beyond the scope of this book, but they are commonly available in the literature (see Barry, 1960; Kessock and Fris, 1965; Cummings, 1977; Dameron, 1977; Hales, 1982). Nielsen and Bucklin (1983) determined that methanol injection offers economic advantages, both relative to activated alumina for incoming gases with 4 $lb_mH_2O$/MMSCF, and relative to molecular sieves for incoming gases with 23 $lb_m$/MMSCF in a turboexpander plant application. For the very low temperature processes, such as turboexpanders with discharge temperatures less than -160$^0$F, however, the normal practice is to use molecular sieves.

### 8.1.3 Hydrate Control Through Heating

On Figure 8-3 consider a trace of the second method of hydrate control through heating. The cooling or formation process may be examined by reversing the processes specified in this section. First consider the case for which hydrate formation has occurred with excess gas, so that all of the free water has been converted to hydrate, and the system consists of only vapor and hydrate in equilibrium at point A. The object of heating, depressurization, and inhibitor injection is to move the system from Point A across the three-phase ($L_W$-H-V) line into the vapor-liquid region. If hydrate formation occurs from a gas and a free water phase, the starting point for the heating process will be on the three-phase line.

Heating may occur at constant pressure in a heat exchanger, with sensible heat input to the vapor and hydrate causing movement from Point A to the equilibrium line at Point B, where the first drop of free water appears as a result of hydrate dissociation. At Point B, all of the energy input is spent on dissociation of the hydrate at constant temperature and pressure. As more hydrate is dissociated the amounts of the gas and liquid phases increase until the complete disappearance of the hydrate. At such time sensible heat causes the remaining vapor and liquid phases to progress toward Point C. The operator may then wish to remove the free water phase using mechanical means, such as a pipeline "pig." Alternatively the operator may choose to dehydrate the equipment through evacuation. As discussed in Chapter 3, hydrates will re-form much more easily if any free water present had previously been converted to hydrates, dissociated, and retained some residual structure.

From a practical standpoint the heating to dissociate a hydrate plug must be done with caution, beginning from the ends and progressing toward the middle of the plug. If a hydrate plug is dissociated in the middle, the undissociated hydrate at either end may act to enclose an exponentially increasing pressure associated with an increase in temperature. Very high pressures

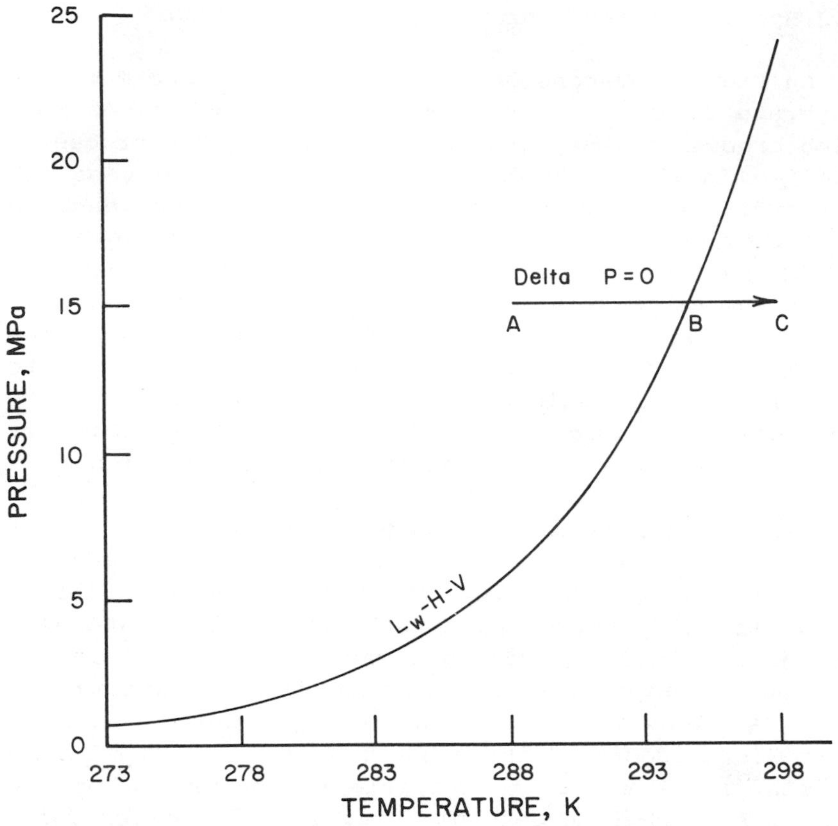


8-3 Isobaric Heating from the Vapor-Hydrate Region (A) Across the Three-Phase Line into the Vapor-Liquid Water Region (C)

in the pipeline can result in equipment wall failure, blowouts, and fires. Heating for hydrate dissociation must be done with some caution, because it is not uncommon for tubing/piping failure to occur.

One possibility of removing hydrate plugs (Hale (1988)) which can be reached by flowing chemicals is to use small amounts of aqueous chemicals with a high exothermic heat of reaction, such as those indicated by Ashton et al. (1986). The reaction combines sodium nitrite ($NaNO_2$) and ammonium nitrate ($NH_4NO_3$) or ammonium chloride ($NH_4Cl$) as follows:

$$NaNO_2 + NH_4NO_3 \rightarrow N_2 + 2H_2O + NaNO_3 \qquad (8.1)$$

A three molar solution of the above reactants produces 400 SCF ($11m^3$) of nitrogen gas and 132,500 BTU's (879 $MJ/m^3$) of heat per barrel of solution. The sodium nitrate produced gives a brine that is equivalent to a 9.7 $lb_m/gal$ ($1152kg/m^3$) sodium chloride solution. If such chemicals can be reacted in the vicinity of a hydrate plug, then the heat generated will act to dissociate the hydrate. As a caution, Dewan (1989) noted also that the reaction generates nitrogen, which may add to the exponential increase in gas pressure, as the temperature increases for hydrate dissociation; this may in turn cause pipeline rupture. While the above method appears promising, it is too recent to have obtained wide-spread industrial use.

### 8.1.4 Hydrate Through Depressurization

There are three means to control hydrates by system pressure reduction: (1) isothermally, as with infinitely slow pressure reduction, (2) isenthalpically, as with pressure reduction in a Joule-Thomson valve expansion, and (3) isentropically, as with pressure reduction through an ideal turboexpander. The three processes are shown for similar pressure drops, beginning from Point A in Figure 8-4a. All three techniques relate to the three-phase ($L_W$-H-V) line, reproduced from that in Figure 4-1, although the last two require the additional calculation of the enthalpy or entropy of the system. As in the previous section, it is assumed that initially, only vapor and hydrate are present at Point A and that hydrate formation has ceased. If hydrate formation is occurring from free water, the system initially exhibits three-phase ($L_W$-H-V) conditions but, with the exception of the starting point, the processes will progress in a similar manner to those described.

As will be shown in the following paragraphs, a common misconception exists about hydrate depressurization. It is commonly thought that depressurization alone will cause dissociation; instead depressurization must be aided by heat input from the surroundings, such as the air or ground. The real depressurization process is usually intermediate between the isothermal and adiabatic cases. From a First Law analysis, since the dissociated hydrate components have more energy in the vapor and liquid than they do in the solid phase, energy to promote hydrate dissociation must come in from the surroundings, be they the ocean floor, a heat exchanger, etc. The fundamental purpose of depressurization is to decrease the pressure at the hydrate so that the equilibrium temperature is below that of the surroundings. The temperature gradient from the surroundings then causes heat input and dissociation of the hydrate. Dissociating hydrates can obtain the heat of dissociation from the hydrate mass itself, thereby lowering its temperature.

From a practical standpoint, it is necessary to depressurize a hydrate plug simultaneously from both ends. If only one end of a plug is depressurized, the pressure gradient across the dislodged plug may cause a hydrate projectile in the pipeline or equipment with resulting damage and safety problems. As a consequence, it is important that there not be a differential pressure across the hydrate plug.

Isothermal hydrate depressurization represents a upper thermal limit to the expansion, shown as a vertical line in Figure 8-4a. Unless the temperature of the surroundings is increased from that prior to hydrate depressurization, the temperature of the dissociating hydrate mass cannot increase beyond its original value. With isothermal pressure reduction the process is extremely slow, so that the surroundings replenish the heat taken from the system as the hydrate mass dissociates. Such an isothermal process may occur for example, with the controlled depressurization of a long ocean pipeline or buried gas transmission pipeline to remove a hydrate plug over a period of days or weeks. As pressure is reduced, sensible heat is supplied from the surroundings to both the hydrate and vapor phases to maintain the temperature while the system moves from Point A to Point B. At Point

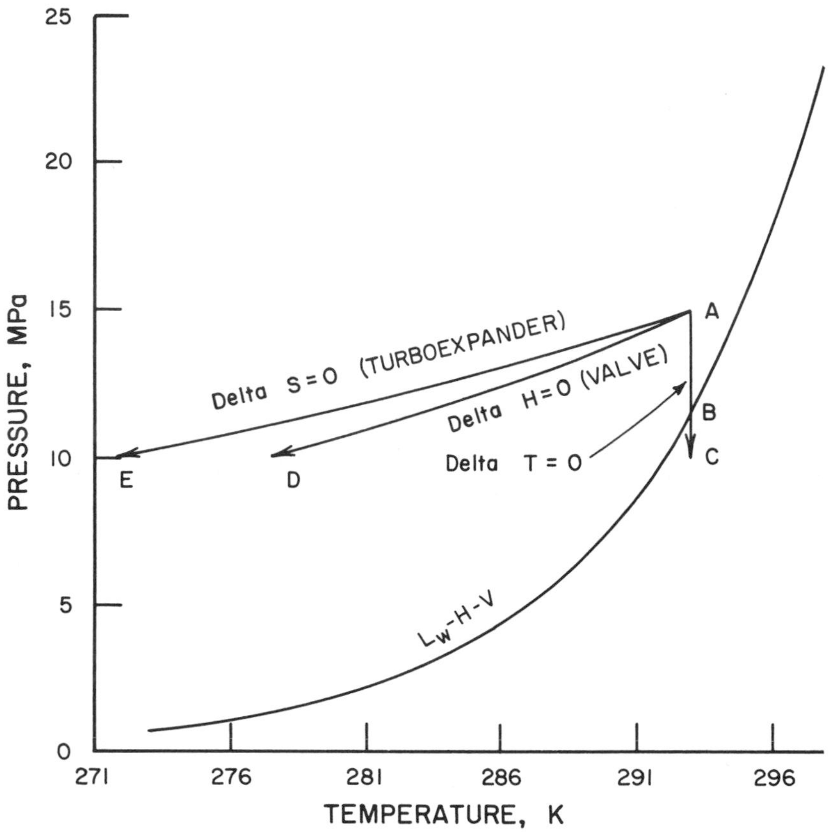


8-4a Three Depressurization Curves from the Vapor-Hydrate Region

B the hydrate begins to dissociate and heat of hydrate dissociation is input to maintain constant temperature until hydrate dissociation is complete, so that further pressure reduction causes the system to progress toward Point C, requiring only sensible heat input to vapor and liquid phases.

A more common pressure reduction for process equipment or smaller pipeline systems is the isenthalpic ($\Delta H=0$) process which occurs with rapid depressurization in the absence of heat exchange, as shown in line $\overline{AD}$ of Figure 8-4a. For rapid hydrate dissociation in a flow

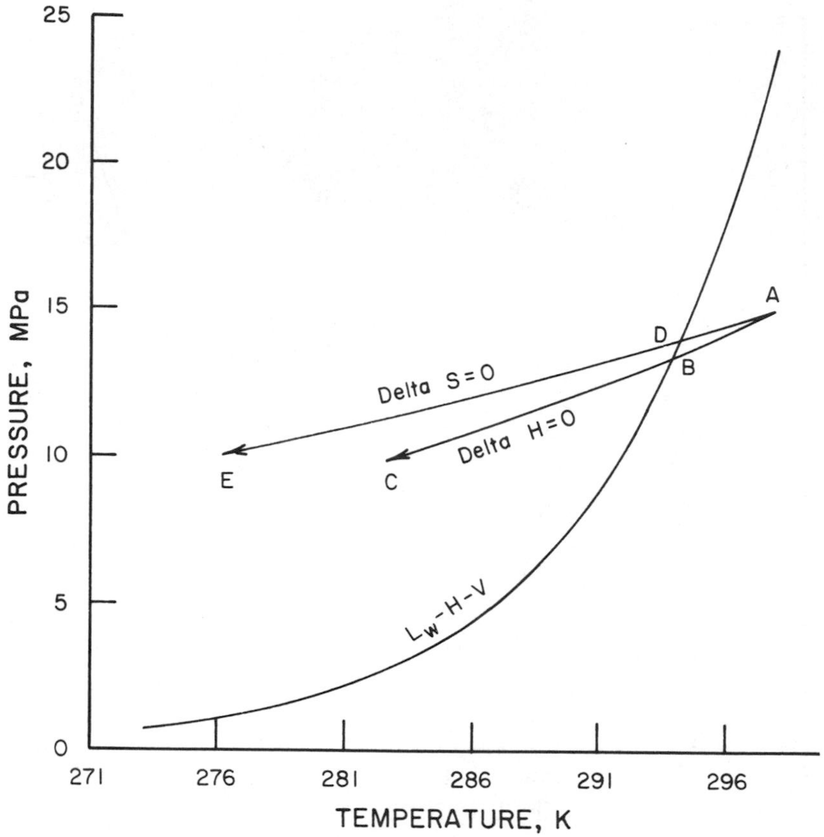


8-4b The Effect of Isenthalpic (Delta H = 0) and Isentropic (Delta S = 0) Depressurization from the Vapor-Liquid Water Region

system, neither heat (Q) nor work ($W_s$) is normally exchanged with the surroundings, so by the First Law of thermodynamics for a system at steady state ($\Delta H=Q-W_s$) the enthalpy change is zero. Relative to the isenthalpic case, any heat input to the system will cause the temperature to increase, while any work removed from the system will cause the temperature to decrease. The isenthalpic limit is normally called a Joule-Thomson expansion.

While the Joule-Thomson expansion characteristics of gases differ, the slopes of the isenthalpic expansion curves are generally much shallower than the three-phase ($L_W$-H-V) equilibrium curves. Therefore, with this type of pressure reduction most of the gases will expand toward (or further into) the hydrate region, i.e. either promoting increased hydrate formation, as shown starting from Point A in Figure 8-4a, or toward the three phase ($L_W$-H-V) equilibrium line from the vapor-liquid region, as from Point A in Figure 8-4b. Isenthalpic expansion, therefore, generally promotes hydrate formation, rather than allowing their removal. As shown in line $\overline{AD}$ of Figure 8-4a, heat from the surroundings will be required to increase the temperature (usually at a constant final pressure such as Point D) to a value beyond the three-phase line. The mechanism of isenthalpic expansion alone will not cause hydrate suppression or removal.

The limit to hydrate-free isenthalpic expansion from the vapor-liquid region, i.e. the point at which hydrates begin to form, is demonstrated in Figure 8-4b by the intersection of the $\Delta H=0$ line $\overline{ABC}$ with the three-phase ($L_W$-H-V) line. These limits are quantified in the original work of Katz (1945) as shown in Figures 4-8 through 4-10 in Section 4.3.1. Later work by Loh et al.(1983) and by Sloan (1984) indicate some limits of those original charts, determined through use of more accurate hydrate data and better equations-of-state.

At very high pressures (8,000 - 10,000 psia) the Joule-Thomson inversion locus is reached for most natural gases. Above those pressures $(dT/dP)_H$ can be negative which causes the temperature to increase as the pressure decreases. Also it is possible for adiabatic vapor expansions to cause both entry and exit of the hydrate phase envelope or the hydrocarbon vapor-liquid phase envelope, such as that shown in Figure 4.5.

The isentropic ($\Delta S=0$) depressurization is less common and represents a lower limit to the adiabatic ($Q=0$) temperature of depressurization, given by lines $\overline{AE}$ in Figures 8-4a,b. The isentropic process generally requires that energy be removed from the system in the form of shaft work, as with a reversible turboexpander, or that the diameter of the flow conduit be changed, as in a nozzle, in order to decrease the temperature below the isenthalpic limit for a given pressure decrease. Flow through such a process is normally too rapid to allow substantial heat exchange with the surroundings. Since the entropy change is nil, these process conditions are an ideal limit rather than a realization.

Here again the isentropic expansion process promotes hydrate formation more rapidly than the isenthalpic expansion, so that at the end of such a process to eliminate hydrates, heat from the surroundings must be used to dissociate hydrates. Line $\overline{ADE}$ in Figure 8-4b demonstrates the limit for an isentropic expansion from the vapor-liquid region at intersection Point D with the three-phase ($L_W$-H-V) boundary.

If either an adiabatic or isentropic expansion is anticipated, there are several alternatives for the operator. He may choose to heat the gas to a temperature considerably to the right of point A in Figure 8-4b, so that the resulting expansion line will not cross the three-phase ($L_W$-H-V) line. He may choose to decrease the expansion from point A, so that the surroundings supply enough heat to ensure that the three-phase line is not crossed. If the expansion cannot be constricted, the operator may even elect to fill the pipeline to prevent rapid depressurization and to provide another heat source. Finally the operator may elect to use inhibitors to shift the three-phase line to the left so that the expansion will not cross the three-phase boundary.

On the other hand, the gas processor frequently chooses to undergo an isenthalpic expansion, combined with refrigeration of the incoming gas, to form hydrates and consequently lower the water dew point of the gas. The solid hydrates are accumulated in a separator immediately after the expansion valve. This technique is called the LTX (low temperature expansion) process. This process may be used when pressure drop is not a concern or when high pressure is economically available. A computer program similar to that in Appendix A may be used to determine the discharge pressure which will be within the hydrate region, which may be supplemented by cooling. As a general rule, lower temperatures and higher pressures will discourage hydrate meta-stability and promote more and faster hydrate formation.

**Example 8.1. Expansion of a Natural Gas**

A simple natural gas consists of 90 mol% $CH_4$, 7% $C_2H_6$, and 3% $C_3H_8$. Free water may occasionally be present in a pipe in which the gas flows. Two initial process inlet conditions are considered: (a) 293K and 15 MPa, or (b) 298K and 15 MPa. For either condition, is hydrate formation a possibility? Are there process limitiations on the means of expansion from either initial condition to 10 MPa?

**Solution:** Before doing any hydrate calculations it is worthwhile to confirm that this gas is not close to the hydrocarbon dew point, to determine the possibility of encountering two fluid hydrocarbon phases. A vapor-liquid equilibria flash calculation indicates that the highest temperature at which a hydrocarbon liquid can occur (the cricondentherm) for this mixture is 231 K, so the process will not form free hydrocarbon liquid.

As a second step, the three-phase ($L_W$-H-V) conditions of the gas should be determined. Since no experimental data are available for a gas of this composition, we may calculate the equilibria either from the gas gravity charts, the K-value charts, or the statistical thermodynamic method, as found in the program CSMHYD, on the floppy disk in the frontpapers of this book. Choosing the latter method, a three-phase line is generated for the subject gas at the following conditions:

| | | | | | | | | | | | |
|---|---|---|---|---|---|---|---|---|---|---|---|
| T,K | 273 | 275 | 278 | 281 | 283 | 286 | 288 | 291 | 293 | 295 | 298 |
| P,MPa | .82 | 1.03 | 1.47 | 2.09 | 2.54 | 3.8 | 4.95 | 7.7 | 11.2 | 17.3 | 27.9 |

A semi-logarithmic interpolation of the above values gives the three phase equilibrium point at 294 K when the pressure is 15 MPa. Therefore the initial condition of

293K and 15 MPa is within the hydrate formation region and the initial conditions of 298K and 15 MPa is in the vapor-liquid (water) region.

If the system at 293 K and 15 MPa has formed hydrates, consider the three means of depressurization. If the system pressure is lowered to 10 MPa slowly and isothermally (with associated heat input) it will cross the three-phase line at 11.2 MPa and will exist in the vapor-liquid water region at 10 MPa. On the other hand, without further heating from the surroundings, the isenthalpic and isentropic depressurization processes give much colder gas at 10 MPa. Using an isenthalpic and isentropic expansion computer program, such as in Appendix A (also on the floppy disk in the frontpapers) the following isenthalpic and isentropic lines are obtained:

| | | | | |
|---|---|---|---|---|
| P, MPa | 13.8 | 12.4 | 11.03 | 10 |
| T,K for $\Delta H=0$ | 289.9 | 285.9 | 281.3 | 277.4 |
| T,K for $\Delta S=0$ | 288.6 | 283.2 | 276.9 | 271.7 |

As plotted in Figure 8-4a, these lines show both expansions to intrude further into the hydrate region. Only with subsequent heating at a constant pressure of 10 MPa, will the system traverse the hydrate three-phase condition at 292.4K.

A similar calculation for the system initially at 298K and 15 MPa shows the jeopardy of expansion by both isenthalpic and isentropic processes. The results, plotted in Figure 8-4b show isenthalpic and isentropic expansion intersections with the three-phase ($L_W$-H-V) boundary at (293.4K, 13.3 MPa) and (293.6K, 13.8 MPa) respectively. To prevent expansion into the hydrate region the system should be initially at a still higher temperature, heat or inhibitor should be added to the stream, or the expansion pressure should be limited.

### 8.1.5 Hydrate Control Through Use of Chemicals

Frequently inhibitors are injected into processing lines as a means of hydrate control by both the breakage of hydrate hydrogen bonds and the competition for available water molecules. The more frequently used inhibitors are strong polar fluids, such as methanol, the ethlyene glycols, and ammonia, arranged in order of decreasing usage frequency. Salts are only used infrequently, because their very low vapor pressures and high melting points prevent vaporization and liquefaction, respectively, needed for flow. As noted in Section 4.5, of the above three chemicals, the uses of methanol and the ethylene glycols are pervasive. While ammonia is more than twice as effective an methanol, it is only used in cases of severe hydrate blockage, (Deaton and Frost, 1946 p. 37f.) due to the reaction of $NH_3$ with $CO_2$ in the gas to form plugs of solid ammonium carbonate, bicarbonate, and carbamate. The handling of ammonia is also an important safety consideration.

The use of inhibitors shifts the three-phase ($L_W$-H-V) equilibrium curve to higher pressures and lower temperatures as shown in Figure 8-5 by relative positions of the solid (inhibited) line to the dashed line. The parallel nature of the two lines indicates the reason for the almost universal neglect of the pressure effect upon hydrate temperature depression. The effect of adding the inhibitor then, is to shift the three-phase equilibrium curve beyond the process conditions so that, in Figure 8-3, Point A would exist in the vapor-liquid region rather than the vapor-hydrate region.

Figure 8-5 may also be used to show that the hydrate temperature depression, ΔT, is always less than the ice temperature depression, ΔT′ at the same inhibitor concentration, regardless of the inhibitor. Nielsen and Bucklin (1983) indicate that the relationship of the two temperature depressions is given by:

$$\frac{\Delta T}{\Delta T'} = \frac{\Delta H(L_W \rightleftharpoons I)}{\Delta H(L_W + V \rightleftharpoons H)} = \frac{\Delta H(L_W + V \rightleftharpoons H) - \Delta H(I + V \rightleftharpoons H)}{\Delta H(L_W + V \rightleftharpoons H)}$$

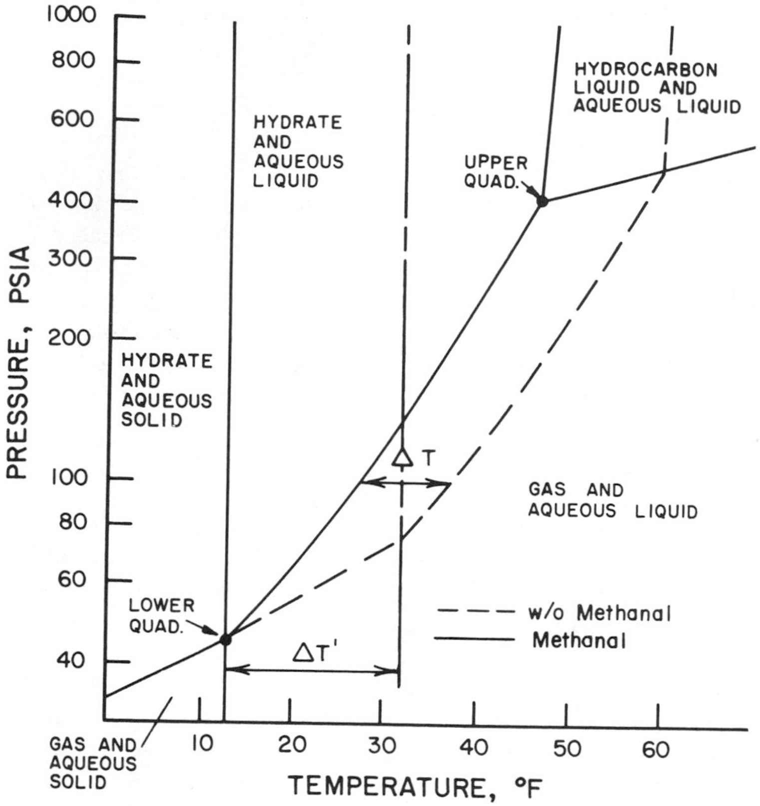


8-5 The Relative Freezing Point Depression of Methanol on Ice (ΔT′) and Hydrate (ΔT) (Reproduced, with permission, from Hydrocarbon Processing, (Nielsen and Bucklin, 1983), by Gulf Publishing Co.

or substituting in the ΔH values for hydrate formation from either ice or water in a ratio of about 0.35:

$$\Delta T = \Delta T'\left(1 - \frac{\Delta H(I+V\rightleftharpoons H)}{\Delta H(L_W+V\rightleftharpoons H)}\right) \approx 0.65\ \Delta T' \qquad (8.2)$$

Equation (8.2) together with the Hammerschmidt Equation (4.7) give a good representation of both glycol and methanol (up to 20 wt%) effects on hydrate depression, relative to the uninhibited three-phase condition at the same pressure:

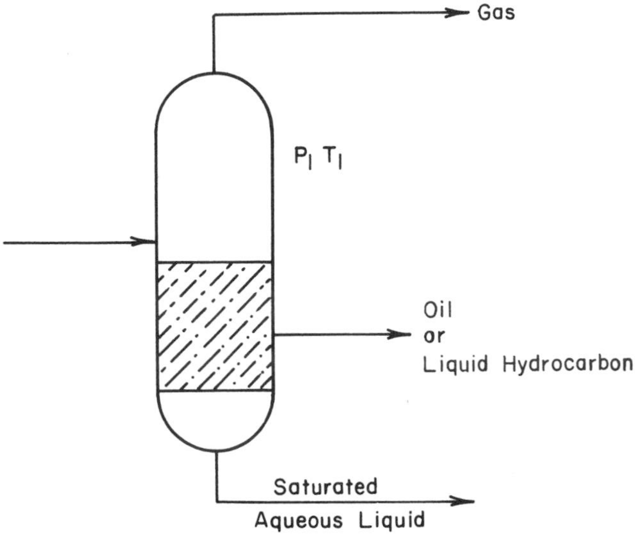


8-6 Three-Phase Separator

$$\Delta T = \frac{2{,}335\ W}{100M - MW} \tag{8.3}$$

where M and W are the molecular weight and the weight per cent of the antifreeze agent, respectively. Nielsen and Bucklin (1983) indicate that for glycols, a cancellation of errors enables the above equation (called the Hammerschmidt Equation) to fit up to a mol fraction of 0.4. Townsend and Reid (1978) suggest that the original Hammerschmidt equation will adequately represent depressions up to 60 wt% for ethylene glycol; if the constant 2,335 is changed to 4,000, it is acceptable for diethylene glycol up to 75 wt%. Ng and Robinson (1983) have presented data to show however, that methanol inhibits hydrate formation more than an equivalent mass fraction of glycol in aqueous liquid. Nielsen and Bucklin (1983) suggest that for methanol the above equation may be corrected for solutions above 20 wt%, for systems operating as low as 166.5K for hydrate control by:

$$\Delta T = -129.6\ \ell n(1 - x_{MeOH}) \quad \text{with } \Delta T \text{ in } {}^0F \tag{8.4}$$

Whether methanol or a glycol is used for hydrate depression, there are three possible phases into which the inhibitor may distribute. Figure 8-6 provides an illustration of a tank which separates the three fluid phases, as follows:

1. the free aqueous phase where the hydrate inhibition occurs,
2. the vapor hydrocarbon phase, where the inhibitor may be lost, as in the usual case of methanol injection, and
3. the liquid hydrocarbon phase, which has appreciable solubility for both methanol and glycols.

Depending upon the temperature and pressure, all three of the fluid phases may be present, or either the vapor or the liquid hydrocarbon may be missing. Typically the vapor hydrocarbon phase is present, but recent practice in subsea pipelines and cryogenic processing suggests that all three phases may be commonly found in the future. Normally calculations are performed assuming that the vapor-liquid hydrocarbon phase envelope is not affected by the presence of a third phase (water) due to the low solubility of water in the hydrocarbon liquid and the low vapor pressure of water. This simplification, which is almost always applicable, allows the calculation of the hydrate from a knowledge of the vapor phase composition. A more detailed discussion of the effect of two hydrocarbon phases on hydrate formation from mixtures is presented in Section 8.1.6.

The concentration of the inhibitor in the aqueous phase is addressed by the Hammerschmidt equation and the more rigorous inhibitor methods of Chapter 5. The loss of inhibitor to the vapor phases may be determined from Figure 8-7, taken from Nielsen and Bucklin (1983) for methanol, or from Figure 8-8 of Polderman (1958) for glycols. The heavier molecular weights of the glycols result in lower vapor pressures, so that their vaporization losses are less than methanol. Townsend and Reid (1978, p.88) indicate that ethylene glycol vaporization losses will be negligible at temperatures below the ice point when the operating pressure is above 5.17 MPa; va-

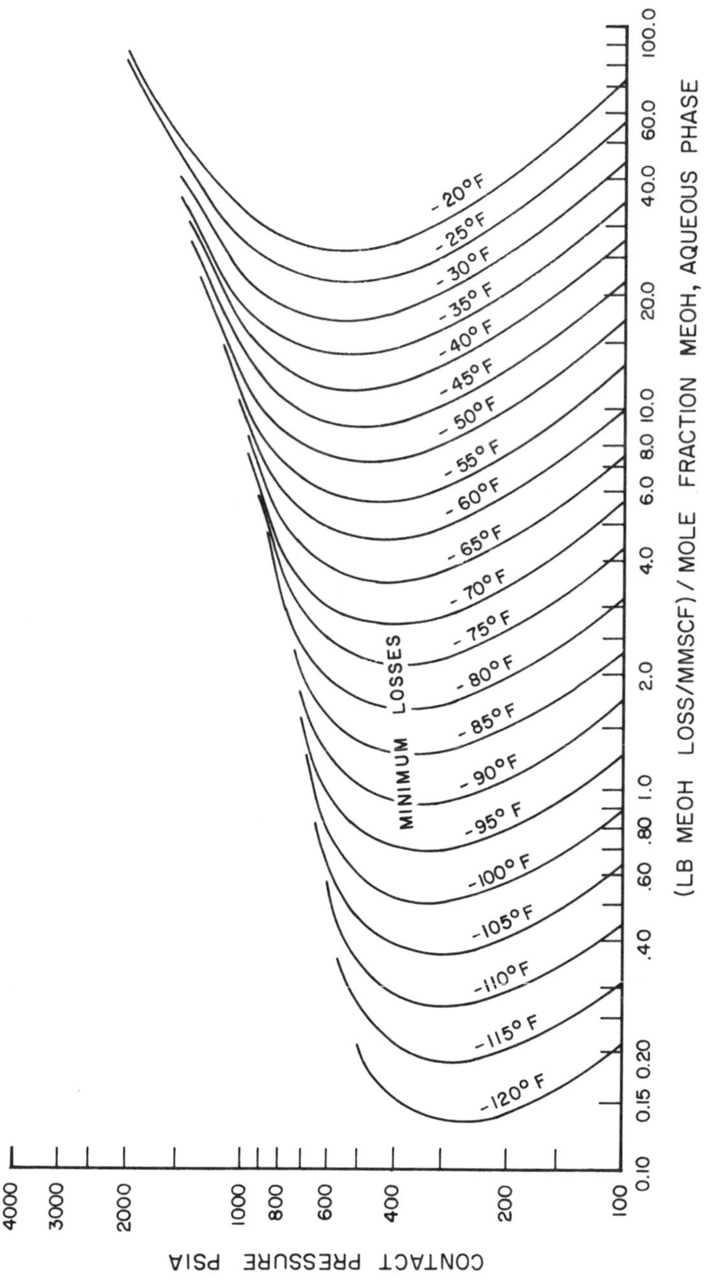


8-7 Loss of Methanol to the Vapor Phase
(Reproduced, with permission, from *Hydrocarbon Processing*, (Nielsen and Bucklin, 1983), by Gulf Publishing Co.

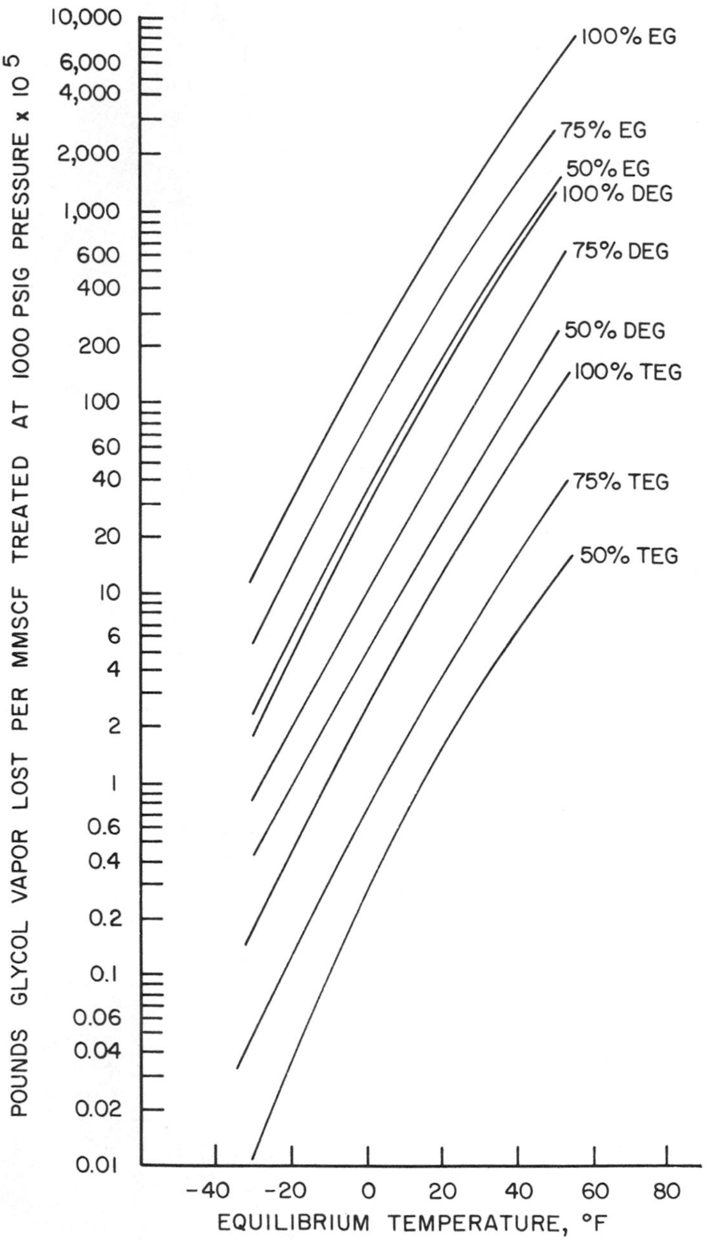


8-8 Loss of Glycols to the Vapor Phase
(Reproduced, with permission, from Proc. Gas Condit Conf (Polderman, 1958), by the University of Oklahoma

porization losses will be still less at these conditions with di- or tri- ethylene glycol.

The solubility of inhibitors in liquid hydrocarbons is ill-defined relative to their solubility in vapor hydrocarbons. Unfortunately, these inhibitors tend to dissolve in the hydrocarbon liquid so that losses here can be appreciable and more phase equilibrium data are needed in this area. The low temperature data which are available for methanol and glycol solubilities in liquid hydrocarbons are presented in Figures 8-9 and 8-10, respectively. Note that each of these figures can only approximate the hydrocarbon liquids normally encountered with natural gases. Chen et al. (1988) measured the high temperature (263K to 323K) solubility of methanol in binary hydrocarbon liquid mixtures of methane with heptane, mythlycyclohexane, and ethylene glycol solubilities were measured in the latter two binaries. In all cases ethylene glycol and methanol show very high solubility if aromatic compounds are present in the liquid hydrocarbon.

The injected methanol concentration is normally ≥ 98 wt%, while the typical ethylene glycol injected into pipelines often falls in the range 65-75 wt%. Most of the ethylene glycol injected will remain in the liquid phase after injection. Nielsen and Bucklin indicate that methanol is injected at a rate such that the concentration in any process/pipeline free water will be 90 wt%, and a typical safety factor is that the minimum recommended hydrate depression is the calculated depressed hydrate point plus $35^0$F.

### Example 8.2. Amount of Ethylene Glycol Injected as Inhibitor

It has been determined to use ethylene glycol to inhibit 28,320 SCM/D ($1\times10^6$SCF/D) of a gas flowing at 6.89 MPa (1000 psia) and 300K ($80^0$F). How much 67 wt% ethylene glycol should be injected to protect the line against a low temperature of 277.6K ($40^0$F). The dry composition of

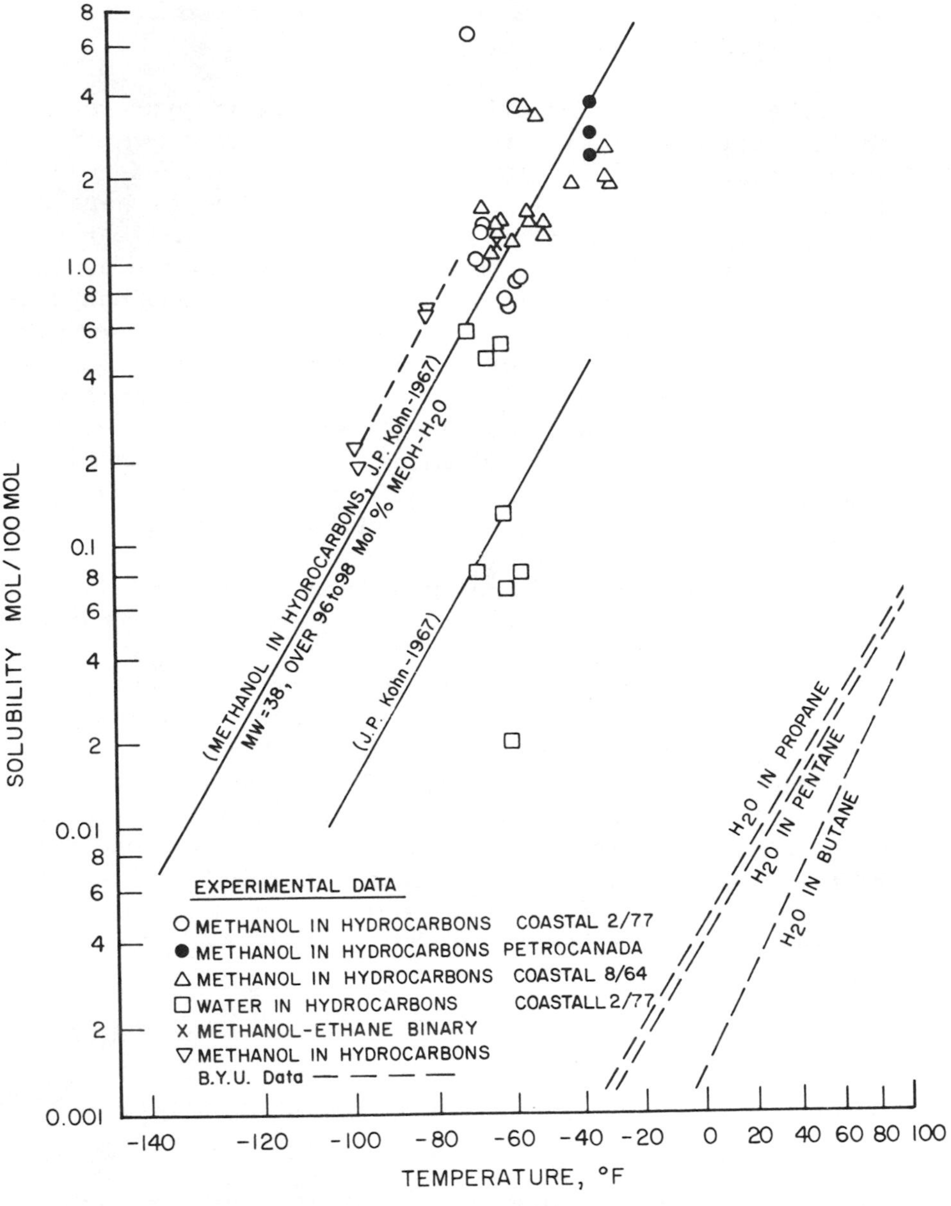


8-9 Solubility of Methanol in Liquid Hydrocarbons (Reproduced, with permission, from Hydrocarbon Processing, (Nielsen and Bucklin, 1983), by Gulf Publishing

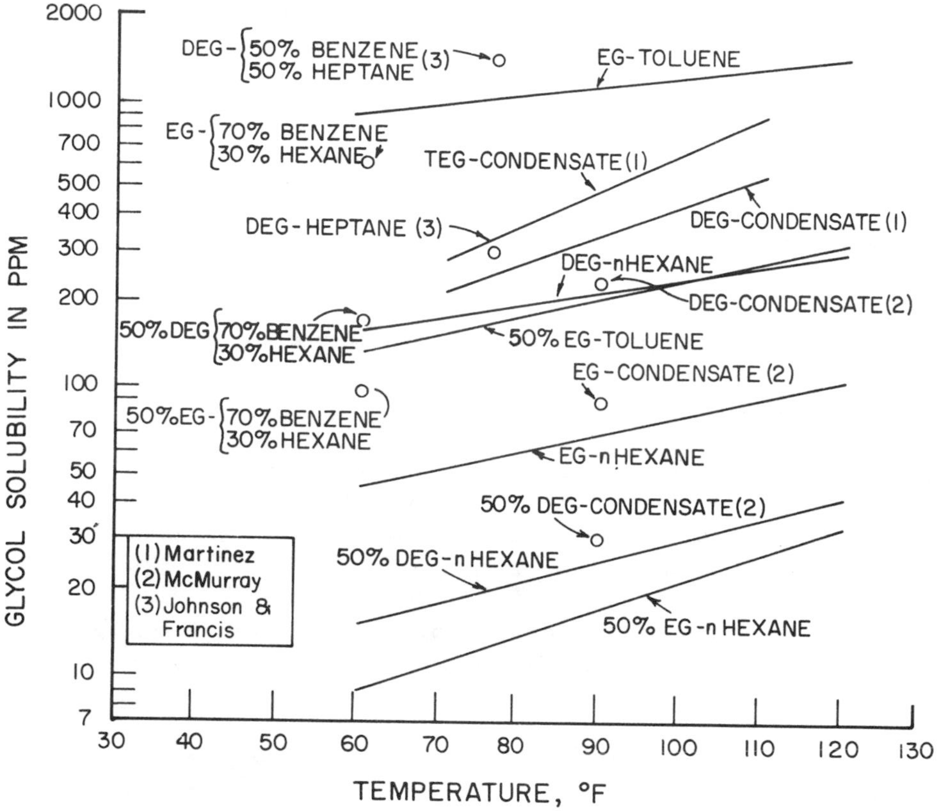


8-10 Solubility of Glycols in Liquid Hydrocarbons (Reproduced, with permission, from Proc. Gas Cond. Conf. (Polderman, 1958), by the University of Oklahoma

the gas is 90% $CH_4$, 7% $C_2H_6$ and 3% $C_3H_8$. The free water has been removed at the wellhead, but the inlet gas is at its dew point.

**Solution:** The amount of ethylene glycol to inject is determined by the amount needed to supply the aqueous liquid phase (steps 1-3), the vapor phase (step 4), and the hydrocarbon liquid phase (not applicable). Since this is the gas composition of the previous example, there is no possibility of obtaining a liquid hydrocarbon phase because the hydrocarbon mixture cricondentherm is far below these temperatures (231 K, as in Example 8.1).

Step 1. Determine the three-phase hydrate formation point (using CSMHYD, the K-value method, or the gas gravity method) as 290.3K at 6.89 MPa.

Step 2. Determine the weight per cent ethylene glycol needed to depress the hydrate temperature from 290.3K to 277.6K. Using Equation (8.3) with $\Delta T = 22.86^{0}F$ (12.7K) and M=62.07 a value of 37.8 wt% ethylene glycol is obtained.

Step 3a. Determine the saturated water content of the gas at 6.89MPa and 300K. Using Equation (4.8) the water content is calculated as 33.1 $lb_m$/MMSCF (697.5 ppm[mol])

Step 3b. Determine the water content of the gas at 277.6K and 6.89 MPa. Using the two-phase V-H option of CSMHYD, or the procedure outlined in Table 4-7, the water content of the gas is 143.37 ppm[mol]

Step 3c. Determine the amount of water which condenses from the gas, as the difference in the above two water concentrations.

$$\left(\frac{(697.5-143.4)\,\text{lbmol}H_2O)}{10^6\,\text{lbmol gas}}\right)\left(\frac{\text{lbmolgas}}{379.3\text{SCF}}\right)\left(\frac{10^6\,\text{lbmolgas}}{\text{day}}\right)\left(\frac{18\,lb_mH_2O}{\text{lbmol}H_2O}\right)$$

$$= 26.3\ lb_mH_2O/D \text{ water which condenses from the gas}$$

Step 4. Determine the amount of ethylene glycol which is lost to the vapor. From Figure 8-8 the loss is determined to be 0.01 $lb_m$/D at 100 psia and $40^{0}F$.

Step 5. Perform an overall mass balance and an ethylene glycol mass balance (assuming steady state and no reaction) to find the amount of ethylene glycol solution injected:

Overall Balance: (assuming unaffected hydrocarbon flows)

$$\text{IN} = \text{OUT}$$

$$(\text{EG Sol})_{in} + (H_2O)_{in} = (\text{EG Sol})_{out} + (\dot{\text{EG}})_{vap}$$

where: $(H_2O)_{in} = 26.3\ lb_m/D$ and $(EG)_{vap} = 0.01\ lb_m/D$

Glycol Balance:

$$0.67(\text{EG Sol})_{in} = 0.378(\text{EG Sol})_{out} + 0.01$$

Solving both of the above balances simultaneously the mass flow of the ethylene glycol solution injected [$(\text{EG Sol})_{in}$] is calculated to be 3.9 $lb_m$/D.

---

The following example for methanol injection was done following the method of Nielsen and Bucklin (1983) which has supplanted the previous method of Jacoby (1953). The example illustrates a calculation of inhibitor for the case in which there is both a vapor and a liquid hydrocarbon, in addition to an aqueous phase.

**Example 8.3. Amount of Methanol Injected as Inhibitor**

Chappelear et al. (1977) present hydrocarbon vapor-liquid calculations for a typical turboexpander gas processing plant, using various equations-of-state for a "bone dry" gas. Consider such a plant in which 70 MMSCF/D inlet gas to the process contains water in the amount of 2.25 $lb_m$/MMSCF. In order to prevent hydrates 98 wt% methanol is injected into the gas prior to refrigeration and subsequent separation of the vapor, liquid hydrocarbon, and aqueous liquid phases. What should be the methanol injection rate?

The separation is done in a flash tank, identical to that in Figure 8-6, at $-80^0$F and 800 psia, with the flows and compositions of the hydrocarbon phases given below, as determined from a two-phase hydrocarbon vapor-liquid flash calculation.

| | Feed to Refrigeration | HC Liquid from Separator | HC Vapor from Separator |
|---|---|---|---|
| Rate, lbmol/day | 184,458 | 41,413 | 143,045 |
| Composition, mol fraction | | | |
| Methane | 0.9016 | 0.7219 | 0.9536 |
| Ethane | 0.0469 | 0.1068 | 0.0296 |

| | | | |
|---|---|---|---|
| Propane | 0.0185 | 0.0630 | 0.0056 |
| i-Butane | 0.0079 | 0.0309 | 0.0012 |
| n-Butane | 0.0051 | 0.0207 | 0.0006 |
| i-Pentane | 0.0027 | 0.0115 | 0.0002 |
| n-Pentane | 0.0018 | 0.0078 | 0.0000 |
| n-Heptane | 0.0063 | 0.0279 | 0.0000 |
| Nitrogen | 0.0040 | 0.0013 | 0.0048 |
| Carbon Dioxide | 0.0052 | 0.0081 | 0.0044 |

**Solution:** The amount of methanol to inject is determined by the amount of methanol needed to supply the three phases leaving the separator, namely (a) the aqueous liquid phase (steps 1 through 4), (b) the hydrocarbon vapor phase (step 5), and (c) the hydrocarbon liquid phase (step 6).

Step 1. Using the CSMHYD program on the floppy disk in the frontpapers (see Appendix B), determine the hydrate point for the vapor phase at the separator pressure of 800 psia to be 53.7$^0$F.

Step 2. Set the methanol concentration in the aqueous liquid leaving the separator. Use the Nielsen and Bucklin criteria that most often 0.835 mol fraction (90wt%) methanol is specified, since it affords the maximum protection without methanol solidification.

Check that 90 wt% methanol enables the Rule-of-Thumb safety factor of having the depressed hydrate temperature at least 35$^0$F below the process point. Since the methanol concentration range is beyond the prediction ability in CSMHYD (≤20wt%) use the modified Hammerschmidt Equation (8.4) to obtain a hydrate depression amount of $\Delta T$ = 233.5$^0$F. The hydrate formation temperature is then:

$$T = 53.7 - 233.5 = -179.8^0F$$

while the separator temperature is -80$^0$F, so that there is a safety factor of -100$^0$F, almost three times the

Rule-of-Thumb amount. In another part of a similar process Chappelear et al. (1977) calculated a turboexpander discharge temperature of $-166^{0}F$, so the safety factor ($-13.8^{0}F$) is not as great there.

Step 3. Determine the amount of water condensed.
Using the G-H option of CSMHYD the water content of the gas leaving the separator is calculated to be 0.167ppm (0.0079 $lb_m$/MMSCF). Effectively all of the water in the inlet gas is condensed by the refrigeration prior to the separator. The water leaving the separator is:

$$\frac{2.25\ lb_m}{MMSCF} \times 70\frac{MMSCF}{day} \times \frac{1\ day}{24hrs} = 6.56\frac{lb_m}{hr}$$

Step 4. Determine the methanol in the aqueous liquid phase via a mass balance

$$0.9\ \frac{lb_m\ MeOH}{lb_m\ soln}\ X\ \frac{lb_m\ soln}{0.1\ lbm\ H_2O} \times 6.56\ \frac{lbm\ H_2O}{hr} = 59.0\ \frac{lb_m\ MeOH}{hr}$$

Step 5. Determine the amount of methanol loss in the hydrocarbon liquid. This is the most uncertain value in the calculation. Nielsen and Bucklin recommend using the highest solubility line of Figure 8-9, and then imposing an additional safety factor of 1.2. The methanol solubility in hydrocarbon liquid is read from Figure 8-9 to be 1.00 mol/(100 mol liquid) at $-80\ ^{0}F$.

$$1.2\left(\frac{1.00\ mol}{100\ mol\ liq}\right)\left(\frac{1{,}726\ mol\ liq}{hr}\right)\left(\frac{32\ lb_m}{mol}\right) = \frac{662.8\ lb_m\ MeOH}{hr}$$

Step 6. Determine the amount of methanol loss in the hydrocarbon vapor. From Figure 8-7 at 800 psia and $-80^{0}F$, we read an abscissa value of 3.4 ($lb_m$ MeOH/(MMSCF•mol fraction MeOH in aqueous phase). The amount of methanol lost to the vapor may be determined as:

$$\left(\frac{3.4lb_mMeOHvaploss}{10^6SCF\bullet mlfrtMeOH}\right)\left(0.835aq.MeOH\right)\left(\frac{379.3SCF}{mol\ gas}\right)\left(\frac{5{,}960\ molgas}{hr}\right)$$

$$= 6.42\ lb_m\ MeOH\ vapor\ loss/hr$$

Step 7. Determine the amount of 98 wt% methanol $M_{in}$ to be input to the system at steady state, by adding the three methanol outputs: (1) to the aqueous phase (step 4), (2) to the hydrocarbon liquid phase (step 5), and (3) to the hydrocarbon vapor (step 6):

$$0.98M_{in} = 59.0 + 662.8 + 6.42$$

$$M_{in} = 743.1 \text{ lb}_m/\text{hr}$$

Note that the amount of methanol in the aqueous liquid for hydrate inhibition is less than 10% of the total methanol injected. The relative amounts of methanol in the three phases suggests that recovery is indicated, at least from the hydrocarbon liquid and aqueous phases, for economic reasons. A process for methanol recovery is indicated by Nielsen and Bucklin (1983).

---

The higher molecular weights of the glycols cause them to condense more easily so that they are more recoverable than methanol. Due to vaporization losses of glycols, Townsend and Reid (1978) recommend that ethylene glycol be used where system temperatures are lower than 255K and that diethylene glycol be used above that temperature. Most frequently the methanol injected into pipelines is taken as a loss, although in gas processing methanol recovery is viable in turboexpander plants (Herrin and Armstrong, 1973; Nelson and Wolfe, 1981; Nielsen and Bucklin, 1983), as well as in absorber plants (Nelson, 1973). The latter work contains operating comparisons of methanol and glycol inhibition for a cryogenic gas processing plant, concluding that methanol is both more efficient and cost effective.

Townsend and Reid (1978 p.99ff) suggest that five other factors should be considered for the use of glycol injection systems: (1) the presence of salt(s) is a serious obstacle due to fouling and corrosion of regenerative exchangers, (2) the pH of glycol solutions should be controlled at 7.0 to mitigate corrosion, (3) that glycol coats all surfaces where hydrates will form, because it is more difficult to atomize and (4) the recognition that glycol-water solutions will freeze if they are too dilute or too concentrated. The relative viscosities of methanol and glycols are a final consideration, particularly where pressure drop is critical. Nelson (1973) indicates that at 233K the viscosity of 80 wt% methanol was only three times that of water at ambient conditions while the viscosity of ethylene glycol was 480 times that value for the same conditions. According to Polderman (1958) the viscosity should be between 100 to 150 centipoise to ensure flow through valves, heat exchangers, etc.

### 8.1.6. Hydrate Formation with Condensed Systems

Hydrate formation in inaccessible pipelines, causes extreme problems; Setliff (1988) reported stoppage of flow for 188 days at one Eldfisk location, and that typically 10% of the overall capital cost of an ocean pipeline is spent on inhibitor injection facilities. Formation of hydrates from liquid hydrocarbons is a topic at the state-of-the-art, particularly in applications such as the North Sea, where dense phase pipelines carrying both hydrocarbon and water phases are common and distances of transport may be as high as 10 km. That hydrates can form in crude oil pipelines has been shown many times by Katz (1971,1972) and his students (see Verma (1974) and Holder (1976)). If hydrates do form from liquid hydrocarbons in ocean pipelines, the opportunities for hydrate dissociation are much more limited than in gas pipelines. One cannot depressurize a liquid pipeline, and it is difficult to get a heat source and chemical inhibitors to the point of a flow stoppage, so a hydrate plug can jeopardize a large investment.

Three decades ago, Scauzillo (1956) reported that the insertion of liquid hydrocarbons with a natural gas lowers the hydrate-formation temperature of the gas. While limited other field data seem to support

Scauzillo's conclusion, later calculations such as those by Katz (1972), and Munck et al. (1988) have been unable to determine a thermodynamic justification for the lowering of the hydrate point. One may infer that fluid mechanics may contribute significantly to the blockage of pipelines in a way that thermodynamics cannot explain.

Behar et al. (1988) and Majeed and Lingelem (1988) discuss the paucity of such flow experiments for hydrate formation in the literature as well as their preliminary experiments with ice and hydrate formation from Freons. Majeed and Lingelem found that ice formation was a reasonably good simulation of hydrate formation. Setliff (1988) reported that Exxon has adopted the strategy of encasing several pipelines in a common conduit in ocean applications, to use a common heating source, or to minimize heat losses. The state-of-the-art of kinetics of hydrate formation, accumulation, and pipeline blockage is discussed in Chapter 3. This may be the single area in which the most major contributions can be made in fundamental hydrate research.

First discussed in Section 4.1.3, the equilibrium of multicomponent hydrocarbons in the four-phase ($L_W$-H-V-$L_{HC}$) region, is of industrial interest, particularly with regard to cryogenic, high $CO_2$, and deep sea applications. In applications with condensed multicomponent hydrocarbons, a vapor phase is very frequently present. The following example, modified slightly from Robinson and Ng (1986) with Chen (1987), is an illustration of multi-component hydrocarbons with vapor and liquid hydrocarbon phases in equilibrium with hydrates:

**Example 8.4. Intersection of Hydrate Conditions with Vapor-Liquid Phase Boundary for a Condensate**

In Figure 8-11, line $\overline{AHFCPEGA'}$ represents the water-free phase envelope of a gas composed of 60 mol% carbon dioxide, 25% methane, 9.5% nitrogen, 5% hydrogen sulfide, and 0.5% ethane. The portion of the phase envelope from point A to the critical point CP ($\overline{AHFCP}$) is the bubble

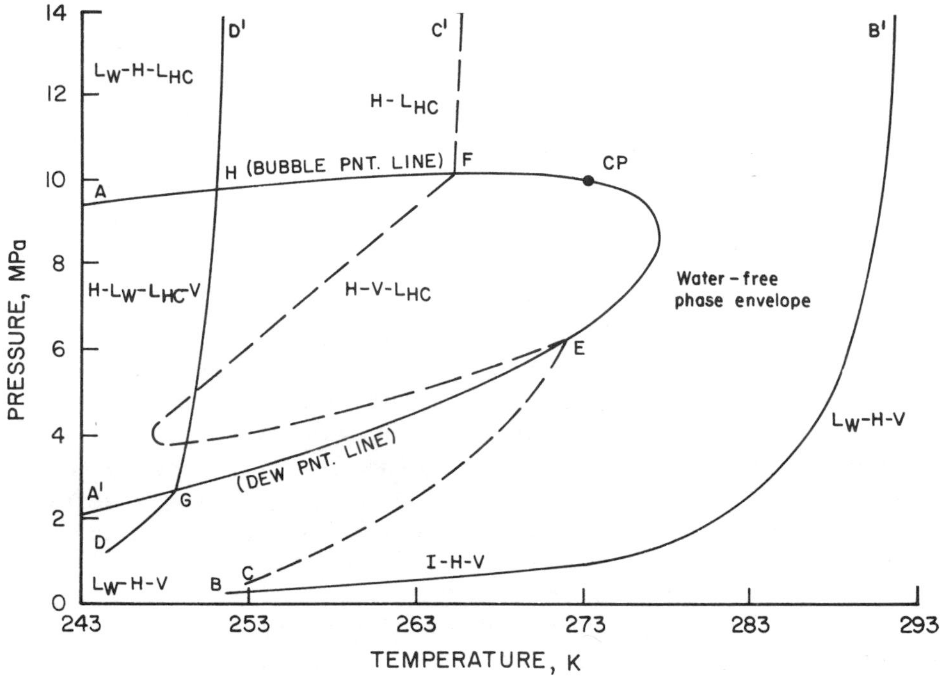


8-11 Calculated Hydrate Forming Conditions in a Gas Mixture Saturated with Water (BB'), Undersaturated with Water(CEFC'), and in the Presence of Methanol (DGHD') (Reproduced, with permission, from J. Can. Petrol Tech. (Robinson and Ng, 1986), by Can. Inst. of Mining and Met.

point curve, while the dew point curve is represented as $\overline{A'GECP}$.

Line $\overline{BB'}$ represents the line of hydrate formation from a gas of the above composition and an aqueous liquid (or ice, below 273K). The position of this three-phase ($L_W$-H-V) or (I-H-V) line represents the maximum temperature at which hydrates can form with a vapor mixture of the above composition at a given pressure. This is the "normal" hydrate line, such as is addressed in Sections 8.1.1 and 8.1.2.

If the above vapor did not have free water or ice present, yet was in equilibrium with hydrates, it would lie to the left of the three-phase line $\overline{BB'}$, on a line such as that marked $\overline{CEFC'}$.[1] On $\overline{CE}$, the lower two-phase (H-V) portion of curve $\overline{CEFC'}$, in addition to pressure and temperature, another variable (typically water concentration) would be needed to specify any state or point. Adiabatic compression of the constant composition system would also cause the temperature to increase along the dashed line until the dew point intersection with the water-free phase envelope at Point E. At Point E the first drop of liquid appears, at a composition in equilibrium with (but different from) the original vapor. Between the lower intersection E and the upper intersection F of these two lines, the relative amounts of the liquid and vapor and their compositions change. Most often a straight line is assumed between Points E and F; however frequently the trace is highly non-linear (as shown) for systems with high $CO_2$ content, causing a dramatic pressure and temperature change in hydrate formation conditions (see Example 8-5). At Point F the last bubble of vapor disappears into a liquid of the original composition, given above. Above the hydrocarbon bubble point F, line $\overline{CEFC'}$ again enters a two-phase (H-$L_{HC}$) region. Note that, by removing the free water phase, the temperature of hydrate formation from either vapor or liquid is greatly reduced from the temperatures at identical pressures along the three-phase ($L_W$-H-V) line.

**Illustration:** Robinson and Ng (1986) with Chen (1987) suggest that the above conditions may be industrially encountered as a process shown in Figure 8-12, in which a vapor and hydrocarbon liquid are separated before drying each phase. The dried phases are then recombined before transporting them in an underground pipeline. One would like to determine the hydrate formation point of the recombined vapor and liquid mixture, relative to the hydrate formation point before drying the vapor and liquid. Consider the composi-

---

[1]Note even this low-water content gas becomes saturated just below point C, joining the I-H-V curve at low T

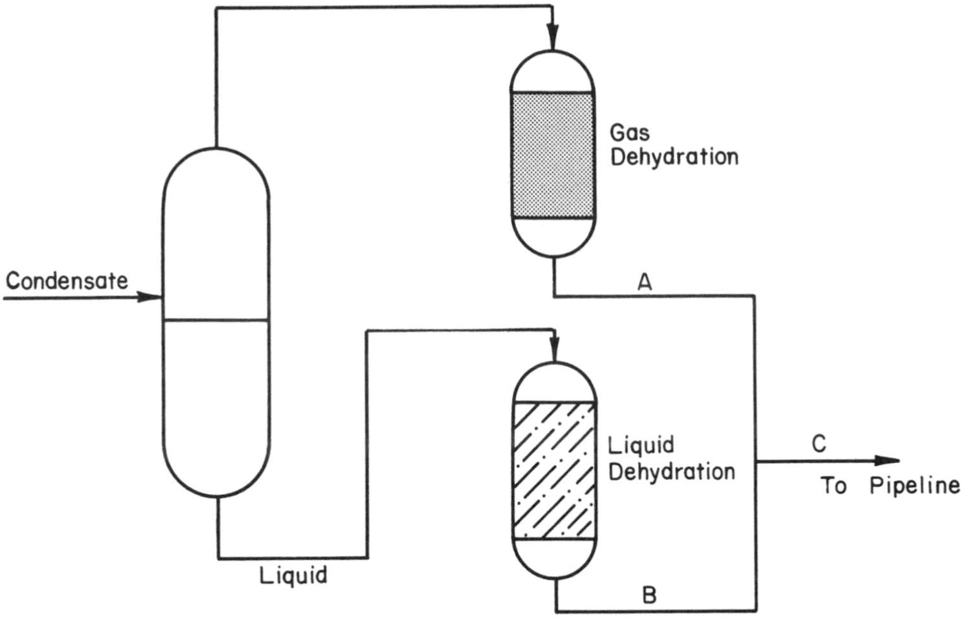


8-12 Schematic Representation of a Process Involving the Drying of Each of Two Phases Before Recombination to Consider Hydrate Formation in a Condensate System
(Reproduced with permission (Robinson et al., 1987) by Gas Processing Association)

tions and flows of three streams A,B,C in Figure 8-12 as given in Table 8-2. Note that the equilibrium separation causes the gas and liquid to be of differing compositions, so they will not fall along the same dashed line, as does vapor (Point E) and liquid (Point F) of identical composition in Figure 8-11.

Before drying, the calculated three-phase ($L_W$-H-V) temperature of the inlet fluid (essentially equal in flow and water-free composition to C) in the presence of hydrate and liquid water at 1400 psia is 84$^0$F. After drying, the gas stream A and the liquid stream B contain $6.5\times10^{-4}$ and $5.7\times10^{-4}$ mol

**Table 8-2**
**Stream Analysis for Figure 8-12, Illustration 8-4**

| Component | Stream A mol fraction | Stream B mol fraction | Stream C mol fraction |
|---|---|---|---|
| $CO_2$ | 0.0324 | 0.0187 | 0.0281 |
| $H_2S$ | 0.2345 | 0.3277 | 0.2635 |
| $CH_4$ | 0.6171 | 0.1806 | 0.4813 |
| $C_2H_6$ | 0.0587 | 0.0562 | 0.0580 |
| $C_3H_8$ | 0.0288 | 0.0605 | 0.0386 |
| $C_4H_{10}$ | 0.0161 | 0.0670 | 0.0319 |
| $C_5H_{12}$ | 0.0072 | 0.0449 | 0.0189 |
| $C_6H_{14}^+$ | 0.0052 | 0.2444 | 0.0797 |
| Relative Flow (lbmol/hr) | 2.22 | 1.00 | 3.22 |

fraction water, respectively. When A and B are recombined, however, the water concentration of C is $6.3\times10^{-4}$ mol fraction. The hydrate formation for the combined fluid at 1400 psia is 59.5°F, so that the drying process has reduced the hydrate point by about 25°F. Robinson Ng, and Chen (1987) represent this temperature reduction schematically for the liquid as in Figure 8-13, in which lines A and B show hydrate formation temperatures with and without a free aqueous phase, at the intersection of the phase envelope for the hydrocarbon.

Strictly speaking, the hydrocarbon phase envelope in Figure 8-13 will shift very slightly when a free water phase is present. However, because the water vapor pressure is relatively low, and because aqueous and hydrocarbon phases are almost totally immiscible, the three-phase P-T envelope is almost identical to the water-free two-phase envelope. However, when high concentrations of miscible components such as carbon dioxide are present, the shift of the phase envelope can be appreciable.

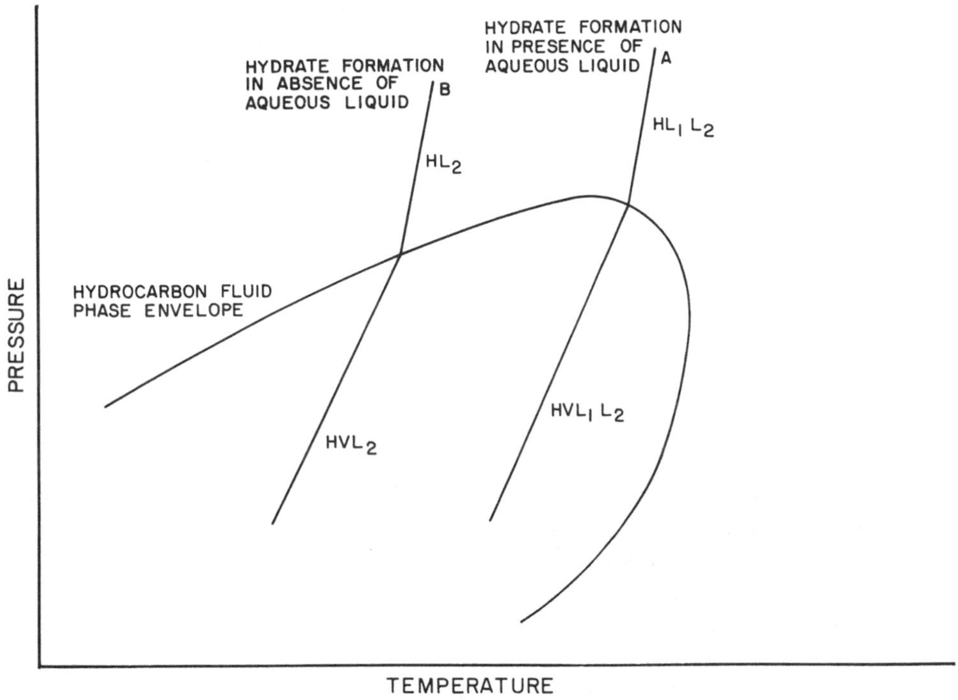


8-13 Schematic Representation of Effect of Dehydration of Incipient Hydrate Formation in a Pipeline Gas Mixture Both With and Without a Free Aqueous Phase (Reproduced with permission (Robinson et al., 1987) by Gas Processing Association)

Thus far in this example an analysis of the relative positions of lines $\overline{BB'}$ and $\overline{CEFC'}$ in Figure 8-11 demonstrated that removing the free water phase will decrease the temperature of hydrate formation. A method of decreasing the temperature of hydrate formation still further is to add an aqueous phase with a substantial amount of inhibitor. In Figure 8-11 the effect of an aqueous inhibitor phase is shown as line $\overline{DGHD'}$, again with the water-free envelope taken as the three-phase envelope. The relative positions of line $\overline{CEFC'}$ and $\overline{DGHD'}$ determine whether dehydration or inhibitor injection is used to prevent hydrate formation.

**Example 8.5. Constant Pressure Cooling of Gas and Intersection with Phase Boundaries**

Case et al. (1985c) provided an example of constant pressure cooling of an unsaturated gas using a line like $\overline{CEFC'}$ of Figure 8-11, to demonstrate an unusual effect within the water-free phase envelope. As the gas was cooled, first the hydrate appeared, then disappeared, then reappeared. The multicomponent gas of the example had a very high $CO_2$ content (63.3 mol% $CO_2$, 24.0% $CH_4$, 0.2% $C_2H_6$, 8.0% $N_2$, 4.5% $H_2S$) and the cooling was done at 700 psia from $70^oF$ to $-50^oF$.

In Figure 8-14 multiphase areas are marked which represent equilibrium between the water-free $CO_2$-rich phases of solid(S), liquid(L), and vapor(V), respectively. Areas I, II, III, and IV thus represent liquid-vapor, $CO_2$ solid-vapor, $CO_2$ solid-liquid-vapor, and $CO_2$ solid-liquid regions, respectively. As in Figure 8-10, line $\overline{BB'}$ represents the three-phase ($L_W$-H-V) hydrate line and line $\overline{CE}$ represents the two-phase (H-V) line for a vapor with a water content of 10 $lb_m$/MMSCF. Note that at point E, the initial liquid formed is very rich in $CO_2$ and consequently it has a higher affinity for water than its associated vapor. The enhanced water carrying capacity of this liquid competes with hydrate for available water, and thus supresses the hydrate forming temperature within the phase envelope. At the upper end of the phase envelope, line $\overline{FC'}$ represents the two-phase (H-$L_{CO_2}$) condition for liquid carbon dioxide with a water content of 10 $lb_m$/MMSCF.

In the cooling process shown in Figure 8-14, consider a gas of the above composition, initially at Point 1 ($70^oF$ and 700 psia), unsaturated in water at 10 $lb_m$/MMSCF. The gas mixture is cooled at constant pressure as a single phase until the intersection of the cooling line with two-phase line $\overline{CE}$ (Point 2) where incipient hydrate formation may occur from the vapor. This hydrate-vapor mixture exists until the mixture is cooled to $20^oF$ (Point 3) where a $CO_2$-rich liquid forms. Hydrate, liquid $CO_2$ and vapor coexist as a three-phase sys-

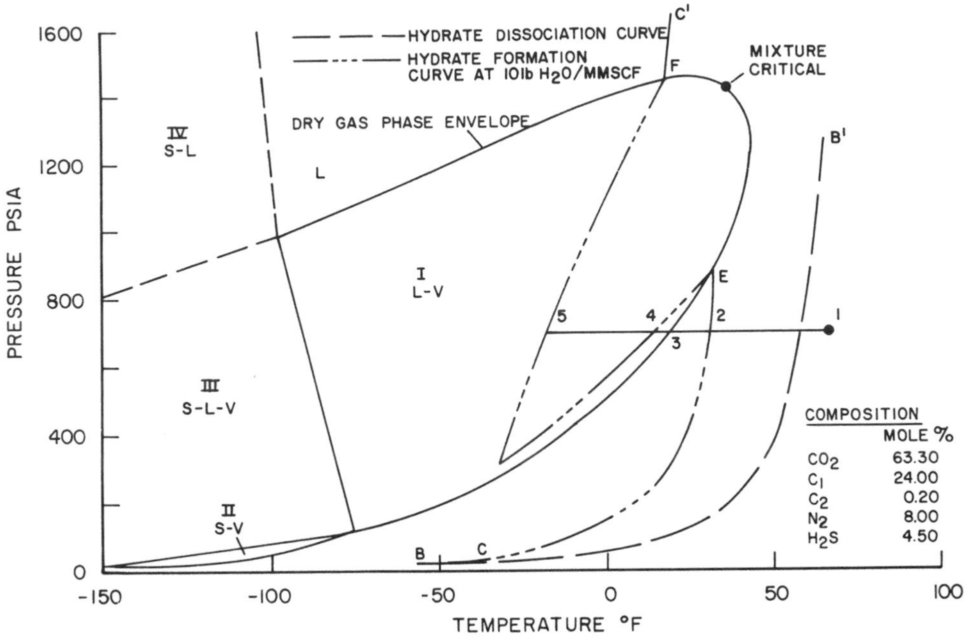


8-14 Isobaric Cooling Charted with Phase Behavior of a Complex Mixture with High Carbon Dioxide and Water Related Behavior
(Reproduced, with permission from Oil Gas Journal, (Case et al., 1985c) by PennWell Publishing Company

tem in the temperature range from 20 to 16$^0$F. At 16$^0$F (Point 4) sufficient liquid $CO_2$ forms to dissolve the hydrate and the system reverts to liquid-vapor equilibrium. As the temperature continues to decrease from 16$^0$F, more liquid is generated to prevent hydrates; however, the water carrying capacity of both the liquid and the vapor phases decreases. At -16$^0$F (Point 5) the water carrying capacity of the $CO_2$ rich phases is decreased to the extent that a hydrate phase reappears.

---

The example problems of Section 8.1.4 concerned inhibitor injection into a system with an aqueous phase coexisting with either (a) vapor (Example 8.2), or (b)

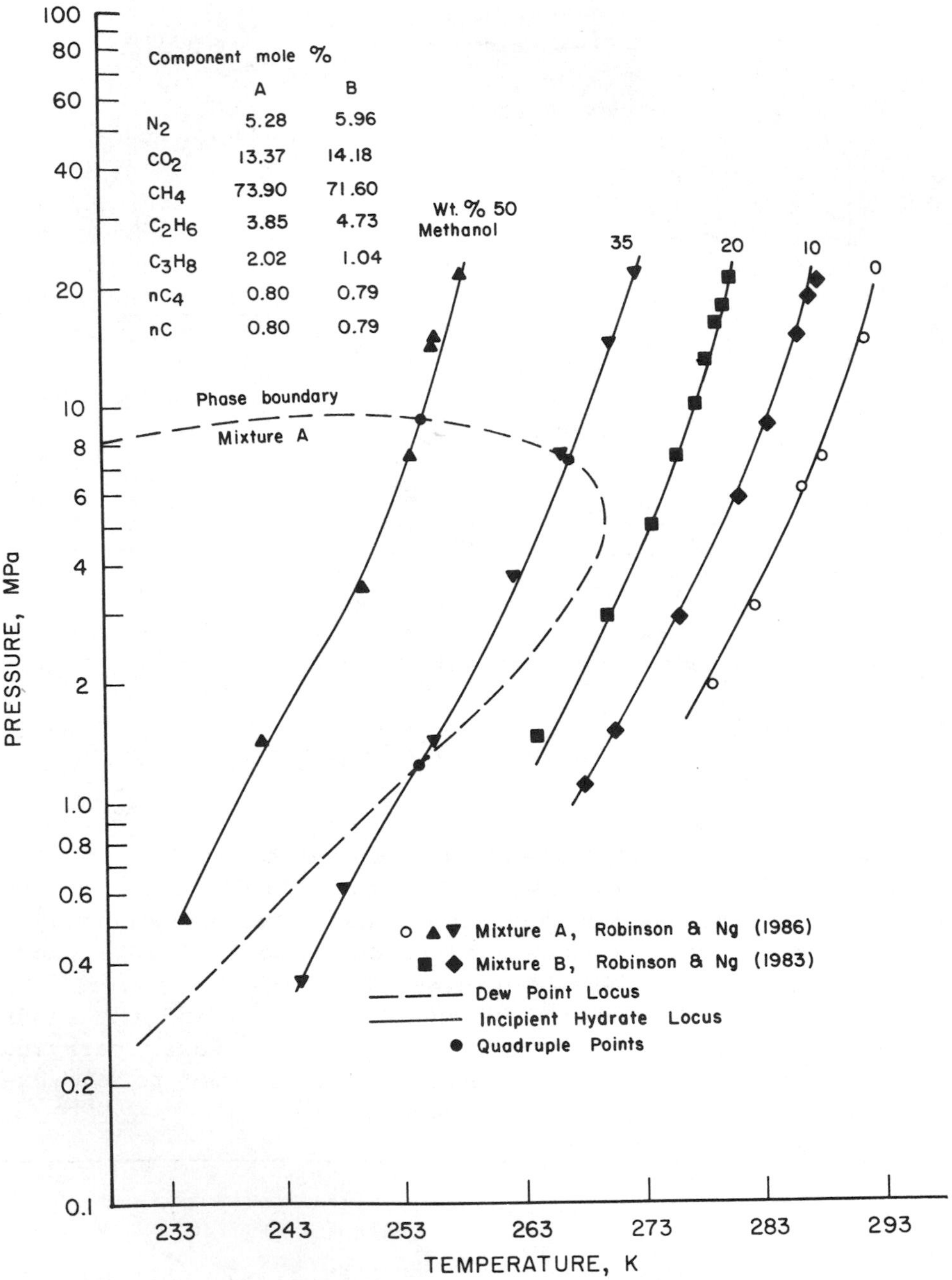


8-15 Influence of Methanol on Hydrate Formation in a Seven-Component Gas Mixture
(Reproduced, with permission, from J. Can. Petrol Tech. (Robinson and Ng, 1986), by Can. Inst. of Mining and Met.

vapor and hydrocarbon liquid (Example 8.3). With the amount of injected inhibitor already established, it is useful to consider the transition between the above two cases for mixtures of hydrocarbons. Figure 8-15 shows Robinson and Ng (1986) hydrate formation data (see Section 6.2.1 for composition and P-T points) for a seven component system with 0 to 50 wt% methanol in the aqueous phase. At methanol concentrations between 20 and 35 wt%, the three-phase ($L_W$-V-$L_{HC}$) envelope is encountered upon cooling from high temperature. Very often natural gas mixtures, including the one shown, have critical points far below the cricondentherm, so that intersection of the $L_W$-H-V line with the $L_W$-V-$L_{HC}$ envelope occurs in the retrograde dew point region, where the three-phase ($L_W$-H-V) mixture becomes four phase ($L_W$-H-V-$L_{HC}$). With further cooling, the mixture will leave the $L_W$-H-V-$L_{HC}$ region at a second dew point, where it again enters the three-phase ($L_W$-H-V) region.

## 8.2 Hydrate Formation and Prevention in Well Drilling

In recent decades, the petroleum industry has expanded drilling operations into the arctic and deep water areas of the world. These operations have resulted in two types of problems: those from the dissociation of **in situ** hydrates in the reservoir around the well during drilling, and those which arise in the well itself due to hydrate formation from drilling fluids. The danger of blowouts, blockage, etc. has caused industry to view these problems from a safety standpoint, with the result of more readily shared information.

Makogon (1974), Bily and Dick (1974), and Franklin (1979) initially reported problems in drilling through **in situ** hydrate formations due to casing collapse, severe mud gassification, and solids in the wellbore. Hydrate reservoirs have low permeability (Patil et al. (1988)) so that drilling mud loss is prevented and an insulating cake buildup results in higher heating. Makogon (1988b) presented a picture of collapsed casing caused by hydrate dissociation, indicating that the heat of drilling through hydrate reservoirs causes decomposition to the extent of causing high pressures and gas influx around the well casing as well as weakening the surrounding rock and sediment. These problems and their solutions are discussed in some detail in Section 8.2.1.

Drilling fluids, which operate as a lubricant, cuttings remover, etc. are restricted to water-based type fluids in many offshore operations by environmental considerations. As drilling depths increase, the potential for formation of hydrates in drilling fluids and equipment also increase due to higher seafloor hydrostatic pressure and lower temperatures. Instances of hydrate-related drilling problems are beginning to come into the literature (Jones and Sherman, (1986); Barker and Gomez, (1987)). Section 8.2.2 treats hydrates in drilling muds.

### 8.2.1. Drilling in In Situ Hydrate Reservoirs

Two techniques have been used for drilling through hydrate reservoirs. The first technique is to prevent the dissociation of hydrates, for controlled gas influx and safe drilling. The second technique, originally proposed by Franklin (1980), is to encourage a controlled dissociation of hydrates while drilling through the hydrate zone, so that safe drilling may be carried out below the hydrate zone. While the former technique predominates in industrial practice, a clear determination has not been made as to which is best. General heuristics have been formulated for both techniques, as below.

To prevent hydrate dissociation:

- reduce the temperature of the drilling fluids (muds),
- increase the mud density, commonly called weight,
- increase the mud circulation, maintaining turbulent flow, and
- run the casing after penetrating hydrate zones.

Various combinations of the above should be used upon obtaining a gas "kick" from hydrates, e.g. closing the blow-out preventer, decreasing the mud temperature, increasing the mud density, and removing the free gas.

To promote hydrate dissociation while drilling through the hydrate zone:

- increase the temperature of the drilling mud,
- decrease the mud weight,
- decrease the penetration rate,
- maintain laminar mud circulation in the well annulus, and
- place the casing at the maximum depth above a potential hydrate zone.

Goodman (1978) and Goodman and Franklin (1982) used computer simulation to specify muds and to determine well temperature and pressure profiles while drilling through hydrates in the arctic offshore. Their model verified that the dissociated gas influx may be controlled by increasing mud density; increased mud density also requires a higher hydrate dissociation temperature by increasing the effective equilibrium pressure adjacent to the hydrates.

If the depth of the hydrate interval is known, one may quantify the gas influx and mud conditions; Roadifer et al. (1987a) combined a hydrate dissociation model with wellbore heat transfer and fluid flow to quantify the effect of a wide range of drilling parameters on well conditions when drilling through hydrates. The combination of parameters chosen for a given application will be a function of economics. Roadifer et al. (1987b) provided the nomogram given in Figure 8-16 to determine optimum temperatures and pressures when drilling through hydrates, either to prevent or to encourage hydrate dissociation, for an assumed gas composition.

The nomogram can also be used to design surface degassing equipment, if the zone of hydrate depth and the hydrate dissociation temperature is known. While this nomogram is the best presently available in the open literature, it must be validated through industrial use over time. Two examples of the nomogram use from Roadifer et al.(1987b) are given below:

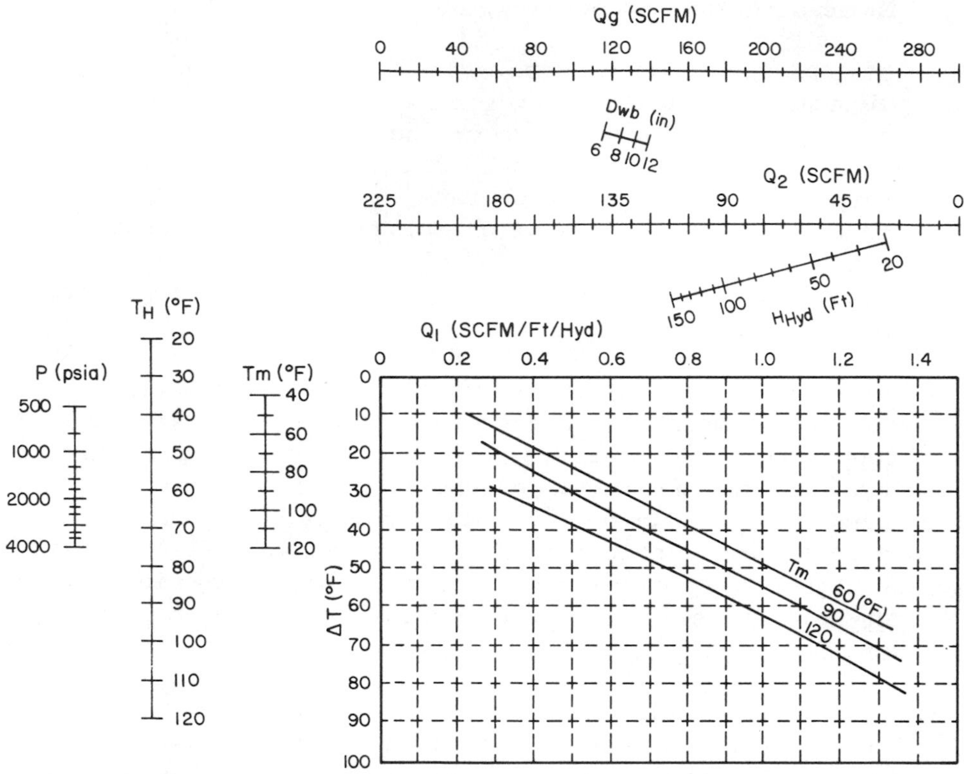


8-16 Nomagram to Size Mud Degasser and to Limit Hydrate Dissociation when Drilling Through Hydrates (Reproduced, with permission (Roadifer et al., 1987b) by the Society of Petroleum Engineers)

### Example 8.6: Sizing a Mud Degasser for Drilling in Hydrate Zones

Given:

| | | |
|---|---|---|
| Hydrate Zone Thickness, $H_{Hyd}$ | 50 | ft. |
| Methane Hydrate Formation Temperature, $T_H$ | 60 | $^oF$ |
| Wellbore Diameter, $D_{wb}$ | 8.375 | in. |
| Minimum Wellbore Pressure, P | 1750 | psia |
| Maximum Annulus Temperature, $T_m$ | 95 | $^oF$ |

Find: For simple hydrates of methane, determine the minimum required gas capacity of the mud degasser

**Solution:** Using the nomogram of Figure 8-16:

Step 1. Construct a horizontal line from the pressure (P) line at 1750 psia to the hydrate dissociation temperature line ($T_H$) at 60$^0$F.

Step 2. Construct a line from $T_H$ = 60$^0$F, through the mud temperature line ($T_m$) at 95$^0$F to the left ordinate of the crosshatched region (the ΔT axis) at ΔT = 35$^0$F.

Step 3. Construct a horizontal line from ΔT = 35$^0$F to the mud temperature within the crosshatched region, at 95$^0$F.

Step 4. Construct a vertical line from the mud temperature to $Q_1$ on the top abscissa, which represents the amount of gas released per foot of hydrate. The value is $Q_1$ = 0.55 SCFM/ ft. hyd.

Step 5. From $Q_1$ construct a line through $H_{Hyd}$, the hydrate zone thickness (50 ft.), to $Q_2$ the amount of gas released. Read $Q_2$ = 28 SCFM.

Step 6. Construct a line from $Q_2$ through the diameter of the wellbore $D_{wb}$ (8.375 in.) to $Q_g$ to account for restrictions in the wellbore. Read the maximum gas rate (minimum capacity of degasser required) at 28 SCFM.

---

**Example 8.7: Wellbore Pressure Required to Limit Dissociation**

Given:

| | | |
|---|---|---|
| | Gas Capacity of Mud Degasser, $Q_g$ | 75 SCFM |
| | Hydrate Zone Thickness, $H_{Hyd}$ | 100 ft. |
| | Wellbore Diameter, $D_{wb}$ | 11 in. |
| | Maximum Annulus Mud Temperature, $T_m$ | 100 $^0$F |

Find: The wellbore pressure required to limit hydrate dissociation

**Solution:** Using the nomogram of Figure 8-16:

Step 1. Construct a line from the top line $Q_g$ (80 SCFM) through the diameter of the wellbore $D_{wb}$ (11 in) to intersect the gas released from hydrate $Q_2$ at 75 SCFM. $Q_2$ lies on the second topmost horizontal line.

Step 2. From $Q_2$ draw a line through the hydrate zone thickness, $H_{Hyd}$ (100 ft.) to $Q_1$, the top horizontal line in the crosshatched region. Read $Q_1$ at 0.74 SCFM/ ft. hyd.

Step 3. Draw a vertical line from $Q_1$ to the mud temperature ($T_m$ = 100 $^oF$) in the crosshatched region. Draw a horizontal line from the mud temperature to the left abscissa ($\Delta T$ = 44$^oF$) on the crosshatched region.

Step 4. From the value of $\Delta T$ = 44$^oF$, draw a line through the mud temperature (100$^oF$ on the vertical $T_m$ line) to the vertical hydrate temperature line $T_H$. Read the intersection at 56$^oF$.

Step 5. Draw a horizontal line from the hydrate temperature to the leftmost vertical line, P, the minimum wellbore pressure required to prevent hydrate dissociation. Read that value as 1400 psia.

---

In closing both of these examples, it is worthwhile to re-emphasize that the nomogram of Figure 8-16 contains many assumptions not explicitly stated. For example, the relative position of the leftmost two lines relating hydrate pressure and temperature are strongly dependent upon gas composition, which has been assumed to be pure methane. Until some track record of success is obtained in the field, the nomogram should be used with caution, as the best literature approximation available.

### 8.2.2 Hydrate Formation in Deep Water Drilling Fluids

Barker and Gomez (1987) cite hydrate experiences with drilling muds in deep water offshore Santa Barbara (1977) and in the Green Canyon of the Gulf of Mexico (1985); losses in drill times were 70 days and 50 days, respectively (Barker, 1988a). Potential problem areas can involve both safety and economic factors including: plugging well choke and kill lines, plugging blowout preventers, and hindrance of drillstring movement.

In a review of the number of drilling experiences in deepwater (>3000 ft.) with water-based drilling fluids, it is apparent that the evidence is so small as to be anecdotal rather than conclusive. Barker (1988a) indicated the following rules-of-thumb used by Exxon in considering hydrate formation with drilling fluids:

- Hydrate problems are fairly pervasive in drilling experiences, but have only come to be recognized as such in recent years.
- Hydrates not only form solids by themselves, but they remove water from the remainder of the mud, leaving a barite plug as a second solid.
- One cannot design a well to operate outside of the hydrate region only if flow conditions are maintained. Flow will stop at some period and operation of the well may be jeopardized, if the well conditions will be in the hydrate formation region at static conditions.
- Several hours may be required for hydrate formation and blockage to occur.
- Hydrate plugs must be removed by thawing from both ends, rather than initiating thawing at mid-plug.
- Exxon currently (10/88) uses salt at the saturation limit range of 150 to 170 mg/l to prevent hydrate formation.
- As general rules-of-thumb about hydrates versus water depth, the summary given below by Barker (1988b) may be used:

Drilling Mud Hydrate Formation in Deep Water

| Water Depth, ft. | Risk of Hydrate Formation Problems |
|---|---|
| ≤1000 | A hydrate problem will probably not occur. |
| ≤1500 | Without inhibition a hydrate problem may occur. |
| ≤2000 | Without inhibition a hydrate problem will occur. |
| ≥3000 | Insufficient experience; salt alone will not suffice. |

Hale (1988) indicated Shell experience with the use of water-based muds in deep water drilling. Shell has drilled 16 wells in the Gulf of Mexico at water depths between 2,000 and 7,500 feet, using 20 wt.% NaCl and PHPA- (partially hydrolyzed polyacrylamide) based muds. In each well Shell has experienced an average of more than one gas kick per well, which signal the possibility of hydrate formation. Only one instance in 2900 ft. of water involved the possibility of hydrate formation when Shell experienced difficulty disconnecting the drill stack.

As discussed in some detail in Section 8.1.3, Hale suggested the use of a highly exothermic chemical reaction to dissociate downhole hydrate plugs; while this method has not been put into practice to date, it appears to have potential in this application. Hale indicated that a preventative goal should be to develop a water-based mud to protect against hydrates in 7,500 ft. of water using 14 $lb_m$/gal drilling mud which is environmentally safe and non-flammable. A mud formulated with 20 wt% NaCl and 30 % glycerol can be used to protect against hydrate formation to 5,000 ft.

There exists the need for a method which may be used with the following objectives:

(a) to prevent hydrate formation in a drilling fluid,

(b) to predict which fluids are susceptible to form hydrates under given drilling conditions, such as pressure and temperature,

(c) to dissociate hydrates in the fluid once they are formed, and

(d) to determine the kinetics, or the time required and possible extent of hydrate formation.

Unfortunately, drilling fluids are very complex and do not approximate the fluids normally encountered in hydrate formation. In relation to drilling fluids the normal hydrate forming fluids (found in other parts of this book) appear to be pristine. It should be noted that while the fluids mentioned here are water-based, the oil-based fluids also contain sufficient water for hydrate formation.

A typical deep-sea, water-based drilling fluid may contain the following components, in addition to water, each given with their function.

1. Bentonite/lignosulfonate/sodium hydroxide: - Bentonite (sodium montmorillonite) acts to increase the density and viscosity of the drilling fluid for lubrication, prevention of drill fluid loss, and cuttings transport/suspension. The sodium hydroxide and chrome-lignosulfonate act to disperse the clay and to maintain the degree of hydration of the clay. The sodium hydroxide acts to convert the acidic lignosulfonate to a more water-soluble sodium salt, which in turn adsorbs onto the clay surface as a negatively charged ion, and causes dispersion through repulsion of like charges.

2. Barite: Barite is barium sulfate, with smaller portions of various minerals, such as quartz, calcite, and silicates. The size of the barite particles should be between 2 and 100 micrometers to give a higher density (blowout protection) and to suspend the cuttings.

3. Sodium Chloride acts as a freezing point depressant or inhibitor.

4. Methanol acts as a freezing point depressant or inhibitor.

5. Various Polymers such as xanthan gum and partially hydrolyzed acrylamide (PHPA) which may hydrogen bond with

water or form ions. These act as viscosifiers and fluid loss preventers.

6. Drill Solids as representative of well hole cuttings, which are highly dependent on the geology surrounding the well.

With the diversity of the many and ill-defined components, it is very difficult to construct an **a priori** prediction technique. Such phenomena as surface effects, ionic interactions, and interactions between components dictate that any correlative or predictive model be based upon experiment. As indicated in Chapter 6 a typical single hydrate formation experiment may require 36-60 hours. For the effect of single component additions at one concentration, at least three such data points will be required to establish a straight line on a $\ell nP$ vs. $1/T$ trace. A methodical, thorough test of the composition effect of all of the above components and their interactions would require a substantial experimental effort. A statistically designed experimental program is suggested by the complexity of the problem.

Hydrate formation measurements were performed in this laboratory on twenty drilling fluids by Cha (1988) to determine the effect of various combinations of the above components. In general the hydrate formation was primarily affected by the amounts of salt and methanol in the water-based fluid. One of the surprising experimental results was that in several cases the drilling fluids formed hydrates both at higher temperatures and in larger amounts, relative to the same amount of pure water. Since this effect had not been recorded in the literature, simpler fluids were measured to discern the causes.

### 8.2.2.a. Hydrate Formation from Simple Fluid Components

Cha et al. (1988) measured hydrate formation thermodynamics and the extent of formation using binary and ternary liquid mixtures of water (Fluid 1) with clay (bentonite - Fluid 2), dispersed clay (bentonite/

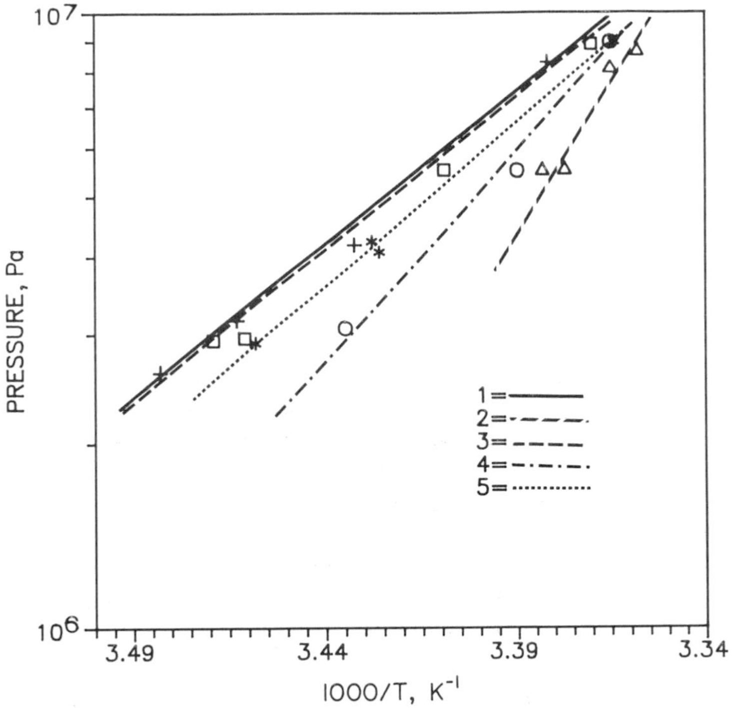


8-17 The Promotion of Hydrate Formation by Third Surfaces (Reproduced, with permission from J. Phys. Chem. (Cha et al., 1988), by American Chemical Society.

lignosulfonate/caustic - Fluid 3), and a hydrophillic polymer (30% partially hydrolyzed polyacrylamide - PHPA - Fluid 4). The thermodynamic conditions of hydrate formation are shown in Figure 8-17 for the four simple fluids relative to water (Fluid 1). Both Fluid 3 with bentonite/lignosulfonate/caustic and Fluid 5 with all of the components, formed hydrates at essentially the same thermodynamic conditions as water, However, Fluids 2 and 4 which contained only bentonite or PHPA, respectively, formed hydrates at substantially higher temperatures than did pure water.

In Figure 8-17, two effects can be observed concerning the slopes of the lines, relative to that of pure water, namely a displacement at the same slope, or a change of slope with a displacement. Both effects may be analyzed by the Gibbs Free Energy change on formation of hydrates, using the expression:

$$\Delta G = \Delta H - T\Delta S \tag{8.5}$$

where $\Delta H$ is determined from P-T data by $d(\ell n\ P)/d(1/T) = -\Delta H/zR$, the Clausius-Clapeyron relation.

Via Equation (8.5) one can hypothesize the cause for the change in Gibbs Free Energy, or the ease of hydrate formation relative to water, either by (1) a displacement from the water line at the same slope, indicating an entropic ($\Delta S$) or ordering effect of water the third surface, and (2) a change in slope, indicating an energetic change ($\Delta H$) on the water molecules by the third surface. It should be noted that a change in slope may contain both ordering and energetic effects.

Cha et al. also presented evidence that some fluids containing a third surface such as water + bentonite/lignosulfonate/caustic (Fluid 3) formed hydrates substantially faster than did pure water. In addition, Fluid 3 formed more hydrates than did pure water. These effects have also not been observed in the literature before, and are related to kinetic phenomena. The results of Fluid 5 with all of the components showed that both thermodynamic and kinetic promotion could be observed in a single fluid.

While detailed investigations continue on these thermodynamic and kinetic phenomena, it is thought that they may be caused by hydrogen-bonding adsorption and surface nucleation, respectively. Bentonite, a swelling clay provides a large surface for strong, ordered adsorption of water (Low, 1961) while the PHPA, a hydrophyllic polymer (Gray and Darley, 1980, p. 269), acts to solvate the water. The chrome-lignosulfonate and caustic are agents that are commonly used to seal and disperse clay platelets (Gray and Darley, 1980, p. 269); these dispersants may act to seal some of the strong hydrogen-bonding sites, as well.

As a result of the experiments, a hypothesis was generated that thermodynamic promotion may be caused by the ordered adsorption of water molecules onto a hydrogen bonding surface, which approximates part of a hydrate lattice, and this surface becomes part of the hydrate structure, with the following mechanism.

$$\text{water} + \text{adsorbing compound} \rightarrow \text{adsorbed water} \quad (8.6a)$$

$$\text{adsorbed water} + \text{gas} \rightarrow \text{adsorbed hydrate} \quad (8.6b)$$

The kinetic promotion may be caused by the presence of an external surface which would provide nucleation sites, with the mechanism.

$$\text{water} + \text{nucleating surface} \rightarrow \text{adsorbed water} \quad (8.7a)$$

$$\text{adsorbed water} + \text{gas} \rightarrow \text{hydrate} + \text{nucleating surface} \quad (8.7b)$$

The difference between the mechanism for thermodynamic promotion and kinetic promotion is that the latter requires only a surface adsorption into random clusters which can encourage the formation of hydrate crystal nuclei, leading to faster and greater crystal growth. Thermodynamic promotion, however, requires that the surface adsorb water molecules into ordered layers which are closer in structure to that of potential hydrate cages. Therefore water can reorganize into hydrates more easily and at higher temperatures for a given pressure.

Many other factors such as surface area, cation exchange, etc. necessitate a very complex model for the formation of hydrate in drilling fluids. This is an area of current research which will yield new insights into hydrate thermodynamics and kinetics.

## 8.3. Summary of Hydrate Formation in Production and Processing

Processes for the production and processing of natural gas are in a state of evolution and as such, the technology is constantly changing. As a production example, only a short while ago there were essentially no wells drilled in water depths greater than 3000 ft. In processing, the last few decades have witnessed the evolution of the turboexpander in gas plants, with the resulting practical example of a process which approached the isentropic expansion limit. Doubtless, technology modifications will continue to occur and older technologies will be replaced. With evolving technology even simple processes such as the gathering/processing flowsheet in Section 8.1.1 and the drilling examples in Section 8.2 will become dated.

However, as the technologies and processes change the physics and the chemistry of hydrates will remain relatively constant, with few perturbations. An exception to this, noted previously, will be the rapidly changing domain of hydrate kinetics, which will enable us to consider time-dependent processes in addition to thermodynamic considerations. Yet the thermodynamic principles will still hold. As such, the principles of chemistry and physics expressed in Chapters 2, 3, 4, and 5 will be applicable to most new processes, and newer physical principles will serve to complement (rather than supplant) them. The classical principles of hydrates form the foundation for Sections 8.1.2 through 8.1.6, and should hold for future processes and technologies.

# Appendix A

# Calculation of Fugacity Coefficients and Enthalpy and Entropy Departures

## A.1 Fugacity Coefficient Derivation for a Component in a Mixture

To perform hydrate equilibria calculations in Chapter 5, it is necessary to be able to determine the fugacity of a vapor or liquid component either in the pure state or in a mixture, for use in equations such as (5-22). In thermodynamic calculations for the vapor phase, fugacity may be considered as a corrected form of partial pressure. In the liquid phase the fugacity may be considered as the quantity which relates the liquid mol fraction of a component in a mixture to the pressure. The fugacity of component i in either phase z ($f_i^Z$) is related to the mol fraction of that component and the pressure through the equation:

$$f_i^Z = w_i \, \phi_i^Z \, P \qquad \text{(A-1)}$$

where $w_i$ is the mol fraction of component i in either phase ($y_i$ for vapor or $x_i$ for liquid) and the superscript Z may be either vapor (V) or liquid (L), as appropriate.

The calculation of the fugacity then resolves into the calculation of the fugacity coefficient $\phi_i^Z$, from an equation-of-state.

### A.1.1 Two Typical Equations-of-State

An equation-of-state may be of any order in volume to fit a single phase, but if it is to fit both the vapor and liquid phases, the equation-of-state must be cubic in volume, as a minimum. The simplest cubic equation-of-state is that of van der Waals (1890); however the use of that equation yields large inaccuracies in the properties of very dense gases and liquids. For purposes of illustration, this appendix deals only with the Soave (1972) modification of the Redlich-Kwong equation-of-state, and with the Peng-Robinson (1976) equation-of-state. While these two equations are among the most popular, higher order equations may be used, but at some expense in computational time. The two equations-of-state are presented in Table A-1, together with mixing rules for their constants.

Each equation-of-state together with its mixing rules may be used to predict the P-V-T properties of mixtures. Given the mol fraction of the components in a phase, the volume of that phase may be predicted at a given pressure and temperature. Since both of the equations in Table A-1 are cubic, three volume roots are obtained for each set of constants. When the liquid phase mol fractions "$x_i$" are used to obtain constants "a" and "b" in either equation-of-state, then the smallest volume root must be taken for that of the liquid. In contrast, when the vapor phase mol fractions "$y_i$" are used to obtain the mixture constants "a" and "b", then the largest volume root must be taken as that of the vapor.

### A.1.2 Fugacity Coefficients from an Equation-of-State

With the ability to obtain the liquid or vapor volume, given the phase compositions, one may also obtain the fugacity coefficients, using the equation (Prausnitz, p. 41 (1969)):

$$\ell n\ \phi_i^Z = \frac{1}{RT}\int_{V^Z}^{\infty}\left[\left(\frac{\partial P}{\partial n_i}\right)_{T,V,n_{j\neq i}} - \frac{RT}{V}\right]dV - \ell n\ \frac{PV^Z}{n_T RT} \qquad \text{(A-8)}$$

where superscript Z represents either the vapor or the liquid phase, $n_i$ is the number of moles of component i, $n_T$ is the total number of moles in the phase, and $V^Z$ represents the total volume of the phase.

When equation (A-8) is applied to the Soave-Redlich-Kwong equation an expression for the fugacity coefficient is obtained as:

$$\ell n\ \phi_i^Z = \frac{b_i}{b_M}(z-1) - \ell n\left(z - \frac{b_M P}{RT}\right) - \frac{a_M}{b_M RT}\left[\frac{2\sum_{j=1}^{n} w_j a_{ij}}{a_M}\right]\ell n\left(1+\frac{b_M P}{zRT}\right) \qquad \text{(A-9)}$$

where $z \equiv Pv/RT$

When equation (A-8) is applied to the Peng-Robinson equation a slightly different expression for the fugacity coefficient is obtained.

$$\ell n\phi_i^Z = \frac{b_i}{b_M}(z-1) - \ell n\left(z-\frac{b_M P}{RT}\right) - \frac{a_M}{2\sqrt{2}\, b_M RT}\left[\frac{2\sum_{j=1}^{n} w_j a_{ij}}{a_M} - \frac{b_i}{b_M}\right]\ell n\frac{v+(1+\sqrt{2})b}{v+(1-\sqrt{2})b} \qquad \text{(A-10)}$$

Note that in both (A-9) and (A-10) the molar volume is required, implying that the mol fractions of each component should be known, both to get the molar volumes "v" and to use in equations (A-9) and (A-10) themselves.

The use of equations (A-9) and (A-10), together with equation (A-1) enable the calculation of the fugacity of component i in the phase of interest, for use in the equations of Chapter 5 such as (5-22).

## A.2 Calculation of Enthalpy and Entropy Changes of a Mixture

In order to perform many of the production and processing calculations in Chapter 8 for some cases of hydrate formation and dissociation it is necessary to determine the enthalpy change ($\Delta H$) or entropy change ($\Delta S$) of a gas or liquid as it goes through a process. Typical examples of the need for these quantities are: flows in a well or pipeline, heat exchanger duties, flow across a Joule-Thomson valve, flow through a turboexpander, non-adiabatic flashes, etc.

Because it is easy to determine the ideal gas changes from one thermodynamic state to another, it is worthwhile to calculate the differences between the actual state and the hypothetical ideal gas state at the same temperature and pressure. Since both enthalpy and entropy are thermodynamic state functions, one may choose any path to obtain the changes from state 1 to state 2. For ease of calculation, one might choose a path which contains the ideal gas as in the diagram below; the actual path is shown as a dashed line while the solid lines indicate the path most convenient for calculations:

**Actual Path and Path of Calculation Convenience**

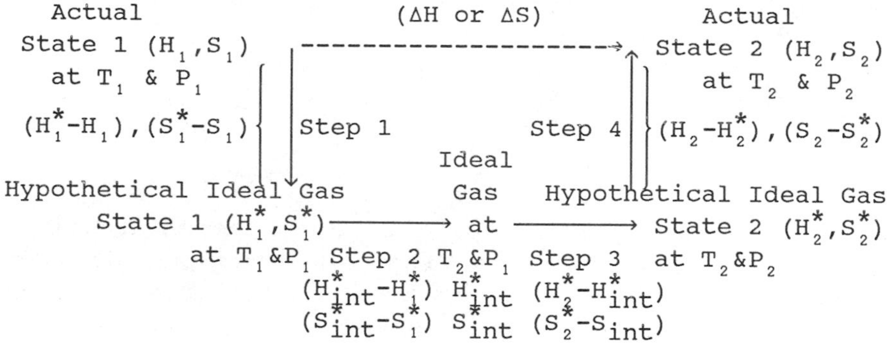


where: int = intermediate value

The above path determines the desired enthalpy or entropy change from state 1 to state 2 as may be shown by adding Steps 1 through 4 (in reverse order) to obtain:

$$(H_2-H_1) = (H_2-H_2^*) + (H_2^*-H_{int}^*) + (H_{int}^*-H_1^*) + (H_1^*-H_1) \tag{A-11}$$

$$(S_2-S_1) = (S_2-S_2^*) + (S_2^*-S_{int}^*) + (S_{int}^*-S_1^*) + (S_1^*-S_1) \tag{A-12}$$

### A.2.1 Ideal Gas Enthalpy and Entropy Changes

The advantage of using the above path is that the sum of the middle two terms or steps may be obtained from ideal gas properties of the mixture as:

$$(H_2^*-H_1^*) = \int_{T_1}^{T_2} C'_{p_{mix}}\, dT \tag{A-13}$$

$$(S_2^*-S_1^*) = \int_{T_1}^{T_2} \frac{C'_{p_{mix}}}{T}\, dT - R\, \ell n \frac{P_2}{P_1} \tag{A-14}$$

where the ideal gas heat capacity $C'_{pmix}$ is obtained from component ideal gas heat capacities $C'_{pi}$ by a linear average:

$$C'_{p_{mix}} = \sum_{n=1}^{c} y_i\, C'_{p_i} \tag{A-15}$$

and the component ideal gas heat capacities $C'_{p_i}$ are tabulated as a power series expansion in temperature for many components in standard thermodynamic references (see for example Passut and Danner (1972)). Note that none of the three equations (A-13,14, or 15) would be correct if the fluid were not an ideal gas.

### A.2.2 Deriving Enthalpy and Entropy Departures

To calculate the total change in enthalpy (ΔH) and entropy (ΔS) via the thermodynamic path shown in the above diagram, the first and last terms of Equations (A-11) and (A-12) require the departure of each quantity (H and S) from the ideal gas conditions ($H^*$ and $S^*$) at the specified temperature and pressure. These expressions may be derived for any equation-of-state using the departure derivations from Hougen, Watson, and Ragatz (1959 pp. 565-566):

$$(H^*-H) = RT - Pv + \int_{\infty}^{V}\left[P - T\left(\frac{\partial P}{\partial T}\right)_{V,n_T}\right] dv \quad \text{(A-16)}$$

and

$$(S^*-S) = \int_{\infty}^{V}\left[\frac{R}{V} + \left(\frac{\partial P}{\partial T}\right)_{V,n_T}\right]dv - R\,\ell n\, z \quad \text{(A-17)}$$

The derivations of the above residual properties for the Soave-Redlich-Kwong and for the Peng-Robinson equations-of-state are lengthy but straightforward, and can be found in modern thermodynamic texts (see for example Walas, p 509, (1985)). For the Soave-Redlich-Kwong equation the residual properties are:

$$(H^*-H) = RT - Pv + \frac{1}{b_M}(a_M+D)\,\ell n(1+b_M/v) \quad \text{(A-18)}$$

and

$$(S^*-S) = -R\ell n\left[\frac{Pv}{RT}\left(1 - \frac{b_M}{v}\right)\right] + \frac{D}{b_M T}\,\ell n\left(1 + \frac{b_M}{v}\right) \quad \text{(A-19)}$$

where: $$D \equiv \sum_i \sum_j \left[y_i y_j \beta_j (1-\delta_{ij})(a_i a_{c_j} T_{r_j})^{1/2}\right] \quad \text{(A-20)}$$

For the Peng-Robinson equation, The residual enthalpy and entropy are derived to be:

$$(H^*-H) = RT - Pv + \frac{a_M}{2.828 b_M}\left(1 + \frac{D}{a_M}\right)\ell n\left(\frac{v + 2.414\ b_M}{v - 0.414\ b_M}\right) \quad \text{(A-21)}$$

and

$$(S^*-S) = -R\ell n\left(\frac{Pv}{RT} - \frac{b_M P}{RT}\right) + \frac{b_M R^2 TD}{2.828 a_M{}^2}\ell n\left(\frac{v + 2.414\ b_M}{v - 0.414\ b_M}\right) \quad \text{(A-22)}$$

where D is given as in Equation (A-20).

The above equations are used with some of the examples in Chapter 8 to determine the hydrate formation and dissociation conditions in production and processing facilities. Note that in Equations (A-18) through (A-22) the volumes v are those of the phase of interest, implying that the mol fractions of the components of that phase are known and that the correct root for v is used. As mentioned in Section A.1, this sort of calculation normally requires the use of a computer. An example of such a computer program is given in the following pages.

### A.2.3 Computer Program for Changes in Enthalpy and Entropy

**A Word of Caution**

**While it is hoped that the programs accompanying this book will be of use in estimating thermodynamic conditions related to hydrate formation, the author should not and cannot be held totally accountable for the use of the predictions which these programs provide. If there is a safety consideration or an important process decision to be made, based upon the program's predictions, the user is cautioned to obtain a second opinion from someone knowledgeable in thermodynamics and phase equilibria, before proceeding.**

A computer program, written in the BASICA language, is included on the disk in the frontpapers for the Soave-Redlich-Kwong equation-of-state. The computer program SRKH&S.BAS was written to solve Example 8.1 in Section 8.1.3, but may be easily modified for other cases. The case in the program was written explicitly for expansion of a three component gas containing 90 mol% $CH_4$, 7% $C_2H_6$, and 3% $C_3H_8$. The program follows the flow pattern indicated by the equations in this section, as may be determined from the "REMark" statements in the program itself.

To use the program:

1. load DOS BASICA into the IBM-PC compatible computer,
2. load the program SRKH&S.BAS from the floppy disk, and
3. run the program, following the instructions given.

4. Input the initial P(psia), T(R), the final P and an estimate of the final T
5. If the expansion is isenthalpic, iterate with various values of $T_{final}$ until the total $\Delta H$ is approximately nil at the endpoint of the calculation
6. If the expansion is isentropic, iterate with various values of $T_{final}$ until the total $\Delta S$ is approximately nil at the endpoint of the calculation.

The program is written to be easily extended to include gases with methane, ethane, propane, hydrogen sulfide, carbon dioxide, and nitrogen, because constants for these components are already included. Note that the input data are set for six components (M=6), so that the first three mol fractions are set to the values in Example 8.1 and the preceding paragraph. The final three components have mol fractions set to 0.0 (Line 910) to achieve the values of the example with an input of six components. To modify the program to include any of the above six components, simply change Statement 910 to reflect the mol fractions of the components present in the same order as the previous line (900), setting absent component mol fractions as nil, and input other components as requested by in program execution.

In order to modify the program for other components, change the program in the following way:

| Statements | Change |
|---|---|
| 310-360 | Modify the dimension statements to the number of components |
| 400-450 | Change critical constants and acentric factors $\omega$ (W) for the components |
| 460-570 | Change ideal heat capacity expansion parameters from a power series expansion such as given by Passut and Danner (1972) |
| 580-690 | Change the interaction parameters $\delta_{ij}$ (called here $k_{ij}$) from a reference such as Graboski and Daubert (1978) |
| 900-930 | Change the input gas fractions and molecular masses |

Before substantial changes are made in the program however, it may be advisable to duplicate the results of Example 8.1 to ensure that the copy of the program is operating correctly.

```
10 REM SRKH&S
20 REM
30 REM THIS IS A PROGRAM TO CALCULATE
40 REM THE ENTHALPY AND ENTROPY CHANGE
50 REM
60 REM THE EQUATIONS FOR ENTHALPY & ENTROPY DEPARTURE
70 REM WERE DERIVED FROM THE SOAVE EQUATION
80 REM USING THE EQUATIONS IN THE APPENDIX A OF THE BOOK
90 REM THE CRITICAL CONSTANTS LISTED HERE ARE TO BE
100 REM FOUND IN THE APENDIX OF THE
110 REM 1981 EDITION OF HENLEY AND SEADER
120 REM "EQUILIBRIUM STAGE OPERATIONS IN
130 REM CHEMICAL ENGINEERING" MCGRAW-HILL
140 REM EXCEPT IDEAL GAS CP VALUES FROM
150 REM PASSUT AND DANNER I&ECPDD, VOL11, 543 (1972)
160 REM
170 REM CURRENTLY THE PROGRAM IS SET TO RUN
180 REM MIXTURES OF CH4,C2H6,C3H8,N2,H2S & CO2
190 REM TO CHANGE THE MIXTURE COMPOSITION
200 REM SIMPLY CHANGE THE VALUES IN LINE 910 IN
                     THE ORDER INDICATED
210 REM IN THE PREVIOUS LINE
220 REM
230 REM TO CHANGE THE COMPONENTS, ONE MUST INPUT
240 REM THE FOLLOWING FOR THE NEW COMPONENT
250 REM TCRIT (LINE 410), PCRIT (LINE 430),
                     ACENTRIC FACTOR (LINE 450),
260 REM HEAT CAPACITY CONSTANTS (LINE 460 - LINE 570),
270 REM INTERACTION KIJ VALUES (LINE 580 - LINE 690),
280 REM CHANGE THE MOL FRACTIONS IN LINE 910,
290 REM THE MOLECULAR WEIGHT IN LINE 930,
                     AND DIMENSION LINES (320-360)
300 REM
310 REM DIMENSION STATEMENTS
320 DIM PC(6),TC(6),W(6),PR(6),TR(6)
330 DIM ALF(6),MK(6),CA(6),CB(6),KIJ(6,6)
340 DIM CC(6),CD(6),CE(6)
350 DIM P(2),T(2),AA(6),BB(6)
360 DIM Y(6),MW(6)
370 REM 1=CH4, 2=C2H6, 3=C3H8
380 REM 4=H2S, 5=CO2, 6=N2
390 REM
400 REM DATA FOR TC IN R
410 DATA 343.9,550.0,665.9,672.4,547.6,227.3
```

```
420 REM DATA FOR PC IN PSIA
430 DATA 673.1,709.8,617.4,1306.5,1070.5,492.9
440 REM DATA FOR ACENTRIC FACTOR
450 DATA 0.0,0.1064,0.1538,0.0868,0.177,0.0206
460 REM HEAT CAPACITY CONSTANTS FOR CH4
470 DATA 0.564834,-.565946E-03,1.252197E-06,
                     -6.102304E-10,9.794285E-14
480 REM HEAT CAPACITY CONSTANTS FOR C2H6
490 DATA 0.273088,0.085912E-03,.938445E-06,
                     -5.55956E-10,10.035115E-14
500 REM HEAT CAPACITY CONSTANTS FOR C3H8
510 DATA 0.179733,1.32916E-04,7.52994E-07,
                     -4.989844E-10,9.467545E-14
520 REM HEAT CAPACITY CONSTANTS FOR H2S
530 DATA 0.238575,-0.048914E-03,0.123201E-06,
                     -.520504E-10,.72426E-14
540 REM HEAT CAPACITY CONSTANTS FOR CO2
550 DATA 0.114433, 0.202264E-03,-.079482E-06,
                     0.138824E-10,-.02628E-14
560 REM HEAT CAPACITY CONSTANTS FOR N2
570 DATA 0.253664,-.029098E-03,.037632E-06,
                     -.068424E-10,-.041195E-14
580 REM KIJ VALUES FOR C1 W/ C2, C3, ETC.
590 DATA 0.0,0.0,0.0,0.095,0.103,0.037
600 REM KIJ VALUES FOR C2 W/ C1,C2, ETC.
610 DATA 0.0,0.0,0.0,0.092,0.135,0.02
620 REM KIJ VALUES FOR C3 W/ C1,C2 ETC
630 DATA 0.0,0.0,0.0,0.079,0.112,0.07
640 REM KIJ VALUES FOR H2S W/ C1,C2,ETC.
650 DATA .095,.092,.079,0.0,.094,.135
660 REM KIJ VALUES FOR CO2 W/ C1,C2, ETC.
670 DATA 0.103,.135,.112,.094,0.0,-0.04
680 REM KIJ VALUES FOR N2 W/ C1, C2, ETC.
690 DATA 0.037,0.02,0.07,0.135,-0.04,0.0
700 REM
710 PRINT "INPUT NUMBER OF COMPONENTS"
720 INPUT M
730 FOR I = 1 TO M
740 READ TC(I)
750 NEXT I
760 FOR I= 1 TO M
770 READ PC(I)
780 NEXT I
790 FOR I = 1 TO M
800 READ W(I)
810 NEXT I
820 FOR I = 1 TO M
830 READ CA(I),CB(I),CC(I),CD(I),CE(I)
840 NEXT I
850 FOR I = 1 TO M
860 FOR J = 1 TO M
870 READ KIJ(I,J)
```

```
880 NEXT J
890 NEXT I
900 REM INPUT GAS MOLE FRACTIONS
910 DATA 0.9,0.07,0.03,0,0,0
920 REM COMPONENT MOLECULAR MASSES
930 DATA 16.04,30.07,44.1,34.08,44.01,28.01
940 FOR I= 1 TO M
950 READ Y(I)
960 NEXT I
970 FOR I = 1 TO M
980 READ MW(I)
990 NEXT I
1000 PRINT "INPUT P1(PSIA),T1(R),P2(PSIA)"
1010 INPUT P(1),T(1),P(2)
1020 PRINT "INPUT T2GUESS"
1030 PRINT "INPUT T2=0 IF YOU WISH TO DISCONTINUE PROGRAM"
1040 INPUT T(2)
1050 IF T(2)=0 GOTO 1770
1060 KOUNT =1
1070 R=10.73
1080 REM
1090 IF KOUNT > 1 GOTO 1120
1100 P=P(1)
1110 T=T(1)
1120 IF KOUNT =1 GOTO 1150
1130 P=P(2)
1140 T=T(2)
1150 FOR I = 1 TO M
1160 MK(I)= .48+1.574*W(I)-.176*W(I)*W(I)
1170 TR(I)=T/TC(I)
1180 PR(I)=P/PC(I)
1190 ALF(I)=(1+MK(I)*(1-SQR(TR(I))))^2
1200 AA(I)=.42747*ALF(I)*R*R*TC(I)*TC(I)/PC(I)
1210 BB(I)=.08664*R*TC(I)/PC(I)
1220 NEXT I
1230 A = 0!
1240 B = 0!
1250 FOR I = 1 TO M
1260 B = B + Y(I)*BB(I)
1270 NEXT I
1280 FOR I = 1 TO M
1290 FOR J = 1 TO M
1300 A = A + (1-KIJ(I,J))*(SQR(AA(I)*AA(J)))*Y(I)*Y(J)
1310 NEXT J
1320 NEXT I
1330 GOSUB 1780
1340 REM NOW BEGIN TO CALCULATE H & S DEPARTURES
1350 SUM=0!
1360 FOR I = 1 TO M
1370 FOR J = 1 TO M
1380 SUM = SUM + Y(I)*Y(J)*MK(J)*(1-KIJ(I,J))*
                        SQR(AA(I)*(AA(J)/ALF(J))*TR(J))
```

```
1390 NEXT J
1400 NEXT I
1410 H1 = - 1.987*T*(Z-1-(1/(R*T*B))*(LOG(1+B/V))*(A+SUM))
1420 S1 = -1.987*(LOG(Z*(1-B/V))+(SUM/(B*R*T))*LOG(1+B/V))
1430 IF KOUNT > 1 GOTO 1480
1440 D1 = H1
1450 D3 = S1
1460 REM
1470 REM D2 = 2ND ENTHALPY DEPARTURE;
                        D4 = 2ND ENTROPY DEPARTURE
1480 IF KOUNT = 1 GOTO 1520
1490 D2=H1
1500 D4=S1
1510 REM NOW CALCULATE IDEAL GAS ENTHALPY
                        AND ENTROPY CHANGES
1520 T3 = T(1)
1530 T4 = T(2)
1540 HT = 0!
1550 ST = 0!
1560 FOR I = 1 TO M
1570 HA = CA(I)*(T4-T3)+CB(I)/2*(T4*T4-T3*T3)+
                         CC(I)/3*(T4^3-T3^3)
1580 HB = HA + CD(I)/4*(T4^4-T3^4)+CE(I)/5*(T4^5-T3^5)
1590 HT=HT+Y(I)*MW(I)*HB
1600 SA = CA(I)*LOG(T4/T3)+CB(I)*(T4-T3)+
                         CC(I)/2*(T4*T4-T3*T3)
1610 SB = SA + CD(I)/3*(T4^3-T3^3)+CE(I)/4*(T4^4-T3^4)
1620 ST=ST+Y(I)*MW(I)*SB
1630 NEXT I
1640 STOT = ST - 1.987*LOG(P(2)/P(1))
1650 KOUNT = KOUNT + 1
1660 IF KOUNT = 2 GOTO 1090
1670 TTH = (D1+HT-D2)
1680 TTS = (D3+STOT-D4)
1690 PRINT"P1(PSIA),T1(DEG R), P2,T2 FOLLOW"
1700 PRINT P(1),T(1),P(2),T(2)
1710 PRINT "1ST DEL H =";D1;"BTU/LBMOL  IDEAL GAS DEL H =";
                             HT;" 2ND DEL H =";D2
1720 PRINT "TOTAL DELTA H = ";TTH;"BTU/LBMOL"
1730 PRINT "1ST DEL S =";D3;"BTU/LBMOL-R; IDLGAS DEL S=";
                             STOT;" 2ND DEL S =";D4
1740 PRINT "TOTAL DELTA S = ";TTS;"BTU/LBMOL-R
1750 PRINT"IF THE ABOVE VALUES ARE UNSATISFACTORY";
                   " GUESS ANOTHER OUTLET T IN R"
1760 GOTO 1020
1770 END
1780 REM SOL'N TO V^3+D*V^2+E*V+F=0
1790 REM
1800 D=-R*T/P
1810 E=A/P-R*T*B/P-B*B
1820 F=-A*B/P
1830 G=(3!*E-D*D)/3!
```

```
1840 H=-(9!*D*E-27!*F-2!*D*D*D)/27!
1850 GH=G^3!/27!+H*H/4!
1860 IF GH<=0! GOTO 2050
1870 S = -H/2!+SQR(G^3!/27!+H*H/4!)
1880 TT = -H/2!-SQR(G^3/27!+H*H/4!)
1890 IF S>=0! GOTO 1920
1900 S=-((-S)^(1!/3!))
1910 GOTO 1930
1920 S=S^(1!/3!)
1930 IF TT>=0! GOTO 1960
1940 TT=-(-TT)^(1!/3!)
1950 GOTO 2000
1960 TT=(TT)^(1!/3!)
1970 REM
1980 REM SINGLE REAL ROOT
1990 REM
2000 V=S+TT-D/3!
2010 GOTO 2200
2020 REM
2030 REM *** 3 REAL ROOTS
2040 REM
2050 XA=(-.5*H/SQR(-G^3/27!))
2060 THETA = (3.14159265#/2!-ATN(XA/(SQR(1!-XA*XA))))/3!
2070 V1=2!*SQR(-G/3!)*COS(THETA)
2080 V2=2!*SQR(-G/3!)*COS(THETA+2.0944)
2090 V3=2!*SQR(-G/3!)*COS(THETA+4.188)
2100 REM
2110 REM ** TAKE LARGEST V FOR VAPOR
2120 REM
2130 IF V1<V2 AND V2<V3 GOTO 2170
2140 IF V1<V2 AND V3<V2 GOTO 2190
2150 V=V1-D/3
2160 GOTO 2200
2170 V=V3-D/3!
2180 GOTO 2200
2190 V=V2-D/3!
2200 Z=P*V/(R*T)
2210 RETURN
```

# Appendix B

# User's Guide for CSMHYD Hydrate Program

**A Word of Caution**

**While it is hoped that the programs accompanying this book will be of use in estimating the equilibrium conditions of hydrate formation, the author should not and cannot be held totally accountable for the use of the predictions which these programs provide. If there is a safety consideration or an important process decision to be made, based upon the program's predictions, the user is cautioned to obtain a second opinion from someone knowledgeable in hydrate phase equilibria, before proceeding.**

## Executive Summary

This program (reproduced in the IBM-compatible PC floppy disk in the frontpapers) estimates the phase equilibria conditions for the formation of hydrates. The details of the theory of the program are presented in

Chapter 5 of this text. The program will enable the calculation of answers to the following types of questions:

a) How much dissolved water is allowable in a hydrocarbon vapor or liquid phase (no free water present) before hydrate will form at a given temperature and pressure?
b) What is the allowable temperature or pressure before hydrates will form with a free water phase and a hydrocarbon vapor?
c) How much methanol should be used as inhibitor for hydrates formation from a free water phase at a given temperature and pressure?
d) What are the conditions for hydrate formation when three other phases (free water-liquid hydrocarbon-vapor) are present?

The programs uses the method of van der Waals and Platteeuw for the hydrate phase, and the Soave-Redlich-Kwong equation of state for the fluid phases, with the parameters obtained from experimental measurements. Standard hydrate forming components are: methane, ethane, propane, iso-butane, butane, hydrogen sulfide, carbon dioxide, nitrogen, ethylene, and propylene. The following non-hydrate forming components are also treated: pentane, hexane, heptane, octane, nonane, decane, toluene, and benzene. The range of applicability is 233 to 300K, and pressures up to 27 MPa. Twenty-two typical multicomponent, hydrate forming natural gases have been tested with an average accuracy to within 9% of the experimental values for three phase pressure conditions at a given temperature (Question b above).

This appendix is divided into three portions, as follows:
B.1 Overview of the program
B.2 Input Data for Starting the Calculation
B.3 Illustrations of Types of Calculations

## B.1 Overview of the Program

The computer hardware required for the program are: an IBM-compatible personal computer, 512 K Bytes of random access memory (RAM), at least one floppy disk drive, and a numeric data 8087 co-processor. Hardcopy of the results may be obtained through hand recording of the screen, or through the "PRINTSCREEN" function; screen scrolling stops are automatically provided at convenient points in the program output for this purpose.

The program user is encouraged to review Chapter Five for the details of the theory upon which the program is based. However, while Chapter Five serves to point out the limitations of the method, the use of the program does not require comprehension of the theory. The program provides a balance between a rigorous empirical fit of the data, and the maintenance of theoretical meaning of the parameters, with the hope of providing accurate interpolation and a cautious extrapolation of existing experimental data.

The program enables the user to predict the below hydrate formation conditions for a given hydrocarbon composition:

1. Three-phase ($L_W$-H-V or I-H-V)) conditions of either temperature or pressure, given a pure aqueous phase,

2. Three-phase ($L_W$-H-V or I-H-V) conditions of either temperature, pressure, or methanol concentration in the aqueous phase, given two of those three variables,

3. Condensed three-phase ($L_W$-H-$L_{HC}$) and four phase ($L_W$-H-V-$L_{HC}$) conditions temperature and pressure both with and without methanol in the aqueous phase, and

4. Two phase (H-V or H-$L_{HC}$) water concentrations in the hydrocarbon phase which is in equilibrium only with hydrate, at a given temperature or pressure.

While some default calculations are built into the program, it is suggested that the user predict additional hydrate conditions in phase equilibria regions which bound the calculation, as provided in the discussions of Chapter 4. For example, infrequently the program indicates when it has been requested to predict hydrates at conditions at which they cannot occur (e.g. at temperatures appreciably above the upper quadruple point). On the other hand the program will not calculate the three-phase hydrate conditions before performing a two-phase calculation, to ensure that the temperature and pressure will only permit two phases to form. Therefore it is worthwhile to predict the limiting three-phase temperature and pressure conditions (see Figures 4-1 and 4-2 and discussion for a detailed explanation) before attempting a two-phase calculation.

### B.2 Input Data for Starting the Calculation

After the computer is "booted" in the MS-DOS operating system, the floppy disk in the frontpapers should be mounted on the disk drive and the instruction CSMHYD (RETURN) should be typed. The computer should respond with the below series of seven questions, regarding the input data and the type of calculation to be performed. It is assumed that the answers to the questions will be input from the keyboard of the computer.

Question 1. Input the number corresponding to the type of phase equilibrium calculation which is to be performed.

-4 Four-phase ($L_W$-H-V-$L_{HC}$) and three phase ($L_W$-H-$L_{HC}$) prediction with methanol inhibition

-3 Two-phase (H-$L_{HC}$) prediction of water concentration in the hydrocarbon, at a given temperature and pressure

-2 Three-phase ($L_W$-H-V) prediction with methanol inhibition, with temperature as the unknown, and with pressure and mass per cent methanol as the known parameters.

-1 Three-phase ($L_W$-H-V) prediction with methanol inhibition, with pressure as the unknown, and with temperature and mass per cent methanol as the known parameters.

0 Three-phase ($L_W$-H-V) prediction with methanol inhibition, with the mass per cent methanol as the unknown, and with temperature and pressure as the known parameters.

1 Three-phase ($L_W$-H-V or I-H-V) prediction with the temperature given and pressure as the unknown.

2 Three-phase ($L_W$-H-V) prediction with pressure given and temperature as the unknown.

3 Two-phase (H-V) prediction of the water content of the hydrocarbon, at a given temperature and pressure.

4 Four-phase ($L_W$-H-V-$L_{HC}$) and three-phase ($L_W$-H-$L_{HC}$) predictions of temperature and pressure.

Question 2. Input the corresponding number for the units in which the temperature and pressure will be entered and calculated

| Option | Temperature | Pressure |
|---|---|---|
| 0 | $^0$F | psia |
| 1 | $^0$C | mm Hg |
| 2 | K | kilopascals |
| 3 | K | atmospheres |

Question 3. Input the total number of components (both hydrate-formers and non-hydrate formers) that are present in the natural gas mixture, not including water. The maximum number of the components is 20.

Question 4. Input the numbers corresponding to which components are present, listing hydrate-formers first. If the components are listed in the wrong order, the program will stop execution if input is from a file, or will ask for the components numbers to be re-entered in the correct order if input is from the keyboard. The component numbers are:

Hydrate-Formers

| | |
|---|---|
| 1 Methane | 14 Hydrogen Sulfide |
| 2 Ethane | 15 Nitrogen |
| 3 Propane | 16 Carbon Dioxide |
| 4 iso-Butane | 18 Ethylene |
| 5 normal-Butane | 19 Propylene |

Non-Hydrate-Formers

| | |
|---|---|
| 6 Pentane | 10 Nonane |
| 7 Hexane | 11 Decane |
| 8 Heptane | 12 Toluene |
| 9 Octane | 13 Benzene |

Question 5. Input the mol fractions of the components, on a water-free basis, listing them in the same order as in Question 4. The program will warn the user if the mol fractions do not sum to unity; the user is given the opportunity to re-input the mol fractions, or to choose that they be normalized.

Question 6. Since different input data are required depending on which type calculation is being performed, the specifics of each are given. Refer to Question 1 for the appropriate case numbers, given on the left below:

-4 Enter an estimate of the temperature and pressure and the desired weight fraction of methanol (maximum is 0.35) in the aqueous phase

-3 Enter the desired temperature and pressure and an estimate of the water concentration in ppm (mol basis)

-2 Enter an estimate of the temperature and the desired pressure and mass fraction of methanol (maximum is 0.35)

-1 Enter the desired temperature, an estimate of the pressure and the desired mass fraction of methanol (maximum is 0.35)

0 Enter the desired temperature and pressure (the program calculates the methanol concentration with no estimate required).

1 Enter the desired temperature and an estimate of the pressure.

2 Enter the desired pressure and an estimate of the temperature.

3 Enter the desired temperature and pressure and a estimate of the water concentration in ppm (mol)

4 Enter the best estimate of both the temperature and pressure

Question 7. The program asks if the user would like to continue execution (No=1, Yes =2). If the answer is Yes and a four-phase calculation is being performed, execution continues at Question 5. If the answer is Yes and a two-phase or three-phase calculation is being performed, execution continues at Question 6. If the answer is No(1), execution ceases.

Examples of each of the nine option cases for the calculation are provided in the following pages. If questions arise concerning the use of the program, it may be helpful to duplicate some of these examples.

## B.3 Illustrations of Types of Calculation

### Example A
Four Phase ($L_W$-H-V-$L_{HC}$) Prediction
with Methanol Inhibition

| Component | Mol Fraction |
|---|---|
| $CH_4$ | 0.07 |
| $C_2H_6$ | 0.23 |
| $C_3H_8$ | 0.28 |
| i-$C_4H_{10}$ | 0.22 |
| n-$C_4H_{10}$ | 0.11 |
| $C_5H_{12}$ | 0.09 |

Weight Fraction Methanol 0.15

Guess for Temperature 30°F

Guess for Pressure 100 psia

```
THIS IS THE COLORADO SCHOOL OF MINES HYDRATE PREDICTION PROGRAM
ALL INPUT DATA SHOULD BE SEPARATED BY COMMAS WITH 10 PIECES OF DATA PER LINE
WHAT IS THE TYPE OF CALCULATION TO BE PERFORMED?
     -4= FOUR PHASE(LW-G-H-LH) PREDICTION WITH METHANOL INHIBITION
     -3= TWO PHASE(LH-H) PREDICTION
     -2= THREE PHASE(LW-G-H) TEMPERATURE PREDICTION WITH METHANOL INHIBITION
     -1= THREE PHASE(LW-G-H) PRESSURE PREDICTION WITH METHANOL INHIBITION
      0= THREE PHASE(LW-G-H) WEIGHT % METHANOL NEEDED PREDICTION
      1= THREE PHASE(LW-G-H) PRESSURE PREDICTION
      2= THREE PHASE(LW-G-H) TEMPERATURE PREDICTION
      3= TWO PHASE(G-H) PREDICTION
      4= FOUR PHASE(LW-G-H-LH) PREDICTION

-4
WHAT ARE THE UNITS ON
         TEMPERATURE  AND  PRESSURE?
     0=   FAHRENHEIT        PSIA
     1=   CENTIGRADE        mm Hg
     2=   KELVIN            KPASCALS
     3=   KELVIN            ATM
```

```
0
HOW MANY COMPONENTS ARE PRESENT?
6
WHICH COMPONENTS ARE PRESENT? LIST HYDRATE FORMERS FIRST!

              HYDRATE FORMERS
     1= METHANE            2= ETHANE
     3= PROPANE            4= ISO-BUTANE
     5= N-BUTANE          14= HYDROGEN SULFIDE
    15= NITROGEN          16= CARBON DIOXIDE
    18= ETHYLENE          19= PROPYLENE

              NON HYDRATE FORMERS
     6= PENTANE            7= HEXANE
     8= HEPTANE            9= OCTANE
    10= NONANE            11= DECANE
    12= TOLUENE           13= BENZENE

1,2,3,4,5,6
WHAT ARE THE MOLE FRACTIONS OF THE COMPONENTS?
    LIST IN THE SAME ORDER AS ABOVE !

0.07,0.23,0.28,0.22,0.11,0.09
WHAT IS THE GUESS FOR TEMPERATURE AND PRESSURE
AND WEIGHT FRACTION METHANOL?

30,100,.15
          RESULTS FOR QUADRAPOLE PREDICTIONS
               WITH METHANOL INHIBITION
```

| COMPONENT NAME | LIQUID COMPOSITION | VAPOR COMPOSITION | HYDRATE COMPOSITION | VLE K FACTOR | HYDRATE K FACTOR |
|---|---|---|---|---|---|
| METHANE | .07000 | .53051 | .5599 | 7.5788 | .9475 |
| ETHANE | .23000 | .31460 | .0363 | 1.3678 | 8.6603 |
| PROPANE | .28000 | .10640 | .2386 | .3800 | .4459 |
| I-BUTANE | .22000 | .03282 | .1615 | .1492 | .2032 |
| N-BUTANE | .11000 | .01200 | .0036 | .1091 | 3.3289 |
| PENTANE | .09000 | .00376 | | .0418 | |

press RETURN to continue ...

FRACTION OF CAVITIES OCCUPIED BY HYDRATE FORMERS

STRUCTURE 2

| COMPONENT NAME | SMALL CAVITY | LARGE CAVITY |
|---|---|---|
| METHANE | .627739 | .006796 |
| ETHANE | .000000 | .081898 |
| PROPANE | .000000 | .538025 |
| I-BUTANE | .000000 | .364135 |
| N-BUTANE | .000000 | .008126 |

press RETURN to continue ...

```
                    TEMPERATURE =  37.17 F

                    PRESSURE =   251.8634 PSIA

                    WEIGHT % METHANOL IN WATER = 15.00 %

                    THE FREEZING POINT OF THE METHANOL-WATER MIXTURE=  12.94 F

                    THE SLOPE OF THE THREE PHASE(LW-G-LH)(DP/DT)=   518.7 PSIA/F


DO YOU WISH TO DO THIS CALCULATION FOR OTHER MOLE FRACTIONS (N=1 Y=2)?
```

## Example B

Two Phase Prediction

| Component | Mol Fraction |
|---|---|
| $C_2H_6$ | 0.06 |
| $C_3H_8$ | 0.92 |
| $i\text{-}C_4H_{10}$ | 0.03 |

Temperature 30 $^{0}$F

Pressure 100 psia

Guess for Mol Fraction of Water 20 ppm (mol basis)

```
THIS IS THE COLORADO SCHOOL OF MINES HYDRATE PREDICTION PROGRAM
ALL INPUT DATA SHOULD BE SEPARATED BY COMMAS WITH 10 PIECES OF DATA PER LINE
WHAT IS THE TYPE OF CALCULATION TO BE PERFORMED?
     -4- FOUR PHASE(LW-G-H-LH) PREDICTION WITH METHANOL INHIBITION
     -3- TWO PHASE(LH-H) PREDICTION
     -2- THREE PHASE(LW-G-H) TEMPERATURE PREDICTION WITH METHANOL INHIBITION
     -1- THREE PHASE(LW-G-H) PRESSURE PREDICTION WITH METHANOL INHIBITION
      0- THREE PHASE(LW-G-H) WEIGHT % METHANOL NEEDED PREDICTION
      1- THREE PHASE(LW-G-H) PRESSURE PREDICTION
      2- THREE PHASE(LW-G-H) TEMPERATURE PREDICTION
      3- TWO PHASE(G-H) PREDICTION
      4- FOUR PHASE(LW-G-H-LH) PREDICTION

-3
WHAT ARE THE UNITS ON
         TEMPERATURE  AND  PRESSURE?
     0-   FAHRENHEIT         PSIA
     1-   CENTIGRADE         mm Hg
     2-   KELVIN             KPASCALS
     3-   KELVIN             ATM

0
HOW MANY COMPONENTS ARE PRESENT?
3
WHICH COMPONENTS ARE PRESENT? LIST HYDRATE FORMERS FIRST!

              HYDRATE FORMERS
     1- METHANE           2- ETHANE
     3- PROPANE           4- ISO-BUTANE
     5- N-BUTANE         14- HYDROGEN SULFIDE
    15- NITROGEN         16- CARBON DIOXIDE
    18- ETHYLENE         19- PROPYLENE
```

```
                NON HYDRATE FORMERS
     6= PENTANE            7= HEXANE
     8= HEPTANE            9= OCTANE
    10= NONANE            11= DECANE
    12= TOLUENE           13= BENZENE

2,3,4
WHAT ARE THE MOLE FRACTIONS OF THE COMPONENTS?
    LIST IN THE SAME ORDER AS ABOVE !

0.06,0.92,0.03
THE MOLE FRACTIONS SUM TO  1.010000
DO YOU WISH TO REINPUT THE MOLE FRACTIONS(N=1 Y=2)?
IF NO THE PROGRAM WILL NORMALIZE THE MOLE FRACTIONS
1
WHAT IS THE TEMPERATURE AND PRESSURE AND A GUESS FOR THE MOLE FRACTION WATER IN PPM?

30,100,20
         RESULTS FOR TWO PHASE(LH-H) CALCULATIONS

COMPONENT     LIQUID         HYDRATE         HYDRATE K
  NAME     COMPOSITION    COMPOSITION         FACTOR
 ETHANE        .05941         .01155          5.1431
 PROPANE       .91089         .96200            9469
 I-BUTANE      .02970         .02645          1.1231

 press RETURN to continue ...


         FRACTION OF CAVITIES OCCUPIED BY HYDRATE FORMERS
                        STRUCTURE 2
              COMPONENT    SMALL       LARGE
                NAME       CAVITY      CAVITY
               ETHANE      .000000     .011545
               PROPANE     .000000     .961512
               I-BUTANE    .000000     .026434

 press RETURN to continue ...

                     TEMPERATURE =  30.00 F

                     PRESSURE =   100.0000 PSIA

                     THE MOLE FRACTION OF WATER =    129.320 PPM

                     THE BUBBLE POINT PRESSURE =     79.2472 PSIA

DO YOU WISH TO DO ANOTHER CALCULATION AT THESE SAME COMPOSITIONS (N=1 Y=2)?
```

## Example C

### Three-Phase ($L_W$-H-V) Temperature Prediction with Methanol Inhibition

| Component | Mol Fraction |
|---|---|
| $CH_4$ | 0.93 |
| $C_2H_6$ | 0.04 |
| $C_5H_{12}$ | 0.005 |
| $N_2$ | 0.015 |
| $CO_2$ | 0.01 |

Pressure 200 psia

Weight Fraction Methanol 0.20

Guess for Temperature $30^0$F

```
THIS IS THE COLORADO SCHOOL OF MINES HYDRATE PREDICTION PROGRAM
ALL INPUT DATA SHOULD BE SEPARATED BY COMMAS WITH 10 PIECES OF DATA PER LINE
WHAT IS THE TYPE OF CALCULATION TO BE PERFORMED?
     -4= FOUR PHASE(LW-G-H-LH) PREDICTION WITH METHANOL INHIBITION
     -3= TWO PHASE(LH-H) PREDICTION
     -2= THREE PHASE(LW-G-H) TEMPERATURE PREDICTION WITH METHANOL INHIBITION
     -1= THREE PHASE(LW-G-H) PRESSURE PREDICTION WITH METHANOL INHIBITION
      0= THREE PHASE(LW-G-H) WEIGHT % METHANOL NEEDED PREDICTION
      1= THREE PHASE(LW-G-H) PRESSURE PREDICTION
      2= THREE PHASE(LW-G-H) TEMPERATURE PREDICTION
      3= TWO PHASE(G-H) PREDICTION
      4= FOUR PHASE(LW-G-H-LH) PREDICTION

-2
WHAT ARE THE UNITS ON
        TEMPERATURE  AND  PRESSURE?
     0=   FAHRENHEIT         PSIA
     1=   CENTIGRADE         mm Hg
     2=   KELVIN             KPASCALS
     3=   KELVIN             ATM

0
HOW MANY COMPONENTS ARE PRESENT?
5
WHICH COMPONENTS ARE PRESENT? LIST HYDRATE FORMERS FIRST!
```

```
                 HYDRATE FORMERS
   1- METHANE            2- ETHANE
   3- PROPANE            4- ISO-BUTANE
   5- N-BUTANE          14- HYDROGEN SULFIDE
  15- NITROGEN          16- CARBON DIOXIDE
  18- ETHYLENE          19- PROPYLENE

                 NON HYDRATE FORMERS
   6- PENTANE            7- HEXANE
   8- HEPTANE            9- OCTANE
  10- NONANE            11- DECANE
  12- TOLUENE           13- BENZENE

1,2,15,16,6
WHAT ARE THE MOLE FRACTIONS OF THE COMPONENTS?
    LIST IN THE SAME ORDER AS ABOVE !

.93,.04,.015,.01,.005
INPUT THE GUESS FOR TEMPERATURE AND THE SPECIFIED PRESSURE
AND WEIGHT FRACTION METHANOL?

30,200,0.2
          RESULTS FOR THREE PHASE (LW-G-H) TEMPERATURE PREDICTIONS
               WITH METHANOL INHIBITION

COMPONENT      VAPOR         HYDRATE        HYDRATE K
  NAME      COMPOSITION    COMPOSITION       FACTOR
 METHANE       .93000         .71277         1.3048
 ETHANE        .04000         .26729          .1497
 NITROGEN      .01500         .00114        13.1833
 C-02          .01000         .01880          .5319
 PENTANE       .00500

 press RETURN to continue ...


          FRACTION OF CAVITIES OCCUPIED BY HYDRATE FORMERS
                        STRUCTURE 1
               COMPONENT     SMALL      LARGE
                 NAME        CAVITY     CAVITY
                METHANE     .853902    .620113
                ETHANE      .000000    .339282
                NITROGEN    .002558    .000591
                C-02        .007568    .021340

 press RETURN to continue ...

                   TEMPERATURE -    6.67 F

                   PRESSURE -    200.0000 PSIA

                   WEIGHT % METHANOL IN WATER - 20.00 %

                   THE FREEZING POINT OF THE METHANOL-WATER MIXTURE-    4.78 F

                   THE TEMPERATURE IS ABOVE THE CRITICAL POINT

DO YOU WISH TO DO ANOTHER CALCULATION AT THESE SAME COMPOSITIONS (N=1 Y=2)?
```

## Example D

### Three Phase ($L_W$-H-V) Pressure Prediction with Methanol Inhibition

| Component | Mol fraction |
|---|---|
| $CH_4$ | 0.60 |
| $C_2H_6$ | 0.30 |
| $H_2S$ | 0.10 |

| | |
|---|---|
| Temperature | 40°F |
| Weight Fraction Methanol | 0.15 |
| Guess for Pressure | 50 psia |

```
THIS IS THE COLORADO SCHOOL OF MINES HYDRATE PREDICTION PROGRAM
ALL INPUT DATA SHOULD BE SEPARATED BY COMMAS WITH 10 PIECES OF DATA PER LINE
WHAT IS THE TYPE OF CALCULATION TO BE PERFORMED?
     -4= FOUR PHASE(LW-G-H-LH) PREDICTION WITH METHANOL INHIBITION
     -3= TWO PHASE(LH-H) PREDICTION
     -2= THREE PHASE(LW-G-H) TEMPERATURE PREDICTION WITH METHANOL INHIBITION
     -1= THREE PHASE(LW-G-H) PRESSURE PREDICTION WITH METHANOL INHIBITION
      0= THREE PHASE(LW-G-H) WEIGHT % METHANOL NEEDED PREDICTION
      1= THREE PHASE(LW-G-H) PRESSURE PREDICTION
      2= THREE PHASE(LW-G-H) TEMPERATURE PREDICTION
      3= TWO PHASE(G-H) PREDICTION
      4= FOUR PHASE(LW-G-H-LH) PREDICTION

-1
WHAT ARE THE UNITS ON
        TEMPERATURE  AND  PRESSURE?
     0=  FAHRENHEIT        PSIA
     1=  CENTIGRADE        mm Hg
     2=  KELVIN            KPASCALS
     3=  KELVIN            ATM

0
HOW MANY COMPONENTS ARE PRESENT?
3
WHICH COMPONENTS ARE PRESENT? LIST HYDRATE FORMERS FIRST!
```

```
            HYDRATE FORMERS
   1= METHANE            2= ETHANE
   3= PROPANE            4= ISO-BUTANE
   5= N-BUTANE          14= HYDROGEN SULFIDE
  15= NITROGEN          16= CARBON DIOXIDE
  18= ETHYLENE          19= PROPYLENE

            NON HYDRATE FORMERS
   6= PENTANE            7= HEXANE
   8= HEPTANE            9= OCTANE
  10= NONANE            11= DECANE
  12= TOLUENE           13= BENZENE

1,2,14
WHAT ARE THE MOLE FRACTIONS OF THE COMPONENTS?
    LIST IN THE SAME ORDER AS ABOVE !

0.6,0.3,0.1
INPUT THE SPECIFIED TEMPERATURE AND THE GUESS FOR PRESSURE
AND WEIGHT FRACTION METHANOL?

40,50,0.15
         RESULTS FOR THREE PHASE (LW-G-H) PRESSURE PREDICTIONS
              WITH METHANOL INHIBITION

COMPONENT    VAPOR          HYDRATE        HYDRATE K
  NAME    COMPOSITION    COMPOSITION        FACTOR
 METHANE     .60000         .12856          4.6670
 ETHANE      .30000         .41401           .7246
 H2-S        .10000         .45743           .2186

 press RETURN to continue ...


         FRACTION OF CAVITIES OCCUPIED BY HYDRATE FORMERS
                        STRUCTURE 1
              COMPONENT    SMALL      LARGE
                NAME       CAVITY     CAVITY
               METHANE    .161312    .112717
               ETHANE     .000000    .536145
               H2-S       .756341    .340248

 press RETURN to continue ...


                    TEMPERATURE =  40.00 F

                    PRESSURE =   208.1696 PSIA

                    WEIGHT % METHANOL IN WATER = 15.00 %

                    THE FREEZING POINT OF THE METHANOL-WATER MIXTURE=  12.94 F

                    THE TEMPERATURE IS ABOVE THE CRITICAL POINT

DO YOU WISH TO DO ANOTHER CALCULATION AT THESE SAME COMPOSITIONS (N=1 Y=2)?
```

## Example E

### Three Phase ($L_W$-H-V) Weight Percent Methanol Needed Prediction

| Component | Mol Fraction |
|---|---|
| $CH_4$ | 0.901 |
| $CO_2$ | 0.099 |

| | |
|---|---|
| Temperature | 31.6 $^oF$ |
| Pressure | 935 psia |

```
THIS IS THE COLORADO SCHOOL OF MINES HYDRATE PREDICTION PROGRAM
ALL INPUT DATA SHOULD BE SEPARATED BY COMMAS WITH 10 PIECES OF DATA PER LINE
WHAT IS THE TYPE OF CALCULATION TO BE PERFORMED?
     -4= FOUR PHASE(LW-G-H-LH) PREDICTION WITH METHANOL INHIBITION
     -3= TWO PHASE(LH-H) PREDICTION
     -2= THREE PHASE(LW-G-H) TEMPERATURE PREDICTION WITH METHANOL INHIBITION
     -1= THREE PHASE(LW-G-H) PRESSURE PREDICTION WITH METHANOL INHIBITION
      0= THREE PHASE(LW-G-H) WEIGHT % METHANOL NEEDED PREDICTION
      1= THREE PHASE(LW-G-H) PRESSURE PREDICTION
      2= THREE PHASE(LW-G-H) TEMPERATURE PREDICTION
      3= TWO PHASE(G-H) PREDICTION
      4= FOUR PHASE(LW-G-H-LH) PREDICTION

0
WHAT ARE THE UNITS ON
         TEMPERATURE  AND  PRESSURE?
     0=   FAHRENHEIT         PSIA
     1=   CENTIGRADE         mm Hg
     2=   KELVIN             KPASCALS
     3=   KELVIN             ATM

0
HOW MANY COMPONENTS ARE PRESENT?
2
WHICH COMPONENTS ARE PRESENT? LIST HYDRATE FORMERS FIRST!

              HYDRATE FORMERS
     1= METHANE            2= ETHANE
     3= PROPANE            4= ISO-BUTANE
     5= N-BUTANE          14= HYDROGEN SULFIDE
    15= NITROGEN          16= CARBON DIOXIDE
    18= ETHYLENE          19= PROPYLENE
```

```
              NON HYDRATE FORMERS
     6= PENTANE           7= HEXANE
     8= HEPTANE           9= OCTANE
    10= NONANE           11= DECANE
    12= TOLUENE          13= BENZENE

1,16
WHAT ARE THE MOLE FRACTIONS OF THE COMPONENTS?
    LIST IN THE SAME ORDER AS ABOVE !

.901,.099
INPUT THE SPECIFIED TEMP AND PRESS

31.6,935
          RESULTS FOR THREE PHASE (LW-G-H) WEIGHT % METHANOL NEEDED

COMPONENT     VAPOR          HYDRATE         HYDRATE K
  NAME     COMPOSITION     COMPOSITION        FACTOR
 METHANE      .90100          .81941          1.0996
 C-02         .09900          .18059           .5482

 press RETURN to continue ...


          FRACTION OF CAVITIES OCCUPIED BY HYDRATE FORMERS
                          STRUCTURE 1

               COMPONENT    SMALL      LARGE
                 NAME       CAVITY     CAVITY
                METHANE    .872802    .776165
                C-02       .065280    .213418

 press RETURN to continue ...

                    TEMPERATURE =  31.60 F

                    PRESSURE =    935.0000 PSIA

                    WEIGHT % METHANOL IN WATER = 20.00 %

                    THE FREEZING POINT OF THE METHANOL-WATER MIXTURE=   4 78 F

                    THE TEMPERATURE IS ABOVE THE CRITICAL POINT

DO YOU WISH TO DO ANOTHER CALCULATION AT THESE SAME COMPOSITIONS (N=1 Y=2)?
```

## Example F

### Three Phase ($L_W$-H-V) Pressure Prediction

| Component | Mol Fraction |
|---|---|
| $CH_4$ | 0.906 |
| $C_2H_6$ | 0.066 |
| $C_3H_8$ | 0.018 |
| i-$C_4H_{10}$ | 0.005 |
| $H_2S$ | 0.005 |

Temperature 68.8 $^oF$

Guess for Pressure 800 psia

```
THIS IS THE COLORADO SCHOOL OF MINES HYDRATE PREDICTION PROGRAM
ALL INPUT DATA SHOULD BE SEPARATED BY COMMAS WITH 10 PIECES OF DATA PER LINE
WHAT IS THE TYPE OF CALCULATION TO BE PERFORMED?
     -4= FOUR PHASE(LW-G-H-LH) PREDICTION WITH METHANOL INHIBITION
     -3= TWO PHASE(LH-H) PREDICTION
     -2= THREE PHASE(LW-G-H) TEMPERATURE PREDICTION WITH METHANOL INHIBITION
     -1= THREE PHASE(LW-G-H) PRESSURE PREDICTION WITH METHANOL INHIBITION
      0= THREE PHASE(LW-G-H) WEIGHT % METHANOL NEEDED PREDICTION
      1= THREE PHASE(LW-G-H) PRESSURE PREDICTION
      2= THREE PHASE(LW-G-H) TEMPERATURE PREDICTION
      3= TWO PHASE(G-H) PREDICTION
      4= FOUR PHASE(LW-G-H-LH) PREDICTION

1
WHAT ARE THE UNITS ON
         TEMPERATURE  AND  PRESSURE?
     0=   FAHRENHEIT        PSIA
     1=   CENTIGRADE        mm Hg
     2=   KELVIN            KPASCALS
     3=   KELVIN            ATM

0
HOW MANY COMPONENTS ARE PRESENT?
5
WHICH COMPONENTS ARE PRESENT? LIST HYDRATE FORMERS FIRST!
```

```
                HYDRATE FORMERS
     1= METHANE            2= ETHANE
     3= PROPANE            4= ISO-BUTANE
     5= N-BUTANE          14= HYDROGEN SULFIDE
    15= NITROGEN          16= CARBON DIOXIDE
    18= ETHYLENE          19= PROPYLENE

                NON HYDRATE FORMERS
     6= PENTANE            7= HEXANE
     8= HEPTANE            9= OCTANE
    10= NONANE            11= DECANE
    12= TOLUENE           13= BENZENE

1,2,3,4,14
WHAT ARE THE MOLE FRACTIONS OF THE COMPONENTS?
    LIST IN THE SAME ORDER AS ABOVE !

0.906,0.066,0.018,0.005,0.005
INPUT THE SPECIFIED TEMPERATURE AND THE GUESS FOR PRESSURE?

68.8,800
          RESULTS FOR THREE PHASE (LW-G-H) PRESSURE PREDICTIONS

COMPONENT     VAPOR         HYDRATE         HYDRATE K
  NAME     COMPOSITION    COMPOSITION         FACTOR
 METHANE      .90600          .65448          1.3843
 ETHANE       .06600          .05236          1.2605
 PROPANE      .01800          .16135           .1116
 I-BUTANE     .00500          .07330           .0682
 H2-S         .00500          .05851           .0855

 press RETURN to continue ...


          FRACTION OF CAVITIES OCCUPIED BY HYDRATE FORMERS
                        STRUCTURE 2
               COMPONENT    SMALL      LARGE
                 NAME       CAVITY     CAVITY
                METHANE    .820452    .187302
                ETHANE     .000000    .146268
                PROPANE    .000000    .450728
                I-BUTANE   .000000    .204750
                H2-S       .078831    .005770

 press RETURN to continue ...


                  TEMPERATURE =  68.80 F

                  PRESSURE =  1598.7130 PSIA

                  THE TEMPERATURE IS ABOVE THE CRITICAL POINT


DO YOU WISH TO DO ANOTHER CALCULATION AT THESE SAME COMPOSITIONS (N=1 Y=2)?
```

## Example G

### Three Phase ($L_W$-H-V) Temperature Prediction

| Component | Mol Fraction |
|---|---|
| $CH_4$ | 0.906 |
| $C_2H_6$ | 0.066 |
| $C_3H_8$ | 0.018 |
| $i$-$C_4H_{10}$ | 0.005 |
| $H_2S$ | 0.005 |

Pressure 100 psia

Guess for Temperature 50 $^0$F

```
THIS IS THE COLORADO SCHOOL OF MINES HYDRATE PREDICTION PROGRAM
ALL INPUT DATA SHOULD BE SEPARATED BY COMMAS WITH 10 PIECES OF DATA PER LINE
WHAT IS THE TYPE OF CALCULATION TO BE PERFORMED?
      -4= FOUR PHASE(LW-G-H-LH) PREDICTION WITH METHANOL INHIBITION
      -3= TWO PHASE(LH-H) PREDICTION
      -2= THREE PHASE(LW-G-H) TEMPERATURE PREDICTION WITH METHANOL INHIBITION
      -1= THREE PHASE(LW-G-H) PRESSURE PREDICTION WITH METHANOL INHIBITION
       0= THREE PHASE(LW-G-H) WEIGHT % METHANOL NEEDED PREDICTION
       1= THREE PHASE(LW-G-H) PRESSURE PREDICTION
       2= THREE PHASE(LW-G-H) TEMPERATURE PREDICTION
       3= TWO PHASE(G-H) PREDICTION
       4= FOUR PHASE(LW-G-H-LH) PREDICTION

2
WHAT ARE THE UNITS ON
         TEMPERATURE  AND  PRESSURE?
      0=  FAHRENHEIT         PSIA
      1=  CENTIGRADE         mm Hg
      2=  KELVIN             KPASCALS
      3=  KELVIN             ATM
```

```
0
HOW MANY COMPONENTS ARE PRESENT?
5
WHICH COMPONENTS ARE PRESENT? LIST HYDRATE FORMERS FIRST!

              HYDRATE FORMERS
     1= METHANE           2= ETHANE
     3= PROPANE           4= ISO-BUTANE
     5= N-BUTANE         14= HYDROGEN SULFIDE
    15= NITROGEN         16= CARBON DIOXIDE
    18= ETHYLENE         19= PROPYLENE

              NON HYDRATE FORMERS
     6= PENTANE           7= HEXANE
     8= HEPTANE           9= OCTANE
    10= NONANE           11= DECANE
    12= TOLUENE          13= BENZENE

2,1,3,4,14
WHAT ARE THE MOLE FRACTIONS OF THE COMPONENTS?
   LIST IN THE SAME ORDER AS ABOVE !

.066,.906,.018,.005,.005
INPUT THE GUESS FOR TEMPERATURE AND THE SPECIFIED PRESSURE?

50,1000
          RESULTS FOR THREE PHASE (LW-G-H) TEMPERATURE PREDICTIONS

COMPONENT     VAPOR          HYDRATE         HYDRATE K
  NAME     COMPOSITION    COMPOSITION         FACTOR
 ETHANE       .06600         .04743           1.3914
 METHANE      .90600         .61691           1.4686
 PROPANE      .01800         .17603            .1023
 I-BUTANE     .00500         .09053            .0552
 H2-S         .00500         .06909            .0724

 press RETURN to continue ...


          FRACTION OF CAVITIES OCCUPIED BY HYDRATE FORMERS
                          STRUCTURE 2
               COMPONENT     SMALL      LARGE
                 NAME        CAVITY     CAVITY
                ETHANE      .000000    .129584
                METHANE     .776599    .132112
                PROPANE     .000000    .480903
                I-BUTANE    .000000    .247320
                H2-S        .091895    .004955

 press RETURN to continue ...

                    TEMPERATURE =  64.02 F

                    PRESSURE =   999.9999 PSIA

                    THE TEMPERATURE IS ABOVE THE CRITICAL POINT

DO YOU WISH TO DO ANOTHER CALCULATION AT THESE SAME COMPOSITIONS (N=1 Y=2)?
```

## Example H

### Two Phase (H-V) Prediction

| Component | Mol Fraction |
|---|---|
| $CH_4$ | 0.60 |
| $C_2H_6$ | 0.30 |
| $N_2$ | 0.10 |

Temperature 30 $^0$F

Pressure 120 psia

Guess for Water Concentration 200 ppm (mol basis)

```
THIS IS THE COLORADO SCHOOL OF MINES HYDRATE PREDICTION PROGRAM
ALL INPUT DATA SHOULD BE SEPARATED BY COMMAS WITH 10 PIECES OF DATA PER LINE
WHAT IS THE TYPE OF CALCULATION TO BE PERFORMED?
     -4= FOUR PHASE(LW-G-H-LH) PREDICTION WITH METHANOL INHIBITION
     -3= TWO PHASE(LH-H) PREDICTION
     -2= THREE PHASE(LW-G-H) TEMPERATURE PREDICTION WITH METHANOL INHIBITION
     -1= THREE PHASE(LW-G-H) PRESSURE PREDICTION WITH METHANOL INHIBITION
      0= THREE PHASE(LW-G-H) WEIGHT % METHANOL NEEDED PREDICTION
      1= THREE PHASE(LW-G-H) PRESSURE PREDICTION
      2= THREE PHASE(LW-G-H) TEMPERATURE PREDICTION
      3= TWO PHASE(G-H) PREDICTION
      4= FOUR PHASE(LW-G-H-LH) PREDICTION

3
WHAT ARE THE UNITS ON
         TEMPERATURE  AND  PRESSURE?
     0=   FAHRENHEIT         PSIA
     1=   CENTIGRADE         mm Hg
     2=   KELVIN             KPASCALS
     3=   KELVIN             ATM

0
HOW MANY COMPONENTS ARE PRESENT?
3
WHICH COMPONENTS ARE PRESENT? LIST HYDRATE FORMERS FIRST!
```

```
            HYDRATE FORMERS
    1= METHANE          2= ETHANE
    3= PROPANE          4= ISO-BUTANE
    5= N-BUTANE        14= HYDROGEN SULFIDE
   15= NITROGEN        16= CARBON DIOXIDE
   18= ETHYLENE        19= PROPYLENE

            NON HYDRATE FORMERS
    6= PENTANE          7= HEXANE
    8= HEPTANE          9= OCTANE
   10= NONANE          11= DECANE
   12= TOLUENE         13= BENZENE

1,2,15
WHAT ARE THE MOLE FRACTIONS OF THE COMPONENTS?
    LIST IN THE SAME ORDER AS ABOVE !

.6,.3,.1
WHAT IS THE TEMPERATURE AND PRESSURE AND A GUESS FOR THE MOLE FRACTION WATER IN PPM?

30,120,200
          RESULTS FOR TWO PHASE(G-H) CALCULATIONS

COMPONENT     VAPOR         HYDRATE        HYDRATE K
  NAME     COMPOSITION   COMPOSITION        FACTOR
 METHANE      .60000        .29283         2.0489
 ETHANE       .30000        .69998          .4286
 NITROGEN     .10000        .00719         13.9114

 press RETURN to continue ...


          FRACTION OF CAVITIES OCCUPIED BY HYDRATE FORMERS
                        STRUCTURE 1
               COMPONENT    SMALL      LARGE
                 NAME       CAVITY     CAVITY
                METHANE     .575117    .153248
                ETHANE      .000000    .824575
                NITROGEN    .020088    .001772

 press RETURN to continue ...

                   TEMPERATURE =  30.00 F

                   PRESSURE =   120.0000 PSIA

                   THE MOLE FRACTION OF WATER =    648.255 PPM

                   THE TEMPERATURE IS ABOVE THE CRITICAL POINT

DO YOU WISH TO DO ANOTHER CALCULATION AT THESE SAME COMPOSITIONS (N=1 Y=2)?
```

## Example I

Four Phase ($L_W$-H-V-$L_{HC}$)

| Component | Mol Fraction |
|---|---|
| $C_2H_6$ | 0.20 |
| $C_3H_8$ | 0.50 |
| i-$C_4H_{10}$ | 0.30 |

Guess for Temperature 32 $^0$F

Guess for Pressure 100 psia

```
THIS IS THE COLORADO SCHOOL OF MINES HYDRATE PREDICTION PROGRAM
ALL INPUT DATA SHOULD BE SEPARATED BY COMMAS WITH 10 PIECES OF DATA PER LINE
WHAT IS THE TYPE OF CALCULATION TO BE PERFORMED?
     -4= FOUR PHASE(LW-G-H-LH) PREDICTION WITH METHANOL INHIBITION
     -3= TWO PHASE(LH-H) PREDICTION
     -2= THREE PHASE(LW-G-H) TEMPERATURE PREDICTION WITH METHANOL INHIBITION
     -1= THREE PHASE(LW-G-H) PRESSURE PREDICTION WITH METHANOL INHIBITION
      0= THREE PHASE(LW-G-H) WEIGHT % METHANOL NEEDED PREDICTION
      1= THREE PHASE(LW-G-H) PRESSURE PREDICTION
      2= THREE PHASE(LW-G-H) TEMPERATURE PREDICTION
      3= TWO PHASE(G-H) PREDICTION
      4= FOUR PHASE(LW-G-H-LH) PREDICTION

4
WHAT ARE THE UNITS ON
         TEMPERATURE  AND  PRESSURE?
     0=   FAHRENHEIT         PSIA
     1=   CENTIGRADE         mm Hg
     2=   KELVIN             KPASCALS
     3=   KELVIN             ATM

0
HOW MANY COMPONENTS ARE PRESENT?
3
WHICH COMPONENTS ARE PRESENT? LIST HYDRATE FORMERS FIRST!

              HYDRATE FORMERS
     1= METHANE          2= ETHANE
     3= PROPANE          4= ISO-BUTANE
     5= N-BUTANE        14= HYDROGEN SULFIDE
    15= NITROGEN        16= CARBON DIOXIDE
    18= ETHYLENE        19= PROPYLENE
```

```
            NON HYDRATE FORMERS
     6= PENTANE           7= HEXANE
     8= HEPTANE           9= OCTANE
    10= NONANE           11= DECANE
    12= TOLUENE          13= BENZENE

2,3,4
WHAT ARE THE MOLE FRACTIONS OF THE COMPONENTS?
    LIST IN THE SAME ORDER AS ABOVE !

.2,.5,.3
WHAT IS THE GUESS FOR TEMPERATURE AND PRESSURE
32,100
          RESULTS FOR QUADRAPOLE PREDICTIONS
```

| COMPONENT NAME | LIQUID COMPOSITION | VAPOR COMPOSITION | HYDRATE COMPOSITION | VLE K FACTOR | HYDRATE K FACTOR |
|---|---|---|---|---|---|
| ETHANE | .20000 | .54967 | .0474 | 2.7483 | 11.6055 |
| PROPANE | .50000 | .36673 | .6288 | .7335 | .5832 |
| I-BUTANE | .30000 | .08358 | .3238 | .2786 | .2581 |

press RETURN to continue ...

FRACTION OF CAVITIES OCCUPIED BY HYDRATE FORMERS
STRUCTURE 2

| COMPONENT NAME | SMALL CAVITY | LARGE CAVITY |
|---|---|---|
| ETHANE | .000000 | .047328 |
| PROPANE | .000000 | .628354 |
| I-BUTANE | .000000 | .323592 |

press RETURN to continue ...

```
TEMPERATURE =  39.66 F

PRESSURE =   113.4250 PSIA

THE SLOPE OF THE THREE PHASE(LW-G-LH)(DP/DT)=-34654.9 PSIA/F

DO YOU WISH TO DO THIS CALCULATION FOR OTHER MOLE FRACTIONS (N=1 Y=2)?
```

# References

Afanaseva, L.D. and Groisman, A.G., Siberian Branch of the USSR Academy of Science, **100**, 103 (1973)

Ahmad, N. and Phillips, W.A. Solid State Commun, **63**, 167 (1987)

Alexander, D.M., and Hill, D.J.T., Aust. J. Chem., **22**, 347 (1969)

Alexeyeff, W., Ber. **9** 1025 (1876)

Allen, K.W., J. Chem. Phys., **41** 840 (1964)

Allen, K.W. and Jeffrey, G.A., J. Chem. Phys., **38**, 2304 (1963)

Al-Ubaidi, B., Personal Communication, February 26, 1988

American Petroleum Institute, Technical Data Book, Chapter 9, (1981)

Anderson, F.E. and Prausnitz, J.M., AIChE J, **32** 1321 (1986)

Andersson, P. and Ross, R.G., J. Phys. C: Solid State Phys.,**16**, 1423 (1983)

Andryushchenko, F.K., and Vasilchenko, V.P., Gasov. Prom.,(10), 4 (1969)

Angell, C.A., in Water: A Comprehensive Treatise (V.7), F. Franks, ed., Plenum Press, New York, 1982.

Aoki, Y., Tamano,T., and Kato, S., Studies in Continental Margin Geology, J.S. Watkins, and C.L. Drake, Eds., Amer. Assn. of Petrol. Geol. Mem. No. 34, pp 309-322, 1983

Aoyagi, K. and Kobayashi, R., "Report of Water Content Measurement of High Carbon Dioxide Content Simulated Prudhoe Bay Gas in Equilibrium with Hydrates," Proc. 57th Annual Convention, Gas Processors Association, p.3, New Orleans, La. March 20-22, 1978

Aoyagi, K., Song, K.Y., Kobayashi, R., Sloan, E.D., and Dharmawardhana, P.B., Gas Processors Assn. Rsch Rpt No 45 Tulsa, Ok, (December, 1980)

Arnett, E.M., Kover, W.B., Carter, J.V., J. Am. Chem. Soc., **91**, 4028 (1969)

Asher, G.B., Development of a Computerized Thermal Conductivity Measurement System Utilizing the Transient Needle Probe Technique: An Application to Hydrates in Porous Media, Dissertation T-3335, Colorado School of Mines, 1987

Asher, G.B., Sloan, E.D., Graboski, M.S., Int. J. Thermophysics, **7**(2), 285 (1985)

Ashworth, T., Johnson, L.R. and Lai, L.-P., High Temperatures-High Pressures, **17**, 413 (1985)

Ashton, J.P., Kirspel, L.J., Nguyen, H.T., Credeur, D.J., Proc. 61st. Ann. SPE Tech. Conf., New Orleans, La., Oct 5-8, 1986 SPE 15660

Avlonites, D., "Multiphase Equilibria in Oil-Water Hydrate Forming Systems," M.Sc. Thesis, Heriot-Watt University, Edinburgh, Scotland (1988)

Avlonites, D., Danesh, A., Todd, A.C., "Measurement and Prediction of Hydrate Dissociation Pressure of Oil -Gas Systems," presented at BHRA Conference on OperationalConsequences of Hydrate Formation and Inhibition Offshore, Cranfield, UK, Nov. 3, 1988

Babe, G.D., Bondarer, E.A., Groisman, A.G., Kanibolotskii, Z., Inz-Eiz 879 (1972)

Baker, P.E., in Natural Gases in Marine Sediments, I.R. Kaplan, Ed., 227, Plenum, New York (1974)

Barduhn, A.J., Towlson, H.E. and Hu,Y.-C, AIChE J, **8** 176 (1962)

Barkan, Y.S. and Voronov,A.N. Sovietskaia Geologiia **7** 37 (1982)

Barker, J.W., and Gomez, R.K., "Formation of Hydrates During Deepwater Drilling Operations," SPE/IADC 16130, SPE/IADC Drilling Conference, New Orleans, LA, March 15-18, 1987

Barker, J.W., "Formation of Hydrates in Deep Water Drilling Operations" Presented at N.L. Baroid Meeting on Hydrate in Deep Water Drilling Muds, Houston, TX, October 20, 1988a

Barker, J.W., Personal Communication, October 20, 1988b.

Barrer, R.M., Nature, **183**, 463 (1959)

Barrer, R.M., and Edge, A.V.J., Proc. Roy. Soc. (London), **A300**, 1 (1967)

Barrer, R.M., and Ruzicka, D.J., Trans. Faraday Soc., **58**, 2253 (1962a)

Barrer, R.M., and Ruzicka, D.J., Trans. Faraday Soc., **58**, 2262 (1962b)

Barrer, R.M. and Stuart, W.I. Proc. Roy Soc. (London) A. No 1233 **243** 172 (1957)

Barry, H.M., Chem. Eng., p. 105, Feb. 8, 1960

Bayles, G.A., Sawyer, W.K., Anada, H.R., Reddy, S., and Malone, R.D., Chem. Eng. Commun., **47**, 225 (1986)

Behar, E., Sugier, A., and Rojey A., "Hydrate Formation and Inhibition in Multiphase Flow," presented at BHRA Conference Operation consequences of Hydrate Formation and Inhibition Offshore, Cranfield, UK, Nov. 3, 1988,

Benedict, W.S., Gailar, N., Plyler,E.K., J. Chem. Phys. **24**, 1139 (1956)

Ben-Naim, A., Hydrophobic Interaction, Plenum Press, New York, 1980

Ben-Naim, A., and Yaacobi, M., J. Phys. Chem, **78**(2), 170 (1974)

Berecz, E., and Balla-Achs, M., Research Report No. 37 (185-XI-1-1974, OGIL),NME, (Tech. Univ. Heavy Ind.), Miskolc (1974)

Berecz, E., and Balla-Achs, M., Research Report No. 56 (232-XI-3-1978, OGIL) NME (Tech. Univ. Heavy Ind.), Miskolc, (1980)

Berecz, E., and Balla-Achs, M., "Gas Hydrates, Studies in Inorganic Chemistry Vol 4", Elsevier, New York, 343 pp (1983)

Bernal, J.D. and Fowler, R.H., J. Chem. Phys. **1**, 515 (1933)

Bernard, B.B., Brooks, J.M., Sackett, W.M., Earth Planet. Sci. Lett.,**31**, 48 (1976)

Bernard, W.R., Wilkie, D.I., and Cooke, W.J., Can Patent 1,056,716, c.6/19/79 f. 7/29/77 (1979)

Berthelot, P.E.M., Ann. Chim. Phys. Ser. 3 **46**, 490 (1856)

Bertie, J.E., Bates, F.E., Hendricksen, D.K., Can. J. Chem. **53**, 71 (1975)

Bertie, J.E., and Devlin, J. P., J. Chem. Phys., **78**, 6340 (1983)

Bertie, J.E. and Jacobs, S.M., Can. J. Chem., **55**, 1777 (1977)

Bertie, J.E. and Jacobs, S.M., J. Chem. Phys., **69**, 4105 (1978)

Bertie, J.E. and Jacobs, S.M., J. Chem. Phys., **77**, 3230 (1982)

Bily, C., and Dick, J.W.L., Bull. Can. Petr. Geol. **22** 340 (1974)

Bishnoi, P.R. and Vysniauskas, A., Kinetics of Gas Hydrate Formation Pt. II.,Final Report to Gas Research Institute, Chicago, (1980a)

Bishnoi, P.R., The Kinetics of Natural Gas Hydrate Formation, Final Rpt. on Contract DSS F 14SU 23235 -0-0497, Dept Energy, Mines, & Resources, Canada, (1980b)

Bjerrum, N., Science, **115**, 385 (1952)

Bolin, B., in Oceanography, The Present and Future, P.G. Brewer (Editor), Springer-Verlag, New York, pp. 305 -326, 1983

Bond, D.C., and Russell, N.B., Trans AIME, **179** 192 (1949)

Bourrie, M.S., and Sloan, E.D., Gas Proc. Assn. Res. Rpt. 100, Tulsa, Ok, June, 1986

Brooks, J.M., Personal Communication, June 29, 1988

Brooks, J.M., and Bryant, W.R., Geological and Geochemical Implications of Gas Hydrates in the Gulf of Mexico, U.S. Department of Energy, DOE/MC/21088-1964 (1985)

Brooks, J.M., Cox, H.B., Bryant, W.R., Kennicutt, M.C., Mann R.G., and McDonald, T.J. Org Geochem **10** 221 (1986)

Brooks, J.M., Jeffrey, A.W.A., McDonald, T.J., Pflaum, R.C., Init. Repts Deep Sea Drilling Project, Vol LXXXIV, 699, R. von Huene, J. Auboin, et al., eds. Washington, U.S. Govt Printing Office, (1985a)

Brooks, J.M., Kennicutt, M.C., Bidigare, R.R., Fay, R.A., EOS Trans. Amer. Geophys. Union, **66(10)**, 106 (1985b)

Brooks, J.M., Kennicutt, M.C., Bidigare, R.R., Wade, T.L.,Powell, E.N., Denoux, G.J., Fay, R.R., Childress, J.J., Fisher, C.R., Rossman, I. and G. Boland, EOS, Trans Amer Geophys Union, **68(18)** 498 (1987)

Brooks, J.M., Kennicutt, M.C., Fay, R.R., McDonald, T.J., Sassen, Science **225** 409 (1984)

Brooks, W.B., Gibbs, G.B., and McKetta, J.J., Pet. Refiner., **30**(10), 118 (1951)

Brown, G.G., Trans AIME, **160**, 65 (1945)

Brownstein, S., Davidson, D.W., and Fiat, D., J. Chem. Phys., **46**, 1454 (1967)

Bucklin, R.W., Toy, K.G., and Won, K.W., "Hydrate Control of Natural Gas Under Arctic Conditions Using TEG", Proc. Gas Conditioning Conference, Norman, Ok (1985)

Bugge, T., Befring, S., Belderson, R.H., Eidvin, T., Jansen, E., Kenyon, N.H., HOltedahl, H., and Sejrup, H.P., Geo-Marine Letters, **7**, 191 (1987)

Burshears, M., O'Brien, T.J., and Malone, R.D., "A Multi-Phase Multi-Dimensional, Variable Composition Simulation of Gas Production fron a Conventional Gas Reservoir in Contact with Hydrates," Proc. SPE Unconventional Gas Technol. Symposium, SPE 15246, Louisville, Ky, May 18-21, 1986.

Butler, J.A.V., Trans. Faraday Soc., **33**, 229 (1937)

Byk, S.S., and Fomina, V.I., Russ. Chem. Rev. **37** (6), 469 (1968)

Cady, G.H., J. Phys. Chem., **85**, 3225 (1981)

Cady, G.H., J. Phys. Chem., **85**, 4437 (1983a)

Cady, G.H., J. Chem. Ed, **60**, 915 (1983b)

Cady, G.H., J. Phys. Chem., **89**, 3302 (1985)

Cailletet, L., Compt. Rend., **85**, 851 (1877)

Cailletet, L. and Bordet, R., Compt. Rend. **95**, 58 (1882)

Callanan, J. and Sloan, E.D., "Calorimetric Studies of Clathrate Hydrates" Paper presented at 1983 Internat. Gas Rsch. Conf. London, May 1983

Calvert, L.D. and Srivastava, P., Acta Crystallogr., **A25**, S131 (1967)

Carslaw, H.S., and Jaeger, J.C.: Conduction of Heat in Solids, Clarendon Press, Oxford (1959)

Carson, D.B. and Katz, D.L., Trans. AIME, **146**, 150 (1942)

Case, J.L., Ryan, B.F., and Johnson, J.E., Proc. 64th Annual Convention of Gas Processors Association, p.258, Houston, Tx, (1985a)

Case, J.L., Ryan, B.F., and Johnson, J.E., Oil Gas Journal, **83**, p 175, May 6, (1985b)

Case, J.L., Ryan, B.F., and Johnson, J.E., Oil Gas Journal, **83**, p 103, May 13, (1985c)

Cha, S.B., Formation of Hydrates in Water-Based Drilling Muds and Mud-Like Fluids, M.S. Thesis, T-3506, Colorado School of Mines, Golden, CO (1988)

Cha, S.B., Ouar, H., Wildeman, T.R., and Sloan, E.D., J. Phys. Chem., **92**, 6492 (1988)

Chappelear, P.S., Chen, R.J.J., Elliot, D.G., Hydrocarbon Proc. p 215. Sept., 1977

Chancel, G., and Parmentier, F., Compt Rend **100**, 27 (1885)

Chen, C.-J., Ng, H.-J., and Robinson, D.B., Gas Proc. Assn. Rsch. Rpt. 117, March 1988

Chen, T.-S., A Molecular Dynamics Study of The Stability of Small Prenucleation Water Clusters, Dissertation, U. Missouri-Rolla, 1980, Univ. Microfilms No. 8108116, Ann Arbor, Mi.

Cherskiy, N.V., Tsarev, V.P., and Nikitin, S.P., Petrol. Geol., **21**, 65 (1985)

Child, W.C., Jr., J. Phys. Chem. **68**, 1834, (1964)

Ciesnik, M. and Krason, J., Geological Evolution and Analysis of Confirmed or Suspected Gas Hydrate Localities: Basin Analysis, Formation and Stability of Gas Hydrates in the Black Sea, U.S. Department of Energy,DOE/MC/21181-1950,Vol 11,(1987)

Claussen, W.F., J. Chem. Phys. **19** 259, 662 (1951a)

Claussen, W.F., J. Chem. Phys. **19** 1425 (1951b)

Claussen, W.F. and Polgase, M.F., J. Am. Chem. Soc., **74**, 4817 (1952)

Claypool, G.E. and Kaplan, I. R., Natural Gases in Marine Sediments, Vol. 3, 99 I.R. Kaplan, ed., Plenum, New York (1974)

Claypool, G.E. and Kvenvolden, K.A. Ann Rev. Earth Planet. Sci. **11** 299 (1983)

Collett, T.S., Detection and Evaluation of Natural Gas Hydrates from Well Logs, Prudhoe Bay, Alaska, M.S. Thesis, U. Alaska, Fairbanks, AL. (1983)

Collett, T.S., Bird, K.J., Kvenvolden, K.A., and Magoon, L.B., Geologic Interrelations Relative to Gas Hydrates Within the North Slope of Alaska, Final Report DE-AI21-83MC20422, U.S. Department of Energy, 1988

Collett, T.S., Ehlig-Economides, C., "Detection and Evaluation of the In-Situ Natural Gas Hydrates in the North Slope Region, Alaska," SPE 11673, Proc. 1983 SPE California Regional Meeting, March 23-25, 1983

Collett, T.S., Godbole, S.P. and Ehlig-Economomides, C., "Quantification of Gas Hydrates on North Slope of Alaska, Proceedings of the 35th Annual Meeting of Canadian Institute of Mining, Calgary, Canada (June 10-12, 1984)

Collett, T.S. and Kvenvolden, K.A., U.S. Geol. Survey. Open File Report 87-225, 1987.

Collier, W.B., Ritzhaupt, G., and Devlin, J.P., J. Phys. Chem., **88**, 363 (1984)

Cook, J.G. and Laubitz, M.J., Proc 17th Internat. Thermal Conduc. Conf., Gaithersburg, Md, 1981

Cook, J.G. and Leaist, D.G., Geophys. Rsch Lett. **10** 397 (1983)

Cottrell, T.L., The Strengths of Chemical Bonds, Butterworths, London (1958)

Culberson, O.L. and McKetta, J.J., Petrol Trans AIME, **189**, 319 (1950)

Culberson, O.L. and McKetta, J.J., Petrol Trans AIME **192**, 223 (1951)

Culberson, O.L., Horn, A.B., and McKetta, J.J., Petrol Trans AIME **189**, 1 (1950)

Cummings, W.P., "Natural Gas Purification with Molecular Sieves,"paper presented at the Canadian Chemical Engineering Conference, Calgary, Alberta, October 23-26, 1977

Dameron, E., in "G.P.A. Cryogenics Panel Conclusion: Cryogenics Plant Control Optimized," Oil and Gas Journal, July 25, 1977

Dang, L.X., Molecular Dynamics Simulations of Non-Polar Solutes Dissolved in Liquid Water or Trapped in Water Clathrate, Dissertation, U. California, Irvine, 1985, Univ. Microfilms No. 8603239, Ann Arbor, Mi

Daubert, T.E., Personal Communication, June 8, 1987.

Davidson, D.W., Can. J. Chem., **49**, 1224 (1971)

Davidson, D.W., "Clathrate Hydrates," in Water: A Comprehensive Treatise, Vol 2., Chptr 3, 115 (1973)

Davidson, D.W., "Gas Hydrates as Clathrate Ices ," in Natural Gas Hydrates: Properties, Occurrence and Recovery, J.L. Cox, ed., Butterworths, Boston (1983)

Davidson, D.W., El-Defrawy, M.K., Fuglen, M.O., and Judge, A.S., Proc Third International Conference on Permafrost, H.M. French, Ed., 938, (1978)

Davidson, D.W., Garg, S.K., Gough, S.R., Hawkins, R.E., and Ripmeester, J.A., Can. J. Chem., **55**, 3641 (1977)

Davidson, D.W., Gough, S.R., Ripmeester, J.A., and Nakayama, H., Can. J. Chem., **59**, 2587 (1981)

Davidson, D.W., Handa, Y.P., Ratcliffe, C.I., Tse, J.S., and Powell, B.M., Nature, **311**, 142 (1984)

Davidson, D.W., Handa, Y.P., Ratcliffe, C.I., Ripmeester, J.A., Tse, J.S., Dahn, J.R., Lee, F., and Calvert, L.D., Mol. Cryst. Liq. Cryst., **141**, 141 (1986)

Davidson, D.W., Leaist, D.G., and Hesse, R., Geochimica et Cosmochimica Acta, **47**, 2293, (1983)

Davidson, D.W. and Ripmeester, J.A., J. Glaciology, **21** 33 (1978)

Davidson, D.W. and Ripmeester, J.A.,in Inclusion Compounds, Vol 3 Chapter 3, J.L. Atwood, J.E.D. Davies, and D.D. MacNichol,eds, Academic Press (1984)

Davy, H., Phil. Trans. Roy. Soc. London, **101**, 1 (1811)

de Boer, R.B., Houbolt, J.J.H.C. and Lagrand, J., Geol. en Mijnbouw, **64**, 245 (1985)

de Forcrand, R., Compt. Rend. **95**, 129 (1882)

de Forcrand, R., Compt. Rend., **135**, 959 (1902)

de Forcrand, R., Compt. Rend. **176**, 355 (1923)

de Forcrand, R., Compt. Rend. **181**, 15 (1925)

de Forcrand, R. and Thomas, S., Compt. Rend. **125**, 109 (1897)

de Forcrand, R. and Villard, P., Compt. Rend. **106**, 849 (1888a)

de Forcrand. R. and Villard, P., Compt. Rend. **106**, 1357 (1888b)

de la Rive, A., Ann. Chim. Phys. Ser. 2, **40**, 405 (1829)

de Roo, J.L., Peters, C.J., Lichtenthaler, R.N. and Diepen, G.A.M., AIChE J., **29**, 651 (1983)

Deaton, W.M. and Frost, E.M., Jr., Oil & Gas J., **36** (1),75 (1937)

Deaton, W.M. and Frost, E.M., Jr. U.S. Bureau of Mines Monograph 8, 1946.

Decharme, C., Ann. Chim. Phys. Ser 4, 415 (1873)

Delsemme, A.H., Colloq. Inst. C.N.R.S., **207**, 305 (1974)

Delsemme, A.H. and Miller, S.L., Planet. Space Sci.,**18**, 709 (1970)

Dendy, J.W., "Trouble Shooting and Preventative Maintenance of Off Shore Gas Systems," Proc. 5th Annual Gas Proc. Assn., p.161, Lafayette, LA (1969)

Devlin, J.P., and Richardson, H.H., J. Chem. Phys, **81**, 3250 (1984)

Dewan, A.K.R., Personal Communication, January 9, 1989

Dharma-wardana, M.W.C., J. Phys. Chem., **87**, 4185 (1983)

Dharmawardhana, P.B., Parrish, W.R., and Sloan, E.D., Ind. Eng. Chem. Fundam., **19**(4), 410 (1980)

Dillon, W.P., Grow, J.A., Paull, C.K., Oil & Gas J., **78** 124 (1980)

Dillon, W.P. and Paull, C.K., in Natural Gas Hydrates: Properties, Occurrence and Recovery, J.L. Cox, Ed., p.73 (1983)

Ditte, A., Compt. Rend., **95**, 1283 (1882)

Dobrynin, V.M., Korotajev, Yu.P., and Plyuschev, D.V., Long Term Energy Resources, R.G. Meyer, Ed., 727, Pitman, Boston, 1981

Dodson, C.R., and Standing, M.B., Drill. and Prod. Prac., 173-179 Amer. Petrol. Inst., Dallas, Tx. (1944)

Dore, J., Physics World, **1**, 25, December 1988

Du, Y. and Guo, T., Proc. Int. Symp. in Thermo in Chemical Eng. and Industry, p. 486, May 30-June 2, 1988, Beijing, China

Dubinin, V.M. and Zhidenko, G.G., Transp. Khranenie Gaza (Ref. Inf.) **6** 20 (1979)

Duclaux, E., Compt Rend., **64**, 1099 (1867)

Durrant, P.J. and Durrant, B., Introduction to Advanced Inorganic Chemistry, John Wiley and Sons, New York (1962)

Edgar, N.T., Gulf Coast Assoc Geol Soc, Trans, **23** 1 (1973)

Englezos, P., A Model for the Formation Kinetics of Gas Hydrates from Methane, Ethane, and Their Mixtures, M.S. Thesis, U. Calgary, 1986

Englezos, P., and Bishnoi, P.R., AIChE J., **34**, 1718 (1988a)

Englezos, P., and Bishnoi, P.R., Fluid Phase Equil., **42**, 129 (1988b)

Englezos, P., Kalogerakis, N., and P.R. Bishnoi, "Formation and Decomposition of Gas Hydrate of

Natural Gas Components, J. Inclus. Phenom., D.W. Davidson memorial volume (in press) (1989)

Englezos, P., Kalogerakis, N., Dholabhai, P.D., and Bishnoi,P.R., Chem. Eng. Sci., **42**(11), 2647 (1987a)

Englezos, P., Kalogerakis, N., Dholabhai, P.D., and Bishnoi,P.R., Chem. Eng. Sci., **42**(11), 2659 (1987b)

Erickson, D.D., Development of a Natural Gas Hydrate Prediction Computer Program, M.S. Thesis T- 2715, Colorado School of Mines, Golden, CO, (1980)

Ertl, H., Khoury, F., Sloan, E.D., and Kobayashi, R., Chem.-Ing.-Tech.,**48**, 811 (1976)

Euclid, Elements Book XIII. Concluding Scholium (fl.c. 300 B.C.), in Great Books of the Western World, Vol 11., R.M. Hutchins, et al. eds., Encyclopaedia Britannica, Inc. Chicago, IL (1952)

Evrenos, A.I., Heathman, J., Ralston, J., J. Petrol Tech., **23**, 1059 (1971)

Falabella, B.J., A Study of Natural Gas Hydrates, Dissertation, U. Mass., 1975, Univ. Microfilms. No. 76-5849, Ann Arbor, MI

Falabella, B.J., and Vanpee, M., Ind. Eng. Chem. Fund., **13**, 228 (1974)

Faraday, M., Phil. Trans. Roy Soc. London, **113**, 160 (1823)

Field, M.E., and Kvenvolden, K.A., Geology, **13**, 517 (1985)

Finjord, J., Hydrates of Natural Gas: A State-of-the-Art Study with Enphasis on Fundamental Properties, report no. T 11/83 Rogaland Research Instutute, Stavanger, Norway, (June 1983)

Finley, P., and Krason, J., Geological Evolution and Analysis of Confirmed or Suspected Gas Hydrate Localities: Basin Analysis, Formation and Stability of Gas Hydrates in the Middle America Trench, U.S. Department of Energy, DOE/MC/21181-1950, Vol 9, (1986a)

Finley, P., and Krason, J., Geological Evolution and Analysis of Confirmed or Suspected Gas Hydrate Localities: Basin Analysis, Formation and Stability of Gas Hydrates of the Colombia Basin, U.S. Department of Energy,DOE/MC/21181-1950, Vol 7, (1986b)

Finley, P.D., Krason, J., and Dominic, K., Am. Assoc Petrol Geol Bull, **71(5)** 555 (1987)

Fleyfel, F., Devlin, J.P., J. Phys. Chem., **92**, 631 (1988)

Frank, H.S., Proc. Roy. Soc. London, **A247**, 481 (1958)

Frank, H.S., Science, **169**, 635, (1970)

Frank, H.S. and Evans, M.W., J. Chem. Phys., **13**, 507 (1945)

Frank, H.S., and Franks, F., J. Chem. Phys., **48**, 4746 (1968)

Frank, H.S., and Quist, A.S., J. Chem. Phys., **34**, 604 (1961)

Frank, H.S., and Wen, W.-Y. Discuss. Faraday Soc., **24**, 133 (1957)

Franklin, L.J., Proc. USGS Clathrates in the National Petroleum Reserve of Alaska Workshop, Bowsher,A.L. ed., USGS Open File Report No. 81-2198 (1982) (Workshop held July 16-17, 1979)

Franklin, L.J., Petrol. Eng. Internat., 112 (Nov. 1980)

Franklin, L., in Natural Gas Hydrates: Properties, Occurrence, and Recovery, J.L. Cox, Ed., p. 115, Butterworths, Boston, 1983.

Franks, F., Editor, Water: A Comprehensive Treatise, (7 Volumes), Plenum, New York, (1973).

Franks, F., in Water: A Comprehensive Treatise, (Vol. 2, Chptr. 1), F. Franks. ed., Plenum, New York, (1973)

Franks, F., in Water: A Comprehensive Treatise, (Vol. 4, Chptr. 1), F. Franks. ed., Plenum, New York, (1975)

Franks, F. and Reid, D.S., in Water: A Comprehensive Treatise, (Vol. 2, Chptr. 5), F. Franks. ed., Plenum, New York, (1973)

Frost, E.M. Jr., and Deaton, W.M., Oil & Gas J., **45** (12), 170 (1946)

Galloway, T.J., Ruska, W., Chappelear, P.S. and Kobayashi, R., Ind. Eng. Chem. Fundam., **9**, 237 (1970)

Gas Processors Association Handbook, Tulsa, Ok. 1981, 1987

Geiger, A., Stillinger, F.H., and Rahman, A., J. Chem. Phys., **70**, 4185 (1979)

Gibbs, J.W., The Collected Works of J. Willard Gibbs, Vol I, Thermodynamics., pp. 55-353, Yale University Press, New Haven, Connecticut, 1928.

Gjaldbaek, J.C., and Niemann, H., Acta Chem. Scand., **12**, 1015 (1958)

Glew, D.N., Nature, **184**, 545 (1959)

Glew, D.N., J. Phys. Chem., **66**, 605 (1962)

Glew, D.N. and Rath, N.S., J. Chem. Phys., **44**, 1710 (1966)

Godbole, S.P., Dissociation Pressures of n-Propane and i-Butane Hydrates Below the Ice Point, Master of Science Thesis, University of Pittsburgh, 1981

Goodman, M.A., World Oil, pp 13-15 (1978)

Goodman, M.A., and Franklin, L.J., Proc. Fourth Canadian Permafrost Conf., H.M. French, Ed., 349, Can. Natl. Rsch. Cncl., (1982)

Goodman M.A., and Guissani, A.P., Detection and Evaluation Methods for In Situ Gas Hydrates, SPE/DPE 10831 (1982)

Gough, S.R., Garg, S.K., Ripmeester, J.A. and Davidson, D.W., Can. J. Chem., **52**, 3193 (1974)

Gough, S.R., Garg, S.K., Ripmeester, J.A. and Davidson, D.W., Can. J. Chem., **53**, 2215 (1975)

Graboski, M.S. and Daubert, T.E., Ind. Eng. Chem. Process Des. and Dev, **17**, 443 (1978)

Grantz, A., Boucher, G., and Whitney, O.T., U.S. Geol Survey Circ. 733, 1976

Gray, G.R. and Darley, H.C.H., Composition and Properties of Oil Well Drilling Fluids; 4th ed., Gulf, Houston, 1980

Hafemann, D.R. and Miller, S.L., J. Phys. Chem., **73**, 1392 (1969)

Hale, A., "Shell's Strategy for Gas Hydrate Inhibition," presented at N.L. Bariod Meeting, "Hydrates in Deep Water Drilling," Houston, TX, October 20, 1988

Hale, A.H., and Dewan, A.K.R., "Inhibition of Gas Hydrates in Deep Water Drilling," Presented at SPE Drilling Meeting, February 27-March 3, 1989, New Orleans, LA.

Hales, G.E., "Drying of Fluids with Adsorbents" in Encyclopedia of Chemical Processind and Design, Vol 16, pp 416-442, J.J. McKetta, Ed., Marcel Dekker (1982)

Hall, G.D. and Polderman, L.D., Chem.Eng., **54**, 52 (1960)

Halleck, P.M., Byrer, C.W., McGuire, P.L., Judge, A.S., Corlett, R.C., Barraclough, B., Proc. Methane Hydrate Workshop, Morgantown, WV, 63 Mar 29-30, 1982.

Handa, Y.P., Can. J. Chem, **62**, 1659 (1984)

Handa, Y.P., Can. J. Chem, **63**, 68 (1985)

Handa, Y.P., J. Phys. Chem., **90**, 5497 (1986a)

Handa, Y.P., J. Chem. Thermo., **18**, 891 (1986b)

Handa, Y.P., Calorimetric Studies of Laboratory Synthesized and Naturally Occurring Gas Hydrates, paper presented at AIChE 1986 Annual Meeting, Miami, Fl, Nov 2-7, 1986c

Handa, Y.P., J. Phys. Chem., **90**, 5917, (1986d)

Handa, Y.P., Ind. Eng. Chem. Res., **27**, 872 (1988)

Handwerk, G.E., Personal Communication, August 3, 1988.

Hamblin, W.K., The Earth's Dynamic Systems, Macmillan, New York, 1985

Hammerschmidt, E.G., Ind. Eng. Chem. **26** 851 (1934)

Hammerschmidt, E.G., Gas, **15** (5), 30 (1939)

Hammerschmidt, E.G., Oil Gas J, **29**(2), 61 (May 23, 1940)

Harmens, A., Sloan, E.D., Can. J. Chem. Eng., (in press) (1989)

Harrison, W.E. and Curiale, J.A., Initial Reports DSDP Vol 67 591 (1982)

Hawkins, R.E. and Davidson, D.W., J. Phys. Chem., **79**, 1889,(1966)

Hedberg, H.D., Problems of Petroleum Migration, Amer. Assn. of Petrol Geol. Studies in Geology No. 10, W.H. Roberts and R.J. Cordell, eds., 179 (1980)

Herrin, J.P. and Armstrong, R.A., "Methanol Injection and Recovery in a Turboexpander Plant," Proc. Gas Condition. Conference, Norman, Ok, 1972.

Hesse, R., Geoscience Canada, **14** (3) 165 (1986)

Hesse, R., "Pore Water Anomalies in Gas-Hydrate Bearing Sediments of the Deeper Continental Margins," J. Inclus. Phenomena, D.W. Davidson Memorial volume (in press) (1989)

Hesse, R., Lebel, J., Gieskes, J.M., Initial Reports of The Deep Sea Drilling Project, Vol. 84, von Huene, R., Aubouin, J., et al. Eds, 727, U.S. Government Printing Office, Washington, D.C., (1985)

Himmelblau, D.M., J. Phys. Chem., **63**, 1803 (1959)

Hitchon, B., Natural Gases in Marine Sediments, I.R. Kaplan, ed., 195, Plenum Press, New York, 1974.

Holder, G.D., Multi-Phase Equilibria in Methane-Ethane-Propane-Water Hydrate Forming Systems, Ph.D. Thesis, U. Michigan, University Microfilms No. 77-7939 (1976) Ann Arbor, Mi 48106

Holder, G.D., Personal Communication, June 7, 1988.

Holder, G.D., Angert, P.F., Proc. Soc. Petr. Eng. Meet., New Orleans, LA, Sept 26-29, 1982, SPE 11105

Holder, G.D., Angert, P.F., John. V.T., and Yen, S., J. Petr. Tech. **34** 1127 (1982)

Holder, G.D., and Godbole, S.P., AIChE J., **28**, 930 (1982)

Holder, G.D., Grigoriou, G.C., J. Chem. Thermo., **12**, 1093 (1980)

Holder, G.D. and Hand, J.H., AIChE J., **28**, 44 (1982)

Holder, G. D. and Kamath, V.A.,"The Hydrates of Propane/n-Butane, Propane/trans-2Butane, Propane/Isobutane, Cis-2-Butene and Trans2-Butene Below the Ice Point," presented at AIChE Meeting, Los Angeles, CA November, 1982

Holder, G.D., Kamath, V.A., and Godbole, S.P., Ann. Rev. Energy, **9**, 427 (1984)

Holder, G.D., Katz, D.L. and Hand, J.H., Am. Ass. Petrol. Geol. Bull., **60**, 981 (1976)

Holder, G.D. and Manganiello, D.J., Chem. Eng. Sci., **37**, 9 (1982)

Holder, G.D., Zetts, S.P. and Pradhan, N., Reviews in Chemical Engineering, **5**, 1 (1988)

Holland, P.M., and Castleman, A.W., J. Chem Phys., **72** 5984 (1980)

Hougen, O.A., Watson, K.M., and Ragatz, R.A., Chemical Process Principles: Part II. Thermodynamics, (Second Edition), Wiley & Sons, New York (1959)

Hunt, J.M., Petroleum Geochemistry and Geology, W.H. Freeman, San Francisco, 1979

Hwang, M.J., Wright, D.A., Kapur, A., and Holder, G.D., "An Experimental Study of Crystallization and Crystal Growth of Methane Hydrates from Melting Ice," J. Inclus. Phenom. D.W. Davidson Memorial volume (in press) (1989)

Jacoby, R.H., "Vapor-Liquid Equilibrium Data for Use of Methanolin Preventing Gas Hydrates," Proc.

Gas. Hydr. Contr. Conf., U. Oklahoma, Norman, Ok (1953)

Jeffrey, A.W.A., Pflaum, R.C., McDonald, T.J., Brooks, J.M., and Kvenvolden, K.A., Initial Reports of The Deep Sea Drilling Project, Vol. 84, von Huene, R., Aubouin, J., et al. Eds, 719, U.S. Government Printing Office, Washington, D.C., (1985)

Jeffrey, G.A. and McMullan, R.K., Prog. Inorg. Chem., **8** 43 (1967)

Jeffrey, G.A., Inclusion Compounds, Vol 1, Acad. Press., J.L. Atwood, J.E.D. Davies, D.D. MacNichol, eds. 135 (1984)

Jhaveri, J., and Robinson, D.B., Can. J. Chem Eng. **43** 75 (1965)

John, V.T. and Holder, G.D., J. Phys. Chem., **85**, 1811 (1981)

John, V.T. and Holder, G.D., J. Phys. Chem., **86**, 455 (1982a)

John, V.T. and Holder, G.D., J. Chem. Eng. Data, **27** 18 (1982b)

John, V.T. and Holder, G.D., J. Phys. Chem., **89**, 3279 (1985)

Johnson, J.J., The Measurement of Two Phase Liquid Propane-Hydrate Equilibrium Conditions Using the Dielectric Constant, Master of Science Thesis T2493, Colorado School of Mines, Golden, CO, 1981.

Jones, R.D., and Sherman, J., Offshore, **46**, 27 (July 1986)

Judge, A.S., Proc. Fourth Can Permafrost Conf, H.M. French ed., National Research Council of Canada, 320 (1982)

Judge, A.S., Personal Communication, August 19, 1987

Kamath, V.A., *Study of Heat Transfer Characteristics During Dissociation of Gas Hydrates in Porous Media*, Ph. D.Dissertation, U. Pittsburgh, Univ. Microfilms No. 8417404, Ann Arbor, Mi (1984)

Kamath, V.A. and Godbole, S.P., *J. Petrol Tech*., **39**, 1379 (1987)

Kamath, V.A., Holder, G.D., and Angert, P.F. *Chem Eng. Sci*. **39** 1435 (1984)

Kamath, V.A. and Holder, G.D. *AIChE J*, **33** 347 (1987)

Kamath, V.A., Holder, G.D., and Angert, P.F. *Chem Eng. Sci*. **39** 1435 (1984)

Katz, D.L., *Trans AIME* **160** 140 (1945)

Katz, D.L., *J. Petrol Tech*, **23**, 419 (1971)

Katz, D.L., *J. Petrol Tech*, **24**, 557 (1972)

Katz, D.L., Personal Communication, November 14, 1983

Katz, D.L., Cornell, D., Kobayashi, R., Poettmann, F.H., Vary, J.A., Elenbaas, J.R., and Weinaug, C.F. *Handbook of Natural Gas Engineering*, McGraw-Hill, New York, 802 pp, 1959

Katz, H.R., *J. Petrol. Geol*., **3**, 315 (1981)

Kennicut, M.C., Brooks, J.M., Bidigare, R.R., McDonald, S.J., Adkinson, D.L., Macko, S.A., *Limnology and Oceanography*, (in press) (1989)

Kessock, A. and Fris, J.P., *Hydrocarbon Proc*., **44**, 123 (1965)

Kiefte, H., Clouter,M.J., and Gagnon, R.E., *J. Phys. Chem*. **89** 3103 (1985)

Kim, H.C., A Kinetic Study of Methane Hydrate Decomposition, Dissertation, U. Calgary, CAlgary, Alberta, 1985.

King, J.D., Chemical Engineering Separation Processes, 737, McGraw-Hill, New York, 1980

Knox, W.G., Hess, M., G.E. Jones, and Smith, H.B., Chem. Eng. Prog., **57** (2), 66 (1961)

Kobayashi, R., Vapor-Liquid Equilibrium in Binary Hydrocarbon-Water Systems, Ph.D. Dissertation, U. Mich.,(1951), University Microfilms No. 3521, Ann Arbor, Mi.

Kobayashi, R., and Katz, D.L., Trans AIME, **186**, 66 (1949)

Kobayashi, R., and Katz, D.L., Ind. Eng. Chem., **45**, 440 (1953)

Kobayashi, R., and Katz, D.L., Trans AIME, **204**, 262 (1955)

Kobayashi, R., Song, K.Y., and Sloan, E.D., "Petrol. Eng. Hndbk." H.B. Bradley, Ed., Chapter 25, Society of Petrol Eng., Richardson, Tx, (1987)

Kobayashi, R., Withrow, H.J., Williams, G.B., and Katz, D.L. Proc. NGAA, **1951**, 27 (1951)

Kollman, P., J. Am. Chem. Soc., **99**, 4875 (1977)

Krason, J., Personal Communication, July 29, 1987

Krason, J. and Ciesnik, M., Geological Evolution and Analysis of Confirmed or Suspected Gas Hydrate Localities: Gas Hydrates in the Russian Literature, U.S. Department of Energy, DOE/MC/21181-1950, Vol 5, (1985)

Krason, J. and Ciesnik, M., Geological Evolution and Analysis of Confirmed or Suspected Gas Hydrate

Localities: Basin Analysis, Formation and Stability of Gas Hydrates in the Panama Basin, U.S. Department of Energy,DOE/MC/21181-1950, Vol 6, (1986a)

Krason, J. and Ciesnik, M., Geological Evolution and Analysis of Confirmed or Suspected Gas Hydrate Localities: Basin Analysis, Formation and Stability of Gas Hydrates in the Northern California Offshore, U. S. Department of Energy, DOE/MC/21181-1950, Vol 8, (1986b)

Krason, J. and Ciesnik, M., Geological Evolution and Analysis of Confirmed or Suspected Gas Hydrate Localities: Basin Analysis, Formation and Stability of Gas Hydrates of the Aleutian Trench and the Bering Sea, U.S. Department of Energy,DOE/MC/21181-1950, Vol 10, (1987)

Krason, J., Ciesnik, M.,and Finley, P.D., U.S. Department of Energy, DOE/METC-86/6037, 23 (1986)

Krason, J., Finley, P.D., and Rudloff, B., Geological Evolution and Analysis of Confirmed or Suspected Gas Hydrate Localities: Basin Analysis, Formation and Stability of Gas Hydrates in the Western Gulf of Mexico, U.S. Department of Energy, DOE/MC/21181-1950, Vol 3 (1985)

Krason, J., Ridley, W.I., Geological Evolution and Analysis of Confirmed or Suspected Gas Hydrate Localities: Blake-Bahama Outer Ridge-U.S. East Coast, U.S. Department of Energy, DOE/MC/21181-1950, Vol. 1 (1985a)

Krason, J., Ridley, W.I., Geological Evolution and Analysis of Confirmed or Suspected Gas Hydrate Localities: Baltimore Canyon Trough and Environs - U.S. East Coast, U.S. Department of Energy, DOE/MC/21181-1950, Vol. 2 (1985b)

Krason, J., Rudloff, B., Geological Evolution and Analysis of Confirmed or Suspected Gas Hydrate

Localities: Offshore of Newfoundland and Labrador, U.S. Department of Energy, DOE/MC/21181-1950, Vol. 4 (1985)

Krishnan, C.V., and Friedman, H.L., J. Phys. Chem., **73** 1572, 3934, (1969)

Kvenvolden, K.A., Fourth Canadian Permafrost Conference, R.J.E. Brown Memorial Vol. 305 (1982)

Kvenvolden, K.A., Natural Gas Hydrates, 63, J.L. Cox, Ed., Butterworths, Boston (1983)

Kvenvolden, K.A., Marine and Petrol Geol **2** 65 (1985a)

Kvenvolden, K.A., AAPG Bull, **69(2)** 276 (1985b)

Kvenvolden, K.A., "Methane Hydrate - A Major Reservoir of Carbon in the Shallow Geosphere," Chemical Geology, **71**, 41 (1988a)

Kvenvolden, K.A., Personal Communcation, October 19, (1988b.)

Kvenvolden, K.A., and Barnard, L.A., Initial Reports DSDP Vol 76, R.E. Sheridan, F. Gradstein, et al., Eds., 353 U.S. Government Printing Office, Washington, D.C.,(1983)

Kvenvolden, K.A. and Claypool, G.E., Trans. Amer. Geophys. Union, **62(45)**, 900 (1981)

Kvenvolden, K.A., and Claypool, G.E., Proceedings of the Unconventional Gas Recovery : Gas Hydrates Peer Review Session, Bethesda, Md., April 24-25, 1985.

Kvenvolden, K.A., and Claypool, G.E., U.S. Geol. Survey Open File Report 88-216, 50 pp. (1988)

Kvenvolden, K.A., Claypool, G.E., Threlkeld, C.N., and Sloan, E.D., Org. Geochem. **6** 703 (1984)

Kvenvolden, K.A., Golan-Bac, M., and Rapp, J.B., The Antarctic Contineneta1 Margin: Geology and Geophysics of Offshore Wilkes Land and the Western Ross Sea, S.L. Eittreim, M.A. Hampton, A.K. Cooper, and F.J., Davey (Eds.), Circum-Pacific Council for Energy and Mineral Resources, Earth Science Series, **5A**, 205 (1987)

Kvenvolden, K.A., and Grantz, A., The Artic Ocean, The Geology of North America,, A. Grantz, L. Johnson, and J.F. Sweeney Eds. The Geological Society of America, 50 (in press) (1988)

Kvenvolden, K.A., and Kastner, M., Initial Reports ODP, Part B., E. Suess, R. von Huene, K-C., Emeis, et al., Eds., 112, U.S. Government Printing Office, Washington, D.C., (in press)(1988)

Kvenvolden, K.A. and McMenamin, M.A., U.S. Geological Survey Circular 825 (1980)

Kvenvolden, K.A., and McDonald, T.J., Initial Reports of The Deep Sea Drilling Project, Vol. 84, von Huene, R., Aubouin, J., et al. Eds, 667, U.S. Government Printing Office, Washington, D.C., (1985)

Kvenvolden, K.A., and von Huene, R., Tectonostratigraphic Terranes of the Circum Pacific Region, D.G. Howell, Ed., Circum-Pacific Council for Energy and Minerals, Earth Sciences Series, **1**, 31 (1985)

Ladd, J., Westbrook, G., and Lewis, S., Lamont-Doherty Geological Observatory Yearbook 1981-1982, **8**, 17 (1982)

Larson, S.D., Phase Studies of the Two-Component Carbon Dioxide-Water System, Involving the Carbon Dioxide Hydrate, Univ. Illinois, 1955

Latimer, W.M. and Rodebush, W.H., J. Am. Chem. Soc., **42** 1419 (1920)

Laughlin, A.R., "Water Content in Natural Gas, The Thermodynamics of its Determination, Comparison of Methods and its Value to the Industry," Proc. Gas Conditioning Conference, Norman, OK (1969)

Lawson, W.F., Reddy, S.M., McCarthy, L.A., Gregoire, C.E., Vassallo, K.L., Barlow, D.A., "Acoustic Velocity and Electrical Property Measurements in the DSDP Methane Hydrate Core," paper presented at 1984 Winter Meeting of AIChE, Atlanta, GA., March 12, 1984

Leaist, D.G., Murray, J.J., Post, M.L. and Davidson, D.W., J. Phys. Chem., **86**, 4175 (1982)

Lewin and Associates, Inc. and Consultants, Handbook of Gas Hydrate Properties and Occurrence, U.S. Department of Energy, DOE/MC/19239-1546, (1983)

Lewis, G.N. and Randall, M., Thermodynamics, McGraw-Hill Book Co., New York, 1923

Lievois, J.S., Development of an Automated, High Pressure Heat Flux Calorimeter and its Application to Measure the Heat of Dissociation of Methane Hydrate, Ph.D. Thesis, Rice University, Houston, TX., 1987.

Lingelem, M., and Majeed, A., "Challenges in Areas of Multiphase Transport and Hydrate Control for a Subsea Gas Condensate Production System," Proc. of the 68th Ann. Gas Proc. Assoc. Convention, San Antonio, TX., March 13-14, 1989

Loh, J., Maddox, R.N., Erbar, J.H., Oil & Gas J., 96, May 16, 1983

Lonsdale, H.K., Proc. Roy Soc. Ser. A, **247**, 424 (1958)

Loomer, J.A. and Welch, J.W., "Some Critical Aspects of Designing for High Dewpoint Depression with Glycols," Proc. Gas Condition. Conf., Norman, OK 1961

Low, P.F., Advances in Agronomy, E.G. Norman, Ed., **13**, p. 269 Academic Press, New York, 1961

Löwig, G., Mag. Pharm., **23**, 12 (1828)

Lubas, J., Nafta (Pol), **34**, 380 (1978)

Luck, W.A.P., in Water: A Comprehensive Treatise, Volume 2, Chapter 4, F. Franks, ed., Plenum Press, New York, 1973

Lyusternik, L.A., Convex Figures and Polyhedra, Dover, New York, 1963

Majeed, A., and M.N. Lingelem, "Hydrate Formation Kinetics in Gas Condensate Pipelines," presented at BHRA Conference on Operational Consequences of Hydrate Formation and Inhibition Offshore, Cranfield, UK, November 3, 1988

Majid, Y.A., Garg, S.K. and Davidson, D.W., Can. J. Chem., **47**, 4697 (1969)

Mak, T.C.W. and McMullan, R.K., J. Chem. Phys., **42** 2732 (1965)

Makogon, Y.F., Gazov. Promst., **5**,14 (1965)

Makogon, Y.F., Hydrates of Natural Gas, Moscow, Nedra, Izadatelstro, 208 pp (1974), Translated from the Russian by W.J. Cieslesicz, PennWell Books, Tulsa, Oklahoma 237 pp (1981)

Makogon, Y.F., Gas Hydrates: Prevention of Gas Hydrates Formation and Their Utilization, Moscow, Nedra, (In Russian) (1985)

Makogon, Y.F., La Recherche, **18**, 1192 (1987)

Makogon, Y.F., Personal Communication, January 13, 1988a

Makogon, Y.F., "Natural Gas Hydrates: The State of Study in the USSR and Perspectives for Its Use," paper presented at the Third Chemical congress of North America, Toronto, Canada, June 5-10, 1988b.

Makogon, Y.F. and Koblova, I.L., E.I. Gas Prom, **22** 3 (1972)

Malenko, E.V., Respublikanskia. Nauchno-Tekgnicheskaia Kenferentoria po Neftekhimii, Zd. Guryar (1974)

Malone, R.D., Gas Hydrates Topical Report, DOE/METC/SP-218, U.S. Department of Energy, April, 1985

Mann, S.L., McClure, L.M., Poettmann, F.H., and Sloan, E.D., "Vapor-Solid Equilibrium Ratios for Structure I and Structure II Natural Gas Hydrates," Proc. 68th Ann. Gas Proc. Assoc. Conv., San Antonio, TX, March 13-14, 1989

Markl, R.G., Bryan, G.M., and Ewing, J.I., J. Geophys. Res., **75**, 4539 (1970)

Marlow, M.S., Carlson, P., Cooper, A.K., Karl, H., McLean, H., McMullin, R., and Lynch, M.B., U.S. Geol. Survey Open File Report 81-252, 83pp (1981)

Marshall, D.R., Saito, S., and Kobayashi, R., AIChE J, **10** 202, 723,(1964)

Marx, J.W., and Langenheim, R.H., Trans AIME, **216**, 312 (1959)

Mathews, M.A., The Log Analyst, 26 (May-June 1986)

Mathews, M.A., and von Huene, R., Initial Reports of The Deep Sea Drilling Project, Vol. 84, von Huene, R., Aubouin, J., et al. eds, 773, U.S. Government Printing Office, Washington, D.C., (1985)

Mauméné, E., Chem. N., **47**, 154 (1883)

McAuliffe, C., Nature, **200**, 1092 (1963)

McAuliffe, C., J. Phys. Chem, **70**(4) 1267 (1966)

McCarthy, J., Stevenson, A.J., Scholl, D.W., and Vallier, T.L., Marine and Petrol Geology,**2**,713 (1981)

McClelland, A.L., Dipole Moments, W.H. Freeman & Co., San Francisco (1963)

McGuire, P.L., Los Alamos Sci. Lab. Rept. LA-91-MS, Los Alamos, NM (1981)

McGuire, P.L., Proc. Fourth Can Permafrost Conf., H.M. French, ed. 356 (1982)

McIver, R.D., Initial Reports of the Deep sea Drilling Project, Hayes, D.E., Frakes, L.A., Eds., **28**, 815, U.S. Government Printing Office, Washington, D.C., (1975)

McIver, R.D., Long-Term Energy Resources, R.F. Meyer and J.C. Olson, eds., Pitman, Boston, **1**, 713, (1981)

McKirdy, D.M., and Cook, P.J., Am. Assn. Petrol. Geol. Bull., **64**, 2118 (1980)

McKoy, V. and Sinanoglu, O., J. Chem. Phys, **38** 2946 (1963)

McLeod, H.D. and Campbell, J.M., J. Petrol Tech., **13**, 590 (1961)

McMullan, R.K. and Jeffrey, G.A., J. Chem. Phys., **42** 2725 (1965)

Mendis, D.A., Nature, **249**, 536 (1974)

Menten, P., Parrish, W.R., and Sloan, E.D., Ind Eng Chem Proc Des and Dev., **20**, 399 (1981)

METC Topical Report: Gas Hydrates Project, Morgantown Energy Technology Center, Topical Report (Feb 1984)

Meyer, R.F., Long-Term Energy Resources, R.F. Meyer and J.C. Olson, eds., 49 Pitman, Boston, MA (1981)

Miller, K.W., and Hildebrand, J.H., J. Am. Chem. Soc., **90** 3001 (1968)

Miller, B. and Strong, E.R., Proc. Am. Gas. Assoc., **27**, 80 (1945)

Miller,B. and Strong,E.R., Am Gas Assn Mon,**28**(2),63(1946)

Miller, S.L., Proc. NSA, **47**, 1515 (1961)

Miller, S.L., Natural Gases in Marine Sediments, I.R. Kaplan ed. (Marine Science, **3**) 151, Plenum (1984)

Miller, S.L. and Smythe, W.D., Science, **170**, 531 (1970)

Millon, H., Compte Rend. **51**, 249 (1860)

Moore, B., III., and Bolin, B., Oceanus, **29** (4), 9 (1987)

Morrison, T.J., J. Chem. Soc., 3814 (1952)

Morrison, T.J. and Billett, F.,J. Chem. Soc., 3819 (1952)

Mozurkewich, M., and Benson, S.W., J. Phys. Chem., **88**, 6429 (1984)

Müller, H.R., and von Stackelberg, M., Naturwiss, **39** 20 (1952)

Müller-Bongartz, B., Personal Communication "On the Accuracy of HYDK", February 6, 1989

Munck, J., Skjold-Jørgensen, S., Rasmussen, P., Chem. Eng. Sci., **43**, 2661 (1988)

Nadem, S.L., Patil, S.L., Mutalik, P.N., Kamath, V.A., Godbole, S.P.,"Measurement of Gas Permeability in Hydrate Saturated Unconsolidated Cores," paper presented at the Third Chemical Congress of North America, Toronto, June 5-10, 1988

Nagata, I. and Kobayashi, R., Ind. Eng. Chem. Fundam. **5**, 344, (1966a)

Nagata, I. and Kobayashi, R., Ind. Eng. Chem. Fundam. **5**, 466, (1966b)

Nakayama, H., Hashimoto, M., Bull. Chem. Soc. Japan, **53**, 2427 (1980)

Narten, A.H., Levy, H.A., Science, **165**, 447 (1969)

Nelson, K., Hydrocarbon Processing, p. 161, Sept. 1973

Nelson, K. and L. Wolfe, "Methanol Injection and Recovery in a Large Turboexpander Plant," Proc. Gas Condition. Conference, Norman, Ok, 1981

Nemethy, G., and Scheraga, H.A., J. Chem. Phys., **36**, 3382 (1962a)

Nemethy, G., and Scheraga, H.A., J. Chem. Phys., **36**, 3401 (1962b)

Ng, H.-J. and Robinson, D.B., I&EC Fundam., **15**, 293 (1976a)

Ng, H.-J. and Robinson, D.B., AIChE Journal., **22**, 656 (1976b)

Ng, H.-J. and Robinson, D.B., AIChE Journal., **23**, 477 (1977)

Ng, H.-J. and Robinson, D.B., Gas Proc. Assn. Rsch Rpt 66, April 1983

Ng, H.-J. and Robinson, D.B., Gas Proc. Assn. Rsch Rpt 74, March 1984

Ng, H.-J. and Robinson, D.B., Fluid Phase Equilibria, **21**, 145 (1985)

Ng, H.-J., Chen, C.-J. and Robinson, D.B., Gas Proc. Assn. Rsch Rpt 87, March 1985

Ng, H.-J., Chen, C.-J. and Robinson, D.B., Gas Proc. Assn. Rsch Rpt 92, September 1985

Ng, H.-J., Chen, C.-J., and Saeterstad, T., Fluid Phase Equilibria **36**, 99 (1987a)

Ng, H.-J., Chen. C.-J., and Robinson, D.B., Gas Proc. Assn. Rsch Rpt 106, April 1987b

Ng, H.-J., Petrunia, J.P. and Robinson, D.B., Fluid Phase Equilibria, **1**, 283 (1978)

Nielsen, R.B., and Bucklin, R.W., Hydrocarbon Processing, p.71, April 1983.

Nikitin, B.A., Z. Anorg. Allg. Chem. **227**, 81 (1936)

Nikitin, B.A., Nature, **140**, 643 (1937)

Nikitin, B.A., Zh. Obshch. Khim. **9**, 1167, 1176 (1939)

Nikitin, B.A., Izv. Akad. Nauk SSSR, Otd. Khim. Nauk **1**, 39 (1940)

Noaker, L.J. and Katz, D.L., Trans AIME, **201**, 237 (1954)

Nolte, F.W., Robinson, D.B., and Ng, H.-J.,Proc. 64th Ann Conv Gas Process Assn, **64** 137 (1985)

Nosov, E.F., and Barlyaev, E.V., Zh. Obshch. Khim, **38**(2), 211 (1968)

Olds, R.H., Sage, B.H., and Lacey, W.N., Ind. Eng. Chem., **34**, 1223 (1942)

Onsager, L., and Runnels, L.K., J. Chem. Phys., **50**, 1089 (1969)

Otto, F.D. and Robinson, D.B., AIChE J., **6**, 602, (1960)

Owicki, J.C., and Scheraga, H.A., J. Am. Chem. Soc, **99**, 7413, (1977)

Pandit, B.I., and King, M.S., Fourth Can Permafrost Conf, H.M. French ed., 335 (1982)

Pang, K.D., Voga, C.C., Rhoads, J.W. and Ajello, J.M., Proc. 14th Lunar and Planet Sci Conf, Houston, **14** 592 (1983)

Paranjpe, S.G., Patil, S.L., Kamath, V.A., and Godbole, S.P. "Hydrate Equilibria for Binary and Ternary Mixtures of Methane, Propane, Isobutane, and n-Butane: Effect of Salinity," Paper SPE 16871, 379, Proc. 62nd SPE Annual Conf. Sept 27-30, 1987, Dallas, TX

Paranjpe, S.G., , Patil, S.L., Kamath, V.A., Godbole,S.P., "Hydrate Formation in Crude Oils and Phase Behavior of Hydrates in Mixtures of Methane, Propane, Isobutane, and n-Butane," paper presented at the Third Chemical Congress of North America, Toronto, Canada, June 5-10, 1988

Parent, J.D., Inst. of Gas Techn. Res. Bull., **1** (1948)

Parrish, W.R. and Prausnitz, J.M., Ind. Eng. Chem. Proc. Des & Dev. **11** 26 (1972)

Parsonage, N.G., and Staveley, L.A.K., in Inclusion Compounds, Vol 3, Chptr 1, J.L. Atwood, J.E.D. Davies, and D.D. MacNichol, eds., Academic Press (1984)

Passut, C.A., and Danner, R.P., Ind. Eng. Chem. Process Des. Develop., **11**, 543 (1972)

Patil, S.L., Measurements of Multiphase Gas Hydrates Phase Equilibria: Effect of Inhibitors and Heavier Hydrocarbon Components, M.S. Thesis, U. Alaska, Fairbanks, AL, 1987.

Pauling, L., J. Amer. Chem. Soc., **57**, 2680 (1935)

Pauling, L. The Nature of the Chemical Bond, Cornell U. Press, Ithaca, NY 1945

Pauling, L., in Hydrogen Bonding, D. Hadzi, ed., Pergamon Press, New York, 1957

Pauling, L., Science, **134**, 15 (1961)

Pauling, L., and Marsh, R.E., Proc. Nat. (U.S.A.) Acad. Sci, **38** (1952)

Paull, C.K. and Dillon, W.P., U.S. Geological Survey: Miscellaneous Field Studies Map MF-1252, Washington, Washington, D.C., 1981

Pearson, C., Murphy. J.R., Mermes, R.E., and Halleck, P.M., LosAlamos Scientific Laboratory Report LA-9972-MS **9** (1984)

Peng, D.-Y. and Robinson, D.B., Ind. Eng. Chem. Fundam., **15**, 59 (1976a)

Peng, D.-Y. and Robinson, D.B., Canad. J. Chem. Eng., **54**, 595 (1976b)

Perry, C.R., Oil & Gas Journal, **58**, 71 (1960)

Pieroen, A.P., Rec. Trav. Chim., **74**, 995 (1955)

Pierre, J., Ann. Chim. Phys. Ser. 3, **23**, 416 (1848)

Pimentel, G.C. and McClellan, A.L., The Hydrogen Bond, W.H. Freeman and Co., San Francisco, (1960)

Pinder, K.L., Can. J. Chem. Eng., **6**, 132, (1964)

Pinder, K.L., Can. J. Chem. Eng., **10**, 271 (1965)

Platteeuw, J.C. and van der Waals, J.H., Rec. Trav. Chim. Pays-Bas **78**, 126 (1959)

Plummer, P.L.M. and Chen, T.S., J. Phys. Chem., **87**, 4190, (1983)

Plummer, P.L.M. and Chen., T.S., J. Chem. Phys., **86**, 7149, (1987)

Poettmann, F.H., Hydrocar. Proc., **63** (6) 111, (1984)

Poettmann, F.H., Personal Communication, July 20, 1988.

Polderman, L.D., "The Glycols as Hydrate Point Depressants in Natural Gas Systems," Proc. Gas Conditioning Conference, Norman, OK, 1958

Pople, J.A., Proc. Roy. Soc., **205**, 163 (1951)

Powell, H.J.M., J. Chem. Soc. (London) **1948**, 61 (1948)

Prausnitz, J.M., Molecular Thermodynamics of Fluid-Phase Equilibria, Prentice-Hall, Inc., Englewood, Cliffs, NJ, 1969

Rahman, A., and Stillinger, F.H., J. Am. Chem. Soc., **95** (24), 7943 (1973)

Rahman, A., and Stillinger, F.H., J. Chem. Phys., **60**, 1545 (1974)

Razaryan, T.S. and Ryabtsev, N.I., Neft. Khoz, **47**(10),54 (1969)

Reamer, H.H. Selleck F.T., and Sage, B.H., J. Petrol Tech,. **4**(8), 197 (1952)

Records, J.R., and Seely, Jr., D.H., Trans AIME, **192**, 61 (1951)

Redus, F.R., "Fundamentals of Natural Gas Dehydration," Proc Gas Condition Conf, Norman, OK, 1965.

Rice, P.A., Gale, R.P., and Barduhn, A.J., J. Chem. Eng. Data **21**, 204 (1976)

Richardson, H.H., Woolridge, P.J., and Devlin, J.P., J. Phys. Chem., **89**, 3552 (1985a)

Richardson, H.H., Woolridge, P.J., and Devlin, J.P., J. Chem. Phys., **83**, 4387 (1985)

Ripmeester, J.A., Personal Communication, May 2, 1988

Ripmeester, J.A. and Davidson, D.W., Mol. Cryst. Liq. Cryst. **43**, 189 (1977)

Ripmeester, J.A. and Davidson, D.W., J. Mol Struct., **75**, 67 (1981)

Ripmeester, J.A., Tse, J.A., Ratcliffe, C.I., Powell, B.M., Nature **325** 135 (1987)

Roadifer, R.D, Godbole, S.P., Kamath, V.A., "Thermal Model for Establishing Guidelines for Drilling in the Arctic in the Presence of Hydrates," SPE 16361, presented at the SPE California Regional Meeting, Ventura, CA, April 8-10, 1987a

Roadifer, R.D, Godbole, S.P., Kamath, V.A., "Estimation of Parameters for Drilling in Arctic and Offshore Environment in the Presence of Hydrates," SPE 16671, presented at the SPE 62nd Annual Tech. Conference, Dallas, TX, September 27-30, 1987b

Roberts, O.L., Brownscombe, E.R., Howe, L.S., Oil & Gas J., **39**, (30), 37 (1940)

Roberts, O.L., Brownscombe, E.R., Howe, L.S. and Ramser, H. Petrol. Eng., (3), 56, (1941)

Roberts, R.S., Andrikidis, C., Tainsh, R.J. and White, G.K. Proc. 10th Internat. Cryogen. Eng. Conf., 499, Helsinki, H. Callan. P. Bergland, M. Krusius, eds., Butterworths, 1984

Robinson, D.B. and Hutton, J.M., J. Can. Petr. Tech., **6**, 6 (1967)

Robinson, D.B. and Mehta, B.R., J. Can. Petr. Tech., **10**, 33 (1971)

Robinson, D.B. and Ng, H.-J., Hydrocarbon Processing, 95 (December 1976)

Robinson, D.B. and Ng, H.-J.,J. Can. Petrol Tech., 26, July-August 1986

Robinson, D.B., Ng, H.-J. and Chen, C.-J., Proc. 66th, Annual GPA Convention, March 16-18, 1987, Denver, CO.

Robinson, R.A., Stokes, R.H., Electrolyte Solutions, Butterworths Scientific Publications, London, (1959)

Roozeboom, H.W.B., Rec. Trav. Chem. Pays-Bas, **3**, 26 (1884)

Roozeboom, H.W.B., Rec. Trav. Chem. Pays-Bas, **4**, 65 (1885)

Rose, W., and Pfannkuch, H.O., Proc SPE/DOE Unconv. Gas Recov. Symp (Pittsburgh), 5/16-18/82, 417 (1982a)

Rose, W., and Pfannkuch, H.O., Proc. 57th Ann. SPE Fall Tech. Conference, Sept. 26-29, 1982b. SPE 1106

Ross, R.G., Andersson, P. and Backström, G., J. Chem. Phys. **68**, 6967 (1978)

Ross, R.G., Andersson, P. and Backström, G., Nature **290** 322 (1981)

Ross, R.G., and Andersson, P., Can. J. Chem. , **60**, 881 (1982)

Rouher, O.S. and Barduhn, A.J., Desalination, **6** 57 (1969)

Rouher, O.S., The n-Butane and Iso-Butane Hydrates, M.S. Thesis, Syracuse University, June 1968.

Rozov, V.N., Bukgalter, E.B., Pushov, V.N., Tunkel, L.E., Gazov Promst **10** 11 (1986)

Rueff, R.M., The Heat Capacity and Heat of Dissociation of Methane Hydrates: A New Approach, Dissertation T-2985, Colorado School of Mines, Golden, CO, 1985

Rueff, R.M. and Sloan, E.D., Ind. Eng. Chem. Proc. Des. Dev. **24**, 882 (1985)

Rueff, R.M., Sloan, E.D., and Yesavage, V.F., AIChE J., **34**, 1468 (1988)

Saito, S., Marshall, D.R. and Kobayashi, R., AIChE J., **10**, 734 (1964)

Saito, S. and Kobayashi, R., AIChE J., **11**, 96 (1964)

Sapir, M.H., Khramenkov, E.N., Yefremov, I.D., Ginzburd, G.D., Beniaminovich, A.E., Lenda, S.M.,and Koslova, V.I., Geol. Nefti i Gaza, **6** 26 (1973)

Scauzillo, F.R., Chem. Eng. Prog., **52**,324 (1956)

Scheffer, F.E.C. and Meyer, G., Proc. Roy Acad. Sci. Amsterdam **21** 1204, 1338 (1919)

Schneider, G.R. and Farrar, J., U.S. Dept. of Interior, Rsch. Dev. Rpt No 292, 37pp. (January 1968)

Schoenfeld, F., A. Ch., Neue Reihe, **19**, 19 (1855)

Schroeder, W., Die Geschichte der Gas Hydrate, Sammlung Chem. Chem. Tech. Vortrage, **29**, F. Enke, Stuttgart, 98 pp (1927)

Schroeter, J.P., Kobayashi, R., and Hildebrand, M.A., Ind. Eng. Chem. Fundam., **22**, 361 (1983)

Selim, M.S. and Sloan, E.D., "Modeling and Dissociation of an In-Situ Hydrate," SPE 13597, 75, Proc. 1985 SPE California Regional Meeting, Bakersfield, CA March 27-29 1985

Selim, M.S. and Sloan, E.D., Unpublished Work on Modeling Hydrate Nucleation, March 19, 1987.

Selim, M.S. and Sloan, E.D. "Hydrate Dissociation in Sediment," SPE 16859, 243, Proc. 62nd SPE Annual Tech. Conf., Dallas,TX, Sept, 27-30, 1987

Selim, M.S. and Sloan, E.D., AIChE J., **35**, 1049, (1989)

Selleck, F.T., Carmichael, L.T. and Sage, B.H., Ind. Eng. Chem. **44**, 2219 (1952)

Selzneu, A.P. and Stupin, D.Y., Kolloidnyi Zhurnal, **39** (1) 86 (1977)

Setliff, S.R., "Hydrate Mitigation in Esso's Deep Water Integrated Production System," presented at BHRA's Conference on Operational Consequences of Hydrate Formation and Inhibition Offshore," Cranfield, UK, November 3, 1988

Sharma, G.D., Kamath, V.A., Godbole, S.P., Patil, S.L., Paranjpe, S.G., Mutalik, P.N., Nadem, N., Development of Alaskan Gas Hydrate Resources, DOE/FE/6114-2608 (DE88010270), October 1987.

Shepard, G.L., Shallow Crustal Structure and Marine Geology of a Convergence Zone, Northwest Peru and Southwest Ecuador, Ph.D. Dissertation, pp157-165, University of Hawaii (1979)

Sheshukov, N.L., Gazov. Delo., **6**, 8 (1972)

Shipley, T.H., Houston, M.H., Buffler, R.T., Shaub, F.J., McMillen, K.J., Ladd, J.W., and Worzel, J.L., Amer Assn Petrl Geol Bull. **63** 2004 (1979)

Shoji, H. and Langway, C.C., Nature **298** 548 (1982)

Sivalls, C.R., "Glycol Dehydrator Design Manual," Proc. Gas Condition. Conf., Norman, OK, (1974)

Skinner, W., Jr., The Water Content of Natural Gas at Low Temperatures,M.S.Thesis, U. Oklahoma, Norman,OK,1948

Slattery, J.C., Momentum, Heat, and Mass Transfer in Continua," R.E. Krieger, Co., Melbourne, FL, 1978

Sloan, E.D. Proc 63th Ann Conv - Gas Process Assn **63** 163 (1984)

Sloan, E.D. Proc 64th Ann Conv - Gas Process Assn **64** 125 (1985a)

Sloan, E.D., Initial Reports of The Deep Sea Drilling Project, Vol. 84, von Huene, R., Aubouin, J., et al. eds, 695 U.S. Government Printing Office, Washington, D.C., (1985b)

Sloan, E.D., Khoury, F.M. and Kobayashi, R., Ind. Eng, Chem. Fundam. **15** 318 (1976)

Sloan, E.D. and Parrish, W.R., in Natural Gas Hydrates: Properties, Occurrence, and Recovery, J.L. Cox ed., Butterworths, 1983.

Sloan, E.D., Sparks, K.A. and Johnson, J.J., Ind. Eng. Chem. Res. **26** 1173 (1987)

Sloan, E.D., Sparks, K.A., Johnson, J.J. and Bourrie, M.S., Fluid Phase Equilibria **29** 233 (1986)

Snell, L.E., Otto, F.D., and Robinson, D.B., AIChE J., **7**, (3), 482 (1961)

Song, K.Y. and Kobayashi, R., Ind. Eng. Chem. Fundam. **21** 391 (1982)

Song, K.Y. and Kobayashi, R., Gas Processors Assn Rsch Rept 80, Tulsa, OK, May 1984

Song, K.Y. and Kobayashi, R., SPE Form. Eval., 500, (December 1987)

Sparks, K.A., The Water Content of Liquid Ethane and Propane in Two Phase Equilibrium with Hydrate, Master of Science Thesis T-2809, Colorado School of Mines, 1983.

Sparks, K.A., and Sloan, E.D., Gas Processors Assn. Resch Rept 73, Tulsa, OK, September 1983

Speedy, R.J., J. Phys. Chem., **91**, 3354 (1987)

Speedy, R.J., Madura, J.D., and Jorgensen, W.L., J. Phys. Chem., **91**, 909 (1987)

Stange, E., Majeed, A., and Overa, S., "Experiments and Modeling of the Multiphase Equilibrium of Inhibition of Hydrates," Proc. 68th Ann. Gas Proc. Assn Conv., San Antonio, Tx, March 13-14, 1989

Stillinger, F.H., Science, **209**, 451 (1980)

Stillinger, F.H., and Rahman, A., J. Chem. Phys., **60**, 1545 (1974)

Stillinger, F.H., and Rahman, A., J. Chem. Phys., **68**, 666 (1978)

Stoll, R.D. and Bryan, G.M., J. Geophys Rsch **84**, 1629 (1979)

Stoll, R.D., Ewing, J. and Bryan, G.M., J. Geophys Rsch, **76**,2090,(1971)

Sumetz, V.I., Gaz. Prom., **2**, 24 (1974)

Summerhayes, C.P., Bornhold, B.D., and Embley, R.W., Marine Geology, **31**, 265 (1979)

Svartas, T.M., "Overview of Hydrate Research at Rogalands-Forskning," presented at BHRA Conference on Operational Consequences of Hydrate Formation and Inhibition Offshore, Cranfield, UK, November 3, 1988

Swaminathan, S., Harrison, S.W., and Beveridge, D.L., J. Am. Chem. Soc., **100**, 5705, (1977)

Tabushi, T., Kiyosuke, Y., and Yamamura, K., Bull. Chem. Soc. Japan, **54**, 2260 (1981)

Tailleur, I.L. and Bowsher, A.L., USGS Clathrates in the NPRA Workshop, (July 16-17, 1979) USGS Open File Rept 81-1298, 147 (1981)

Taylor, A.E., Wehmiller, R.J., and Judge, A.S., Symposium on Research in the Laborador Coastal and Offshore Region, W. Denver, ed., Memorial University of Newfoundland, 91 (1979)

Tanret, C., Compt. Rend. **86**, 765 (1878)

Thakore, J.L., Hydrate Azeotrope Formation in Ternary (Gas+Gas+Water) Systems, Master of Science Thesis, University of Pittsburgh, Pittsburgh, PA, 1984

Thakore, J.L. and Holder, G.D., Ind. Eng. Chem. Res.,**26**, 462 (1987)

Thies, M.C., Personal Communication, December 20, 1988

Topham, D.R., Chem. Eng. Sci. **39** 821 (1984)

Topham, D.R., Chem. Eng. Sci. **39** 1613 (1984)

Touloukian, Y.S., Kirby, R.K., Taylor, R.E., and Lee, J.Y.R., Thermophysical Properties of Matter Vol 13., 263-266, Plenum Press, New York (1977)

Townsend, F.M. and Reid, L.S., Hydrate Control in Natural Gas Systems, Laurance Reid Associates, Inc., P.O. Box 1188, Norman, OK 73070 (1978)

Trofimuk, A.A., Cherskiy, N.V., Lebedev, V.S., Semin V.I., et al. Geol. Geofiz. **2** 3 (1983)

Trofimuk, A.A., Cherskiy, N.V., and Tsarev, V.P., Future Supply of Nature-made Petroleum and Gas, R.F.Meyer, ed., 919, Pergamon Press, New York, 919 (1977)

Trofimuk, A.A., Cherskiy, N.V., Makogon, Y.F., and Tsarev, V.P., "Possible Gas Reserves in the Continental and Marine Deposits and the Methods of Their Prospecting and Development," presented at the 11th ASA Converence on the Conventional and Unconventional World Natural Gas Resources, Laxenburg, Austria, June 3 - July 4, 1980.

Trofimuk, A.A., Makogon, Y.F., and Tolkachev, M.V., Geologiia Nefti i Gaza, **10** 15 (1981)

Trofimuk, A.A., Cherskii, N.V., Tsarev, V.P. and Nikitin, S.P., Geologiya i Geofizika, **23** 3 (1982)

Tsarev, V.P. and Savvin, A.Z., Gazov Prom-st. **2** 46 (1980)

Tse, J.S., J. de Physique, Coll C1, Suppl 3, **48**, 543 (1987)

Tse, J.S., Personal Communication, June 6, 1988

Tse, J.S. and Davidson, D.W., Proc. Fourth Canadian Permafrost Conf, p 329 Calgary, Alberta, March 2-6 1981, National Research Council of Canada, Ottawa, 1982

Tse, J.S., Handa, Y.P., Ratcliffe, C.I., and Powell, B.M., J. Incl. Phenom., **4**, 235 (1986)

Tse, J.S., Klein, M.L., and McDonald, I.R., J. Chem. Phys. **78** 2096 (1983a)

Tse, J.S., Klein, M.L., and McDonald, I.R., J. Phys. Chem. **87** 4198 (1983b)

Tse, J.S., Klein, M.L., and McDonald, I.R., J. Chem Phys. **81** 6146 (1984)

Tse, J.S., Klein, M.L., and Marchi, M., J. Phys. Chem. 91 4188 (1987)

Tse, J.S., and Klein, M.L., J. Phys. Chem, **91**, 5789, (1987)

Tse, J.S., McKinnon, W.R., and Marchi, M., J. Phys. Chem., **91**, 4188 (1987b)

Tucholke, B.F., Bryan, G.M., and Ewing, J.I., Amer. Assn. of Petrol. Geol. Bull., **61**, 698 (1977)

Uchida, T. and Hayano, I., *Repts. of the Govt. Chem. Ind. Res. Inst, Tokyo*, **59**, 382 (1964)

Uchida, T. and Hayano, I., *Desalination*, **3**, 373 (1967)

Ullerich, J.W., Selim, M.S. and Sloan, E.D., *AIChE J*, **33** 747 (1987)

Unruh, C.H. and Katz, D.L., *Trans AIME*, **186**, 83 (1949)

van Cleeff, A., *Gashydraten van Stikstof en Zuurstof*, Dissertation Delft University, 1962

van Cleeff, A. and Diepen, G.A.M., *Rec. Trav. Chim.*, **79**, 582 (1960)

van der Waals, J.H., *On the Continuity of the Gaseous and Liquid State*, Dissertation, Leiden (1890)

van der Waals, J.H. and Platteeuw, J.C., "Clathrate Solutions," *Adv. Chem. Phys. Vol 2*, 1 (1959)

van Welie, G.S.A., and Diepen, G.A.M. *Rec. Trav. Chim.*, **80**, 666 (1961)

Venart, J.E.S., Prasad, R.S., and Stocher, D.G. in *Water and Steam, Their Properties and Current Industrial Application*, Straub and Scheffler, eds., Pergamon Press, New York, NY 392 (1979)

Verigin, N.N., Khabibullin, I.L., and Khalikov, G.A., *Izv. Akad. Nauk. SSSR, Mekhanika Zhidkosti Gaza No. 1*, 174 (1980)

Verma, V.K., *Gas Hydrates from Liquid Hydrocarbon-Water Systems* Ph.D. Thesis, University of Michigan, 1974, Univ. Microfilms No. 75-10,324, Ann Arbor, MI

Verma, V.K., Hand, J.H., and Katz, D.L., "Gas Hydrates from Liquid Hydrocarbons (Methane-Propane-Water System," *AICHE -VTG Joint Meeting*, p. 10, Munich (Sept. 1974)

Verma, V.K., Hand, J.H., Katz, D.L. and Holder, G.D., J. Pet. Tech., **27**, 223 (1975)

Villard, P., Compt. Rend. **106**, 1602 (1888)

Villard, P., Compt. Rend. **111**, 302 (1890)

Villard, P., Compt. Rend. **123**, 337 (1896)

Volmer, M. and Weber, A., Z. Phys. Chem. (Leipzig), **119**, 277 (1926)

Volpe, A.M., Shipley, T.H., and Moore, G.F., Initial Reports of The Deep Sea Drilling Project, Vol. 84, von Huene, R., Aubouin, J., et al. eds, 851, U.S. Government Printing Office, Washington, D.C., (1985)

von Herzen, R.P. and Maxwell, A.E., J. Geophys. Rsch., **64**, 1557 (1959)

Von Huene, R., Aubouin, J., Azema,J., Blackinton, G., Carter, J.A., Coulbourn, W.T., Cowan, D.S., Curiale, J.A., Dengo, C.A., Faas, R.W., Harrison, W., Hesse, R., Hussong, D.M., Laad, J.W., Muzylov, N, Shiki, T., Thompson, P.R., and Westberg, J., Geol Soc Amer Bull Part I, **91**, 421 (1980)

von Stackelberg, M., Naturwiss **36**, 327, 359 (1949)

von Stackelberg, M., Z. Electrochem., **58**, 104 (1954)

von Stackelberg, M., Rec. Trav. Chim. Pays-Bas, **75**, 902 (1956)

von Stackelberg, M. and Frübuss, H., Z. Electrochem., **58**, 99, (1954)

von Stackelberg, M. and Jahns, W., Z. Electrochem., **58**, 162 (1954)

von Stackelberg, M. and Meinhold, W., Z. Electrochem., **58**, 40, (1954) von Stackelberg, M. and Meuthen, B., Z. Electrochem. **62**, 130 (1958)

von Stackelberg, M. and Müller, H.R., Naturwiss **38**, 456 (1951a)

von Stackelberg, M. and Müller, H.R., J. Chem. Phys. **19**, 1319 (1951b)

White, R.S., Earth and Planetary Science Letters, **42**, 114 (1979)

White, M.A. and MacLean, M.T., J. Phys. Chem., **89**, 1380 (1985)

Wilcox. W.I., Carson, D.B., and Katz, D.L., Ind. Eng. Chem., **33**, 662 (1941)

Wilms, D.A. and van Haute, A.A., 4th International Symposium on Fresh Water From the Sea, **3**, 477 (1973)

Woolfolk, R.M., Oil Gas J. **50**(50) 124 (1952)

Woolridge, P.J., Richardson, H.H., and Devlin, J.P., J. Chem. Phys., **87**, 4126 (1987)

Wright, D.A., A Kinetic Study of Methane Hydrate Formation from Ice, M.S. Thesis, U. Pittsburgh Pittsburgh, PA (1985)

Wroblewski, S.V., Compt. Rend., **94**, 212, 1355 (1882)

Wu., B.-J., Robinson, D.B. and Ng, H.-J., J. Chem. Thermodyn.,**8**, 461, (1976)

Yefremova, A.G., and Gritchina, N.D., Geologiya Nefti i Gaza, **2**, 32 (1981)

Yefremova, A.G., and Zhizhchenko, B.P., Doklady Akad. Nauk. SSSR, **214**(5) 1179 (1972)

Yousif, M.H., Abass, H., Selim, M.S., and Sloan, E.D., "Experimental and Theoretical Investigation of

Methane Gas Hydrate Dissociation in Porous Media," SPE 18320, presented at the 63rd Annual Technical Conference of SPE, October 2-5, 1988, Houston, TX

Yousif, M.H., Li, P.M., Selim, M.S., and Sloan, E.D., "Depressurization of Natural Gas Hydrates in Berea Sandstone Cores," J. Inclus. Phenom., D.W. Davidson Memorial volume (in press) (1989)

von Stackelberg, M. and Müller, H.R., Z. Electrochem. **58**, 25 (1954)

von Stackelberg, M. and Neumann, F., Z. Phys. Chem. Abt B, **19**, 314 (1932)

Vysniauskas, A., A Kinetic Study of Methane Hydrate Formation, Dissertation, U. Calgary, Calgary, (1980)

Vysniauskas, A. and Bishnoi, P.R., in Natural Gas Hydrates: Properties Occurrence and Recovery, J.L. Cox, ed., Butterworths, 1983a

Vysniauskas, A. and Bishnoi, P.R., Chem Eng. Sci. **38**,1061 (1983b)

Vysniauskas, A. and Bishnoi, P.R., Chem Eng. Sci. **40**, 299 (1985)

Wagner, J., Erbar, R., and Majeed, A., Proc 64th Ann. Conv, Gas Process. Assn., **64**, 129 (1985)

Walas, S.M., Phase Equilibria in Chemical Engineering, Butterworth Publishers, Stoneham, Ma (1985)

Wall, T.T., Hornig, D.F., J. Chem. Phys., **43**, 2079 (1965)

Weaver, J.S. and Stewart, J.M., Proc Fourth Canadian Permafrost Conf, H.M. French Ed., National Research Council of Canada 312 (1982)

Wendlandt R., and W. Harrison, Personal Communication, October 13, 1988.

Wetlaufer, D.B., Malik, S.K., Stoller, L., and Coffin, R. L., J. Am. Chem. Soc., **86**, 508 (1964)

Whalley, E., J. Geophys. Res., **85**, 2539 (1980)

Whiffen, B.L., Kiefte, H., and Clouter, M.J., Geophys. Res. Lett., **9**, 645, (1982)

# Index